general ecology

general ecology

S. J. McNaughton • Larry L. Wolf
Syracuse University

HOLT, RINEHART AND WINSTON, INC.
New York   Chicago   San Francisco   Atlanta   Dallas
Montreal   Toronto   London   Sydney

To Margaret and Janet

preface

Our primary objectives in writing this book have been (a) to approach ecology through a consideration of the evidence on which ecological principles are based, (b) to integrate various aspects of ecology (that is, plant and animal ecology, basic and applied ecology, empirical and theoretical ecology) into a single unified body of knowledge, (c) to emphasize the importance of evolutionary adaptations in ecological phenomena, and (d) to provide sufficient flexibility in subject matter and approach to allow teachers and students freedom to develop their own approaches to the science.

Data from the scientific literature are extensively incorporated into our discussion of ecological topics so that the relationship between hypotheses and their tests is apparent. We have tried to avoid presenting ecology as "revealed truth," and instead have relied on carefully chosen examples of actual studies to document the current state of the science. Many of the reviewers during the manuscript stage, both students and professional ecologists, indicated that there are certain instances in which their conclusions from the evidence depart from our own. We value this freedom of interpretation highly since it is the way ecology "works" as an active, growing science.

Our intent in writing the book was to remove the barriers between plant and animal ecology, pure and applied ecology, or empirical and theoretical ecology. For instance, rather than including a separate section on ecology and man, we have incorporated applied aspects at suitable places in the text where they relate to the ecological principles being developed. Thus a chapter on the movement of biocides in the ecosystem immediately follows a chapter on materials flow because the principles of materials flow help to further the understanding of biocides. Similarly, we draw freely on examples from plant, animal, or microbial ecology whenever they are significant for developing certain points and strengthening the understanding of principles.

An evolutionary viewpoint is maintained throughout the book, from a separate chapter on the ecological aspects of evolution to frequent evolutionary interpretations of ecological phenomena. We believe that this avoids developing ecology in static terms as a series of more or less disconnected phenomena divorced from their biological origins. In many cases, a consideration of adaptive adjustments is mandatory for an adequate understanding of present patterns and reliable predictions of future patterns.

We have avoided relying on a single approach to the science, for instance, through community ecology, or ecosystem ecology, or population ecology, although all occur where appropriate. We have found in our own teaching experience that overemphasis of a particular organization forced us to develop many subjects in an artificial way.

Another major objective was to develop a versatile book within the limits of a sound approach to the science. This is an introductory college ecology text. However, there are several levels at which students can be introduced to the science, depending on their backgrounds. Hence we have aimed at providing faculty the utmost freedom in selection of topics. For a one-term survey course at the sophomore level, we suggest the following as particularly appropriate for developing a basic introduction to the way ecosystems are organized, their diversity in nature, and the relationship of ecology to man: Parts 1 and 2, Chapter 14, the biome chapters of Part 4, and Chapter 20. In a one term course for students with a strong background in quantitation and biological fundamentals, Parts 1, 2, and 3, Chapter 19, and Part 5 provide a more rigorous introduction. In its entirety the book can be used in one year sequences of ecology for biology majors. Although the book is structured in what we believe is a logical development for such a sequence, the sections are sufficiently self-contained that they are adaptable for accommodating alternative arrangements.

Basically the book starts (Parts 1 and 2) with a broad overview of ecology and progresses to a consideration of the patterns of structural and functional organization in complex ecosystems. The middle section (Part 3) breaks up these comprehensive patterns into a series of intensive considerations of how the complex phenomena arise out of relationships among the biological components of ecosystems. These intensive examinations are then reintegrated (Parts 4 and 5) into higher levels of organization to reestablish an overview of ecological patterns. Within a chapter, we begin with basic principles, and then develop an increasingly rigorous treatment of these principles. Through this approach all students should become acquainted with ecology's basic ideas, and even the best students should be challenged by the more difficult material. In the book as a whole, the most challenging material is in Part 3, in which we examine ecosystem functioning in considerable detail and with as much precision as is reasonable for an introductory text. Thus, depending on the objectives of a given course, the book may be used either to provide students with a broad overview of ecology or, with different emphasis, to prepare students to go further into ecology at an advanced level.

An appendix of introductory quantitative methods is provided for students unfamiliar with statistical interpretations of data sets or for students who feel the need for a quick review. The appendix is not meant to be intensive or comprehensive. It is intended to provide background for the quantitative methods most often employed in the text.

The book, like the science itself, relies heavily on data to test old ideas and generate new ones. Data and ideas, after all, are the foundations of science. We rely heavily on graphical material to document and summarize ecological principles. We are particularly pleased by the skill with which the artists made these data an integral part of the book's teaching function.

We want to express considerable appreciation to these professional ecologists who reviewed the entire manuscript: Daniel B. Botkin, Peter W. Frank, Richard S. Miller, Robert B. Platt, and Robert H. Whittaker. Their comments were very helpful, and the book profited from them. We also appreciate Peter Skaller's review of the entire manuscript as a first year graduate student in ecology, and the many undergraduates at Syracuse University who reviewed the manuscript. The faults of the book are our own, but many of its virtues originated from good reviews. In addition, we want to express our appreciation to Dorothy Garbose Crane, whose assistance as our editor was superb.

Syracuse, New York
Gilgil, Kenya
January 1973

S. J. McNaughton
Larry L. Wolf

contents

ECOLOGICAL FUNDAMENTALS

Ecology concentrates on phenomena that are among the most inclusive of any in the natural sciences. Since it seeks to understand how organisms function in their natural surroundings, it must draw from various sciences that describe organismal function and physical factors. Part 1 provides a general introduction to the organism in nature. Chapter 1 furnishes a general definition of the limits of ecological interest. Chapter 2 focuses our approach to nature on the relationships among organisms and their surroundings. Chapter 3 considers the evolutionary responses which organisms may make to their surroundings, and some of the consequences of these responses for our interpretations of ecological phenomena. Chapter 4, concluding this part of the book, deals with some patterns of structural organization in nature, both in the conventional sense of geometry and in the less apparent sense of the commonness and rarity of species. Taken together, these chapters represent an essential background for other ecological principles which follow.

introduction to ecology

Some 2.5 to 3 million years ago in East Africa, a small portion of the tree-inhabiting ape population began to move out of the forest as it dwindled from drought. From small bands of hunters and scavengers roaming the savannas and lakeshores of the East African highlands, we have gradually become modern man, *Homo sapiens.* Man presently is a successful species. His success is attested by the variety of habitats he occupies and the size of his population (Fig. 1-1). Our large brain, erect posture, and grasping hand allow us to manipulate objects, pass knowledge from generation to generation in written form, and combine the knowledge with manipulation to modify the earth more extensively than any other living thing. Astronauts can see lumbering roads in Canadian forests and smog clouds over our major cities. Surveying our massive impact upon the earth has led some of us to conclude that we are the *most* successful species. But we should temper such a conclusion with the considerable evidence that success is often impermanent. The fossil record is filled with once successful, but now extinct, species. Perhaps it might be helpful to ask: In what resides the success of man, or of any species? Charles Darwin and A. R. Wallace

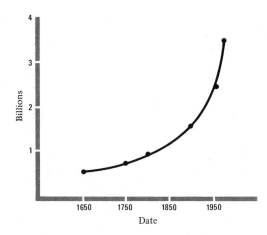

FIGURE 1-1 Growth of the world's human population from 1650 to the present. (Data from Deevey, 1960 and United Nations, 1969.)

(1859) gave us the answer more than a century ago: The success of a species resides in its ability to obtain resources in sufficient quantities to increase or maintain its numbers.

The relationship between organisms and resources is within the subject matter of the science of ecology. *Ecology* is the scientific study of the relationships between organisms and their environments. The word is derived from the Greek roots *oikos,* meaning "home," and *logos,* "the study of." As first used by Reither and Haeckle in 1865 (Kormondy, 1965), it meant the study of the organism in its home, or its natural surroundings.

Ecology is one of the biological sciences. Modern biology can be traced to the sixteenth and seventeenth centuries when it arose out of natural history through systematic studies of the circulatory system of humans, studies of birds and insects, and the invention of the microscope. Modern ecology is a product of the present century; only over the last few decades has it emerged from natural history into a systematic, experimental, and objective study of organisms and their environments.

Because of its recent emergence from natural history, and the complexity of ecological phenomena, there is no comprehensive theory of ecology. At the same time, public awareness of deterioration of the human environment has placed demands upon the science for answers to pressing practical problems. As a consequence, much of what has previously been known as the "conservation movement," deriving from such naturalists as John Muir and Aldo Leopold, has become identified in the public mind as ecology. We wish to emphasize that ecology is not synonymous with conservation. Ecology seeks to understand the way organisms function in nature. Out of this understanding we may derive less destructive approaches to managing our environment, but the objective of ecology is understanding, not necessarily management techniques.

OBJECTS OF INTEREST

Ecology is concerned with phenomena at several levels of biological organization. First, it is concerned with the interaction of the *individual* organism and its environ-

ment. *Environment* is the sum of substances and forces external to the organism
that influences the organism in such a way that it affects the organism's existence
(Mason and Langenheim, 1957). A somewhat more inclusive term is *habitat,* which
refers to the area where the organism occurs. It includes the organism's environment
but, in addition, encompasses those external substances and forces which may not
affect the organism directly. A difficult problem in ecology is separating direct and
indirect causes in nature. Direct causes are what constitute an organism's environment, indirect causes are part of the habitat.

Consider a violet growing on the forest floor. Its environment consists of a certain
combination of temperature, sunlight, water, gases, and minerals which constitute
its life requirements, plus such plant-eating animals as affect it. The combination
of environmental factors, however, may be regulated to a considerable extent by
the trees that constitute the forest. These trees will intercept much of the sunlight
and may utilize many of the minerals in the soil, thereby modifying the violet's
environment. The trees are an indirect cause, therefore constituting part of the
habitat. The environment consists of those direct causes, such as sunlight, predation, and minerals, that regulate organism processes.

The distinction between environment and habitat is an essential one if we are
to make sense of the causal factors in natural situations. What, for instance, are
we to make of the accelerating deterioration of many lakes and streams in the
industrialized nations? It is generally agreed that the habitat modification which gives
rise to this deterioration is the utilization of these waterways for waste disposal. But
what are the causative agents in this sewage? About this there is very little agreement.
Some people feel phosphates are among the most important factors contributing
to waterway deterioration, others feel that other factors may be more important.
This is a case where everyone agrees about the indirect cause, but there is considerable disagreement about direct causes. Part of ecology's task in understanding the
function of nature is to identify direct causes, that is, to define habitat factors which
function as environmental factors.

Ecology is particularly concerned with groups of organisms. A *population* is one
or more individuals of the same species co-occurring in time and space (McMillan,
1959*a*). Populations are often defined in functional terms by the degree of interbreeding among co-occurring individuals. Members of the same population generally
interbreed relatively freely with one another and less freely with members of other
populations. Ecology is also concerned with groups of populations occupying a
common area. A *community* is one or more populations co-occuring in time and
space. Community is usually applied to organisms with similar life habits; thus we
will refer to tree communities, insect communities, or bird communities, rather than
lumping such diverse organisms as a single community (Warming, 1909; MacArthur,
1971).

If we return for a moment to our violet growing on the forest floor, we discover
that it grows intermingled with other violets—a patch of similar individuals. This
group of violets constitutes a population. A variety of other herbs, such as May apple
and jack-in-the-pulpit, also may grow interspersed with the violets. These herbs
constitute a forest floor community, a group of populations co-occurring in time and
space. It should be recognized that these ecological units cannot be rigidly delimited.

There are, for instance, several different species of violets and these may interbreed to varying extents. And violets in a given spot may only occasionally breed with other individuals of the same species several meters away. The delimitation of a population therefore is somewhat arbitrary. Similarly, we have referred to a forest floor herb community. We might extend this community to include the shrubs and tree seedlings growing under the shade of the trees. Or we might extend the concept even further, to include the trees also. Conversely, some communities may consist of a single species. Herbicide-treated agricultural fields, or more naturally, unusual habitats such as marshes, may support single species communities. By and large it is most useful to restrict the community concept to organisms of similar size and ecological characteristics, and to restrict population to individuals of the same species in a local area.

The *fundamental unit of study* in ecology, which cuts across all three levels of biological organization, is the *ecosystem*. As first defined by Tansley (1935), it includes "not only the organism-complex, but the whole complex of physical factors forming what we call the environment." The ecosystem therefore has a biotic component and an abiotic component. The *biotic component* may consist of an individual, a population, a community, or several communities. The concept is often applied in studies of several interacting communities, for instance, to all the plants, animals, and microbes in a pond or length of stream. The *abiotic component* consists of all the substances and forces in the habitat which affect the organisms; that is, the abiotic component is the sum of all the environments.

FIGURE 1-2 In this aerial view of coastal ecosystems, a series of distinct ecological zones are evident from the open ocean on the left through surf and sand dunes to shrubs and trees on the protected land area. (U.S. Dept. of Conservation and Economic Development)

FIGURE 1-3 Other ecological transitions are less distinct than those of Figure 1-2, such as this contact between tundra on the left and coniferous forest on the right. (Steve and Dolores McCutcheon. Alaska Pictorial Service)

The ecosystem approach to ecology is both natural and artificial. Treating the organisms and their environments as a single functional entity is a natural way of organizing our study. But the difficulty of objectively delimiting the borders between ecosystems requires arbitrary decisions about scale in applying the concept. Although the boundaries between ecosystems often may be defined relatively sharply (Fig. 1-2), the definition is much less objective in other cases (Fig. 1-3). Even a distinct boundary such as that between aquatic and terrestrial systems is confounded by amphibians and many other animals that may move freely across the boundary.

In our forest containing violets, we would most often apply the ecosystem concept in such a way that it included several different communities. We might consider the forest floor ecosystem and would therefore concentrate on the herbs and shrubs of the understory plus whatever insects, birds, rodents, and other animals feed upon them. Or we might consider the forest as a whole as an ecosystem, examining the relationships among all the organisms occurring in that forest. One of the important features of this ecological concept is its flexibility. It is defined by function, not by some arbitrary criterion of scale, and therefore may be useful at a variety of levels. We can apply the ecosystem concept from a carnation growing in a pot to an entire forest. Our objectives, of course, will vary radically with the level of our application. It is important to recognize that this application does not make the definition of ecosystem arbitrary. Quite the contrary, the definition maintains its essential functional meaning regardless of the scale of application.

A higher level of ecological organization than the ecosystem is the biome. A _biome_ is an abstraction of many separate ecosystems, with similar characteristics, into a single ecosystem-type. For instance, when we refer to the prairie biome, we are referring to a huge variety of specific ecosystems that are similar in that the principal plants are grasses and the principal animals are grazers.

The largest level of biological organization is the _biosphere,_ that region of the earth from the lower atmosphere to the upper layer of the earth where organisms occur. We will occasionally refer, in a general way, to this level of organization, but its scope is all-inclusive and hence it can be of little use in formulating most ecological ideas.

PROCESSES OF INTEREST

Ecosystems are dynamic functional entities. If we examined our violet population over a growing season and studied the relationship of the population to the habitat, we would find that the violets stored energy, and that this energy was later dissipated; similarly, we would find that minerals were taken up and later returned to the habitat. If we counted the violets from year to year as a measure of population size, we would find that the number was rarely the same every year, but that it fluctuated. Some years, when conditions were particularly favorable, there might be many violets. Other years, the population might be relatively small. Ecologists are particularly interested in the dynamics of ecosystems, the changes that occur with time and the processes which govern, and are governed by, ecosystem composition.

Among the processes of interest to ecologists are _population growth_ and _regulation._ The inception of ecology as a science can be traced to a paper in 1838 by a 34-year-old Belgian, P. F. Verhulst. Verhulst developed a mathematical description of population growth that still represents one of the fundamental insights of ecology:

$$dp/dt = mp - np^\alpha \qquad (1\text{-}1)$$

This equation says that the growth rate of the population (dp/dt = change (d) in population size (p) with a change (d) in time, (t)) is equal to the birth rate of the population (m) times population size (p), minus the death rate of the population (n) times population size (p) raised to some power (α) which describes the increase in death rate as the population approaches a limit on size. This is one of the most interesting propositions in ecology, with substantial implications for the human population. We return to it in considerable detail later in the book. Its essential point is that population growth rate will decline, and finally become zero, as the population reaches some size limit. This limit is determined by the availability of resources and the efficiency of resource utilization by members of the population.

Verhulst's idea was derived directly from T. R. Malthus' famous essay in 1798. Malthus and his father had argued repeatedly the proposition widely espoused by eighteenth century Romantics that man is perfectible through the modification of his social institutions. The senior Malthus shared this viewpoint, but his son argued its futility with such persuasiveness that the father urged him to publish an essay on the topic. In this essay Malthus stated that the achievement of utopia was unlikely because "the power of the population (to increase) is indefinitely greater than the power of the earth to produce subsistence for man." This dropped a bombshell into turn-of-the-century intellectual circles which the younger Malthus spent the rest of his life defending. The rhetorical polemic surrounding his thesis continues to reverberate in the popular press. From this essay, certainly one of the most influential in the history of Western thought, we can trace directly Darwin's (1859) great theory and much of the resulting conceptual base of biology. Verhulst's more precise formulation of the argument was lost in the polemic and never utilized until Pearl and Reed (1920), unaware of Verhulst's forgotten paper, developed the same concept 82 years later.

Growth and regulation of populations are important to man, with a world-wide population of 3.65 billion and a growth rate that is constantly accelerating. Man, for all of his recorded history, seems to have been running a race with the earth's ability to supply him with food; he has depended largely on the opening of previously uncultivated lands for much of it (Russell, 1969). Food is, in fact, a problem of another process of ecological interest, that of *energy flow* in the ecosystem. Organisms maintain themselves and reproduce by utilizing energy from the environment. This movement of energy from organism to organism is called *energy flow*. Energy flow in ecosystems occurs through the medium of the feeding process. An oak leaf "fixes" the energy of the sun in the form of sugar and is subsequently eaten by a caterpillar. The caterpillar is then eaten by a bird. This constitutes energy flow. Lindeman (1942) laid the conceptual framework for its study by emphasizing the idea of *trophic levels* or, as they might be called, feeding levels. He pointed out that organisms obtain their energy at different feeding "distances" from the primary energy source, the sun.

A third process of interest is *materials flow* in the ecosystem. In addition to energy, organisms require a variety of chemical substances in order to maintain themselves and reproduce. These substances are obtained from the environment and eventually are returned there. Although energy is dissipated as it moves from organism to organism, materials are returned to the environment in a form that eventually may be reused. Harvey (1926), who was interested in the biological chemistry of the sea, was among the first to outline the general nature of such a biogeochemical cycle, that for nitrogen.

Energy and materials flow are coupled processes in the ecosystem. As an organism degrades organic molecules for their energy content and uses this energy content for maintenance, growth, or reproduction, it releases materials into the habitat. Respiration, for instance, does cellular work by releasing energy in organic bonds. As it does so, it releases carbon dioxide to the habitat. The "consumption" of energy and release of matter occur simultaneously.

A fourth process of interest to ecologists is the process of *competition*. As populations grow, dissipate energy, and use chemical nutrients, they exert influences on the environment and upon co-occurring populations. The relationship between organism and environment is reciprocal and the growth, maintenance, or decline of one population affects the availability of energy and materials to other populations. Competition occurs when a resource is present in insufficient quantities to meet the needs of all the organisms present in an ecosystem.

The effects of organisms on their habitat may result in a sequential change in the populations occupying a given area as certain populations appear, grow, and then disappear, to be replaced by other populations. This process of community change is called *succession*. A general conceptual description of succession was first developed by F. E. Clements (1916) from studying the grasslands of the central United States. Most work on succession has been descriptive and intuitive in approach, and there are few data that relate clearly to the mechanisms driving the process.

Information obtained by studying the organism-environment complex at various levels of organization will be developed around many concepts throughout this book. A fundamental concept is *evolution*—change in the genotypic characteristics of a population through the differential reproduction of population members. Ecological properties of organisms arise through evolution. Many of our discussions of ecological phenomena are based upon the assumption that a full understanding of their significance requires that they be placed in an evolutionary context. But this is not an evolution text; we treat evolution, or physiology, or physics where it is appropriate to an understanding of ecology. At many levels, however, ecology and evolution are impossible to disentangle; attempting to do so robs both of much of their power for organizing information.

APPLICATIONS OF ECOLOGY

Ecology might be characterized currently as the relevant science. But relevance, after all, must arise from the interaction between the individual person and the subject matter. It is difficult to imagine that any subject can be decreed universally relevant or irrelevant. Professional ecologists are ecologists because they find the ideas and methods of the science exciting and rewarding. Few of them would be unmindful of how important an understanding of ecosystems is to society. Societies, in fact,

FIGURE 1-4 The earth is littered with the remains of civilizations which disappeared through a lack of balance between the demands of the society and the ecosystem's ability to meet those demands. (Jane Latta)

FIGURE 1-5 The demands which industrial societies place upon their habitats are apparent in the smog-choked air above many major cities. (De Wys)

are biological components of ecosystems. It can be argued that much social failure has arisen from ecological ignorance. A component of the failure of certain tropical civilizations (Fig. 1-4) may have been the inability of tropical soil to sustain the intensive agriculture necessary to support a civilization. The most trivial seeming points which are discussed in subsequent chapters (the patterns of litter decay in ecosystems, for instance) represent components of a conceptual base upon which an understanding of ecosystem function depends. Many views of our current civilization (Fig. 1-5) provide evidence that a lack of such understanding works to our detriment.

Consider an individual in its environment (Fig. 1-6). It takes up matter and energy, and as a consequence, it grows. As it feeds (that is, takes up matter and energy), it must first meet the maintenance costs associated with supporting its existing body size. We call this mass of living material, *biomass*. If its energy and materials uptake is in excess of that required for maintenance, it either may excrete the excess as waste or add biomass through growth. All three of these processes, feeding, growth, and maintenance, represent components of functional loops connecting the individual with its environment. Such loops are called *feedback loops*. A common example of such a loop in an engineered system is a furnace thermostat which senses the room temperature, generates an electrical signal according to whether the temperature is below or above a preset level, and activates or inactivates the furnace accordingly. Feedback loops abound in ecosystems; as a population grows, for instance, its individuals respond to decrease in resource availability by decreases in size or reproductive rate, or in a variety of other ways. Because feedback loops either may amplify or diminish the effect of ecosystem variables (von Foerster, 1958), they may have either destabilizing or stabilizing effects on ecosystems.

As the individual grows, it may attain a sufficient biomass and an appropriate developmental state so that it reproduces. Reproduction, by multiplying the number of individuals making demands upon the environment, generates competition for whatever life requirements may be in short supply relative to the individual's needs. And, of course, the multiplication of individuals gives rise to an expanding population. Competition is a feedback loop connecting population members through their environments.

Reproduction is the source of expression of genetic variation in the population, as certain genes may be assembled in different combinations in the offspring. The increase in genetic variation, combined with competition, can produce a change in the average gene structure of the population as certain individuals become capable of exploiting resource combinations not exploited by previous individuals. This change in population gene structure is evolution. Continued evolution, coupled perhaps with barriers to gene exchange among individuals, may give rise to several different populations and eventually different species. This production of diversity in the ecosystem gives rise to more complex communities. In addition to the competition between members of the same species, there will now be competition among individuals of different species.

The objective of ecology is understanding how the phenomena of Figure 1-6 are organized to generate the natural world. The entire contents of the figure constitute an ecosystem. The primary inputs to the biotic component are energy and matter; the primary outputs are communities of organisms.

Our violet population arose because individuals encountered appropriate conditions on the forest floor for their establishment. That is, they were able to obtain sufficient matter and energy from the habitat not only to meet their maintenance requirements as seedlings but to grow into mature individuals. Upon maturity they

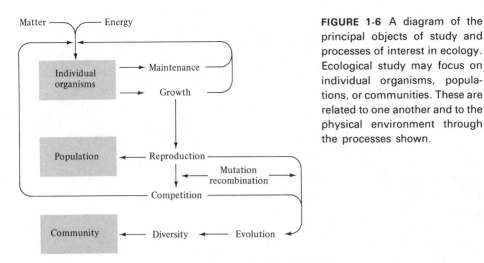

FIGURE 1-6 A diagram of the principal objects of study and processes of interest in ecology. Ecological study may focus on individual organisms, populations, or communities. These are related to one another and to the physical environment through the processes shown.

reproduced. This reproduction gave rise to an increasing population. Perhaps there were *mutations,* changes in the genetic material of the individuals. Or perhaps there were only rearrangements of the existing genetic material, *recombinations,* producing individuals different from the colonists. As the population colonized all sites favorable for violet occupation, there was competition for required resources. Certain new gene combinations were more efficient at this competition; perhaps they were capable of depleting soil phosphorus to a lower level than the initial colonists. The population began to diversify, certain individuals occurring in high phosphorus spots, others in low phosphorus spots. This differentiation in the type of environment exploited could give rise eventually to different species of violets, one particularly efficient at exploiting high phosphorus, another particularly efficient at exploiting low phosphorus. We would then have a more complex violet community.

Insects ate portions of the violets. As the violets evolved, the insects evolved. The feeding process dissipated some of the energy fixed by the violets and returned minerals to the soil with the fecal materials of the insects. This, in turn, may either make the site more or less favorable for colonization by subsequent violets. And, of course, the insects feeding on violets may be eaten by yet other organisms in the forest.

The inputs to the violets were matter and energy, as regulated by the indirect causes—the trees constituting the forest. The outputs were a complex of organisms and ecosystem processes. The objective of ecology is to understand the functional organization of such ecosystems. Our objective in succeeding chapters of this book is to synthesize a partial understanding of what connects the inputs and outputs of such an ecosystem. We start by focusing on the relationships between individuals and their environments and some of the descriptive features of ecological study (Part 1), then proceed to a "big picture" of the organization of ecosystems (Part 2). This overview is then broken up into a detailed consideration of phenomena responsible for the big picture (Part 3), and then reintegrated into a consideration of the breadth of ecosystem diversity and its causes (Part 4). Finally, we place man's evolutionary development in the ecological context and summarize some of ecology's important insights and problems (Part 5).

GENERAL CONCLUSIONS

Ecology: scientific study of the relationships between organisms and their environments.

Objects of study	Processes of interest
1. Organism	1. Energy flow
2. Population	2. Materials flow
3. Community	3. Population growth and regulation
4. Ecosystem	4. Competition
5. Biome	5. Succession
	6. Evolution

the organism in its environment

2

As we saw in the preceding chapter, one of the difficult problems in understanding the functioning of organisms in nature is distinguishing between direct and indirect causes regulating function. The rationale of laboratory experimentation in biology, of course, is to control the many variables which fluctuate constantly in nature. Through this control it is possible to vary one factor at a time and to determine the effect of each upon organism processes. Although laboratory experimentation may be an important part of unraveling ecological problems, it is always imperative that laboratory results in ecology be generalized to the field, since ecology seeks an understanding of field conditions.

It is often extremely difficult to determine what are the actual factors affecting populations under field conditions. A single example can suggest the hazards of intuitive explanations of ecological occurrences. Along rocky seacoasts, there are commonly copious growths of brown algae on rock faces. In sheltered coves in the Northern Hemisphere, rockweeds (*Fucus*) are particularly abundant (Fig. 2-1). These algae are relatively rare, however, in exposed areas subject to direct wave action.

FIGURE 2-1 Rockweeds (*Fucus* spp.) are an abundant algae on sheltered coastlines in the Northern Hemisphere. (American Museum of Natural History.)

It was long supposed, therefore, that rockweeds were poorly adapted to withstanding the physical pounding of direct surf action. Often abundant on the exposed coasts where rockweeds are rare are a variety of marine herbivores called limpets. These marine gastropod molluscs are active grazers on marine algae, and Southward (1964) found that removing them from exposed areas allowed a copious growth of rockweeds to develop. He concluded that the rarity of rockweed on exposed coastlines, rather than being a direct effect of wave action, arose out of the abundance of limpets which crop the rockweeds into near extinction. The direct factor governing rockweed abundance on these coasts was limpet-grazing intensity. Intense surf may be an important environmental factor that permits high-density limpet populations. A principal objective of ecology is understanding those factors which are operative in such distribution patterns. In this chapter we attempt to add some focus to our consideration of organisms in nature through precisely defining the relationship between an organism and its surroundings.

Organisms and their environments form a functional unit with a reciprocal relationship between them. An environment represents a constraining influence within which an organism must operate; the organism, in turn, influences the properties of its environment. The term environment refers to a class of phenomena. It is organism-directed and is determined by the organism's life requirements (Mason and Langenheim, 1957). As defined by ecologists, an organism's *environment* consists of all those substances and forces external to the organism that enter its reaction systems or otherwise directly affect its maintenance, growth, and reproductive functions. For instance, carbon dioxide is taken up by a plant in the process of photosynthesis and thereby becomes an integral component of the plant's reaction systems until released by respiration. Temperature, on the other hand, may regulate the rate at which carbon dioxide is taken up by the plant. Carbon dioxide therefore functions as a *resource*, an environmental factor that enters a reaction system of the organism. Temperature functions as a *regulator*, an environmental factor that affects the organism by modifying the rate or course of a reaction system. The environment of every organism can be subdivided into three classes of phenomena: energy, chemical substances, and other organisms. Components of each class may function as either resources or regulators.

Regulators are particularly important in the practical application of ecology because they cause much of what we refer to as *pollution*, the detrimental modification by an organism of its own environment. Pollutants are inhibitory environmental substances produced by the activities of the organism itself. These do not constitute resources in the same sense as a nutrient element, although they may enter a reaction system of the organism directly and limit uptake of a certain resource. Carbon monoxide, for instance, regulates oxygen uptake in animals by competing with uptake sites on the hemoglobin molecule. All organisms generate environmental modifications which, if they accumulate, arrest the organism's growth. Man is no different and his pollution is exceptional only in its volume, persistence, and pervasiveness. Most organisms depend upon a large habitat volume, or upon the mobility of themselves or their habitat to dissipate these toxic products. When the rate of pollutant production exceeds the rate of dissipation, accumulation to toxic levels may occur. Such accumulations are as common, perhaps more common, in bacterial cultures as in man's cities.

In our definition of environment, how do we consider factors that are potentially environmental factors but which are not directly affecting an organism at present? Pollutants, for instance, may not affect us until they exceed some threshold of tolerance. When do they become environment?: When they exceed the threshold. Until that time they are habitat components, not environmental factors.

Energy

Among the three fundamental constituents of the environment is *energy*, the capacity to do work. This, of course, constitutes the driving force of the ecosystem. A fundamental generalization from physics which describes the limits of this driving force is the second law of thermodynamics (Blum, 1962). In its simplest form, the law states that all processes in an isolated system are accompanied by an increase in the randomness of arrangement of parts of that system. This randomness is called *entropy.* To describe this we can write

$$\Delta S = k \ln P_2/P_1 \tag{2-1}$$

where ΔS is the change in entropy accompanying the process, P_1 is the number of possible arrangements of the system at the beginning of the process, P_2 is the number of possible arrangements of the system at the end of the process, and k is the Stefan-Boltzman constant. If there was only one possible state of the system,

$$\Delta S = k \ln 1 = 0$$

and the entropy change would be zero because nothing would occur, that is, the system could not change state.

The important factor in applying this law to ecosystems is that they are not isolated systems. In particular, ecosystems are connected to the sun through photosynthesis. This energy from the sun, therefore, may be used to drive the entropy of the ecosystem lower. Ecosystems, of course, cannot violate the second law, so they must decrease their own entropy by accelerating the rate of entropy production in the universe as a whole.

Energy occurs in two forms, potential energy and kinetic energy. *Potential energy,* the energy of position, is the form in which energy is transferred from organism to organism in the ecosystem. In the bonds of organic molecules, potential energy represents the energetic currency of the ecosystem. The only significant source of energy for driving ecosystem processes is the sun (Fig. 2-2). The photosynthetic

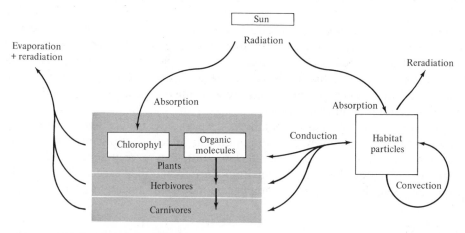

FIGURE 2-2 Main processes of energy exchange in an ecosystem. Exchanges of potential energy among organisms occur through the feeding process inside the box; exchanges involving the habitat are shown by arrows.

apparatus of green plants functions as the ecosystem's energy gateway. Photosynthetic pigments absorb light particles, photons, from the sun and store some of their energy in the bonds of organic acids, which are further converted into carbohydrates, proteins, fats, and other molecules. The potential energy of organic bonds is subsequently transmitted through the ecosystem by the feeding process.

Kinetic energy, the energy of motion, is the form in which energy is transferred between an organism and its abiotic environment. It may occur in radiant or thermal forms. *Thermal energy,* the energy of molecular motion, may be transferred directly by the processes of conduction and convection. *Conduction* is the movement of energy directly between particles as a result of collisions. *Convection* is the movement of thermal energy from one place to another as a result of the actual motion of a given particle type. An example of convection is the motion of gas molecules above road surfaces in the summer which cause "heat waves."

Kinetic energy also may be transferred between objects through the combined processes of radiation and absorption. In *radiation* thermal energy is transformed into the radiant energy of electromagnetic waves. This energy is then reconverted into thermal energy upon *absorption* by an object. At the surface of the sun, the thermal energy of molecules is converted into radiant energy. Some of this energy is reconverted into thermal energy upon reaching the earth 8 minutes later; some is reflected into space by atmospheric particles. Some of the absorbed energy is utilized in photosynthesis to generate the potential energy of molecular bonds that function in organismal energy transfer. Some of the energy is reradiated into space by absorbing objects, some raises the kinetic energy of the absorbing substance.

The *temperature* of a substance is a measure of the kinetic energy of the substance's molecules. This energy can be transferred directly to the organism from the surrounding medium (air, soil, or water) by conduction. This transfer of energy affects organisms directly by influencing the rate of metabolic reactions. These biochemical effects are then translated into effects upon such biological processes as development and reproduction. Organisms occupying distinct thermal climates often show distinct evolutionary adjustments to temperature. For example, cod occur in somewhat colder waters than mackerel; this temperature difference is reflected in the time required for eggs to hatch (Kinne, 1963). Mackerel eggs hatch relatively rapidly at temperatures between 10 and 21 degrees; cod eggs develop at temperatures as low as −1 degree, but fail above 14 degrees (Fig. 2-3).

An interesting property of both patterns is that the most rapid egg development occurs just below lethal temperatures. In this study, only 2 degrees were required to convert the eggs from rapid development to complete failure. This suggests that a relatively minor change in water temperature around 15 degrees would be sufficient to convert from a cod to a mackerel fishery, or vice versa depending upon the direction of the temperature change.

The relatively narrow viability threshold at the upper range of the fish egg temperature response indicates that minor environmental changes might, in certain instances, have profound effects upon the modified ecosystems. For instance, in many areas the electrical power industry is currently installing nuclear power plants that

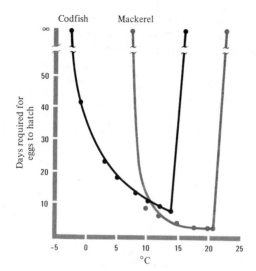

FIGURE 2-3 Relationship between water temperature and the time required for eggs of two fish species to hatch. Cod is a cold water fish and mackerel is a warm water fish. (Data from Kinne, 1963.)

are designed to process huge quantities of water as a reactor coolant and then to return the water to aquatic ecosystems 4 to 5 degrees hotter than previously. Although consumption rates of 100,000 to 200,000 gallons per minute may seem great, they are currently drawn from sources representing billions of gallons, and the results may be minor and localized (Fig. 2-4). Since there is increasing deployment of such generators, however, their impact will increase. The threshold transitional ranges in cod and mackerel temperature response indicate that minor changes could have profound effects upon species in the discharge area. At the same time, the power industry is caught in a conflict among greater power demands by both industrial and domestic users, increasingly stringent air pollution standards, and dwindling high-quality fuel reserves.

The problem of thermal pollution, as it is now called, is a difficult one. Neither the power companies nor their opponents have been able to reconcile the conflict between increasing power demands and the decreasing capacity of man's habitat to accept disruption. Most environmental problems, in fact, revolve around this very point: How can the ultimate ecological cost of degrading our habitat be balanced with the immediate economic cost of arresting the degradation?

Although many biological processes may show rapid transitional ranges between optimal and lethal temperatures, such as the cod and mackerel eggs, others may show a more evenly centered optimum (Fig. 2-5). The abrupt transition suggests a threshold response. Exceeding the threshold results in abrupt simultaneous disruptions of the process examined. A central optimum, on the other hand, suggests a more even balance between promotive and disoperative effects. When the transitions are gradual, environmental temperature changes of a fairly wide range may produce minor ecological effects. One of the important practical applications of ecological research will be accurate differentiation between habitat modifications that can be tolerated and those that generate significant environmental degradation.

FIGURE 2-4 The thermal changes caused by many industrial installations are visible in this aerial view of a nuclear power generating plant. Large open water areas melted in the adjacent frozen river are due to heating of water as it passes through the plant. (Rotkin, PFI.)

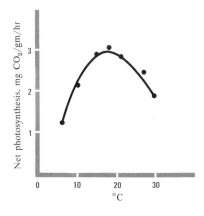

FIGURE 2-5 Relationship between net photosynthesis of a moss and air temperature. (Data from Stafelt, 1937.)

In addition to being organism-directed, environment is ordered temporally by the life cycle of the organism. Environmental effects may be quite different at different times in an individual's life. For instance, the optimum night temperature for growth of green pepper plants (Dorland and Went, 1947) declines as the plants age (Fig. 2-6). Twenty-four-day-old plants grow best at 25 degrees, but 96-day-old plants grow best at 12 degrees. Similarly, population growth rate in relation to temperature may depend on the age structure of the population. A uniform age population would tend to show a somewhat more narrow tolerance to temperature than one of mixed ages, in which decreasing growth by one age group would be compensated by increasing growth by another age group. The temperature zone within which the population may operate can therefore be somewhat wider than that within which any one individual may operate.

Similarly, optimum temperatures for different life processes may differ in the same organism. In populations of the grain beetle, *Calandra oryzae* (Birch, 1953), the maximum number of eggs per adult was produced at 25.5 degrees, whereas the minimum time from egg to adult occurred at 29.1 degrees (Table 2-1). The rate of population growth is affected both by the number of offspring produced per individual and the time required to complete the developmental sequence from egg to reproductive age. Hence the actual growth rate is influenced by the balance between these two population properties. This overall growth rate was maximal at the higher temperature because shorter development time more than compensated for reduced egg production.

A second form of kinetic energy that is an important constituent of the environment of most organisms is *radiant energy*. Radiant energy occurs as particles, called photons, which move in waves from the radiating source. The wavelengths of energy reaching the earth (Gates, 1965) lie largely between 0.3 and 10 μ (Fig. 2-7). Radiant energy drives the ecosystem by (a) providing the sole significant food source through photosynthesis (wavelengths of 0.4 to 0.7 μ), (b) affecting mutation rates in DNA and

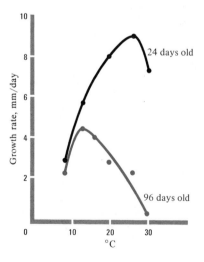

FIGURE 2-6 Effect of night temperature on growth of pepper plants of different ages, showing how growth requirements may change with age. (After Dorland and Went, 1947.)

TABLE 2-1 Effect of temperature upon egg output (eggs per female per week), development time from egg to adult (weeks), and the reproduction rate of the population (females/female/week) for the grain beetle, *Calandra oryzae*

°C	Eggs per female per week	Development time	Females per female per week
13.0	0	∞	0
15.2	1	32.9	0.005
18.2	4	15.5	0.14
23.0	266	6.1	0.43
25.5	384	4.9	0.61
29.1	344	4.0	0.77
32.3	197	4.1	0.50
33.5	27	9.4	0.12
35.0	5	∞	0

Birch, 1953.

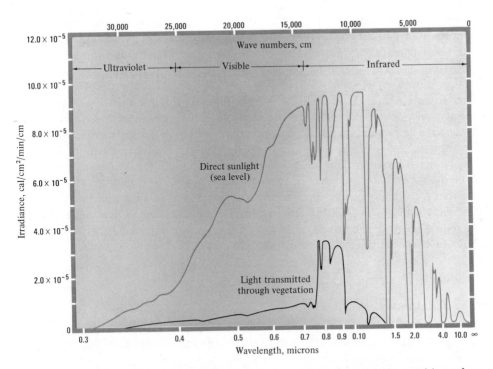

FIGURE 2-7 Spectral distribution of the energy in sunlight reaching the earth's surface and passing through a layer of vegetation. Note that most of the energy in sunlight occurs in the infrared. (After Gates, 1965.)

lethal effects (wavelengths smaller than 0.4 μ), and (c) generating a substantial thermal input upon absorption (wavelengths longer than 0.7 μ). These three classes of radiant energy are visible, short wave, and long wave.

In addition to the direct input of radiant energy to organisms by their own absorption, all objects in the environment absorb radiation, increase their own thermal energy, and thereby affect the thermal energy of organisms by conduction and convection, as we have already seen. These factors also generate movement of the atmosphere (wind), which may be important in producing direct cooling effects. In addition, the absorption of energy by atmospheric molecules, together with conduction from the earth's surface, initiates the large-scale convective patterns that we call weather.

The increase in kinetic energy of an organism's molecules from conduction and absorption of radiant energy is called *heat load*. Part of the behavior of metabolically nonthermoregulatory animals such as lizards is organized around maintaining an appropriate heat load, with the animal often moving from sunny to shady habitats to maintain body temperature within a narrow range. Similarly, the energy requirements of thermoregulatory animals, such as birds and mammals, change substantially according to the heat load. As the heat load either falls below or rises above the level which passively maintain the animal's body temperature, the animal expends more energy to either raise or lower its body temperature. The heat load may be dissipated by conduction, convection, reradiation, and also by evaporation of water from body surfaces (Gates, 1965). This last phenomenon is particularly important in terrestrial plants. They liberate huge quantities of water to the atmosphere through the evaporation of water from the leaf; this process is called *transpiration*.

Our discussion of energy in ecosystems treats several properties that apply to other environmental factors besides temperature. (a) Environment is an organism-directed phenomenon that affects the physiological processes of individuals separately. (b) The population response to environment may be somewhat different from the response of single individuals; in particular, the optima for population processes may be much broader than similar optima for individuals. (c) Optima may be close to the lethal limit for an organism for certain processes and may be more evenly centered for others; hence the effects of environmental modification will be quite different for the two optimum patterns. (d) The environment is determined by the life requirements of the organism, so the same environmental input may have quite different effects at different times in the life cycle.

Ions and Molecules

A second major class of environmental factors is *chemical substances*. If energy may be characterized as the driving force of the ecosystem, chemical substances are its structural constituents.

Chemical substances that constitute part of the environments of organisms are generally in the form of *ions* (atoms or radicals with an electrical charge) or *molecules* (uncharged atomic combinations). Many chemical nutrients important in metabolism, such as potassium and magnesium, may be taken up from the environment as ions. Other important chemicals, such as water and oxygen, are obtained from the environment as molecules. Although some substances may be taken up directly in solid form, as deer obtain sodium from salt licks, most ions and molecules are in solution or in gaseous form when available to organisms.

One of the principal differences between aquatic and terrestrial ecosystems is the availability of ions to animals. In aquatic ecosystems most organisms have access to the chemical constituents of the environment through uptake directly from the surrounding medium. In terrestrial ecosystems, however, many of the elements important in animal nutrition enter the ecosystem's biotic component only through plant membranes, where they are concentrated from the surrounding soil solution. Soil particles are typically covered with thin films of water that contain ions and molecules that have weathered from the particle surfaces or been released by decay of organisms. Plants accumulate many ions and molecules to concentrations 10- to 10^4-fold higher than environmental concentrations (Price, 1970). In terrestrial systems animals are dependent upon this plant uptake through their inability to obtain most nutrients directly.

An important constituent of the surrounding aquatic medium, including the soil solution, is the concentration of hydrogen ions (protons). These have a regulatory effect upon the availability of other nutrients (Truog, 1948). The negative logarithm of the concentration of protons is called pH. At concentrations around neutrality (pH = 7), many nutrients are highly available; others may be much less available than at more acid conditions (Fig. 2-8). At lower pHs, at which iron and manganese are more available, the availability of other nutrients, nitrogen and phosphorus, for instance, falls off markedly. The interaction between hydrogen ion concentration and the availability of many essential elements exemplifies the operation of an environmental regulator. The concentration of hydrogen ions in the aquatic medium affects organisms by regulating the supply of essential nutrients, rather than functioning directly in an organism's reaction systems.

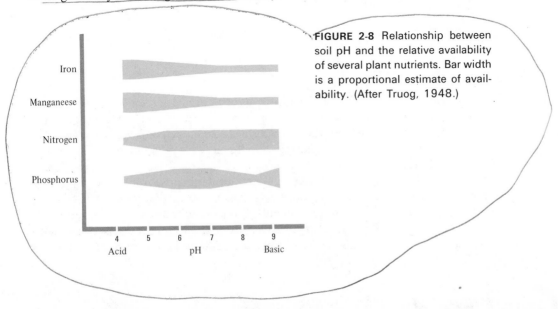

FIGURE 2-8 Relationship between soil pH and the relative availability of several plant nutrients. Bar width is a proportional estimate of availability. (After Truog, 1948.)

Iron

Manganeese

Nitrogen

Phosphorus

4   5   6   7   8   9
Acid   pH   Basic

(a)

FIGURE 2-9 Patterns of carbon monoxide emissions on Manhattan Island: (a) daily emissions for different areas in tons per day; (b) hourly average carbon monoxide concentration and traffic count over a 24-hour period on East 45th Street. (After Johnson et al., 1968.)

Gases are among the chemical constituents of all organisms' environments, either as they occur in the atmosphere and soil in terrestrial systems or in water in aquatic systems. The elements in gases affect organisms as functional molecules in stabilizing light energy (carbon dioxide), releasing stored energy (oxygen), or protein structure (nitrogen). Carbon dioxide acts as an acceptor of hydrogen ions from water to form carbohydrates, oxygen releases the energy of molecular bonds through the respiratory process, and nitrogen is an integral constituent of protein structure. All of these substances occur as significant constituents of the organism's gaseous environment.

In addition, many significant pollutants are gases. Carbon dioxide serves as both a resource, for green plants, and as a regulator, inhibiting the respiratory process

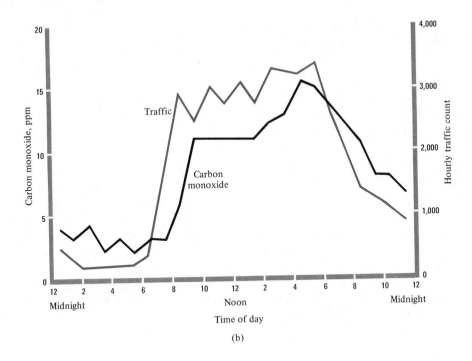

(b)

in both plants and animals. The ambiguous nature of pollutants is, in fact, exempli-fied by the dual function of carbon dioxide in ecosystems. Massive burning of fossil fuels initiated by the Industrial Revolution has increased the atmospheric concen-tration of carbon dioxide by releasing it from the deposits of ancient ecosystems. Some scientists have suggested that this may either warm (Charlson and Pilat, 1969) or cool (McCormick and Ludwig, 1967) the temperature of the earth by either acting to hold thermal radiation within the atmosphere or by keeping sunlight out. One ecologist (Hutchinson, 1970) has even wryly pointed out that agricultural yields since 1900 may have been enhanced somewhat by the increasing atmospheric supplies of carbon dioxide. The conflicting viewpoints regarding the current or potential effects of carbon dioxide upon the world's ecosystems indicates the ambiguous nature of much of the pollution controversy.

Other pollutants are definitely more detrimental. One is carbon monoxide, a by-product of burning petroleum. Motor vehicles are a principal source of atmos-pheric carbon monoxide. The estimated emission from automobiles in New York City in 1968 was 4140 tons per day (Johnson, Dworetzky, and Heller, 1968). The distribution of this pollutant varied significantly in different regions (Figure 2-9a), ranging from 93 tons per square mile per day in the midtown area to 14 tons per square mile per day in Central Park (Heller, 1968). The concentration of carbon monoxide varied conspicuously over a 24-hour period, ranging from about 3 parts per million (ppm) during the early morning hours before dawn to 15 ppm during the afternoon rush hour (Fig. 2-9b), with the concentration closely associated with traffic volume (Johnson et al., 1968).

Carbon monoxide has a direct effect upon man because it displaces oxygen from hemoglobin to form carboxyhemoglobin:

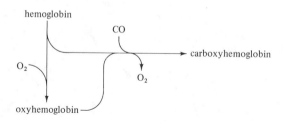

This results in the impairment of choice discrimination and psychomotor ability (Heller, 1968). Both of these capabilities may be extremely important when controlling 3000 pounds of complex, self-propelled machinery under adverse situations. It is ironic that carbon monoxide impairs just those abilities that are most likely to be required when its concentration in the environment is maximal. Dubos (1965) has argued that the ability to tolerate air pollution may be an important evolutionary consequence of urbanization and industrialization. It does not seem improbable that there is substantial natural selection in the human population of urban areas for tolerance to high levels of such pollutants as carbon monoxide.

It has proven extremely difficult to assess the costs of the damage which environmental pollution does to society. The loss to California agriculture alone has been estimated as 7 to 10 million dollars annually (Thomas, 1965). An unusually high concentration of hydrogen sulfide over Poza Rica, Mexico, in 1950 killed 22 persons, hospitalized 320 others, and killed 50 percent of the town's animals. Smelting operations at Copper Hill, Tennessee, destroyed 17,000 acres of forest (Fig. 2-10) and damaged 30,000 acres more (Hursh, 1948). Isolated local disasters such as these are immediately recognizable, but they can also be dismissed as tragic accidents. The increase in automobile deaths and incapacitation resulting from the effects of carbon monoxide upon choice discrimination and psychomotor abilities represents a chronic social effect that is presently impossible to measure. There is little evidence, however, that we are realistically balancing short-term benefits and long-term detriments arising out of environmental pollution.

Other Organisms

The third major class of environmental factors is other organisms. Since ecology focuses much of its interest upon relationships among groups of organisms, this is central to the subject. Much of the material in ecology revolves around how organisms affect one another. Two direct ways are by removing from the environment some life requirement in short supply and by serving as either a food source or a predator. Indirect effects include modifying regulatory environmental properties, for instance, by modifying pH of the water in aquatic systems. Since the mechanisms of these effects constitute much of the subject matter of the science, there are extended discussions of them in subsequent chapters.

FIGURE 2-10 Deforestation of areas around smelters in Tennessee has converted the adjacent ecosystems, once lush forests, into devastated areas. (U.S. Dept. of Agriculture.)

THE ENVIRONMENTAL COMPLEX

Although, for study purposes, we can conveniently categorize the environment into a series of classes of phenomena, the organism perceives its environment as an indivisible complex (Billings, 1952). As we have seen, a change in the environmental concentration of hydrogen ions influences the availabilities of a variety of other chemical substances required by the organism. And since the ecosystem forms a functional unit, a change at one level in the amount of nutrient uptake may be transferred throughout the ecosystem. Suppose, for instance, the supply of phosphorus to an herbivore population was very close to the amount it required. In that case, a shift to more acid soil pH, with a resulting decrease in phosphorus uptake by the plants, could decrease the size of the herbivore population.

There is an important distinction to be made, however, between what could and what must affect the herbivore. Some people concerned with environmental problems have often unconsciously paraphrased Francis Thompson's rhyme:

> All things by immortal power
> Near or far
> Hiddenly
> To each other linked are,
> Thou canst not stir a flower
> Without troubling a star.

Although there is an important element of truth in these lines, ecologists must distinguish between changes which *may* be transferred through the ecosystem and changes which *will* be transferred through the ecosystem. While an organism perceives its environment as a functional complex with multiple interactions, a crucial task for ecologists is to distinguish between important changes and trivial changes. The task is immense because it requires careful definition of the biotic and the abiotic components of ecosystems. A rational policy for such social problems as thermal pollution depends on this knowledge.

LAW OF LIMITING FACTORS

Although it may be traced historically at least as far as the work of Liebig (1840), the fundamental environmental generalization of ecology was most concisely formulated by F. F. Blackman (1905). He pointed out that an organism process that is dependent upon many distinct environmental factors for its operation will be limited by a single factor whose value is farthest from the process requirements. Ecologists call this the *law of limiting factors.* Blackman was interested primarily in physiological processes under experimental conditions in which the principle is more evident in such patterns as the temperature curves of Figures 2-3, 2-5, and 2-6 than it may be under field conditions where the factor farthest from the optimum is difficult to identify.

In Blackman's examinations of the effect of environmental variables upon assimilation, he observed that "when the rate of a function exhibits a transition from rapid increase to a stationary value, it becomes at once probable that another 'limiting factor' has come into play." The law of limiting factors is clearly illustrated by the following experiments that test the effects of light intensity, atmospheric CO_2 concentration, and temperature upon carbon assimilation by plants (Fig. 2-11). (a) When light intensity is low, the assimilation rate is independent of CO_2 and temperature. (b) At high light intensities and low CO_2 concentrations, the rate is temperature independent. (c) At high levels of both CO_2 and light (that is, when neither of these is limiting), the rate becomes temperature dependent. It is clear from the graph (Fig. 2-11), however, that the idea of a single limiting factor is inappropriate, because the effect of low light upon photosynthesis is determined by the levels of carbon dioxide and the temperature. In other words there are certain ranges of environmental factors for which organismal processes may be simultaneously limited by combinations of more than one factor (Smith, 1966).

The concept of limiting factors is central to ecology. The problem of defining limiting factors, and predicting the effect that their alteration will have upon ecosystem organization, is of key importance in formulating a predictive theory of ecology. In addition, this predictive ability will be important in predicting the impact of environmental alteration, such as increasing water temperature by nuclear power plant deployment, upon ecosystems. We might hypothesize, for instance, that a 5 degree increase in water temperature would have less effect in tropical waters, where temperatures may not be an important limiting factor, than in temperate latitudes,

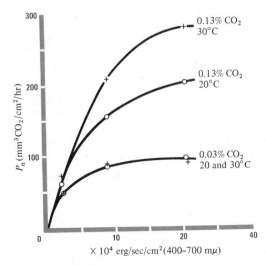

FIGURE 2-11 Relationship between photosynthetic rate of a leaf and light intensity at different combinations of temperature and carbon dioxide supply. The leaves become light saturated at progressively greater intensities as carbon dioxide concentration and temperature are increased; temperature was an important limiting factor only at higher levels of carbon dioxide. (After Gaastra, 1963.)

where temperature may be far from the optimum for certain organisms. Identifying limiting factors in ecosystems depends on a knowledge of the physiological requirements of the biotic component as well as the distribution of factors in the abiotic component. A 5 degree temperature change when water temperatures were 7 degrees would have much less effect on the relative hatching success of mackerel and cod (Fig. 2-3) than the same change when water temperatures were near 12 degrees. The problem of distinguishing between substantial and trivial environmental changes caused by growth of human populations and technologies is essentially a problem of accurately defining limiting factors in affected ecosystems.

ENERGETIC ORGANIZATION OF ECOSYSTEMS

Energy is a common requirement of all organisms for maintaining themselves and for reproducing. One way to organize our information about the kinds of organisms in an ecosystem is by considering the movement of energy within the ecosystem. Patterns of energy movement can be organized at a simple level by listing how many transfers take place. Each time energy is stored in an organism, its movement is temporarily stopped until that organism serves as an energy source for another organism. Each of these temporary stopping points in the movement of energy is called a *trophic level*. The trophic levels can be divided into three major categories depending on the source of energy to support the level and the destination of the energy that is stored at the level. Green plants and some chemosynthetic bacteria convert either the sun's energy or chemical bond energy into types of chemical bond energy that can be used by other organisms. They are called *primary producers*. They serve as energy converters and are the fundamental energy sources for the remainder of the organisms in the ecosystem. All other organisms are incapable of manufacturing their own food and are dependent on the primary producers. The amount of energy available to support organisms depends primarily on the ability of the producers to convert and store the sun's energy.

The two categories of nonproducers are consumers and decomposers. *Decomposers* utilize energy and materials stored in previously dead organisms, while *consumers* eat live prey or kill the prey before eating. The categories tend to merge for organisms such as scavengers. Our major distinction, although somewhat difficult to always use precisely, will be whether the food is killed by the nonproducer (consumers) or is exploited as already dead material (decomposers). Some of the energy and material eaten by consumers and decomposers is stored in their tissues; the rest is released to the environment. Eventually all stored energy is transformed to heat, but this may require millions of years for fossil fuels. The amount of dead organic material present in the environment depends on the balance between rates of production, consumption, and decomposition. Decomposers may increase the potential biomass in an ecosystem by releasing nutrients from dead organisms so they can be reincorporated into living organisms. Decomposers are essential in a functioning ecosystem because of their role in release of stored minerals from organic debris. It is difficult to assign decomposers to specific trophic levels, but to the extent that it can be done, decomposers are included unspecified in discussions of trophic level relationships.

The other nonproducer component of the ecosystem, the consumers, while functionally important, is nonessential in terms of either energy fixation or nutrient release. Some consumers are nutrient and energy traps that slow the release of these ecosystem components to the environment; others may accelerate decomposition. Consumers utilize organic material before it reaches the decomposer portion of the ecosystem. Potentially, then, the consumers and decomposers are both attempting to use the same energy fixed by the primary producers unless other factors in the environment keep populations of either nonproducer from becoming energy limited. Presumably geologic periods of coal formation occurred when some factors in the environment kept decomposer and consumer populations lower than could be supported energetically; thus excess production was stored as fossil fuels (Hairston, Smith, and Slobodkin, 1960).

Consumers live off the energy stored in living organisms, including other consumers. The initial consumer trophic level above the green plants is composed of *herbivores*, or animals that use plants as their main energy source. Trophic levels above the herbivores are all *carnivores* and depend on animal tissue as a source of energy.

Because each trophic level is dependent on energy from the previous level, each successive level can, in general, support no more organisms than the previous level. In fact, in most cases each successive level actually supports fewer organisms than a previous level, primarily because of the inefficiency of energy transfers from level to level. Charles Elton, a British ecologist, defined the structure of the biotic component of an ecosystem by arranging numbers of organisms at each trophic level in a graphic system, called the Eltonian pyramid of numbers (Fig. 2-12). This graphic portrayal gives a quick view of the relative abundance of the organisms comprising each level. However, a mouse-sized herbivore obviously is not functionally

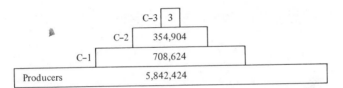

FIGURE 2-12 A pyramid of numbers with three consumer levels. Plant data are from Evans and Cain, 1952; animal data from Wolcott, 1937. (Figure after Odum, 1959.)

equivalent to an elephant. To remedy the problem of size disparity among species that occupy the same or different trophic levels one can convert numbers of individuals in a trophic level to total biomass or weight of individuals at that level, thereby producing a biomass pyramid (Fig. 2-13). In this sense 1 gm of mouse tissue is considered equivalent to 1 gm of elephant tissue.

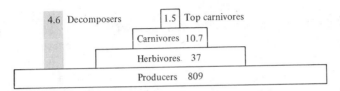

FIGURE 2-13 Pyramid representation of the standing crop biomass (gm/m²) arranged by trophic level for Silver Springs, Florida. (After H. T. Odum, 1957.)

Both types of pyramids share the possibility that at times they may be inverted (Fig. 2-14). This is contrary to the expectation, stated earlier, that each successive trophic level should be relatively smaller. The major problem is that when enumerating organisms either in terms of individuals or grams one can only take into account individuals that are alive when the census is made. This means that organisms with longer life expectancies, but requiring less energy per gram per unit area, can, in fact, be more common at any given time than species at a lower trophic level with shorter life spans, smaller body size, but processing more energy as a population per time period. Populations with a rapid turnover of small individuals can still provide sufficient energy to support more individuals and biomass of the next higher trophic level, if the measurements of numbers or biomass are made at specific times rather than over long time periods.

A trophic level pyramid will always be of the classical pyramidal form, if each level is described in terms of the amount of energy that passes through the level.

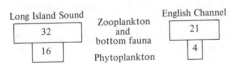

FIGURE 2-14 Inverted pyramids of standing crop biomass (gm/m²) in two marine communities. (After Odum, 1959.)

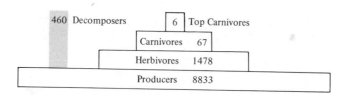

FIGURE 2-15 Net productivity (kcal/m²/yr) of the several trophic levels in Silver Springs, Florida. (After H. T. Odum, 1957.)

in a period of time. Some energy is lost as it moves from level to level, thus reducing the total energy flow at each successive level. An energy flow pyramid (Fig. 2-15) indicates the efficiency of energy transfer between trophic levels and the potential number of levels that can be supported in a system, on the assumption that there is a minimum energy requirement to maintain a trophic level.

FOOD CHAINS AND WEBS

A more detailed approach to the study of energy flow in ecosystems would be to follow the movement of a small, hypothetical packet of energy as it moves from organism to organism and is finally dissipated as heat into the atmosphere (Fig. 2-16). Following a single packet of energy leads to a straight line movement of energy from individual to individual along a so-called *food chain* in which a link is represented by each organism that takes up, stores, then passes along the energy packet. However, there are probably no ecosystems in which the organisms occur only in independent food chains. A food chain presents a detailed, straight line analysis of energy movement in an ecosystem. It, however, reduces information on the interrelated structure of the whole system.

A slightly more difficult, but ecologically more realistic method of charting pathways of energy flow is by following a series of energy packets that start at various points in the first trophic level. Several energy packets starting in a single species of green plant would quickly ramify throughout the ecosystem because many different species of animals probably feed on the plant, several carnivore species feed on the herbivores, and so on (Fig. 2-17). The proliferation of interconnecting relations based on energy utilization leads to the concept of *food webs,* or series of interlocking food chains, as the usual route of energy movement in an ecosystem.

As the number of alternate sources of energy available to a link in a food web increases, the probability that loss of one portion of the web causes extinction of the whole system decreases, at least within certain limits. To an extent, a more complex food web is more stable in terms of buffering against environmental fluctuations that adversely influence components of the system (MacArthur, 1955). However, Paine (1969) has suggested that certain "keystone" species may produce an effect on the stability of an ecosystem disproportionate to their abundance. Removing one species of starfish (*Pisaster ochraceous*) caused a marked change in the species present and their relative numbers; other species of starfish probably would have had much less of an effect (Paine, 1966; 1969).

Trophic
level

I

II

III

FIGURE 2-16 Simple food chain on the arctic tundra near Point Barrow, Alaska.

35

FOOD CHAINS AND WEBS

Several factors tend to set a theoretical upper limit to the complexity of an ecosystem. Most of these revolve around maintaining a minimum-sized population to insure the continued presence of a species. Consumers must find enough prey to meet energy costs. For sexually reproducing forms, minimum numbers may relate to the number of individuals that are necessary so that at least one male and one female find each other during the mating period. There is also a greater risk that a small population will encounter chance environmental changes that cause extinction. For example, the heath hen population on Martha's Vineyard apparently was holding its own and even increasing somewhat in the early 1900s after being drastically reduced when man intruded onto the island. However, the population was sufficiently low that a combination of adverse influences, including a fire and nu-

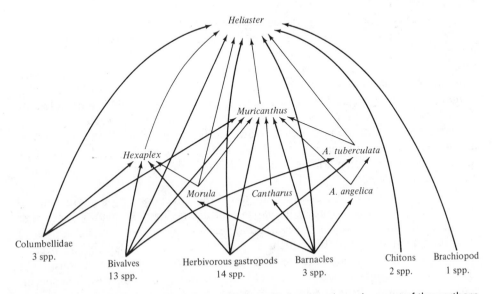

FIGURE 2-17 A portion of the food web in the intertidal region along the coast of the northern part of the Gulf of California, Mexico. The component species are mollusks, a brachiopod, and a starfish. (After Paine, 1966.)

merous predators, apparently reduced the heath hen population to a too small size and it became extinct (Allee et al., 1949).

NICHE

A more detailed way of analyzing the organization of an ecosystem than patterns of energy movement relates to what each organism actually does in the ecosystem. Habitat and environment each characterize the way species relate functionally since each term is defined by the activities of the organism. Another term, essentially the same as our previous definition of environment, is niche. MacFadyen (1957) defined the *niche* of a species as a "certain set of ecological conditions under which a species can exploit a source of energy effectively enough to be able to reproduce and colonize further such sets of conditions." It remained for Hutchinson (1957a) to provide an operational definition of niche. Hutchinson pointed out that for each individual (or species) there are environmental factors that are important in determining whether a species can survive and reproduce in a certain area.

Along each of these environmental parameters an individual or species can survive and reproduce within only a certain range of values. Usually they do best at a certain point in the range and worse at other points until they decline to zero ability to survive and to reproduce at the end of the range. Hutchinson described the niche in terms of all these parameters and the ranges of exploitable values along each parameter. To simplify the concept of the niche, he limited his initial presentation to considering the theoretical situation in which a species either is successful at a certain point along the environmental parameter or it is not—an all-or-none effect.

Graphically these factors can be represented along the axes of a one-, a two-, a three-, to an n-dimensional graph. The first axis might be the range of pH over which the species can survive and reproduce; the second axis the range of temperatures; the third the size of insects that can be eaten. A three-dimensional representation of these three parameters would produce a solid figure. Adding further dimensions, or niche parameters, would produce what Hutchinson called an n-dimensional hypervolume as a representation of the niche of a species. This niche represents the sum of the interactions between the organism and all environmental parameters that are influencing it and is called the *fundamental niche*. The actual description of a fundamental niche requires isolation of the species from nongenetic restrictive influences.

Most organisms do not occur throughout the entire range of environmental conditions that they are genetically capable of occupying. The actual dimensions occupied in nature constitute the *realized niche*. Several species may share parts of their fundamental niches; their niche hypervolumes may overlap. Niche overlap results in simultaneous demands upon the same resource by two populations. If the resource is in insufficient supply to meet the demands of both species, interspecific *competition* ensues. If the species are unequally efficient in the overlap zones, the abundance of the less efficient species will be limited by the interaction with the more efficient species. Restriction of the abundance of one species by a more efficient

competitor is called *dominance*. If the poorer competitor is eliminated from the niche
overlap zone, then its fundamental niche has been restricted by the presence of the
second species. When ecologists discuss niche relations they are usually considering
the realized rather than the fundamental niches.

Certain portions of the niche hypervolumes of two species may overlap and not
lead to displacement of either species from that part of its fundamental niche. For
example, temperature will be included in the fundamental niche of all species, but
one species will not actively displace another from the temperature portion of the
hypervolume. Similarly, pH is a physical factor in the environment that organisms
do not compete for and hence will not be a gradient along which hypervolumes
are actually displaced. It is possible that a species may not occur at some portions
of its temperature or pH range, because it is a less efficient competitor than other
species along another gradient that is associated closely with either temperature or
pH; but the reduction of the niche hypervolume will result from active interactions
along gradients other than temperature and pH. In other words, organisms compete
for resources, not regulators. Thus a species of algae might be excluded from hot
springs above a certain temperature, not because it cannot grow there or because
it loses in temperature competition, but because another species of alga can take
up a limiting nutrient more efficiently at those temperatures.

This definition of niche is organism-oriented, as was the definition of environment
presented earlier. A niche is defined by the species filling it. Certain parameters may
be unutilized or only partly utilized and capable of supporting further species in
the ecosystem. Dixon (1960) showed that the Chestnut-backed Chickadee (*Parus
rufescens*) successfully invaded the San Francisco Bay area of California with no
apparent detrimental effect on the most probable competitor (Root, 1964). This
means some energy sources were unexploited before the arrival of the chickadee,
at least by the resident bird species. New islands also present a series of unexploited
resources, and species can be added to island biotas for some time before niches
begin to overlap along certain critical parameters and competition occurs (MacArthur
and Wilson, 1967; Wilson, 1969; Simberloff and Wilson, 1969). Hence parts or all
of certain potential resources can be unutilized. They are not, however, included
in niches until they enter the reaction system of a species.

It is very difficult to actually delimit the niche of a given species. Many categories
of interactions must be investigated, and it is often difficult to decide just what all
the n parameters might be. However, many ecologists have been interested in some
particular portion of the n-dimensional hypervolume of a species and have provided
data that allow examining certain aspects of the niche parameters.

Among the easiest parameters to measure are physical aspects of the environment.
Each organism has a range of physical conditions that are suitable for life. For many
of these conditions the range for a group of species is similar; some parameters,
however, tend to be associated with specialized forms. The easiest method for
assessing the physical requirements of a species is to examine a range of values
along a given parameter and to ask if the species is able to exist and reproduce
throughout the range or only at certain points. Most of these techniques for describing the niche of a species have the inherent problem that measurements are taken

in nature and thus describe the realized niche, rather than the fundamental niche. Rogers and King (1972), for instance, examined the distribution of grasses in English pastures in relation to soil pH and moisture. The abundances of several species were examined in relation to various combinations of these two soil characteristics. They found that the two species had quite different realized niches in terms of these two niche dimensions (Fig. 2-18). One species of bentgrass (*Argrostis canina*) was most abundant in wet soils at low pH. Another species (*A. tenuis*) was most common at higher pH and drier soils. Although these are two-dimensional representations of the species' realized niches, such an analysis could be expanded indefinitely given sufficient dedication to the task. The important point is that these two closely related species could be segregated ecologically by measuring their distributions in relation to two of the realized niches' *n*-dimensions.

Cairns (1964) obtained data on the occurrence of 20 species of protozoa in 202 aquatic localities and at the same time measured 19 physiochemical characteristics of each habitat. He reported data for only those 20 species which were represented in his samples by six or more individuals. He eliminated the possibility of chance occurrences of species that normally do not survive in a particular habitat. Unfortunately, he did not present abundance data so the niche parameters must be con-

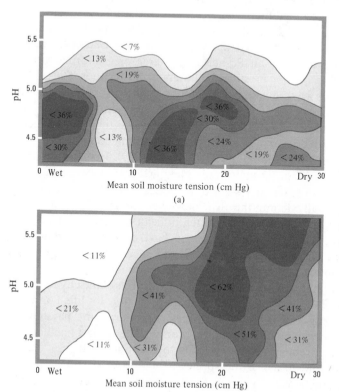

FIGURE 2-18 Distribution of two grass species in a community in relation to combinations of available soil moisture and soil pH. The data are percentage cover that the species contributes to the community at a certain moisture and pH combination. *Agrostis canina* (a) was most abundant at low pHs and high moisture, while *A. tenuis* (b) was most abundant at higher pHs and drier sites. (After Rogers and King, 1972.)

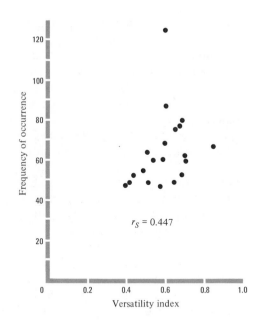

FIGURE 2-19 Relationship between the frequency of occurrence of 20 protozoan species and a versatility index devised by Maguire (1967). Total possible occurrences was 202. There is a general tendency for species with a higher versatility index to be found at more sampling sites. See text for method of calculation of the versatility index.

sidered in terms of presence or absence of each species. Using Cairns' data, Maguire (1967) established two versatility indices that summarize the niche hypervolume of each species in terms of the 19 parameters Cairns measured. One of these indices ranked each physical or chemical measurement from 0 to $N - 1$ in terms of the relative range of values occupied along that parameter; the species with the narrowest tolerance was given a rank of zero and the species with the broadest tolerance the rank of $N - 1$, which is 19 in this example since 20 species were considered. Each rank was then expressed as a relative value of the highest possible rank (19 in this case) for each environmental characteristic. For a single species the ranks were then summed for each parameter and divided by the total number of parameters (19) to give an average rank over all parameters.

Maguire assumed that his versatility index was a good first-order approximation of a portion of the fundamental niche of each protozoan species. The approximation should reflect the relative size of the niche hypervolume, which in turn should reflect the average tolerance of each species to the 19 environmental conditions. Maguire presented other evidence that the frequency of occurrence of each species in the 202 bodies of water was probably not related to their dispersal ability. He then reasoned that the tolerance range of a species, as reflected in its relative fundamental niche size, should relate to the frequency of occurrence in the diverse bodies of water that Cairns sampled. By relating frequency of occurrence to relative niche size (Fig. 2-19), Maguire found that the two variables were indeed positively related, thereby indicating that size of niche hypervolume for the 19 characteristics is a predictor of the relative frequency of occurrence of the species. However, the low correlation coefficient ($r_s = 0.45$) indicates that factors other than those used to generate the versatility index are also responsible for the observed distribution of protozoan species.

This approach to a niche description could be modified meaningfully by reflecting relative abundance of a species at each point along an environmental gradient. This, however, requires some arbitrary division of the gradient. If it can be done in a reasonable fashion (see Colwell and Futuyma, 1971), then we might derive a niche volume index by combining means and variances along all n parameters to arrive at a value that reflects the ability of the species to occupy those environmental conditions. In this case the measure would be in terms of relative success rather than simply presence or absence.

An important dimension of the niche is the energy source for the species. For heterotrophs, food items can appear in a variety of forms, necessitating a series of niche dimensions. The actual dimensions utilized are determined largely by the type of organisms being considered. Herbivores will have quite different niche parameters for food from carnivores, and even within a trophic level it may be difficult to determine appropriate parameters. For insectivorous birds one parameter might be taxonomic categories of insects, which on a broad scale reflect ecological categories; another parameter might be size classes of insects; another might be flying abilities of the prey. Each or all of the potential differences in prey characteristics might be a parameter along which a bird species operates. It probably will be extremely difficult to actually define the parameters associated with certain aspects of the environment, such as the appropriate parameters along which to measure energy exploitation.

Root (1967) examined the niche relationships of a group of bird species that feed on arthropods that inhabit oak trees, primarily oak foliage, in coastal mountains of central California. The five bird species tended to vary in morphological and behavioral characteristics that were more or less reflected in differences in diet. He found that although each bird species was foraging in the same area, each tended to eat insects of somewhat different taxonomic categories and different sizes (Table 2-2 and Fig. 2-20). There was complete overlap in size, but the mean values and variances tended to be somewhat different, although some species ate similar sizes

TABLE 2-2 Arthropods in the diets of sympatric, foliage-gleaning birds. The data are expressed as the percent of total prey individuals (*n*) that could be identified

	Polioptila caerulea	*Vireo gilvus*	*Vireo huttoni*	*Vermivora celata*	*Parus inornatus*
Hemiptera	36.0	10.3	11.9	47.8	13.2
Coleoptera	32.3	15.0	29.8	6.5	55.3
Lepidoptera	7.1	62.0	24.6	37.0	6.5
Hymenoptera	13.8	6.6	22.4	4.3	10.2
Other	10.8	6.1	11.2	4.3	14.8
n	287	213	134	46	81

Root, 1967.

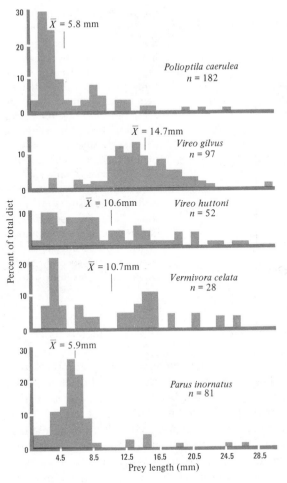

FIGURE 2-20 Length of intact prey in the stomachs of five species of foliage-gleaning birds collected in oak woodland in central coastal California. The birds are the Blue-gray Gnatcatcher (*P. caerulea*), Warbling Vireo (*V. gilvus*), Hutton Vireo (*V. huttoni*), Orange-crowned Warbler (*V. celata*), and Plain Titmouse (*P. inornata*). *n* is the total prey items measured. (After Root, 1967.)

of prey. Taxonomic categories of insects eaten also overlapped, but each bird species tended to specialize on certain taxa. Root also found that the five bird species fell into three general categories with respect to foraging technique (Fig. 2-21). (a) Two species were primarily gleaners, taking their prey from leaf surfaces while the bird walked on a solid substrate through the foliage. (b) One species foraged mostly by gleaning, but also caught some insects on the wing (hawking) and hovered near foliage to capture insects sitting on the foliage. (c) The other two species divided their time about equally between gleaning and hovering with very little hawking. On the assumption that insects differ in behavior, these differences in bird behavior could lead to differences in insect types eaten by bird species in each foraging category and probably within foraging categories. Hence although these five bird species were foraging in about the same place, each was capturing insects with different combinations of sizes, taxonomic types, and behavior. The total available insect fauna was being partitioned among the species in various ways that could

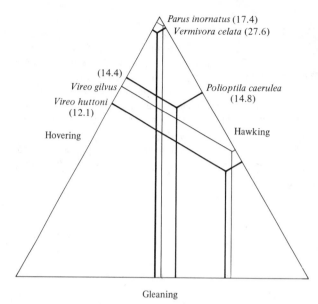

Hovering

Gleaning

Hawking

Parus inornatus (17.4)
Vermivora celata (27.6)

(14.4)
Vireo gilvus
Vireo huttoni
(12.1)

Polioptila caerulea
(14.8)

FIGURE 2-21 Foraging tactics of members of the foliage-gleaning guild of bird species inhabiting oak woodland in central coastal California. The three types of foraging maneuvers used by these birds are represented as the three sides of the triangle. The length of a line perpendicular to the side of the triangle is proportional to the percent of time spent in that foraging maneuver by a bird species. The sum of all three lines for each species equals 100 percent. Numbers in parentheses are attacks on prey per 500 seconds of observations. (After Root, 1967.)

be described as dimensions of the ecological niche—cast as parameters associated with the insects.

There tended to be considerable overlap among the birds in the parameters Root measured, meaning that the realized niches of the species overlapped to some extent. If the availability of these shared resources should become limited, by our earlier definition these species would compete for the resource(s) in short supply. Competitors could have an important influence on the size and shape of the realized niche of a species and by extension might influence characteristics of the fundamental niche. This, however, is more difficult to show experimentally because it involves genetic change in the capabilities of the organism. However, competitive ability can be selected for, thereby reflecting a change in the size and/or shape of the fundamental niche (Moore, 1952; Ayala, 1969a).

Selander (1966) discussed the potential effect of competitors on the range of foraging activities of bird species, primarily woodpeckers. He concluded that where there were few or no other species of bark-foraging birds, woodpeckers tended to expand their foraging capabilities. It was done primarily through sex differentiation in morphological traits associated with foraging (Fig. 2-22) and in foraging behavior of the sexes as measured by the type of foraging maneuvers (Fig. 2-23). Gill (1971) also found "competitive release" in a species of white-eye (*Zosterops*) that colonized Reunion Island in the Indian Ocean. Here the change in diet included a shift toward nectar feeding, presumably related to the absence of other nectar-feeding birds that normally occur with white-eyes in continental situations.

Harper and his associates in Wales have also examined how the realized niche is defined in nature. The primary assumption of their work has been that the abundance of a species will be a function of its own adaptive properties and of the

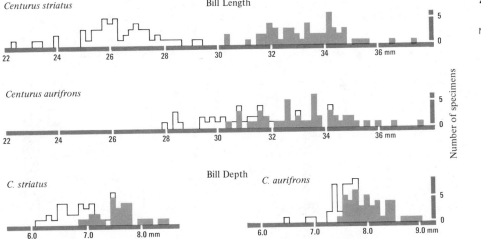

FIGURE 2-22 Variation in bill length and bill depth in two species of *Centurus* woodpeckers. *Centurus striatus* is an island species; *C. aurifrons* is from a mainland population. Colored histograms are males; open histograms are females. (After Selander, 1966.)

distribution of resources in the habitat as these are affected by other organisms in the same community. Rather than asking what is the niche configuration of a species, they have asked how niches overlap in communities.

Putwain and Harper (1970) used herbicides to selectively remove competitors from grassland communities and examined the effect that this removal had on the abundances of two species of sorrel, *Rumex acetosa* and *R. acetosella*. They studied the effects of removing all species but the sorrel (*E* in Fig. 2-24), all non-grasses (*C*), all grasses (*D*), or the effect of removing all sorrel upon subsequent abundance of sorrel (*B*), as compared to a control with no species removed (*A*). The effects of the different treatments were somewhat different for the two species studied.

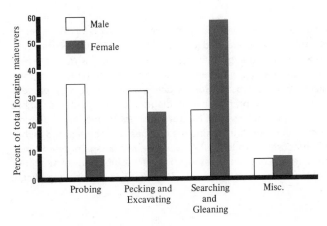

FIGURE 2-23 Comparative foraging maneuvers of male and female woodpeckers of the species *Centurus striatus*. (Data from Selander, 1966).

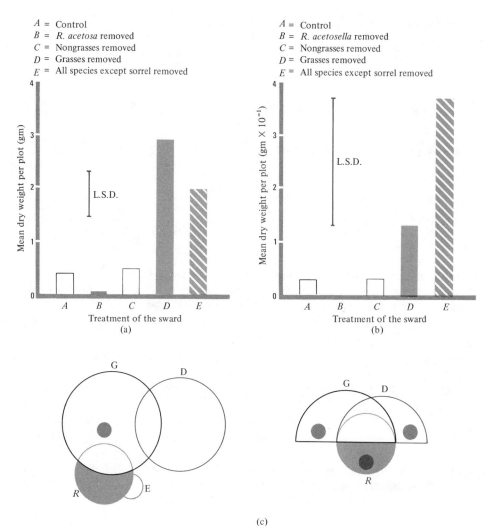

FIGURE 2-24 Effect of removing competitors from a grassland on growth of two species of Sorrel (*Rumex*): (a) *R. acetosa,* (b) *R. acetosella,* and (c) niche diagrams for the two species. Note that *R. acetosa* grew better if only the grasses were removed, while *R. acetosella* showed a significant growth stimulation only if all species were removed. In (c) the diagrammatic representation of the niche relationships of *Rumex acetosa* (left) and *R. acetosella* (right), was deduced from the removal experiments. In each diagram the fundamental niches of grasses (G) and all herbs except the *Rumex* species (D) are indicated as overlapping areas. Fundamental niches of the *Rumex* species (R) are continuous outlines with the realized niches colored. E is that portion of the fundamental niche of *R. acetosa* which is expressed when grasses were eliminated and does not overlap with G and D. Fundamental niches of seedlings are the lightly colored areas. (After Putwain and Harper, 1970.) L.S.D. is the least significant difference between treatments.

For *R. acetosa,* removal of the grasses resulted in a substantial increase in growth (Fig. 2-24a). This indicated that there was niche overlap between the sorrel and the grasses and that the grasses were better competitors in the overlap zone. Removing all nongrass species had no effect, which demonstrated that this sorrel had little niche overlap with other nongrass species. Removing the species itself resulted in no subsequent establishment, thus suggesting that maintenance of the population was dependent upon seed from established plants and that the invasion potential from outside was small. Significantly, the removal of all species except *R. acetosa* resulted in less of a population stimulation than merely removing the grasses, which suggests the interesting possibility of some cooperative effect between *R. acetosa* and non-grasses in suppressing the grasses.

For *R. acetosella,* in contrast, the only treatment that had a significant promotive effect was eliminating all other species in the stand (Fig. 2-24b). There was some promotion on removing the grasses alone, none at all when only the nongrasses were removed, but a substantial increase in yield upon removing all competitors. This suggests that there is a concerted suppressive effect of a large number of species upon *R. acetosella.* From these experiments Putwain and Harper constructed a diagrammatic representation of the niche structure of the two sorrels, relative to the niche structure of other species in the stand (Fig. 2-24c). *Rumex acetosa* has a fundamental niche that overlaps with the grasses, but it does not overlap with nongrasses in limiting niche dimensions. Its realized niche is restricted by the presence of the grasses as a result of unequal efficiency in the overlap zone. The other sorrel, *R. acetosella,* has a fundamental niche that overlaps with both grasses and nongrasses, and both classes of competitors were effective either in concert or when competing separately.

It would be interesting to know also what effect elimination of the sorrels had upon growth of the other species so that we could more accurately assess the comparative efficiencies in the overlap zones. These data, however, were not reported by Putwain and Harper. Their approach does provide considerable insight, however, into the comparative, rather than the absolute niche structures of the species. The niche relations within the overlap zones are clearly complex, with multiple interactions of both a cooperative and disoperative nature within these overlap zones.

Predators (in the broad sense of herbivores and carnivores) can influence the exploitation patterns of prey populations and hence can influence their realized niches. Paine and Vadas (1969) reported that the number of species of algae occurring in an area showed an initial increase after sea urchins (herbivores) were removed or drastically reduced. Apparently the herbivores ate certain species to such an extent that the plants were unable to maintain viable populations in areas of high numbers of herbivores despite their physiological capabilities of inhabiting the area. At a later time the total number of species declined and changed somewhat in specific composition from the initial samples, indicating that competitive interactions reduced the probability of a species occurring in a particular area. Conversely, Harper (1969) showed that in grassland ecosystems the number of plant species initially increased when the grasslands were grazed.

A potential method of escaping from predators and competitors is to utilize the area at times mutually exclusive from the other forms. To reduce the influence of competitors the assumption must be made that the resources taken by a species at one time does not influence what is available for the second species at a later time. This method of predator and competitor escape leads to distinct temporal patterns of utilization of certain resources, which in turn influence the exploited range of other parameters, such as temperature and relative humidity.

Temporal patterning of behavior and the offsetting of time of activity of several species can occur as daily or seasonal (or longer) patterns. Linsley, MacSwain, and Raven (1963) showed that some bee species visiting ephemeral species of desert plants in the genus *Oenothera* tend to have offset times of daily activity even though they visit the same flowers (Table 2-3). The fruiting seasons of 18 species of second growth trees in the genus *Miconia,* which co-occur on the island of Trinidad in the West Indies, tend to be somewhat distinct although overlapping (Fig. 2-25; Snow, 1965). Wolf (1970) found that although there was a major peak of flowering activity of trees in a seasonally dry area in Costa Rica (Janzen, 1967), the flowering period of the numerically predominant species tended to be temporally exclusive, which increased the probability that pollinators would successively visit different individuals of the same plant species. On the other hand, Ricklefs (1966) reported that temporal displacement of breeding activities probably was not an important method of reducing competition in sympatric species of tropical birds in the areas he surveyed.

There are many other potential niche gradients, but the foregoing discussion indicates some dimensions of the *n*-dimensional hypervolume and illustrates techniques for examining the fundamental niche and the realized niche for species. Major difficulties are encountered in describing the niche of a species, because the abundance of the species normally decreases gradually at the extremes of the exploitation

TABLE 2-3 Comparative foraging times of three species of *Andrena* bees visiting the same species of plant, *Oenothera clavaeformis,* west of Austin, Nevada, on June 6 and June 7, 1960[a]

	No. of visits by bee species		
Time of day	A. chylismiae	A. raveni	A. roseni
June 7			
4:50–5:40	2	0	0
5:41–6:20	3	1	0
6:21–7:00	2	19	0
7:01–7:17	3	7	0
June 6			
15:50–16:10	0	2	0
16:11-16:50	0	26	5
16:51–17:30	0	25	17
17:31–18:16	0	10	4

[a] Some individuals arrived with and some arrived without pollen loads. Data from Linsley, MacSwain, and Raven, 1963

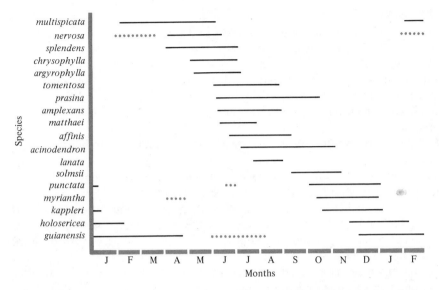

FIGURE 2-25 The seasonal progression of fruiting by 18 species of trees of the genus *Miconia* (Melastomaceae) on the island of Trinidad, British West Indies. Dots indicate occasional out-of-season fruiting. (After Snow, 1965.)

range on a particular environmental factor. For example, at extreme temperatures a species may survive and reproduce, but it reproduces at a much slower rate than it does at more moderate temperatures.

Colwell and Futuyma (1971) pointed out that units used by ecologists to measure environmental gradients may not be equivalent to the units recognized by organisms exploiting the gradient. A change from pH 2 to pH 1 may require much more genetic and physiologic rearrangement before individuals can exploit the lower pH than would be required to change from exploiting pH 7 to pH 8. Viewed another way, a range of 3 pH units near the middle of the tolerance range for life may be equivalent in the necessary genetic and physiologic specializations required to occur there as would a half unit range of pH at either extreme. The ecologist would view these organisms as about six times as different in their abilities to exploit pH gradients, but in fact the physiological and genetic requirements necessary for the species to exploit their particular portions of the gradient are about identical.

The resolution of this dilemma is presently probably beyond the reach of ecologists since there are so few data on the relative adjustments required to exploit portions of niche gradients. A further problem encountered in defining the fundamental niche is the possibility of synergistic effects of co-occurring species. In the studies of Putwain and Harper (1970) mentioned earlier in this chapter, *Rumex acetosa* grew better in the presence of nongrass herbaceous vegetation than when growing alone. Removed from this synergistic effect *R. acetosa* would appear to have a smaller fundamental niche than when grown in the presence of the nongrasses.

Each thing a resource of a regulator (margin handwriting)

GENERAL CONCLUSIONS

1. Environment is the sum of those substances and forces external to an organism that become internalized as resources or regulate the manner of resource utilization.
   (a) Environment is organism-directed, ordered by the life cycle of the organism, and both affects and is affected by organism processes. *weeds grow → change soil*
   (b) Environment may be factored into energy, chemical substances, and other organisms.
   (c) Pollutants are products of an organism's activity that are inhibitory to that organism. *part 17*
2. An organism process will be limited by that factor or factor combination, the value(s) of which is (are) farthest from process requirements.
3. Energy movement between organisms is organized in a series of trophic levels characterized by their feeding distance from the initial energy source, the sun.
4. The amount of energy passing through a trophic level decreases with distance from the initial energy source.
5. An organism's niche is an n-dimensional hypervolume, where n is the number of environmental factors affecting the organism.
   (a) The fundamental niche is determined by the genetic properties of the organism.
   (b) The realized niche is determined, as well, by resource structure of the ecosystem and the occurrence of other organisms.

SOME HYPOTHESES FOR TESTING

Most ecological investigations are pointed toward understanding more inclusive phenomena than those which relate to the functioning of the individual organism in a certain set of ecological conditions. The latter approach is often called environmental physiology. It should be apparent by now, however, that understanding more complex phenomena ultimately may depend upon knowledge of the individual-environment interface.

Since the response of a population to an environmental factor will depend upon the responses of its component individuals, understanding the latter may be imperative in certain instances. We saw, for instance, that the effect of temperature upon growth of pepper plants varied dramatically according to age of individuals. Therefore our ability to predict population responses to temperature would depend upon our knowledge of population age structure and age-dependent physiology. An interesting phenomenon to examine would be the degree to which populations may stabilize their environmental response through different age structure. That is, would a population occupying a habitat subject to pronounced environmental fluctuations tend to have a complex age structure? And, would a population occupying a uniform habitat have a simple age structure synchronized for maximum efficiency at exploiting that uniform habitat?

We also know very little about the interactive effects among environmental factors. To what extent can there be simultaneous limitation of an organism by multiple limiting factors? Is size of a population in nature often limited simultaneously by, say, predation, temperature, and food supply? Or will a single one of these usually be limiting? Understanding the operation of limiting factors in nature, of course, is imperative if we are to develop an ability to predict the effect that environmental modification will have upon ecosystem organization.

Another interesting question that arises out of a consideration of the organism-environment interface is the extent of niche overlap in communities. To what extent are species with different abundance in a community unequally efficient in niche overlap zones? How may niches overlap to the degree documented by Root in the five bird species in California? Does the overlap of prey size taken by these birds indicate that they are not food-limited?

Finally, we may ask whether it would be possible to predict the realized niches of two competing species if we had good estimates of the properties of their fundamental niches. The experiments by Putwain and Harper indicated that species with partially overlapping fundamental niches may have unequal efficiences in the overlap zone. Would it be possible to measure these differential efficiencies in the laboratory so that we could predict the effect that competition would have in modifying the fundamental niche into the realized niche?

adaptation

3

Living organisms occur in diverse habitats, from the ocean depths to the polar ice caps. The key to this ability is *adaptation*. To an ecologist adaptation means both the *genetic process* by which organisms become increasingly better able to exist under prevailing environmental conditions and the specific *genetic trait* that renders an organism more capable of existence. A related, but nongenetic, process is *acclimation,* the modification of an organism's phenotypic traits by the environment. Thus adaptation refers to genetically determined phenotypic traits that generate the organism's ecological properties, whereas acclimation refers to environmental modification of gene expression. The distinction is quite important, as will become apparent in the discussion that follows.

VARIATION BETWEEN AND WITHIN POPULATIONS

In preceding chapters we have often referred to species as if they constituted ecological units. However, it has been known since 1922, from the work of the Swedish ecologist Turresson, that a species consists of local populations that may differ conspicuously in their adaptive properties. Turresson noticed that many plant populations in coastal habitats tended to grow prostrate, whereas inland populations of the same species often grew erect. The coastal populations were exposed to sea winds, salt spray, and shifting sands, and the inland populations were not. Turresson realized that the difference in growth forms might arise either from adaptation or acclimation. To discriminate between the alternatives, he moved samples of both populations to a single inland garden. The result was that both the coastal and inland samples maintained their distinct morphological forms over indefinitely long periods in the same habitat. This experiment provided strong evidence for the existence of genetically distinct populations within the same species adapted to different environments. For such populations Turresson coined the term *ecotype.*

The relative effects of adaptation and acclimation are illustrated in the work of Clausen, Keck, and Hiesey (1958), who examined populations of yarrow (*Achillea*) native to sites ranging from sea level to 3350 meters above sea level in California. Populations within each of the two species examined displayed marked differences in size when grown in the same garden in Stanford, California (Fig. 3-1). There was no significant difference in height between the two species studied. Taken together, the populations formed a continuous distribution, becoming taller from the coastal populations to the central valley and then shorter again toward the high mountain populations. Such a continuously changing character gradient is called a *cline* (Huxley, 1938).

Many other characters were found to vary clinally in yarrow populations, including both morphological and developmental traits. The association of a cline with an environmental gradient, such as the changing climates across California, is regarded as evidence of adaptive function. It is not a simple matter, however, to determine how important adaptation is in explaining population differences observed. For example, the tendency to stay active in the winter, which is a characteristic of many coastal populations, is an appropriate adaptation to the mild winter temperatures and midsummer drought in this part of California, whereas such a tendency is clearly maladaptive in harsh montane winters. But of what importance are characters such as height, which may not be of obvious advantage in different sites? Since height is controlled by a whole array of genes (Clausen and Hiesey, 1958) related to photosynthetic efficiency and growth capacity, such a character is used more as an index of vigor than of site-dependent survival value (as, for example, winter activity would be).

The degree of adaptation, or *selective value,* conferred by a trait is often referred to as *fitness.* Clausen et al. were able to estimate fitness by employing multiple transplant gardens. In addition to the Stanford site, they had gardens at Mather in the Sierra foothills and at Timberline near the Sierra crest. They took several individuals from each population, divided them into separate segments, called *clones,* and transplanted the clones to each site. Since each clone was derived from a single individual, it was then possible to evaluate the growth potential of a given genetic stock in three different climates. Comparison of Sierra foothill and alpine yarrow populations at the Mather and Timberline gardens demonstrated significant differ-

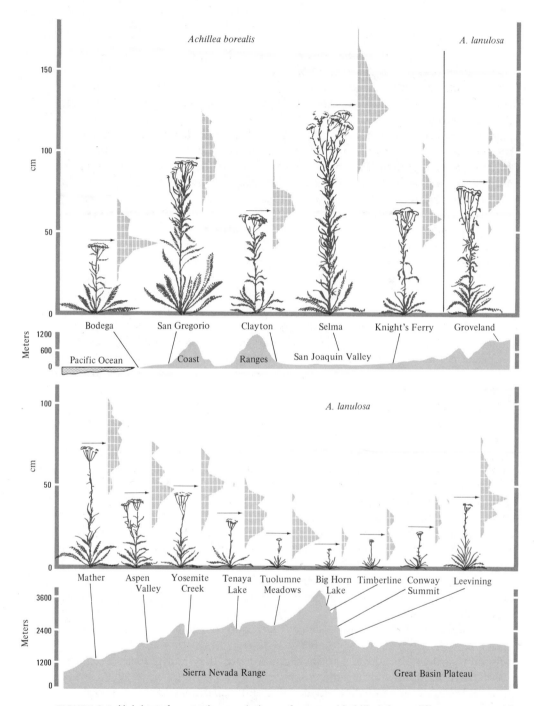

FIGURE 3-1 Heights of ecotypic populations of yarrow (*Achillea*) from different geographic locations across California when all samples were grown in Stanford, California. Above is the western half of the gradient starting at the Pacific Coast and below is the eastern half from the Sierra Nevada foothills to the Great Basin. The plants pictured represent the mean height for each population; frequency diagrams show height variation in the populations; horizontal lines separate 5 cm intervals and the distance between vertical lines represents two individuals. (After Clausen, Keck, and Hiesey, 1958.)

ences in the populations' fitnesses in the two different environments (Fig. 3-2). The
foothill population was taller in the Mather garden and the alpine population was
taller in the Timberline garden. Timberline was a particularly difficult site for the
foothill population and many of the individuals died there. Although alpine individuals were able to grow at Mather, they would be substantially overtopped by
foothill plants if the two were growing together in a natural community. Similarly,
foothill individuals capable of withstanding the winter at Timberline would not fare
well in the shade of the larger alpine plants at the same site.

This work demonstrated both of the processes by which an organism responds
to its environment. Adaptation was seen in the differential survival and growth
of individuals from different climates grown in the same conditions. Differential

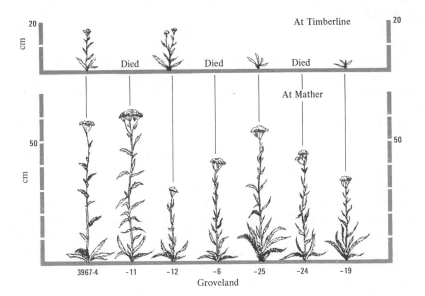

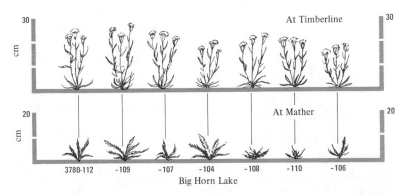

FIGURE 3-2 Size of foothill (Groveland) and timberline (Big Horn Lake)
Achillea individuals when grown at foothill (Mather) or timberline transplant
gardens. (After Clausen, Keck, and Hiesey, 1958.)

survival and growth, which persisted when plants from different climates were grown together in a new environment, were a result of adaptation in the species that suited the two populations to their distinctly different original environments. Different heights of individuals from the same population growing in the different gardens arose out of environmental modification of the expression of genetic traits, that is, acclimation.

Genetic variation in a population, exemplified by the distinct responses of individual yarrows to different climates, is the raw material of adaptive change. Such variation may arise from *mutation,* a change in the basic composition of the genetic substance (DNA), or from *recombination,* a change in the structural organization of the chromosome during meiosis that modifies the gene complement received by the progeny. Recombination is not a new genetic event in the same sense as mutation, but merely the reorganization of existing genes into novel combinations during the reproductive process. Mutation is the ultimate source of all genetic variation, but recombination may allow new expressions of this variation.

An important way in which variation through recombination may occur is by interbreeding among populations. Novel genes may thereby be introduced into populations that have evolved separately. These genes then may serve as the basis for genetic reorganization of the interbreeding populations.

Clausen and Hiesey (1958) performed intensive ecotype interbreeding experiments with common cinquefoil (*Potentilla glandulosa*) using populations from Timberline in the Sierras and Oak Grove in the foothills (Fig. 3-3). The traits of the first offspring generation, called the F_1, were intermediate between the traits of the two parents on an index devised to summarize differences between the parental populations. F_1 individuals were bred to generate a second offspring generation, the F_2. The 521 F_2 individuals ranged in character from individuals very much like the Oak Grove parent to individuals very much like the Timberline parent, with a normally distributed range of variation between. Interbreeding between the Oak

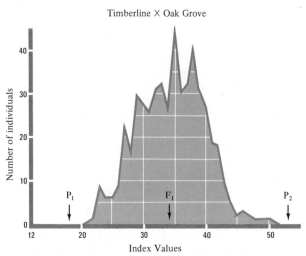

FIGURE 3-3 Segregating phenotypes in the F_2 generation of hybridization between ecotypes of *Potentilla* from high (Timberline, P_1) or low (Oak Grove, P_2) altitudes. (From Clausen and Hiesey, 1958.)

Grove and Timberline populations generated a substantial range of variation in the F_2 from which one could select individuals very much like either parent, or individuals quite distinct from each parent. The exchange of genes among natural populations is called *gene flow*. Such exchange can be an important factor in producing variability upon which natural selection may act.

EVOLUTION

At any instant in time a population will contain both old and novel genes. Old genes are those that have been present in the population for a considerable period of time. During this period, the frequency of the genes in the population depends on the mating patterns and natural selection occurring in the population.

Natural selection was first described by Darwin and Wallace in 1858. It was elaborated in Darwin's great book (1859), *On the Origin of the Species by Natural Selection*, as "survival of the fittest" in the "struggle for existence."

In less romantic terms, natural selection is the process of interaction between phenotypes and environments which determines the survival and reproductive output of individual genotypes. As reformulated since the rediscovery of Mendel's (1866) work on the inheritance of traits, natural selection is an environmentally directed change in gene frequency in a population over a period of time. Certain types of evolution, such as "genetic drift" in isolated populations, may not be a consequence of natural selection. That is, evolution need not be an adaptive response to the environment, although in most cases it is.

Most higher organisms have duplicates, at least, of every gene. Multiple alternative forms of the same gene are called *alleles*. If we imagine just two alleles, p and q, occurring in a population in specified proportions at a certain chromosomal position, we can see that the total frequency is given by $p + q = 1$. Furthermore, $p = 1 - q$ and $q = 1 - p$. If meiosis occurs normally, the alleles will be randomly segregated among gametes according to their frequency in the population. For instance, if p occurs at the chromosomal site in 60 percent of the population, it has a frequency of 0.6 and 60 percent of the gametes will carry p. The frequency of q, the only other allele, is then 0.4, by definition, and 40 percent of the gametes will carry allele q. There will be $p + q$ male gametes combining with $p + q$ female gametes. Reproduction will therefore yield

$$(p + q)(p + q) = (p + q)^2 = p^2 + 2pq + q^2 \tag{3-1}$$

in the subsequent generation (Hardy, 1908). In other words there will be p^2 individuals homozygous for p, q^2 individuals homozygous for q, and $2pq$ heterozygous individuals. For the population example chosen, with $p = 0.6$ and $q = 0.4$, we see that 0.36 of the individuals (36 percent) will be homozygous for p, 0.16 will be homozygous for q (16 percent), and 0.48 will be heterozygous (pq) (48 percent).

Equation (3-1) is called the Hardy-Weinberg law. The law holds for freely interbreeding populations, where gametes are distributed at random and there is neither mutation, selection, nor differential migration. Since we may expect all of these events to occur in any real population, the utility of the law is limited. It does, however, yield the general insight that reproduction tends to generate genotypic stability. In this sense, evolution may be described as the tendency of a population to depart from the gene-frequency equilibrium predicted by the Hardy-Weinberg equation. The mechanisms of evolution, then, are the influences upon the population that tend to generate such departure. Genes will tend to become distributed in a population according to the Hardy-Weinberg equilibrium, but the distribution will be modified by nonrandom mating, mutation, gene flow, recombination, and selection.

Novel genes in a population may originate through mutation, or by gene flow from other populations. In a population isolated from other populations by barriers to reproduction, novel genes must arise primarily by mutation. On the other hand, in a population in reproductive contact with other populations, novel genes may arise primarily by interbreeding. Thus variation in a population, upon which natural selection acts, must arise through mutation or through recombination involving old or alien genes.

Darwin (1859) recognized that the inherent variability of organisms could lead to differences in the abilities of individuals to survive and reproduce. If more individuals of a certain variety survive and reproduce, then the population will contain proportionately more individuals of that variety in the next generation. Differential contribution of individual genotypes to the next breeding generation leads to changes in the relative representation of genotypes in the next generation. A change in the relative gene frequencies is an evolutionary change. The possibility of observing these genotypic changes depends on how prominently they affect the phenotypic characteristics in the population. Not all genotypic variation that leads to differential reproduction need express itself as obvious phenotypic traits. Simple enzyme changes that influence metabolic characteristics would be sufficient to cause a change in energetic efficiency, leading in turn to different amounts of energy being channelled into reproductive effort and, hence, to change in total reproductive output. Whether this change would be a selective advantage or disadvantage depends on the relative contribution of the new genotype to the next breeding generation.

Natural selection in its simplest form involves environmental influences on phenotypes that determine the reproductive output of individuals with those phenotypes. Differential reproduction of genotypes producing a change in the relative frequency of genotypes in the next breeding generation is evolution. According to this concept, natural selection operates at the level of the individual, in terms of genetic contribution to the next generation, and not at the level of a population, community, or ecosystem as has been suggested occasionally.

That differential reproduction can lead to an evolutionary change in the population will be apparent from the following hypothetical example (Table 3-1). Assume that the number of eggs a bird lays is genetically determined and is subject to some degree of genetic variation, both of which assumptions are undoubtedly correct. During the spring, our hypothetical population of birds lays an average of 2.5 eggs per clutch, varying from one to four eggs per nest. Now assume that conditions are good this year, that the birds can easily find food for themselves and their young, and that all the young can find enough food after they leave the nest to survive

TABLE 3-1 Hypothetical distribution of clutch sizes in a population of birds[a]

	First year			Second year	
Clutch size	Number of clutches	Percent total clutches	Total young	Number of clutches	Percent total clutches
1	10	17	10	15	11.1
2	20	33	40	40	29.6
3	20	33	60	50	37.0
4	10	17	40	30	22.2
Average clutch size		2.50			2.70

[a] Assume female determines clutch size genetically; assume no mortality from egg laying to reproduction by the young; assume 50:50 sex ratio of adults and young.

until the next spring. The average clutch size will increase next year through relatively greater production of offspring carrying genes for large clutch size. If conditions remain good for several years, clutch sizes will approach an average of 3.0 rather than 2.5, representing a marked evolutionary change in the course of a few generations. Of course, there are many factors in the environment in addition to food availability that influence whether parent birds successfully raise their young. Each of these factors is a selective agent determining the direction of evolutionary changes in the population.

Certain types of adaptive behavior in populations have led some authors to suggest that differential reproduction of individual genotypes is not the only mode of selection operating in natural populations. A principal behavioral characteristic of this sort appears to be altruism. How, for example, does the caste system of insects evolve in which some individuals do not reproduce, but aid in the reproductive output of another individual? Classic examples of apparent altruism occur in social hymenoptera, such as bees, wasps, and ants.

Hamilton (1964) and Williams and Williams (1957) have shown theoretically that if the altruistic members of the caste system are sufficiently close genetically to the reproductive individual, their genotype actually may be at a selective advantage even though they do not breed themselves. This will be the case if the altruists can leave enough alleles via the reproductive individual to more than make up for their own lack of reproduction. Depending on how many genes the altruists share with the reproductive individual, they may influence the gene frequencies of the next generation more than if they tried to breed alone. The most likely circumstance in which this may occur is when close genetic relatives of the reproductive individual are the nonbreeding individuals. In many social insects, for example, the workers and soldiers are the offspring of the queen and are therefore closely related genetically to her.

Wynne-Edwards (1962, 1965) has postulated that selection at the population level rather than the individual level is primarily responsible for behavioral activities that tend to reduce the reproductive output of the population. Therefore the reproductive output approaches an optimum rather than a maximum value in relation to resource availability. He argues that this phenomenon, *group selection,* determines which populations will survive. These populations eventually colonize areas left vacant by the extinction of populations that have not evolved the appropriate internal regulatory mechanisms and have overrun the capacity of the environment to support them.

Territorial defense by birds is seen by some ecologists as an example of a trait evolved under group selection. They postulate that territorial behavior establishes an upper population size at which certain critical resources in the habitat are never so reduced that the population can no longer survive in the area. Mechanisms of population regulation inherent in the genetic characteristics of the population must have evolved by group selection because it would be disadvantageous for any one individual to reduce its reproductive output if its evolutionary success is measured by reproductive output.

These arguments are weak in two respects. First, reproductive success is not measured by the number of offspring produced but by the number of genes contributed to the next breeding generation. Returning to our previous example of the evolution of clutch size, suppose that during some years food is hard to find. Then pairs attempting to raise four young might produce fewer total young per pair than those trying to raise only three young. This would place the four young pairs at a selective disadvantage compared to pairs trying to raise only three young. Lack (1966) and his co-workers have data for the Great Tit (*Parus major*) which indicate that individuals in larger broods tend to weigh slightly less on the average and that there is a positive correlation between weight at the time of leaving the nest and the probability of survival. Second, territorial behavior can result from individual selection if the time and energy required to maintain a territory enhance the reproductive ability of the territorial individual(s). Although group selection is a very interesting theory of evolutionary mechanisms, most examples that are attributed to group selection can, in fact, be explained on the basis of classical natural selection acting on the individual.

SELECTION TYPES

Although we tend to think of natural selection as a directional process moving a population from one genetic constitution to another, Mather (1953) pointed out that such *directional selection* is just one of three possible ways the environment may act upon a population's genetic structure (Fig. 3-4). Directional selection represents a change in the phenotypic mean for the population over a period of time. *Stabilizing selection* is a reduction in the variation within the population while the mean remains unchanged. *Disruptive selection* is an increase in population variation with the mean unchanged. This increase in variation may merely take the form of greater population variability or it may involve the splitting of an initial population into two or more semidistinct subpopulations.

Bennett (1960) examined the effect of directional selection for resistance to DDT upon populations of the fruit fly, *Drosophila melanogaster*. Part of the progeny of a mating were exposed to DDT-impregnated paper in the food vials. A series of vials varying in DDT concentration was used and the amount required to kill half

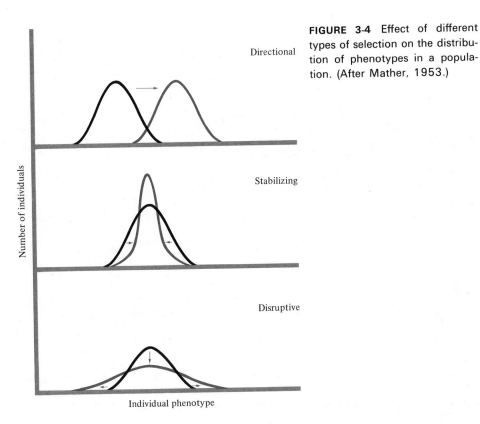

Directional

Stabilizing

Disruptive

Number of individuals

Individual phenotype

FIGURE 3-4 Effect of different types of selection on the distribution of phenotypes in a population. (After Mather, 1953.)

of the test progeny after 18 hours exposure was recorded. From the nonexposed progeny corresponding to the tested progeny, matings were performed to produce resistant and nonresistant lines. (This technique is called *sib-selection.*) By the end of the fourth generation, the two lines had diverged (Fig. 3-5) and they remained distinctly different from each other, and from an unselected line, for the rest of the experiment. When the experiment ended, the resistant line was able to tolerate 125 times the concentration of DDT that the intolerant line could withstand. This experiment, in which sib-selection rather than direct selection was used to isolate resistant genotypes, suggests that the initial population contained the genetic traits necessary to generate low or high tolerance to DDT and that these traits were concentrated in the selected individuals by recombination.

In a sense the population was preadapted to changing environmental conditions. Recombination of the existing genetic complement proved sufficient, in the absence of direct mating of individuals that had survived DDT exposure, to produce populations that were different in their adaptation to high DDT levels. The use of sib-selection in this experiment eliminates the possibility that acclimation to high DDT levels produced the divergence observed.

Thoday (1959) executed an experiment designed to test for the existence of the other two components of Mather's model, disruptive and stabilizing selection. He

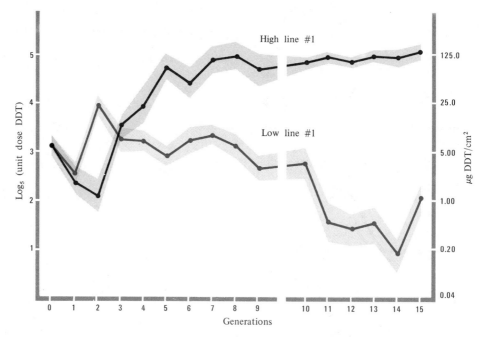

FIGURE 3-5 Change in DDT resistance of fruit fly (*Drosophila melanogaster*) populations exposed to directional selection for high and low resistance. The left-hand scale is the log of the dose tolerated, and the right-hand scale is the actual dose which would kill half a population sample in a standard exposure time. Colored bands are confidence limits on lines. (After Bennett, 1960.)

selected lines of *D. melanogaster* for the number of abdominal bristles, establishing one line (the disruptive one) in which individuals with high and low bristle numbers were always mated to one another and another line (stabilizing selection) in which parents had bristle numbers very close to the population mean. The coefficient of variation of the disruptively selected line showed a significant increase with time, while the value for the line undergoing stabilizing selection showed a significant decrease with time. The difference between the variability of the two populations increased consistently during the course of 35 generations (Fig. 3-6a).

In a later experiment, Thoday examined the ability of populations exposed to different types of selection to respond to subsequent directional selection. Since disruptive selection increases the genetic variability of the population, while stabilizing selection decreases such variability, he predicted a greater variety of genotypes to draw upon in the disruptively selected population subsequently exposed to directional selection. Thoday therefore selected for high or low bristle numbers in populations previously exposed to the two types of selection and compared their responsiveness to that of the wild-type population from which they were derived (Fig. 3-6b). The population exposed to stabilizing selection responded much less rapidly to subsequent directional selection than the wild-type populations. The disruptively selected line, in contrast, was able to respond more rapidly than the wild-type to

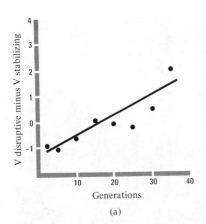

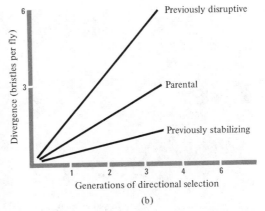

FIGURE 3-6 Effect of different selection types on (a) population variability and (b) ability to respond to subsequent directional selection. In (a) the Y-axis is the difference between coefficients of variation of bristle number in flies exposed to disruptive and stabilizing selection for this trait. In (b) the Y-axis is the difference between initial number of bristles per fly and the number after given periods of directional selection. Disruptive selection increases the ability of a population to respond to directional selection while stabilizing selection diminishes this ability. (After Thoday, 1959.)

such directional selection. In an observation from population genetics that has powerful application to ecology, Thoday concluded that "disruptive selection can promote and stabilizing selection can decrease genetic flexibility, and, therefore, heterogeneity of habitat may be an important cause of genetic diversity in natural populations." We return to this idea many times in subsequent discussions.

SELECTION IN NATURE

In a habitat that is changing from one type to another, what is initially disruptive selection may become directional and finally stabilizing as environmental conditions change. One of the best examples of natural selection under field conditions has been provided by Kettlewell's (1956) studies of a British moth, *Biston betularia*. This moth consists of a series of ecotypic populations occupying different areas. Forms are known that are almost white, while others are very dark gray. One of the consequences of burning coal and oil in Britain since the start of the Industrial Revolution has been that soot deposits are quite heavy in industrial areas. The dark gray ecotype is inconspicuous in polluted woods and conspicuous in unpolluted woods, whereas the converse is true of the light ecotype (Fig. 3-7). The frequency of phenotypes in the two habitats is distinctly different (Table 3-2), with substantially more dark gray forms in polluted habitats and many more light forms in unpolluted habitats. When marked individuals of both ecotypes were released, more white individuals were recaptured in forests with trees lacking soot staining, whereas more gray individuals were recovered in the soot-stained forests.

FIGURE 3-7 *Left:* The peppered moth (*Biston betularia*) and its black form (*B. betularia carbonaria*) at rest on a lichen-covered oak trunk in Dead End Wood, Dorset, England. *Right:* The two moth ecotypes on a soot-covered and lichen free oak trunk near Birmingham, England, (From the experiments of Dr. H. B. D. Kettlewell, University of Oxford.)

Direct observation by Kettlewell and Tinbergen indicated that there was substantially more predation by birds upon maladapted forms than upon the native ecotype. These studies indicate that the distribution of dark, or melanic, ecotypes is associated with the degree of air pollution because of differential predation upon different ecotypes relative to the darkness of tree trunks in the habitat.

If we visualize changes in the populations subsequent to the Industrial Revolution, it seems likely that originally light-colored populations began to be exposed to disruptive selection through predation as they became more conspicuous on the darkening trees. Dark forms of mostly light populations would become favored survivors as pollution became worse. What was initially disruptive selection, and then directional selection, finally would be converted into stabilizing selection as most of the light individuals were eliminated from populations in polluted areas. It has recently been found (Cook, Askey, and Bishop, 1970) that pollution abatement programs are being accompanied by an increase in the frequency of light forms in previously polluted areas, indicating that selection now is being reversed.

TABLE 3-2 Phenotypic patterns in a moth occupying woodlands with dark-colored tree trunks resulting from the death of light-colored lichens on tree trunks (see Fig. 3-7) from air pollution and woodlands with unpolluted, lichen-covered, and light-colored tree trunks[a]

Moth phenotype	Polluted area	Unpolluted area
	Percent of phenotypes in native population	
White	10	95
Gray	85	0
	Percent of phenotypes recaptured	
White	13	13
Gray	28	6
	Observed predation, number taken	
White	43	26
Gray	15	164

[a] Data are numbers of each form, or percentage of each form. After Kettlewell, 1956

SPECIES FORMATION AND BARRIERS TO GENE FLOW

The degree of differentiation among populations of the same species occupying distinct habitats will depend upon the distinctness of occupied habitats and the amount of gene flow between the populations. Aston and Bradshaw (1966) demonstrated that the degree of genetic discontinuity between adjacent population samples may be related directly to environmental discontinuity. In populations of bentgrass, *Agrostis stolonifera,* growing on the coast of Wales, long stolons are at a decided disadvantage in exposed sites where they are subject to severe winds. Samples from the population growing at different positions relative to the sea were transplanted into an inland garden and their growth there was examined. The mean length of stolons changed gradually in population samples originating along a gradually changing environmental gradient, but changed abruptly in populations from an environmental sequence with an abrupt discontinuity (Fig. 3-8). These studies show that genetically distinct plants may be differentiated along an environmental gradient within a single population. Since bentgrass individuals are not self-fertile, differences in stolon length must be maintained in the face of substantial gene flow within the population. Differentiation within a population may be maintained by strong selection even when there is substantial gene exchange.

Maintenance of genetic differentiation in distinct habitats in the presence of substantial gene flow suggests reproductive inefficiency, since many gametes will contribute to offspring not adapted to local environmental conditions. This inefficiency must be balanced against the probability that such offspring will encounter conditions to which they are adapted. If the latter probability is low, the individual may derive an advantage if gene flow from other population members is re-

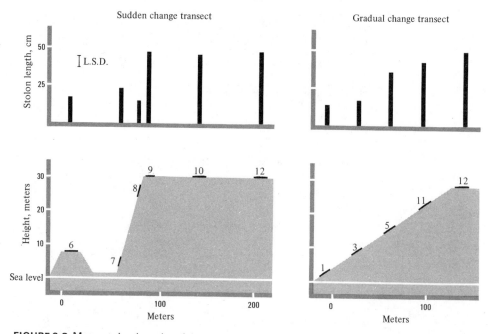

FIGURE 3-8 Mean stolon lengths of *Agrostis* from populations along the seacoast when grown under uniform conditions. The degree of discontinuity between the populations is a function of habitat discontinuity. (After Aston and Bradshaw, 1966.)

stricted. Antonovics (1968) examined self-fertility among grass populations in areas where there are mine spoils containing heavy metals. The metals are toxic to most plants, but Bradshaw (1959, 1960) found that certain species were able to colonize the mine spoil. His work showed that individuals growing on the mine were more tolerant of heavy metals in the soil than members of the same species growing in sites only a few meters from the mine edge. For an individual growing on the mine, the probability that an offspring will encounter a similar habitat increases with the

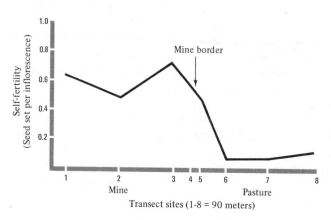

FIGURE 3-9 Relationship between self-fertility of a plant and its position on a contact zone between a heavy metal mine (on the left) and a pasture (on the right). Self-fertility increases strikingly in plants adapted to the specialized mine habitat. (After Antonovics, 1968.)

individual's distance from the edge of the spoil pile. Antonovics examined the distribution of self-fertility in plants growing along a transect from the center of the mine spoil to the adjacent pastures (Fig. 3-9) and found that it increased abruptly in spoil-pile plants.

Self-fertility may serve to restrict gene flow from non-mine individuals by increasing the probability that a gamete will already have been fertilized by another from the same individual. McNeilly and Antonovics (1968) have shown that there also may be displacement of flowering time and other developmental traits related to reproduction along the contact zone between the mine and adjacent pasture. In these species, a single population is differentiating into two separate populations,

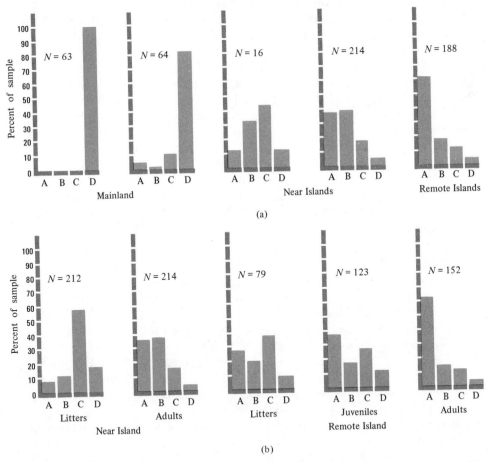

FIGURE 3-10 Frequency of unbanded (A) to darkly banded (D) individuals in water snake populations in Lake Erie: (a) adults along a gradient from the mainland on the left to remote islands on the right; (b) a comparison of litters and adults on a near island and a remote island showing that the population changes as it ages, presumably as a result of greater mortality among darkly banded individuals on the islands. (After Camin and Ehrlich, 1958.)

one of which is heavy-metal tolerant, more self-fertile, and flowers earlier than the other. These changes represent incipient species formation. These populations eventually may become differentiated into distinct species if gene flow is restricted more.

Reproductive barriers of the type documented for the grass populations are one means by which gene flow may be restricted. It also may be restricted merely by physical isolation as a result of geographic barriers, including distance. Camin and Ehrlich (1958) studied the color patterning of water snakes occurring on land in and around Lake Erie. They ranked snakes according to the color patterns from A (gray) to D (darkly banded) and scored the distribution of color patterns in different areas (Fig. 3-10). There was a high frequency of darkly banded snakes in mainland populations, but the gray phenotype increased in frequency on islands until it was the predominant form on the most remote island.

They also compared the coloring of newly produced litters with adult populations and found that there was a higher frequency of the dark types on the island closest to the mainland. Decreasing frequency of dark types with age of the population suggested a low fitness for this phenotype on the island. The greater frequency of light phenotypes in litters, as well as in adults, on the more remote island suggested that selection there was less susceptible to dilution by gene flow from the mainland. Barriers to gene flow, whether physiological as in the plant populations in mine-spoil areas, or geographic as in the snake populations on the Lake Erie Islands, tend to accelerate the differentiation of populations by reducing the incorporation of genes from populations affected by different selective forces. The coastal bentgrass population indicates that genetic differentiation can be maintained by strong selection in the face of substantial gene flow. Barriers to alien genes will accelerate the differentiation of populations in distinct habitats, thereby accelerating the process of species formation.

GENERAL CONCLUSIONS

1. Species are differentiated into local populations that are genetically distinct. This distinctness will depend primarily on environmental distinctness of the native sites.
2. A local population will consist of individuals with different adaptive properties. The amount of variation within a population will be related primarily to local environmental variation.
3. Genetic variation and phenotypic-environmental interaction constitute the bases of evolution, a change in gene frequency within a population over a time period.
4. Natural selection is a consequence of environmental factors which influence changes in population gene frequency with time.
5. Among adjacent populations adapted to distinct environments, reproductive economy may be facilitated by barriers to gene flow.

geographic obstacles, self fertility

SOME HYPOTHESES FOR TESTING

Although the major generalizations of Darwinian natural selection have been elucidated, much remains to be understood about the mechanisms of adaptation in individuals and, hence, populations. Of particular ecological interest are mechanisms of adaptation to spatially and temporally variable environments. An important

question is, what are the relative roles of genetic changes and acclimation in increasing fitness of individuals to variations in the environment? To what periodicity of environmental changes can organisms respond by acclimation and when are genetic changes necessary? How do organisms adapt to nonpredictable environmental fluctuations? What are the genetic strategies that optimize adaptational possibilities (Levins, 1968)?

A significant aspect of the role of genotypic changes in environmental adaptation is the effective size and continuity of breeding populations. These traits of populations determine, in part, the genetic variability present in the population. Breeding continuity among populations will affect the availability of novel genes or genetic combinations from other breeding populations. Evidence is accumulating that many breeding units are much smaller than is apparent from dispersion patterns in nature. Selander (1970) has shown that local populations of housemice (*Mus musculus*) within a single barn may be genetically quite distinct. Ehrlich and Raven (1969) have also noted the existence of local breeding units of butterflies within short distances. Storage and movement of novel gene combinations can be achieved in many ways and each strategy may have maximum importance in one or a few types of environmental variations.

Finally, an important question is, what is the evolutionary role of natural selection at the level of the population rather than the individual? The importance of group selection is presently unclear and needs testing, especially in nature.

ecosystem structure

4

One of the earliest and most persistent interests of ecologists has been ecosystem *structure,* the spatial arrangement of ecosystem components in relation to one another, and the abundances of different populations in the ecosystem. This chapter will be directed primarily toward describing static aspects of ecosystem organization. An understanding of the distribution of biomass in space, and in species, will provide a basis for interpreting more dynamic aspects of ecosystems in succeeding chapters. Initial sections of the chapter discuss the geometry of ecosystems, the distributions of individuals in three-dimensional space. Intermediate sections consider abundance patterns in communities, that is, the patterns of energy partitioning among populations co-occurring in time and space. The final section considers the life form structure of plants in terrestrial ecosystems as it reflects habitat differences.

SPATIAL STRUCTURE

Spatial structure is a universal characteristic of ecosystems and has been particularly thoroughly examined in coral reefs and tropical rainforests. The coral reef (Odum and Odum, 1955) consists of a series of layers varying in biological character and in ecological properties (Fig. 4-1). The outer layer is the living zoological component of the coral, the polyp. These organisms produce hard, rocklike exoskeletons that aggregate, through time, to produce the reef. The reef builds up layer by layer with older portions dying and becoming overgrown. The polyps withdraw into a hard exoskeleton during the day and expose their tentacles at night when they feed on zooplankton swimming over the coral surface.

Primary producers in coral are algae that occur on the surface and within the coral matrix. The majority are green filamentous algae embedded in the reef. A living coral reef eventually consists of an algal layer overlaying the old skeletal remains, a developing skeletal layer above this that is impregnated with algal filaments and honeycombed with chambers containing polyps, and the surface of the reef that comprises a polyp colony. The water immediately above the polyps contains zooplankton communities, which presumably feed upon microbial populations and detritus and, in turn, are fed upon by the polyps. The algae and polyps form a complex interrelated functional unit with each depending on the other for many of their needs. This interdependence is reflected in the structural organization of the coral ecosystem.

The vertical structure of ecosystems also has been examined in tropical rainforests where substantial stratification is evident (Fig. 4-2). A series of canopy layers has been described (Richards, 1952) from the tallest (a) emergents to (b) an upper canopy, (c) a lower canopy, (d) a shrub layer, and (e) a ground layer. There is a spatial hiatus between the shrub layer and the lower canopy that is often occupied by lianas (vines) and epiphytes (nonparasitic plants growing on other organisms). As even the diagrammatic representation indicates, these layers merge into one another, making the divisions somewhat artificial. It is clear, however, that different species tend to occupy different positions in the canopy; the tall emergent species soar above the canopy, while other species are buried within the canopy. Likewise, different animal species may be associated with different levels of the canopy, thus producing a vertical stratification pattern similar to the vegetation structure (Harrison, 1962).

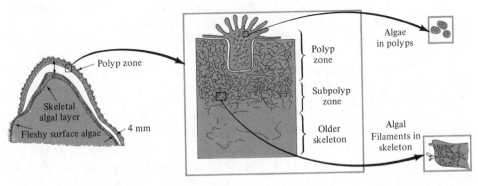

FIGURE 4-1 Spatial structure of a coral reef showing the organized relationship among different ecosystem constituents. (After Odum and Odum, 1955.)

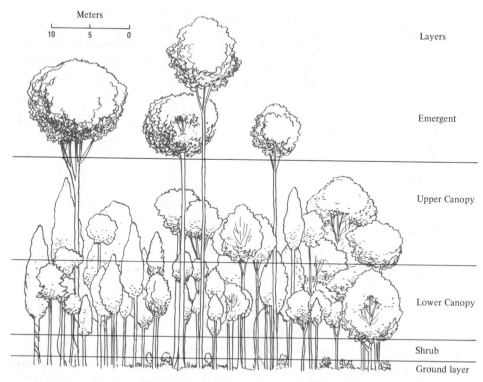

Meters

10   5   0

Layers

Emergent

Upper Canopy

Lower Canopy

Shrub

Ground layer

FIGURE 4-2 Spatial structure of plant canopy in a tropical rainforest. Different tree species occupy different positions, as adults, in the canopy. (After Richards, 1952.)

Studies of bird movement within spruce forests indicate that animal populations show both vertical and horizontal structure with species abundances varying in three-dimensional space. MacArthur (1958) has examined the distribution of warblers in spruce trees by recording the amount of time different species spent in various positions in the trees. He found that different species varied considerably in spatial distribution of their activity (Fig. 4-3). The Cape May warbler spent 70 percent of its time in outer crown layers. The Myrtle warbler, in contrast, spent 50 percent of its time in the region between the lower crown and the soil, and the rest was spread rather evenly over the rest of the tree. The Bay-breasted warbler, however, spent about 50 percent of its time in the central part of the tree, ranging from the trunk to the outer crown.

An ecosystem, then, occupies a volume, and the distribution of individuals varies within that volume. Individuals of the same population tend to occupy similar positions in the ecosystem volume. A tree species in the rainforest, for example, may occur with its roots at the base of depressions where runoff water collects, with the crowns occupying an intermediate position in the crown structure of the forest. Similarly, a bird may spend its time in the upper crown area of spruce trees but only the outer surfaces of this area. One of the interesting problems of ecology is

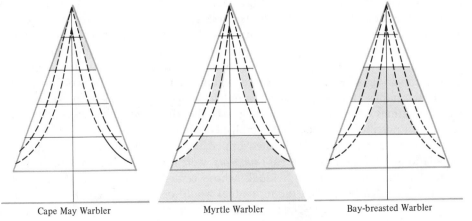

Cape May Warbler   Myrtle Warbler   Bay-breasted Warbler

FIGURE 4-3 Spatial structure of warblers in a coniferous forest with shaded area in trees indicating where a species spends most of its time. (After MacArthur, 1958.)

interpreting the significance of such patterns in terms of the partitioning of resources among co-occurring populations.

DISPERSION

More mental and physical exertion in ecology probably has been expended on measuring the two-dimensional spatial organization, or *dispersion,* of populations than any other phenomenon (Greig-Smith, 1964; Pielou, 1969). The three types of dispersion described (Fig. 4-4) are (a) *random,* in which the presence of an individual at a point does not affect the probability of recording a member of the same population at an adjacent point, (b) *aggregated,* in which the occurrence of an individual at a point increases the probability of recording a similar individual at an adjacent point, and (c) *regular,* in which the occurrence of an individual at a point decreases the probability of occurrence of a similar individual at an adjacent point.

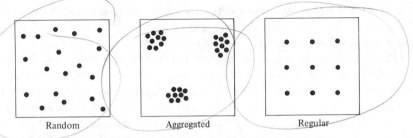

Random   Aggregated   Regular

FIGURE 4-4 Three different types of two-dimensional structure in an ecosystem. The dots could be individuals of the same species, or different species populations, or some other component of the ecosystem that can be assigned a position.

Pielou (1960) proved that regular, aggregated, or random distributions may describe the same dispersion pattern depending on the limits of individual size, the technique used to sample the population, and the sampling area. The model was based on the following biological assumptions: (a) offspring were distributed independently of one another in space and (b) sequentially in time; (c) these offspring grew into various-sized individuals that occupied circular areas in two-dimensional space. To establish the "populations," numbers were assigned to lines on graph paper and individuals were positioned on the grid by drawing random numbers. At a point, the largest possible individual was established without overlapping another individual and within arbitrarily established upper and lower size limits. To sample the populations, square sampling areas (*quadrats*) were randomly thrown onto the graph paper and the number of individuals whose center fell within the quadrat was counted (the idea being that the individual would occupy the center of its area of influence and would be recorded in a natural population only if the sample contained this center). One hundred quadrats were employed. The range of individual size was 0.2 to 0.7 units for the A populations, 0.1 to 1.0 for the B populations, and 0.1 to 2.0 in the C populations. Quadrat sizes of 1.96 and 4.41 square units were used.

A range of densities was established for each of the population sets. As a test for the dispersion pattern, the variance : mean ratio of number of individuals per quadrat was used. This value is one for a random distribution, greater than one for an aggregated distribution, and less than one for a regular distribution. The null hypothesis is that population distribution will be random and with $n = 100$ the 5 percent significance limits on the null are 0.716 to 1.204. Pielou found that the same assumptions generated random, aggregated, or regular distributions depending on the combination of quadrat size and individual size limits involved (Table 4-1).

In set A, with a relatively narrow range of sizes permitted, the population showed regular distribution over the entire range of densities with both quadrat sizes. In set B, with a larger range of individual sizes, some population-quadrat combinations were regular and some were random. And in set C, with the largest range of sizes permitted, most of the combinations showed an aggregated distribution, although the lowest density combination was aggregated when sampled with the large quadrats and random when sampled with the small quadrats. Clearly, the apparent dispersion pattern of these populations is a consequence of the interaction between individual size and sampling area. The degree of apparent aggregation increased with increasing range of individual size in the population and with increasingly large sampling area.

Pielou's model "populations" suggest that care must be exercised when examining the dispersion of individuals in ecosystems. There are, however, sound biological reasons for expecting that nonrandom distributions may be produced within ecosystems by ecological mechanisms. For instance, Feller (1943) has pointed out that an aggregated, or clumped, distribution may be generated by (a) the tendency for offspring to congregate around their parents; in animals this may be manifested as a tendency to remain in the same neighborhood, in plants it may result from seed tending to fall close to the parent or from vegetative reproduction; (b) dispersal, the movement of organisms away from a source point, may be regular or random,

TABLE 4-1 The variance/mean ratio (V/m) obtained by sampling each model with 100 quadrats

Model	4.41 sq unit quadrats		1.96 sq unit quadrats	
	Mean	V/m	Mean	V/m
A_1	2.27	0.527	1.00	0.480
A_2	2.59	0.472	1.16	0.564
A_3	3.13	0.368	1.26	0.343
A_4	3.77	0.429	1.71	0.401
A_5	4.60	0.309	2.18	0.499
B_1	2.66	0.573	1.36	0.669
B_2	3.94	0.745[a]	1.85	0.512
B_3	4.75	0.566	2.10	0.519
B_4	5.76	0.632	2.58	0.746[a]
B_5	6.90	0.601	3.03	0.953[a]
B_6			3.75	0.903[a]
B_7			4.20	1.081[a]
C_1	2.54	1.405[b]	1.25	1.046[a]
C_2	3.40	1.865[b]	1.28	1.423[b]
C_3	4.56	2.304[b]	2.19	1.760[b]

[a] V/m test indicates that the model is random.
[b] V/m test indicates that the model is aggregated.
Pielou, 1960

while habitats favorable for colonization may be clumped so that individuals attempting to colonize outside these clumps are unsuccessful. Although habitat heterogeneity and reproductive clumping may play important roles in producing clumped distributions in nature, Pielou's examples suggest that one should be wary of assigning a causal mechanism in the absence of unambiguous evidence demonstrating the mechanism.

Such evidence was presented by Harper and Sagar (1953) in their study of the distribution of two species of buttercup (*Ranunculus*) in grasslands containing hummocks. *Density* (the number of individuals occurring in a given area) was measured in 0.3 meter × 3.6 meter rectangular strips (called *transects*) placed along a series of ridges and furrows (Fig. 4-5a). *Ranunculus repens* tended to be aggregated in the furrows and to be almost absent from the ridgetops, whereas the opposite was true for *R. bulbosus*. Abundances of the two species were negatively associated (Fig. 4-5b), with an abrupt transitional region where there was a shift from a predominance of one species to a predominance of the other. The ridgetops tend to be freely drained and the furrows often soggy with excess moisture. These field observations suggested that *R. repens* might be adapted to high soil moisture and *R. bulbosus* to drier soils.

The small scale of this pattern suggests that seeds from both buttercup species would be distributed throughout the grassland in sufficient numbers to establish uniform populations and that strong differential habitat acceptability maintained separate distributions for the two species. Because the establishment stage involving seed germination may be a critical step in determining population distribution,

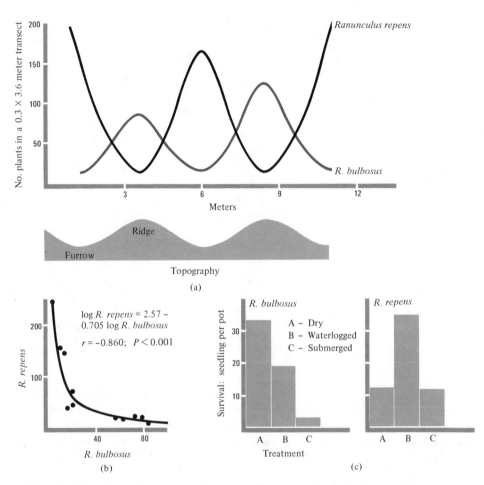

FIGURE 4-5 Spatial structure of two buttercup (*Ranunculus*) species in a field with considerable topographical variation: (a) distribution of the species in relation to topography, (b) relationship between densities of the two species, (c) ability of the species to grow under different soil moisture conditions. Distribution of the two species is a function of the degree of soil waterlogging in the habitat and relative adaptation to waterlogging. (After Harper and Sagar, 1953.)

Harper and Sagar examined the germination responses of seeds under different conditions of soil water (Fig. 4-5c). *R. bulbosus* seeds germinated best in pots with free drainage and *R. repens* in the half-waterlogged soil. The latter species also germinated substantially better than the former in the completely waterlogged soil.

Earlier we examined the maintenance of ecologically distinct subpopulations of bentgrass along the Welsh coast (Fig. 3-8) in response to strong selection gradients and found (Fig. 3-9) that fertility barriers might begin to arise among such subpopulations. The habitat differentiation of these two species of buttercups in a small, but ecologically diverse region suggests that buttercup speciation may be the outcome of ecotypic differentiation in a diverse habitat. The alternative, of course, is that the buttercup species originated in geographically remote areas and then later appeared in the same area. To select the correct alternative we need to know more about the historical development of the two species.

Harper, Williams, and Sagar (1965) performed a series of experiments which indicated that complex dispersion patterns among multiple species may be generated by habitat heterogeneity and distinct growth requirements. They utilized three species of plantain: *Plantago lanceolata*, *P. media*, and *P. major*. Four replications of eight treatments plus a control were randomized in a soil bed. The treatments, designed to alter the soil surface and microhabitat conditions, consisted of (1) a square depression 1.25 cm deep made by pressing a sheet of glass into the soil, (2) a similar depression 2.5 cm deep, (3) a square sheet of glass left laying on the soil, (4a) a sheet of glass inserted into the soil to a depth of 1.25 cm with 3.5 cm above the soil in a north-south orientation, (4b) identical to (4a) but with glass oriented east-west, (5) a square wooden frame without a top or bottom pressed 1.25 cm into the soil with 2.5 cm projecting above, (6) identical to (5) but with only 1.25 cm above the soil, (7) also identical to (5) but with nothing above the soil surface, and (8) the control plots which were untreated. Before treatment, 500 seeds of each species were sown on the soil surface. Thus some seeds were left laying on the surface, others were in depressions of various depths and character, and some were buried.

The experimental production of substantial microhabitat diversity created different distribution patterns among the three species (Fig. 4-6). Each species reached its greatest density under different conditions. The correlations of percent emergence under a given treatment for different species were poor with $r = 0.262$ for *P. lanceolata* and *P. media*, $r = 0.493$ for *P. lanceolata* and *P. major,* and $r = 0.318$ for *P. media* and *P. major.* For *P. lanceolata*, the greatest concentration of seedlings occurred under treatments (2) (23.5 percent of emergent seedlings) and (7) (23.2 percent). Only 0.4 percent of the seedlings occurred on the control plots, indicating that soil microtopography and the associated changes in soil moisture and seed illumination were important factors in the germination of this species. Treatment (3) was by far the most favorable treatment for *P. media*, with 46.7 percent of all seedlings under the glass plates. Although the untreated area comprised a substantial portion (43.5 percent) of the total area sown, significant portions of the seedlings appeared only for *P. major* (27.9 percent). This suggested that it showed less of a preference for topographically disturbed areas than the other species.

These experiments indicate that varied microhabitats can generate varied distribution patterns in a relatively confined area. Different requirements for successful germination by seeds from different species may, when combined with habitat heterogeneity in natural ecosystems, generate substantial clumping of species.

The experiments with *Plantago* and those with *Ranunculus* show that ecosystem patterns may be generated by habitat heterogeneity even when dissemination of offspring is random. On the other hand, dispersal is rarely random. In studies of barn owl dispersal, Stewart (1952) recorded the distance that banded barn owls traveled from the site where they hatched. There was a decided tendency for the banded barn owls to remain close to their home area (Fig. 4-7). Less than one-fourth

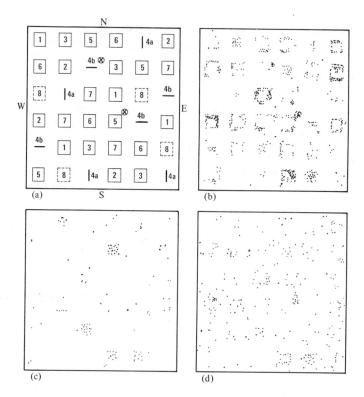

FIGURE 4-6 Effect of different soil microtopography on germination of plantain (*Plantago*) seeds: (a) plan showing the positions
of various objects placed on the soil surface after sowing with
seeds of three species (see text for explanation; ⊗ = worm casts),
(b) distribution of *P. lanceolata* seedlings, (c) distribution of
P. media seedlings, and (d) distribution of *P. major* seedlings.
Each species has a preferred, and distinct germination site.
(After Harper et al., 1965.)

of more than 200 owls moved farther than 100 miles from their hatching place. This
tendency is well known for a large number of animals, and it is common knowledge
to anyone who has walked under an oak tree in the fall that similar reproductive
clustering occurs in plants. Seeds tend to cluster around the parent plant, just as
animal progeny tend to move limited distances from their birthplace. This nonrandom dispersal can generate aggregation among populations even when the habitat
is relatively homogeneous.

The tendency for individuals of a population to remain in that population,
together with the tendency for its members to become aggregated because of habitat
heterogeneity, leads to important evolutionary consequences. Under stable environmental conditions, the population may develop a long evolutionary history in which
the same genomes are selectively favored. This tends to generate genetic homo-
geneity, thereby reducing the population's ability to respond to rapid environmental

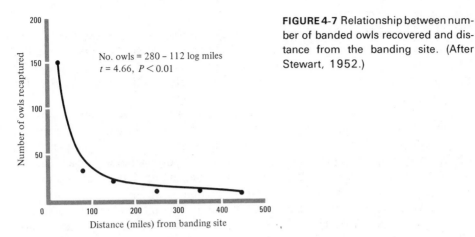

No. owls = 280 − 112 log miles
$t = 4.66$, $P < 0.01$

FIGURE 4-7 Relationship between number of banded owls recovered and distance from the banding site. (After Stewart, 1952.)

77

DISPERSION

change. Dispersal, on the other hand, results in the occasional introduction of novel genes from remote populations which have had a different evolutionary history.

If novel genes are of no selective advantage to the population, they may disappear or more likely may be maintained at a low frequency. A low frequency of nonharmful but nonadvantageous genes may serve as a reservoir of variability that the population can draw upon during periods of environmental fluctuation. We have seen an example of such "preadaptation" in the experiments by Bennett (1960) on the ability of fruit fly populations to respond to directional selection for DDT resistance. The parental population contained genes which were redistributed in the offspring, through recombination, to produce lines that were substantially more resistant to DDT than the parental stock. The tendency for offspring dispersal to decline logarithmically from the point of origin means that there is a decreasing probability of gene flow between populations with interpopulational distance. We can therefore write

$$p = h \log d \qquad (4\text{-}1)$$

where p is the probability that a population will receive genes from another population, d is the distance between populations, and h is a constant that describes the rate at which migration decreases with distance. Clearly, h is a constant of considerable ecological and evolutionary interest. It may vary, for instance, in different directions over the same distance. For example, in the Lake Erie water snakes there was more gene flow from mainland to island populations than vice versa merely because the mainland populations were much larger. A number of things may influence h, including population sizes, migration barriers, and distribution of favorable habitats. The integrity of structure of the local ecosystem must depend upon its component populations not being "swamped" with alien genes. At the same time, the ability of an ecosystem to maintain its structure during periods of environmental flux may depend on a reservoir of novel genes introduced into the component populations through the appearance, from time to time, of foreign individuals (McNaughton, 1966).

SPECIES STRUCTURE

The structure of an ecosystem also can be examined in terms of the species that comprise the system. Simply listing all the species in an area tells very little about the structure of the system, although it permits simple comparisons of species composition of two ecosystems. A more detailed analysis of the abundances of the various species in equivalent units is usually employed.

There are numerous ways to measure the abundances of organisms. Major techniques for population sampling reflect the required precision of the estimate. The most precise technique is to count all the individuals in the sampled area. Obviously in large areas or for mobile organisms this poses a major time and energy problem for the investigator. To surmount this problem population sizes are usually estimated from a subsample of the total population.

If an estimate of the size of the total population is not required, it is sufficient to obtain equivalent sample abundance values that can be used to compare several populations. A useful method is to standardize the sampling procedure so that abundance can be expressed as number of individuals per unit of effort. This might take the form of individuals per steps by the investigator, per hour spent, per unit area sampled, and so on. One might also sample a given number of areas and express abundance as *frequency,* the percent of the total areas in which a given species occurs. Common species normally will occur in a greater percentage of the areas than rare species. These abundance estimates are most useful for investigations in which absolute numbers are not important and when speed of obtaining the measure is critical or a sampling involves very mobile organisms that cannot be identified individually.

For many studies it is important to know how many organisms are in the population, necessitating estimates of absolute abundance from sampling data. For stationary organisms there are several techniques that require randomly selecting a series of areas of known size and counting all individuals of a species within each area. The number of individuals per unit area is called *density.* By knowing what proportion of the total area has been sampled it is easy to calculate the total number of individuals. This technique assumes that individuals are distributed at random over the total area and that sampled areas are determined randomly. As the dispersion of individuals departs from random, the population estimates become less precise.

For mobile organisms other methods may be needed to estimate total population size. Perhaps the most popular, the *mark-recapture* method, is based on capturing and marking an unknown percentage of the population, which is then released. Slightly later, another sample is caught and the percentage of marked individuals noted. The total population can then be estimated from the equation

$$M/N = R/N_2 \qquad (4\text{-}2)$$

where N is the total population, M the number marked the first time, N_2 the total number in the second sample, and R the number in the second sample that was marked at the first capture. The randomness assumptions for this technique include equal ability to catch each individual and random movement of marked individuals throughout the population. For the simple equation (4-2), it must also be assumed that no animals have been added to or subtracted from the population between the time of the marking and the resampling.

More elaborate mathematical procedures are available that modify the results if some assumptions do not hold. The precision of these more elaborate procedures depends on how precisely the assumptions can be modified according to life history characteristics of the species. In each circumstance the absolute numbers are usually presented as density so that comparisons may be made between areas of different sizes.

Once abundances have been estimated, abundances of several species can be compared, either within or between communities. Raunkiaer (1928) described the relative abundances of species in a community by their frequency of occurrence in a series of subsamples from the community. Each species was categorized according to the percentage of the total number of subsamples in which it occurred. Percent occurrences were then grouped into five categories: 0–20 percent, 21–40 percent, 41–60 percent, 61–80 percent, 81–100 percent, and the number of species in each frequency class was plotted (Fig. 4-8). These graphs tended to have a backward J-shape such that class A (0–20 percent) > B > C ≧ D < E. Raunkiaer formalized this apparent relationship into what is now called the *law of frequencies*. In its simplest form this law states that most species are rare and only a few species are abundant.

On a very large scale, Preston (1962) has shown that the relative numbers of groups of species may follow a lognormal distribution. In plotting abundances to see if they fit a lognormal, species are grouped in categories called *octaves,* according to the number of individuals per species; each succeeding octave contains a range of individuals per species that is double the range of the previous category. The first octave contains species that are represented in the sample by one or two individuals; octave two is species represented by two to four individuals; octave three is species represented by four to eight individuals, and so on, until it includes the most abundant species in the sample. Species that fall on the border of categories are divided equally between categories. These categories are plotted as the units of the *X*-axis; since they represent doubling of individuals per species, the axis is a logarithmic plot to the base 2. The ordinate (*Y*-axis) is the number of species that fall into each category (Fig. 4-9). For large heterogeneous samples, the shape of

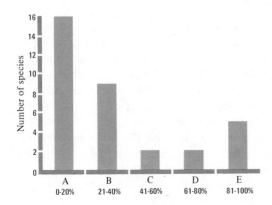

FIGURE 4-8 Relative frequency of occurrence of species in plots in a pine forest area in Luzon, Philippine Islands. Frequency here is the percent of total plots in which a particular species occurred. (Data from Kowal, 1966.)

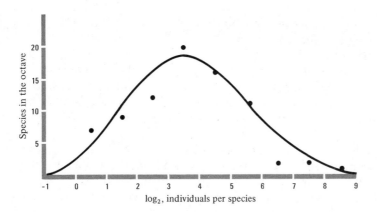

FIGURE 4-9 Lognormal distribution of species abundance in a 10-year breeding bird census at Westerville, Ohio. (After Preston, 1962.)

the curve tends to approximate a normal or bell-shaped curve. The normal distribution of frequencies plotted against a log abscissa leads to the term *lognormal* for this type of distribution. This distribution spreads out the smallest abundance class of the Raunkiaer distribution to show that in large samples most species are neither rare nor common, but of medium abundance.

The entire population of organisms in the sampling area may be called the sampling universe. If the universe is sufficiently large, then usually only samples will be taken to estimate the characteristics of the universe. Species abundances in samples are reduced from their total abundance by a proportion equivalent to the proportion of the total number of individuals represented by the sample, provided samples and species distributions are random. If half the total number of individuals is sampled, the commonest species is represented by half as many individuals as in the universe, in this case the total area under consideration. Sampling only half the total number of individuals thus reduces the maximum observed octave by one in comparison to the universe. Originally the probability of sampling rare species was very small, so some of the rare species may not be included in the samples. This eliminates from the lognormal curve some species that would occur in the octaves.

If the distributions of organisms and the sampling procedures are random, the errors inherent in sampling will shift the mode or peak of the lognormal curve closer to the Y-axis; the shift will occur at the rate of one octave to the left for each decrease by one-half in sample size. This movement will truncate the curve on the left-hand side, thus in effect cutting off part of the left-hand tail of the lognormal since no species can drop below an abundance of one, the lower limit of the first octave. With each doubling of sample size the curve should move one category to the right until eventually the entire universe has been sampled and the curve is complete. If samples are too small, the lognormal curve actually may be shifted so far to the left toward the Y-axis that the mode may not be represented.

Another method of approaching the distribution of relative abundances of species employs *dominance-diversity curves* (Whittaker, 1965) in which the *Y*-axis is the common logarithm of some measure of abundance of a given species, and the *X*-axis is the rank of that species from most abundant (1) to rarest (*S*) species. For samples within a local area, most communities fit a straight line (Fig. 4-10). The mathematical explanation of this distribution is that each succeedingly rarer species makes the same proportional contribution to total abundances of "remaining" species in the sample. For instance, if the major species (species with rank 1) contributes one-half of the total community productivity, the next most important species (rank 2) will contribute one-half of the remaining one-half, that is, one-fourth. The series will continue according to this proportionality factor until, finally, the rarest species (rank *S*) contributes all of the remaining abundance.

A series such as this, with a constant slope when log abundance is plotted against rank, is called a *geometric series*. It differs from the lognormal series in that the rare species are much rarer for a given sample size (Fig. 4-11). We shall consider possible explanations of such series in a later chapter on community organization. At this point, however, you should recognize that most ecosystems will contain a few very abundant species and many relatively rare species. The functional reasons for population abundance and rarity are among the most perplexing in ecology.

These patterns of distribution of relative abundance, such as the law of frequencies, the lognormal, and the geometric series, are empirical, that is, they were derived after plotting actual data. Presently, it is hard to ascribe any sound biological interpretation to the distributions. We do know that as the heterogeneity of the sample is increased, presumably increasing sample randomness and decreasing the influence of species interactions, the distribution of relative abundances tends to approach closer to a lognormal (McNaughton and Wolf, 1970). Although these series all indicate that there are many more rare species than common species, they provide no mechanistic explanation of this phenomenon.

Ecologists would like to develop a theoretical distribution of relative abundances based on meaningful biological assumptions, rather than apparent correlations. MacArthur (1957, 1960) devised a model of relative abundances of species in

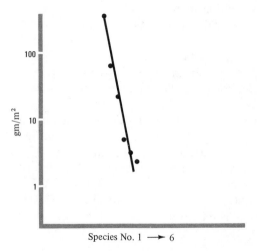

FIGURE 4-10 Dominance-diversity curve for plant species at peak biomass on sandstone soil with a northwest slope exposure in the Jasper Ridge Experimental Area on San Francisco Peninsula, California. (Data from McNaughton, 1968b.)

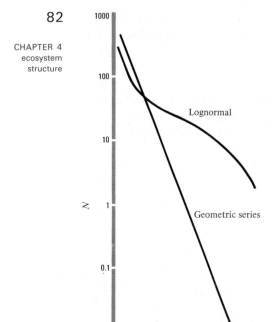

communities that has received much attention. He assumed that the positions of a series of species along a particularly important niche parameter would not overlap, that the portion of the parameter exploited by a species was random in extent and position and that all usable portions of the parameter would be exploited. He then was able to describe the theoretical distribution of relative abundances based on these assumptions (Fig. 4-12). He initially tested this hypothetical distribution against data from bird censuses in the United States. The fit of the observed data to the expected distributions was fairly good. The MacArthur model has instigated a series of further papers, both confirmatory and critical (for example, King, 1964; Hairston, 1969; Pielou, 1969). We discuss the model in some detail in the chapter on community organization.

In addition to describing the abundance distributions of species in single communities, techniques are required for comparing abundances among communities. Each of the empirical measures of the abundance distributions contains one or two constants that might be used as comparative indices between communities. Unfortunately the constants are not the same for each major type of distribution. A more useful index would be one that is comparable in all communities, regardless of species abundance distributions, and which is a good measure of relative abundances

(Margalef, 1957, 1963). Such an index comes from information theory. MacArthur and MacArthur (1961) were among the first to popularize its use in ecology. The index is

$$H = - \sum_{i=1}^{N} p_i \cdot \log p_i \qquad (4\text{-}3)$$

where H is the information content of the collection of species, the units of which depend on the logarithmic base used in the calculations, and p_i is the proportion of the total number of individuals that are in the ith species. The index indicates the uncertainty of predicting the specific identity of the next individual when drawing individuals at random. It provides a single number, allowing for comparisons between samples from distinct communities. The index depends partly on the number of species in the sample and partly on the equality of abundances of individuals per species. Samples with different numbers of species can have the same index depending on equality of relative abundances; in fact, the index can be used as a measure of the concentration of dominance in a group of species (Whittaker, 1965).

Other indices may be required for estimating the evenness of distribution of individuals among species in several communities. Lloyd and Ghelardi (1964) and Pielou (1966a) have proposed somewhat different indices of *equitability,* that is, the evenness of distribution of individuals among species. Lloyd and Ghelardi reasoned that there would be a maximum evenness of distribution of individuals that was

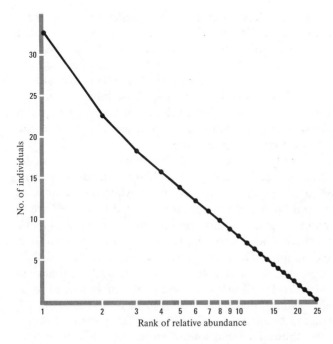

FIGURE 4-12 Distribution of relative abundances of species in a 25 species community that fits the MacArthur random but nonoverlapping niche model.

something less than equal numbers of individuals for each species in the sample. They used as their theoretical maximum the MacArthur distribution of relative abundances on the assumption that this represented the maximum diversity attainable in a community. Any deviations from this diversity measure would be in the direction of abundant species more abundant than expected coupled with associated decreases in the abundance of moderately common and rare forms. Goulden (1969) and Deevey (1969) presented data from communities of aquatic organisms that support this idea. From this theoretical base Lloyd and Ghelardi could calculate an equitability index in one of two ways: (a) compare the number of species distributed according to the MacArthur model that would give the same H value as found for the sample; (b) compare the actual H value with the H value if the species were distributed according to the MacArthur model. In each case the ratio of the observed value to the expected value ranges from one when the sample follows the MacArthur distribution toward zero when there are two species with markedly different abundances.

Pielou (1966b, 1969), on the other hand, argues that maximum diversity is achieved when each species is represented by equal numbers of individuals. Her measure of equitability is

$$E = H/H_{\max} \qquad (4\text{-}4)$$

where $H_{\max} = \log_n S$, where S is the number of species in the sample. This measures how closely the sample distribution approaches the theoretical maximum. The range is from one toward zero. Despite the disagreement as to the theoretical maximum diversity, either equitability index is useful for comparing samples from different communities.

Another potential mechanism for describing diversity is based on the fact that as the numbers of individuals sampled increases, the number of species sampled will also increase until all species in the community have been sampled. Data of this type can be plotted as increases in species present with increases in numbers of individuals sampled or the reverse (Monk and McGinnis, 1966). Frequently, increasing the area of sampling is substituted for increasing the numbers of individuals (Gleason, 1925). In either instance two factors, the rate of increase of species with increasing sample size (area or numbers) and the number of species at the asymptote, provide indices for comparing communities.

In a representation of community structure by relative abundance curves or by diversity indices, the problem of size inequality of different individuals occurs. Schoener and Janzen (1968) used length as a size estimator when they examined size relationships in several insect populations from tropical and temperate areas. They found that the data fitted a lognormal distribution better than an arithmetic normal distribution, but that there tended to be similar patterns of departure from the lognormal in most of the samples. The curves tended to be more peaked than a normal distribution, meaning that the sizes tended to be more nearly equal than in a normal distribution. They also found that the mode of the curve tended to be shifted somewhat to the left, meaning that there tended to be more species in the shorter size classes than expected in the normal distribution. The pattern of occurrence of different-sized insects showed a trend toward larger insects in areas with more moisture and a longer growing season. Physiological differences associated with body size are important in regulating geographical and seasonal distributions of insects. These differences relate largely to growth requirements and ability to resist desiccation. The relative availability of insects of different sizes probably will strongly influence the organization of insectivorous consumer groups, as we shall see when considering energy in ecosystems.

LIFE FORM STRUCTURE

The preceding patterns of ecosystem structure have been somewhat dependent on evaluating the taxonomic character of the ecosystem, breaking it up into species groups. The taxonomic position of ecosystem components performing similar functions may, however, vary from place to place. Different species may constitute ecosystems that are quite similar structurally. For instance, forests are on all the nonpolar continents, but these vary considerably in the trees that compose them and in the consumer species that they support. Although the taxonomic component may change in different geographic areas, Raunkiaer (1934) recognized that similar sorts of species (ecological homologues) occur in similar environmental circumstances. He devised a life form system for describing terrestrial primary producers that eliminated the problems created by taxonomic distinctions. He categorized plant species according to the techniques they have evolved for maintaining themselves during dormant periods, whether caused by cold temperatures, as during winter in temperate climates, or by drought, as in many dry tropical climates.

Annuals, plants which complete their life cycle in one growing season and therefore spend the dormant period as seeds, are called *therophytes* (Fig. 4-13). Perennials and biennials, which produce buds that initiate growth upon termination of the dormant period, are classified according to the position of the dormant bud relative to the soil surface. *Phanerophytes* have buds upon shoot apices some distance above the soil surface. *Chamaephytes* have buds very close to the ground but still aerial. *Hemicryptophytes* have buds at the soil surface and all aerial portions of the plants die back to these buds during the dormant period. Finally, *cryptophytes* are plants with buds that are either buried in the soil or submerged in water during the dormant period. Plant communities occurring in different climates have quite

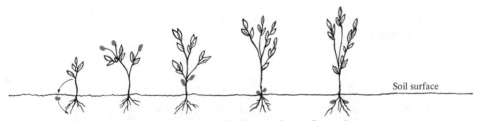

Therophyte Phanerophyte Chamaephyte Hemicryptophyte Cryptophytes

FIGURE 4-13 Life form structure of terrestrial vascular plants according to position of the dormant bud (colored dot). (After Raunkaier, 1934.)

TABLE 4-2 Percentage composition of various life forms occurring in the flora from different climatic areas

Area	Therophytes	Phanerophytes	Chaemaephytes	Hemicryptophytes	Cryptophytes
Tundra	0	0	24	66	12
Desert	42	12	21	20	5
Wet tropical forest	14	59	12	9	4

Raunkiaer, 1934

different life form spectra (Table 4-2). In wet tropical climates there is a predominance of forms with the buds in aerial positions, and in arctic communities there is a predominance of forms with the perennating buds protected from freezing by being buried in the ground. Dry desert communities have a predominance of therophytes which complete their life cycle during the brief wet season and spend the dry season as seeds.

The ability of Raunkiaer's life form spectrum to describe communities with different species composition but similar environments suggests that there has been a strong tendency toward *convergent evolution,* the production of taxonomically distinct but ecologically similar species, in similar environments.

GENERAL CONCLUSIONS

Structure consists of (a) the spatial arrangement of ecosystem components and (b) the abundances of different species in the ecosystem.

1. Spatial structure
   (a) Ecosystem components show definable organization in three-dimensional space. birds in dull parts of trees   self-texture
   (b) Three classes of dispersion patterns are random, aggregated, and regular. Departures from randomness may be generated by nonrandom dispersal or by nonrandom distribution of favorable habitats.   seeds

2. Species structure
   (a) Most species in an ecosystem are neither rare nor common, but are of intermediate abundance.   lack of domination
   (b) Species diversity may be summarized by the index

   $$H = -\Sigma p_i \log p_i$$

   (c) The evenness of distribution of individuals among species, equitability, may be summarized by the index

   $$H/H_{max}$$

3. Life form structure

   Although the species composition of ecosystems will vary from area to area, the similarity in life form in similar habitats suggests a strong tendency toward convergent evolution in similar, though remote, habitats.

ECOSYSTEM PROCESSES

Part 2 considers the processes which give the ecosystem its functional identity: energy and materials flow. The initial step in energy flow is the capture of light energy by plants through photosynthesis. The accumulation of mass by photosynthesizing plants is the primary energy gateway for the ecosystem. The energy of this accumulated mass is subsequently consumed by other organisms in the ecosystem in the process of energy flow within the ecosystem. Most animals obtain chemical substances and energy simultaneously as they feed. In plants, however, energy capture and chemical uptake are largely separated. The movement of chemicals through the ecosystem is called materials flow. Together these two processes define the functional interrelationships among organisms. Chapter 5 concentrates on the entry of energy into the ecosystem. Chapter 6 then considers the transfer of this energy through the feeding process. Chapter 7 examines the uptake and transfer of chemicals through the ecosystem. Finally, Chapter 8 considers an application of the principles of energy and materials flow to increasing concentrations of biological toxins in the biosphere.

primary production

5

A machine may be defined as a set of interacting objects that functions to transmit or transform energy. The ecosystem, like a machine, consists of objects (organisms) which both transform and transmit energy. The kinetic energy of photons in sunlight is transformed into the potential energy of organic compounds. This energy then may be transmitted through food chains. The transmission and transformation of energy are proscribed by the first law of thermodynamics which states that

$$Q = \Delta E + W \qquad (5\text{-}1)$$

where Q is the energy input to the system, ΔE is the change in the energy content of the system, and W is the work done by the system. Thus if an ecosystem receives energy from an outside source, that energy is either stored or translated into work. The universal consequences of work performance are an increase in environmental entropy and either a change in, or preservation of, the entropy level of the system in which the work is performed.

The only significant energy source for ecosystems is the energy of photons from the sun, captured by the photosynthetic pigments of green plants. Kinetic energy of electrons expelled from the photosynthetic pigments is stabilized in the form of chemical bond energy as the electron loses energy on its return to the chlorophyll molecule:

$$\text{electron} \longrightarrow 6CO_2 + 6HOH$$
$$\text{light} \longrightarrow \text{chlorophyll} \longleftarrow C_6H_{12}O_6 + 6O_2$$

The energy of the electron is not translated directly into carbohydrate bond energy, of course, but takes part in many intermediate energy exchange reactions involving adenosine triphosphate and other energy-rich compounds. Some of these compounds may be used directly by the plant in performing work rather than in storing energy. . Energy stored as a variety of organic molecules in plants subsequently may be utilized by the plant or by other organisms in the food chain to do work.

Ecologists have restated the first law of thermodynamics as

$$P_g = P_n + R \tag{5-2}$$

where P_g is *gross productivity* or energy input, P_n is *net productivity* or energy stored, and R is *respiration* or energy used to do work. Biological work may be regarded as maintenance costs that must be met to sustain the system or generate growth. The simple equation (5-2) can be applied to any level of biological organization. As ecologists, however, we are interested in ecosystems, where the equation may be applied from a group of communities to individual organisms. It is applicable to animals and decomposers as well as to plants, although considerable confusing terminology has developed around the process of energy utilization in animals. If gross productivity equals respiration ($P_g = R$), the ecosystem just maintains itself and there is no change in total energy content. Net productivity represents an excess of energy incorporation above maintenance costs and is expressed as an increase in stored energy with time. At any one time, the total stored energy content is called *biomass (B)*. Biomass is the total cumulative net productivity over the period that the ecosystem has been functioning. Thus net productivity may be expressed as an increase in biomass with time. When biomass is stable with time, $P_g = R$. When P_g is less than R, the biomass decreases with time and there is a deterioration of the ecosystem. When P_g is greater than R, the system is accumulating biomass.

At a given trophic level, of course, P_g may be consumed by death and predation as well as by respiratory costs. Therefore

$$dB/dt = P_g - R - H - D \tag{5-3}$$

where dB/dt is the biomass change over a time period, H is the biomass harvested by a higher trophic level over that time period, and D is biomass lost through death. It must be emphasized that net productivity is not biomass, but change in biomass with time.

Since green plants constitute the only significant energy gateway into the ecosystem, the production of this trophic level is often referred to as *primary production*. Assessing the energetic balance of ecosystems is incomplete because direct diversion of high energy compounds from the photosynthetic apparatus into work performance cannot be measured at present. Much of the cellular maintenance function in plants may be met by this direct diversion so that estimates of primary production do not, in fact, represent a complete energy estimate but only set a lower limit.

Gross productivity estimates for plants are measures of their energy accumulation capacities and its subsequent partitioning into energy storage and additional maintenance function not met directly from the photosynthetic reactions. As we shall see later, the proportion of energy input expended in maintenance is generally higher in animals than in plants, but this may arise to an unknown amount from the erroneous bookkeeping procedure. Estimating P_g for plants is further complicated by the fact that a certain proportion of the carbon dioxide incorporated into organic molecules is released very rapidly to the environment, so that carbon incorporation measurements may underestimate P_g by this factor.

Productivity is a rate function. It may be expressed in various ways, but the most common one is change in some unit of energy or mass per unit of ecosystem surface per unit time. It is usually expressed in grams or kilocalories per square meter per day or year. Errors inherent in measuring the gross productivity of plants are not particularly important ecologically because the maintenance energy of plants is lost from the food chain. There is substantial practical significance to this, however, since man's agricultural objective is to preserve as much primary P_g as possible in P_n available for his own use.

ENERGY BUDGETS

An *energy budget* is an estimate of sources of energy incorporation and loss for a system. An estimate of the potential productivity of plants based upon the light energy available from the sun has been made by Loomis and Williams (1963). Of the total energy available in sunlight, a substantial proportion (56 percent) is outside the range of wavelengths absorbed by plant pigments (Table 5-1). It is not likely

TABLE 5-1 Estimated efficiency of primary production under optimal conditions

	Inputs	Loss	Percent
Total solar energy	5000		100
Not absorbable by plant pigments		2780	−55.8
Absorbable by plant pigments	2220		44.2
Reflected by plant surfaces		185	−3.7
Inactive absorption		220	−4.4
Energy available for photosynthesis	1815		36.1
Energy not stabilized in carbohydrate synthesis		1633	−32.5
P_g	182		3.6
R		61	−1.2
P_n	121		2.4

Values in kcal/m²/day
Loomis and Williams, 1963

that pigment absorption ranges might be expanded since most of the energy outside this range is either at too low a level (infrared) to have important energy content or at such a high level (ultraviolet) that it destroys biological molecules. During the evolution of plants, various pigments have appeared that transfer much of the energy in the visible wavelengths to the chlorophyll molecule. Further evolution of this transfer ability is not likely to increase substantially the energy input to ecosystems.

Minor proportions of the potential energy input are lost by reflection and by dissipation in inefficient electron transfers in the photosynthetic apparatus. Under ideal conditions about half of the incident radiation, however, is absorbed by the leaf, and a significant loss (90 percent of absorbed) results from absorption dissipated in water evaporation or in stabilizing the energy as organic bonds. About 10 percent of the absorbed light energy potentially may be stabilized as the bond energy of organic molecules. The maximum estimated amount available for support of subsequent food chains (P_n) is 2.4 percent of total solar energy and 5.2 percent of the energy absorbed by the photosynthetic apparatus (Loomis and Williams, 1963). This, it should be emphasized, is a maximal estimate. In fact, only under very carefully manipulated laboratory conditions with all factors optimized have such values been obtained.

An obvious source of additional ecosystem energy would be to increase the proportion of trapped energy that is stabilized as chemical bond energy. The photosynthetic reactions occur through the physical act of electron transfer in a series of steps involving many intermediate receptor molecules. Part of this energy, as we have already discussed, may be directly diverted into plant maintenance functions. The effective operation of electron transfer depends on the concerted migration of electron pairs rather than single electrons; one electron can do only so much work, regardless of its initial energy content as it is expelled from a pigment by the collision of a photon with the absorbing pigment. More efficient operation of the ecosystem's energy gateway would probably require the evolution of additional high energy compounds beyond the few known to occur in living things. Some of them would be capable of preserving for storage the extremely high energy of an electron produced by, for instance, a photon in the high-energy blue range of the spectrum.

A number of ecosystem energy budgets have been determined. The earliest attempt was Transeau's (1926) study of a cornfield in Ohio (Fig. 5-1). Slightly over half of the energy available to the corn was either unabsorbed or inactively absorbed. Terrestrial plants, such as corn, have a great heat load on the leaves due to wavelengths of light in the infrared. Much of the energy in sunlight occurs in these wavelengths (Fig. 5-2), but there is insufficient energy in the photons to drive the photosynthetic apparatus. The infrared wavelengths serve to heat plant surfaces to temperatures above air temperature, resulting in the evaporation of water from the plant, *transpiration*. This cools the leaves and keeps them at physiological temperatures. In the cornfield, 44 percent of the energy of the sun was dissipated in transpiration. Of the total energy available to the plants, Transeau found that 1.6 percent appeared in the stable form of organic molecules. This value for P_g is substantially less than the 3.6 percent possible if all limiting factors except the light-trapping ability of the photosynthetic apparatus were eliminated.

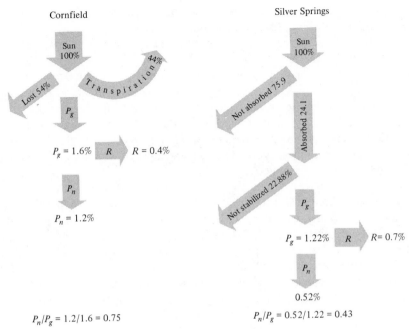

Cornfield

Silver Springs

FIGURE 5-1 Energy budgets for primary production summarizing the way the energy of sunlight is dissipated in an Ohio cornfield and a spring-fed stream in Florida. (After Transeau, 1926, and H. T. Odum, 1957, respectively.)

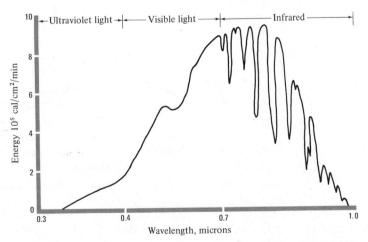

FIGURE 5-2 Distribution of energy in sunlight reaching the earth's surface. The energy per photon decreases from left to right; the high input of energy in the infrared arises out of the abundance of photons at these wavelengths. Light useful in primary production occurs in the range of 0.4 to 0.7 μ. (After Gates, 1965.)

One of the classic studies of ecosystem energetic organization was H. T. Odum's (1957) examination of Silver Springs, Florida. Of the total energy available to the plants from the sun, 75.9 percent was never absorbed and 22.9 percent was absorbed but never appeared as a stable form (Fig. 5-1). The aquatic system was somewhat less efficient than the cornfield, trapping only 1.22 percent of the total available solar radiation as stable molecules. The P_g of the cornfield was 44 percent of the potential if only light were limiting the photosynthetic apparatus, and the aquatic ecosystem's P_g was 34 percent of this potential. Both ecosystems were far from being limited by the innate efficiency of the photosynthetic apparatus. As Eastin (1967) has pointed out, "neither the inherent photosynthetic efficiency nor the CO_2 assimilation potential of leaves currently limits productivity."

Surveys of gross productivity in diverse ecosystems indicate that the upper limit of energy input is about 100 kcal/m^2/day (Table 5-2), which is 55 percent of the maximum possible if only light limits P_g. The greatest energy input occurs in systems such as tropical agriculture, coral reefs, and marshes. At the other end of the gross productivity scale are deserts and open oceans with values of less than 4 kcal/m^2/day. Intermediate values are recorded for grasslands, forests, and most agricultural systems. There has been substantial interest during recent years in utilizing marine systems for greater quantities of human food, but many ecologists feel doubtful about this prospect. Although oceans make up two-thirds of the earth's surface area, their productivity is about one-third that of the total world (Whittaker, 1970b). By far the vast majority of marine productivity occurs on the coastal shelf areas or in certain areas where nutrients are brought into the photosynthetic zone from deep within the sea. In general, the sea has little productivity potential (Ryther, 1969) in the absence of intensive fertilization.

LIMITS ON PRODUCTIVITY

A number of environmental factors in addition to light are important in regulating the energetic potential of the ecosystem. As we have seen, light may be among the least important under natural conditions. There are two important environmental inputs affecting primary production, carbon dioxide and water; two important environmental regulators, temperature and oxygen, affect the final yield of the production process (Fig. 5-3). The size of the photosynthetic biomass also may be affected by the supply of essential mineral nutrients, such as phosphorus and magnesium, and by predation by herbivores. Given a certain photosynthetic biomass, the energy yield of that biomass may be affected by the light available, the carbon dioxide available, and the supply of water from the environment. Carbon dioxide often is an important limiting factor in aquatic environments and may be important in terrestrial systems if other factors are optimized. Water, on the other hand, which is superabundant in aquatic systems, often may be an important determinant of terrestrial productivity. Temperature and oxygen operate as regulatory environmental factors. Temperature affects the rates of chemical reactions, and oxygen affects respiration and loss of fixed carbon dioxide.

Gaastra (1963) has examined limiting factors affecting the energy available from

TABLE 5-2 Survey of primary P_g in diverse types of ecosystems

Ecosystem type	P_g (kcal/m²/day)
Desert	2
Open ocean	4
Oceanic shelf, Grasslands, Cold climate forests	2–12
Most forests, Agriculture, Humid prairies	12–40
Springs, Coral reefs, Tropical agriculture	40–100

Odum, 1959

primary production. As we saw in the chapter on the environment, his studies indicated that the photosynthetic apparatus could be readily saturated with light (Fig. 2-11). If the carbon dioxide concentration of the air was increased above the normal value of 0.03 percent, then more light could be utilized. The rate then could be further increased by raising the temperature. This indicates that P_n is a complex function with at least three interacting environmental limitations upon productivity.

Another way to investigate limitations upon primary productivity is by examining combinations of environmental factors which generate similar P_n. In Figure 2-11, for instance, P_n is the same at a light intensity of 5×10^4 ergs/cm²/sec and 30°C temperatures as it is at 9×10^4 ergs/cm²/sec and 20°C temperatures. Thus over

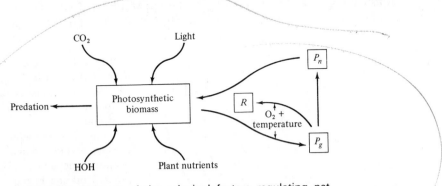

FIGURE 5-3 Diagram of the principal factors regulating net primary productivity of any ecosystem. Efficiency of the photosynthetic biomass is controlled by the availability of carbon dioxide, light, water, and nutrients in the environment. Total yield (P_n) will be determined by the amount of photosynthetic biomass regulated by the previous degree of predation and the rate at which net productivity was transformed into photosynthetic biomass in previous periods.

this range, there is a balance of light intensity and temperature. A given P_n is reached at lower light intensities at the high temperature than at the low temperature. An *isopleth* is a line that shows a constant value for one variable in relation to two other variables. A common example appears in newspaper maps indicating temperature over a geographic area, with constant values of temperature shown over a range of latitude and longitude. Scott and Billings (1964) have presented isopleths of P_n for various combinations of temperature and light intensity (Fig. 5-4). Their study shows the balance between just these two factors in determining P_n. If light intensity is high, temperature becomes an important limitation upon P_n; if light intensity is low, temperature is not an important limitation upon P_n.

Primary production can be regarded as a surface in five-dimensional-environmental space, with the five dimensions being light, carbon dioxide, water, oxygen, and temperature. Various combinations of these will generate equivalent values of photosynthetic yield, but at any one point in the space a single factor may be most important in limiting the process. The degree to which a shift in this limiting factor will move P_n onto another isopleth depends on how far from the optimum the factor is and its balance with the other limiting factors. For instance, if we were to start an aquarium with an excess of mineral elements, high light intensity, and optimum oxygen and temperatures for algal growth, the rate of growth would be limited by the diffusion of carbon dioxide from the atmosphere into the water. By bubbling this gas through the water, we would pass across a series of isopleths of P_n very rapidly at first and then progressively more slowly as we saturated the production system with carbon dioxide. Furthermore, if we provided all environmental factors in excess, productivity ultimately would be limited by the amount of production biomass itself. In addition, production biomass may be cropped significantly by herbivore populations. Although the ultimate upper limit to efficiency is set by energy transfer properties of the photosynthetic apparatus, it is unlikely that the photosynthetic energy input to the ecosystem is frequently limited by light.

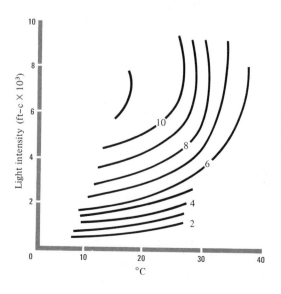

FIGURE 5-4 Isopleths of net productivity (mg CO_2/gm of leaf/hr), averaged for 47 plants, at various combinations of air temperature and light intensity. At low light intensities, P_n is relatively independent of temperature, while, conversely, at high temperatures, P_n is relatively independent of light intensity. (After Scott and Billings, 1964.)

FIGURE 5-5 In this conifer-ous forest, the vegetation is so dense that little light reaches the soil surface. (Redwood Empire Association.)

There is good evidence to indicate that ecosystems are capable of saturating the light supply, up to full sunlight, with photosynthetic biomass. Studies of model communities under laboratory conditions indicate that plants can adjust the number of leaf layers produced so that there is a balance between P_g and R that maximizes P_n. The ratio of leaf surface to soil surface is called the *leaf area index* (LAI). For instance, if the LAI is four, the community has developed a leaf surface four times larger than the soil surface under the community. This results from the community piling canopy level upon canopy level until all of the available light is utilized (Fig. 5-5). Eventually, a point is reached in the lower canopy levels in which maintenance costs are just met by the photosynthetic return ($R = P_g$). The LAI at which this balance occurs increases as light intensity increases. Communities will adjust their LAI to saturate the light available with photosynthetic biomass unless there are other limiting factors. In forests, for instance, almost no light of photosynthetically useful wavelengths reaches the forest floor (Vezina and Boulter, 1966). Because there are ample supplies of other potential limiting factors, many forests are able to utilize much of the available sunlight. In other communities, other limiting factors, such as water in desert systems, may generate very sparse communities in which much of the sunlight is never intercepted by leaves (Fig. 5-6). The energy input to an ecosystem such as this will be extremely low and the ratio, P_n or P_g/light, will be very low.

Jordan (1971) assembled data on the light-trapping efficiency of a wide variety of terrestrial plant communities. There are two main ways we can study this efficiency, either on an annual basis or on a growing season basis. The latter approach expresses light-trapping ability as influenced by limitations other than an intrinsically short growth period. Expressing light-trapping efficiency on an annual basis, of

FIGURE 5-6 In this desert ecosystem where productivity is limited by available water, much of the energy of sunlight is not trapped by the primary producers. (American Museum of Natural History.)

course, does not compensate for the radically different favorable periods of, say, moist tropical sites and near-polar arctic sites. In addition, perennial plants may channel much of their energy storage into organs like woody stems and roots which persist for years and are not readily accessible to either consumers or decomposers. Therefore we can consider energy-trapping ability on a total basis, including long-term storage tissues and tissues which are transitory in duration. Or we can consider the different storage forms separately.

On both an annual and a growing season basis, the minimum efficiency of long-term storage, excepting annual plants, was in desert shrubs (Table 5-3). Maximum efficiency was in a deciduous forest in the southeastern United States. Annual efficiency varied over 600-fold between these extremes. Growing season efficiency varied almost 400-fold. For ecosystems with a continuous growing season, such as tropical rainforest, annual and growing season efficiencies, of course, are identical. Where climate is distinctly seasonal, growing season efficiency will be somewhat higher than annual efficiency.

Minimum short-term energy storage efficiency also occurred in the desert shrubs (Table 5-3). Maximum efficiency, however, was in ecosystems where the major primary producers were annual plants. The range of variation in channeling of light energy into leaves and other short-lived organs was much less than the variation in long-term storage. The range between maximum and minimum varied only 30-fold on an annual and 16-fold on a growing season basis.

With total annual P_n plotted against annual light-trapping efficiency, the result is a straight line (Fig. 5-7) such that

$$P_n = 60 + 1170 \, E \tag{5-4}$$

TABLE 5-3 Light energy input and net productivities of plant communities

Community No.	Description	Age	Location	Incident light [cal/m²(10⁷)] between 0.4 and 0.7 μm — Year	Incident light — Growing season	Net productivity (10⁴ cal/m²/yr) — Leaves	Net productivity — Storage
1	Annual herbs	1 year	Georgia	65.8	37.6	206	0
2	Ragweed	1 year	Oklahoma	70.5	42.3	601	0
3	Annual herbs	1 year	New Jersey	60.9	35.0	1081	0
4	Maize	1 year	Minnesota	56.4	21.6	629	0
5	Perennial grass	8 years	Georgia	65.8	37.6	180	58
6	Tallgrass prairie	?	Missouri	65.8	36.0	196	184
7	Perennial herbs	?	Japan	56.4	30.5	181	127
8	Old field	14 years	Michigan	57.5	47.1	272	416
9	Old field, upland	30 years	Michigan	65.8	37.6	135	0
10	Old field, swale	30 years	Michigan	57.5	47.1	620	0
11	Desert shrubs	Varied	Arizona	90.4	30.9	59	2.7
12	Alpine tundra, xeric	?	Mts. of Wyoming	–	15.8	32	0
13	Alpine tundra, mesic	?	Mts. of Wyoming	–	15.8	87	144
14	Tropical rainforest	1–3 years	Puerto Rico	48.7	48.7	75	140
15	Tropical rainforest	Mature (?)	Puerto Rico	48.7	48.7	226	197
16	Tropical seasonal forest	?	Ivory Coast	75.2	37.6	164	377
17	Tropical rainforest	Mature	Thailand	75.2	75.2	866	223
18	Broadleaf evergreen forest	?	Japan	56.4	30.5	442	383
19	Mixed evergreen broadleaf	?	Japan	56.4	30.5	285	–
20	Average from forests of four equatorial regions	Varied	Varied	70.5	70.5	422	–
21	Average from forests of seven warm temperate regions	Varied	Varied	61.1	37.6	261	–
22	Average of 10 stands of angiosperms	Varied	Mts. of Tennessee	61.1	35.2	152	466
23	Stand dominated by *Liriodendron tulipifera*	?	Mts. of Tennessee	61.1	35.2	195	1108
24	Average of 13 stands of gymnosperms	Varied	Mts. of Tennessee	61.1	35.2	131	421
25	Angiosperm forest	Young	Japan	56.4	30.5	257	–
26	Pine forest	17 years	Virginia	61.0	35.2	234	449
27	Oak–pine forest	40 years	New York	56.0	35.0	190	368
28	Average of 10 angiosperm forests	Varied	Europe	47.0	35.2	133	354
29	Average of 7 gymnosperm forests	Varied	Europe	47.0	35.2	133	498
30	Ash plantation	12 years	Denmark	42.3	32.9	157	188
31	Beech plantation	8 years	Denmark	42.3	32.9	128	221
32	Beech plantation	25 years	Denmark	42.3	32.9	186	442
33	Beech plantation	46 years	Denmark	42.3	32.9	186	442
34	Beech plantation	85 years	Denmark	42.3	32.9	186	340
35	Savanna	?	Minnesota	56.4	32.5	–	238
36	Oakwood	?	Minnesota	56.4	32.5	–	378
37	Average from 22 forests of cool temperate regions	Varied	Varied	56.4	35.2	172	–
38	Two subalpine conifer forests	?	Varied	56.4	30.5	172	–
39	Average from forests of three arctic-alpine regions	?	Varied	37.6	32.9	53	–

See Jordan, 1971 for original data sources

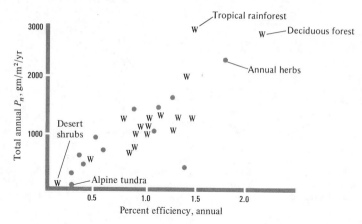

FIGURE 5-7 Relationship between total annual net primary productivity and annual efficiency of converting light energy into biomass. W = woody communities, • = herbaceous communities. (Data from Jordon, 1971.)

where P_n is annual P_n in gm/m²/year and E is annual percentage efficiency of transforming visible light energy into biomass. By far the bulk of these ecosystems fall into an efficiency range between 0.5 and 1.5 percent, with a corresponding P_n of 650 to 1800 gm/m²/year.

When light-trapping efficiencies are converted to a growing season basis, the pattern is essentially the same (Fig. 5-8), but efficiencies are higher:

$$P_n = 198 + 677\,G \tag{5-5}$$

where P_n is annual productivity in gm/m²/day and G is percentage efficiency of energy transformation on a growing season basis. The tropical forest, which had a relatively high annual efficiency, was somewhat less efficient on a growing season basis. The least efficient were, again, the desert shrubs and, in addition, arctic-alpine forests.

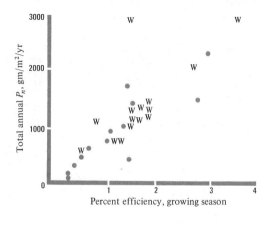

FIGURE 5-8 Relationship between total annual net primary productivity and the efficiency with which light is converted into biomass during the growing season. W = woody communities, • = herbaceous communities. (Data from Jordon, 1971.)

The difference between the relationships of equations (5-4) and (5-5) arise out of
several ecosystems where day-to-day efficiency, as expressed in (5-5), is not particularly high, but the growing seasons are so long that they compensate. Daily productivities of several tropical forests, for instance, are around 3 gm/m²/day. Since
the growing seasons are continuous, annual productivities exceed 1000 gm/m²/year.

Annual herbs and maize, in contrast, have average growing season productivities
around 11 gm/m²/day. The growing seasons of these temperate stands are less than
200 days, so annual primary productivities of the two radically different ecosystems
are essentially equivalent. We can conclude, therefore, that daily productivity during
the active period will be directly related to the efficiency of light trapping. Annual
productivity, however, will also reflect variations in the number of days per year
that light trapping is sustained.

Rosenzweig (1968) examined environmental factors that might limit net primary
productivity in terrestrial ecosystems and found that an environmental index combining available moisture and length of growing season was a good predictor of
P_n (Fig. 5-9). Ecosystems with the highest total annual yield were tropical rainforests,
where the growing season is continuous and there is heavy rainfall. Any environmental modification that tends to decrease either of these climatic parameters will
result in lower ecosystem yield. The lowest net productivities occur in arctic tundra
and deserts. The limitations upon growth in these two systems, of course, are fundamentally different. Yield in the tundra is limited by the extremely short growing

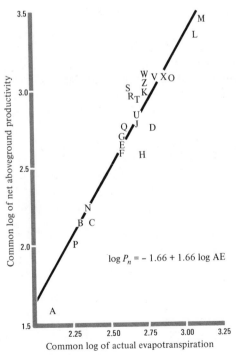

FIGURE 5-9 Relationship between net primary productivities of plant communities
and the evapotransporation indices of the
sites. Communities are: A = creosote bush
desert, B = arctic tundra, C = alpine tundra, D = tall grass prairie, E to J = mountain shrublands, K = beech-maple forest, L
and M = tropical forests, N = desert sand
dunes, O = oak forest, P = annual grassland, Q = fir forest, R and S = spruce-fir
forests, T and U = beech forests, V = hemlock forest, W and X = lowland deciduous
forest, and Z = hemlock forest. (After
Rosenzweig, 1968.)

$$\log P_n = -1.66 + 1.66 \log AE$$

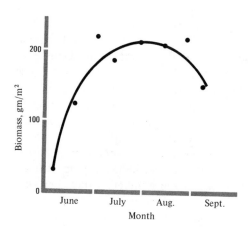

FIGURE 5-10 Growing season change in plant biomass on abandoned fields in New Jersey. (After Malone, 1968.)

season and in the desert by severe water shortage. Net productivity is frequently limited by combinations of low temperatures and short growing season with low available moisture. The causal nature of such a correlation should not be over-emphasized, however. Other limitations, such as the supply of nutrients from sub-strate weathering, may be associated with the climatic index. Although these are not direct limitations on P_n, they do regulate it through their effect upon the ability of the ecosystem to support a photosynthetic biomass.

PRODUCER BIOMASS

In temperate ecosystems the producer biomass goes through annual cycles. Although total community biomass may reach a peak early in the growing season (Fig. 5-10) and remain constant until winter, there will usually be a series of production waves in which different species reach peak biomass at different times (Table 5-4). The ecosystem has some upper limit to the supportable biomass that is fairly constant from July to September, but it is able to support waves of P_n as one species declines

TABLE 5-4 Maximum biomasses (B = gm/m²) of individual producer species in a New Jersey herbaceous community and the dates of B_{max}

Species	B_{max} = gm/m²	Date
Ambrosia artemisifolia	123	July 30
Chenopodium album	69	July 16
Ipomoea pandurata	83	July 2
Raphanus raphanistrum	90	July 11
Sitavia glauca	6	August 27
Digitaria sanguinalis	8	July 30
18 infrequent species	19	July 30
	P_n = 398 gm/m²/yr	

Malone, 1968

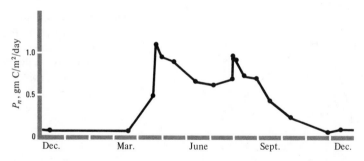

FIGURE 5-11 Annual pattern of net primary productivity in a Danish lake. (After Jonasson and Mathieson, 1959.)

in biomass and another increases. When estimated by peak biomass, the net productivity of this ecosystem was only 55 percent of the actual P_n expressed in the sum of separate species productivities.

In a Danish lake where P_n was measured directly by radioactive carbon incorporation, the ecosystem went through two waves of high P_n, one in the spring and one in the late summer (Fig. 5-11). Different species were predominant at different times during the year, suggesting that there was a replacement of one species by another in productivity waves similar to those occurring in the terrestrial system.

Ovington and Lawrence (1967) examined changes in biomasses of four terrestrial systems during a single annual cycle in Minnesota. Although their estimates of productivity in the natural systems are undoubtedly low since they did not assess the productivity waves as species replaced species, some interesting insight is provided into the organization of net primary production in different systems. Biomass varied over several orders of magnitude in the four ecosystems (Fig. 5-12). The proportional biomass change during the growing season was conspicuous only for the cultivated field.

Peak productivity, for the one month period showing the maximum change in biomass, was 3.04×10^2 cal/ha/day for the prairie, 6.19 for the maize field, 5.63 for the savanna, and 16.63 for the forest. The maize estimate is more accurate than the others because it is a species poor system and was not complicated by the species replacement waves occurring in species rich systems. The P_n values must be regarded as minimal estimates for the other systems. Nevertheless the agricultural field is not particularly productive. All of the natural systems compared favorably, with the forest P_n peak over two-and-a-half times the maize P_n. The most conspicuous property of the maize, of course, is that it is able to generate a peak biomass greater than the prairie, although it starts at near zero biomass at planting time. As an annual plant it is able to generate a high productivity from an extremely small early season biomass.

Dividing the peak biomasses by the peak productivities gives values with the dimensions of time. This ratio is called *turnover time* of the producer component of the ecosystem, which is the time required to replace the peak biomass completely given the maximum productivity potential of the system. The turnover time is 75

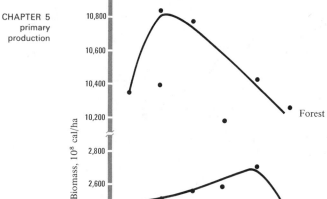

FIGURE 5-12 Annnual patterns of plant biomasses in four different types of ecosystems in Minnesota. Note the breaks in the Y-axis. (After Ovington and Lawrence, 1967.)

days for maize, 126 for the prairie, 461 for the savanna, and 652 for the forest. The much higher turnover time for the forest indicates that the ecosystem will be less susceptible to structural modifications as a result of environmental perturbations than will the maize field. A long turnover time indicates that biomass maintenance is less likely to be dependent on chance environmental fluctuations that influence productivity. In a system with a short turnover time, an environmental perturbation that affects P_n will be more rapidly translated into a change in biomass.

A distinctive difference among the four ecosystems is the biomass burst of the agricultural ecosystem, but this ecosystem depends on substantial energy inputs from fossil fuels, via man, to generate a rather meager net productivity. Self-maintenance is one of the important properties of nonagricultural systems. The ability of these ecosystems to maintain their integrity with energetic inputs solely from sunlight makes them important models to understand in fashioning minimal maintenance agricultural systems. A self-maintaining ecosystem with a high food yield should be the objective of agricultural practices rather than continued reliance on the fundamentally unstable systems that are currently exploited for food production. In later discussions of community organization and succession we consider some of the properties of uncultivated ecosystems that may influence their stabilities.

Another important difference among the four ecosystems is the partitioning of energy among various production compartments. Biomass can be separated into

TABLE 5-5 Partitioning of energy among producer compartments in four different ecosystems in Minnesota. Values are mean ± 95 percent confidence interval for four way comparison

	Proportion of energy in	
Community	Roots	Litter
Prairie	0.603 ± 0.051	0.342 ± 0.051
Savanna	0.205 ± 0.036	0.225 ± 0.027
Forest	0.067 ± 0.013	0.202 ± 0.011
Maize	0.158 ± 0.178	–

Ovington and Lawrence, 1967

shoot, root, and litter components; one of the significant differences among these ecosystems is the partitioning of energy among these compartments (Table 5-5). In the maize field little productivity was in litter at the end of the growing season. Subsequent to harvest, of course, a substantial fraction of this system's P_n will become litter. A conspicuous property of the maize and forest was that a minor proportion of their energy was in roots. The prairie, however, channeled 60 percent of its energy into the root compartment. Although the absolute size of the root and litter compartments increases from prairie to forest, the percentage contribution declines significantly. One of the most outstanding differences between the forest and the prairie is the major contribution of shoots to the total energy content of the forest (73 percent) and its minor contribution to that of the prairie (5.5 percent). Most of the energy of the prairie occurs in those portions of the community associated with the soil.

In a survey of litter accumulation patterns in prairies in different climatic regions, Koelling and Kucera (1965) found that litter accumulation increased from south to north, while P_n decreased over the same range (Table 5-6). As a result, the ratio of litter to P_n (litter turnover time) increased almost twofold over a few hundred mile latitudinal range. These data indicate that the litter compartment becomes progressively larger as climate becomes cooler, and that its proportional size increases even faster than its absolute size. Weaver and Rowland (1952) found that litter accumulation could decrease primary productivity. P_n was about 25 percent larger in prairies of different types when the litter layer was removed. Although litter

TABLE 5-6 Primary P_n (gm/m^2/yr) and size of the litter compartment (gm/m^2) in prairies in different geographic areas

	Geographic location of prairies		
	Southwest Missouri	Central Missouri	Northeast Iowa
P_n	544	508	390
L	382	471	510
L/P_n	0.70	0.93	1.31

Koelling and Kucera, 1965

represents an important storehouse of energy and materials that the ecosystem may draw upon under certain circumstances, its immediate effect may be an inhibition of energy flux into the system.

COMPONENTS OF PRIMARY PRODUCTION

Primary production supports the ecosystem. Positive biomass increments by the producers, even if instantaneously cropped by consumers, still must be sustained over extended time periods if the ecosystem is to persist. If the balance of equation (5-3) goes negative for producers, they lose biomass. Furthermore, for any time period, the balance of equation (5-2) must be positive, or the producer component is consuming its own biomass to meet maintenance and/or reproductive requirements. On a diurnal basis, of course, plants usually will store energy during the day and then consume some of it during the night in growth and maintenance. Thus for the dark period of every day, plants are on a negative balance for equation (5-2). There also may be short time periods immediately following dormancy and during fruit ripening when producers consume previously stored biomass to produce a canopy rapidly or to produce mature seeds. Support of consumers depends on these negative energy periods being relatively insignificant over long times. Some consumers, however, may overharvest producer biomass as their own populations grow. We shall see later, when considering aquatic ecosystems, that this often generates seasonal cycles in which producers have a high early season P_n that is subsequently consumed by herbivores with a resulting decline in producer biomass.

If we ignore herbivore consumption, we can get some insight into the ability of plant communities to trap light from studies of experimental "communities." McCree and Troughton (1966) studied the properties of white clover plants grown inside cylinders which confined the leaves within a known area. As LAI built up, they examined the balance between daytime P_n and nighttime R. Early in the experiment, when LAI was small, the rate of photosynthesis per unit ground area was low (Fig. 5-13). As the canopy developed, of course, community photosynthesis and respiration increased. The gross photosynthetic rate reached a constant value at an LAI of 3, while R continued to increase. There was, therefore, an optimum LAI for net photosynthesis. As a consequence of declining P_n and increasing R, the proportion of daily P_g that was subsequently consumed in respiration increased from about 28 to 60 percent.

In these model communities the area of living leaves reached a plateau after about 40 days (Fig. 5-14), and the LAI of dead material was essentially a linear function of time for the duration of the experiment. As the lower canopy levels developed an increasingly negative energy budget by virtue of shading, they died. The rate of leaf production, however, remained relatively constant throughout the experiment. This phenomenon is probably a consequence of the type of plant used for the experiment—a pasture species which has been under strong selection for high biomass yield in leaves. The situation is somewhat different, we may assume, for other producers. In deciduous forest, for instance, LAI probably increases very rapidly in the spring and then declines over the rest of the growing season through death and herbivory. These experiments do indicate, however, that LAI will become

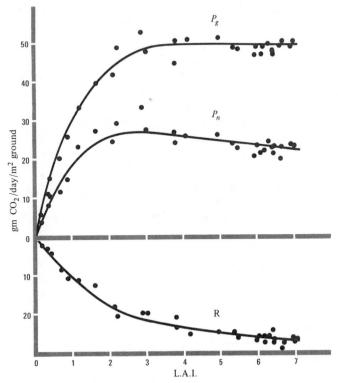

FIGURE 5-13 Relationship between leaf area index (L.A.I. = leaf surface per unit ground surface) and the balance of P_g, P_n, and R in synthetic clover communities. (After McCree and Troughton, 1966.)

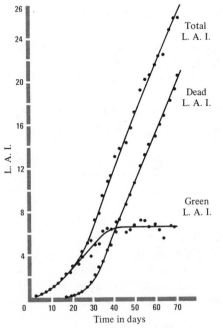

FIGURE 5-14 Development of leaf area index (L.A.I.) in synthetic clover communities with time and the partitioning of leaves between living and dead biomass. (After McCree and Troughton, 1966.)

stabilized at some R/P_g balance less than one. Sustained yield to higher trophic levels depends on such a balance.

Primary producers in aquatic ecosystems may have an even more pronounced biomass "turnover" than occurred in the clover plants. In highly productive aquatic ecosystems, a thick layer of producers near the surface may trap most of the light so that light penetration into the water is minor. Producers in these systems can be heavily cropped by herbivores with little effect upon primary P_n. We have emphasized terrestrial primary productivity in this chapter and will consider aquatic ecosystems in more detail in Chapters 14, 15, and 16.

GENERAL CONCLUSIONS

1. Entry of energy into the ecosystem is dependent on the conversion of light energy into chemical bond energy by green plants.
2. The transformation and transmission of energy in the ecosystem are proscribed by

$$P_g = P_n + R$$

where P_g is gross productivity, P_n is net productivity, and R is respiratory rate.
   (a) If $P_g - R > 0$, biomass increases.
   (b) If $P_g - R < 0$, biomass decreases.
   (c) If $P_g = R$, biomass is constant.
3. For a given trophic level biomass may be consumed by death and higher trophic levels as well as respiration, so that

$$dB/dt = P_n = P_g - R - H - D$$

where H is rate of biomass harvest and D is rate of biomass death.
4. Of the solar energy in wavelengths absorbed by plant pigments (0.4 to 0.7 μ), the maximum trapped as primary P_g will be 5.4 percent. Since the maximum values recorded for nonlaboratory ecosystems are near 4 percent, few (if any) ecosystems are limited by the efficiency of the light-trapping apparatus.
5. Primary productivity of terrestrial ecosystems is limited primarily by water shortage and/or short frost-free periods.
6. The partitioning of biomass among different plant organs will vary conspicuously in different types of ecosystems. This undoubtedly will influence the nature of the predominant consumers.
7. As the photosynthetic layer becomes thicker the efficiency of light trapping will decline due to shading of deeper portions. Yield to consumers depends upon R/P_g smaller than one.

SOME HYPOTHESES FOR TESTING

Among the areas of active ecological interest which we have not touched upon in this chapter is the relationship between canopy geometry and productive efficiency. The clover communities we considered are composed of flat leaves which are generally horizontal to the soil surface. Most producers, of course, have a somewhat more complex canopy structure, ranging from the conical needle-leaved canopies of conifers to the erect lanceolate leaves of grasses. Since the ability of light to penetrate a canopy depends to a considerable extent on the relationship between the sun's angle and the leaf angle, understanding the geometry of foliage structure and its relation to light-trapping efficiency is among the most complex problems in ecology.

Another interesting problem is the relationship between foliage production patterns and the producer's potential for supporting subsequent trophic levels. The ability of white clover to produce foliage continuously is probably related to its importance as a grazing and hay crop. It would be interesting to compare the foliage producing capabilities of, say, deciduous trees, with plants subjected to substantial intensive grazing. Can oak trees recover from repeated denudation as rapidly as grasses?

Since there are abundant data to indicate that few, if any, ecosystems are limited by the inherent light-trapping efficiency of the photosynthetic apparatus, it would be interesting to find out the degree to which primary P_n may be increased by, say, irrigation and fertilization. We have seen that agricultural ecosystems are not particularly productive when compared with natural ecosystems, even though the former often receives huge inputs of fertilizer and water. To what degree could the efficiency of natural ecosystems be stimulated by comparable environmental enrichment?

energy flow   6

All organisms above the level of primary producers are ultimately dependent on these producers as their source of stored energy. This chapter considers the movement of energy among the populations of an ecosystem and has two primary goals: (a) to examine the energy interrelationships of organisms in an ecosystem and (b) to understand the sources and effects of energy losses from an ecosystem. The first goal provides a view of the ecosystem's functional organization. The second goal is related to predicting and controlling the partitioning of energy output from ecosystems. From the human viewpoint, control of output of usable energy is important in terms of feeding the growing human population. Much of what we will discuss is dependent on the relationships among energy input, energy storage, and work similar to those already discussed for plant populations.

    There are three general approaches to the study of energy flow in ecosystems and each has its own inherent strengths and shortcomings. The largest scale approach views energy flow on an ecosystem basis. Each species is assigned to one or more trophic levels and energy input and output values are measured for each level. Such an analysis of a complete ecosystem necessarily reduces the detailed information of energy flow in individual populations. By reducing the scope of the investigation to movements of energy in single species at each trophic level the ecologist can examine more specific mechanisms influencing energy flow. This type of investigation is called a *food chain* analysis, because each species is considered a link in the movement of energy from the primary producers to the highest level of consumers and decomposers. To examine in detail the factors that influence the amount of energy moving between populations, the factors influencing these movements, and the amount of energy stored by a population, the ecologist usually studies one or two populations in detail in the laboratory, where variable influences can be controlled. Each level of investigation generates different data and none is better than another; the type of investigation must be determined by the type of information desired. We examine three studies that illustrate the three specific levels of investigation on energy flow.

ANALYSIS OF ENERGY FLOW

The broadest view of energy flow in ecosystems is obtained by an analysis of the trophic levels of the entire ecosystem. All organisms in the ecosystem are assigned to one or more trophic levels, and energy movement is estimated by summing values for populations that operate at the same level. The most difficult problem inherent in this approach is that not all organisms can be easily assigned to a trophic level. Man, as a good example, is an omnivore and operates at trophic levels two, three, four, and probably five. Many assumptions, some of which cannot be easily tested, must be made about losses and sources of energy in the ecosystem to construct a balance sheet of energy flow. Unless all energy flow values can be determined independently, in all three types of analyses a primary assumption is that the system is stable; that is, the input of energy is just balanced by output, including respiration loss and net productivity at each level. Short-term changes in net productivity or respiration will decrease the accuracy of estimates of these parameters. Technically, this is a difficult method of analyzing energy flow because it requires knowledge of, or assumptions about, the energies of all organisms in the system and for many small species this can be extremely difficult. Juday (1940), Lindeman (1942), H. T. Odum (1957), and Teal (1957, 1962) are among the few workers to attempt a complete ecosystem analysis of energy flow.

Investigations of energy flow in complete ecosystems have been limited primarily to aquatic systems. A major advantage of many aquatic systems is that temperatures are less variable compared to terrestrial systems. Odum (1957) worked on an aquatic system at Silver Springs, Florida, with a continual input of fresh water from an underground source. Water temperature varied from 22.2 to 23.3°C in 19 measurements during a two-year period. Respiration values for organisms could be calculated at a single temperature and used as an annual estimate of the total respiration for the sample. Even in a terrestrial tropical climate in which there is less climatic variation than in temperate areas, daily temperature fluctuations may have a marked influence on energy budgets compared to uniform temperature ecosystems such as those at Silver Springs.

A second major advantage of an aquatic ecosystem for energetic analysis is the circumscribed nature of the system. No ecosystem is completely bounded and divorced from outside influences, but aquatic systems allow more precise estimates of outside influences than do terrestrial systems. The incoming energy that continually supplies the system must cross the boundary from one medium to another. Once in an aquatic system, the energy leaves as heat energy or stored energy that again must cross a distinct boundary; it can be fairly easily estimated compared to the estimation problems in terrestrial ecosystems in which boundaries are indistinct. To estimate the biomass lost from a cold spring pond, Teal (1957) used emergence traps for adult insects that matured from larvae living in the pond (Table 6-1). These emigrating adults were a net loss of stored energy from the pond that could be more easily estimated than can migration losses in most terrestrial systems.

Finally, a system bounded by another medium can be useful for estimating the amount of energy lost as maintenance energy before it is dissipated beyond the bounds of the system. Carbon dioxide released in respiration is difficult to measure in a terrestrial system because it quickly becomes part of the general atmospheric supply by diffusion and mixing. Gases are dissolved in the water in an aquatic system, and changes in concentration are not difficult to estimate by standard techniques. Some errors will be introduced by movement of gases across the water-air interface, but these can be assumed and correction factors applied.

Ecosystem analyses of energy flow depend at present on numerous assumptions, estimates, and potential sources of error. However, the problems inherent in aquatic analyses are less severe than those in terrestrial analyses. Some data on terrestrial systems are available from analyses of a rainforest ecosystem at the Rio Yanque Experimental Station in Puerto Rico, in which the system was defined artificially by enclosing a section of the rainforest in a huge plastic cylinder (H. T. Odum, 1970). Other workers (for example, Whittaker and Woodwell, 1968; Whittaker and Wood-

TABLE 6-1 Population of larvae and emerging adults of *Calopsectra dives* in Root Spring, Concord, Mass., in 1954. Energy content is estimated on the basis of 1.58 cal/mg as determined for *Analopynia dyari*

	Jan.–April	May	June	July	Aug.	Sept.	Oct.–Nov.
Larvae							
Number in thousands/m²	–	1.7	89.5	65.0	57.0	0.2	–
Weight in gm/m²	–	3.0	58.5	82.4	127.0	0.2	–
Energy content in kcal/m²	–	2.1	40.4	56.8	87.6	0.1	–
Adults							
Number/m²	–	13	170	953	3464	13250	533
Weight in mg/m²	–	12	156	876	3170	12200	190
Energy content in kcal/m²	–	0.019	0.246	1.38	5.00	19.30	0.775

Teal, 1957

well, 1969; Woodwell and Botkin, 1970) have attempted to measure energy flow in portions of an ecosystem and then to correlate these estimates with easily measured aspects of the system such as growth rates or sizes of individuals. These regression equations then can be used to estimate total energy flow in the ecosystem once the component parts have been recognized and appropriately measured. This total ecosystem approach, as in the aquatic system method, is confounded by the gross level of the measurements of energy flow and energy storage, especially since it is difficult to measure the component populations accurately enough to warrant applying the precise regression equations.

Ecosystem Analysis of Energy Flow

The Silver Springs, Florida study (H. T. Odum, 1957) exemplifies the ecosystem approach to energy flow analysis. The results of the energy flow analysis of Silver Springs are presented in Figure 6-1. The total amount of energy stored at successive consumer trophic levels decreases dramatically, partly as a result of respiration costs and partly because so much of the energy (57 percent) is not used by consumers but passes directly into the decomposer portion of the ecosystem. A small amount is also lost as export. Odum did not quantify the amount of energy entering the decomposer populations from each trophic level but gave a total figure for the decomposers. He assumed that all decomposer energy requirements were met by organic material from within the system. There is very little energy flow at the highest trophic levels. Since these are usually made up of fairly large organisms (fish and turtles in this case), the density per square meter is low and by the fourth trophic level the potential energy has been reduced to such an extent that a fifth level cannot be supported. None of the aquatic ecosystems studied to date had more than four clearly defined trophic levels in addition to the decomposers. The limit is set by the amount of energy initially fixed by the plants and the losses of energy as it flows in the system. To increase the number of consumer trophic levels, P_n/P_g ratios could be increased, more energy could be channeled into consumers and less into decomposers, or the size of the primary producer base could be increased.

Teal (1957) also used the trophic level analysis of an ecosystem to study energy flow in Root Spring, a cold spring in Massachusetts. One major difference from the Silver Springs data was that 76 percent of the total net primary productivity was imported from primary producers living outside the system. Most of the energy supporting the higher trophic levels was not produced in the cold spring and presumably the future production of the primary producers could not be influenced by the grazing activities of the first level consumers. Most of what the consumers ate had no impact on what would be added to the system during and after the next growing season. Interestingly, Teal found that 76 percent of the total net primary productivity was assimilated by the herbivores. Odum reported only about 32 percent in Silver Springs and most values are about one-fifth that reported by Teal (Table 6-2). The cold spring herbivores thus were very efficient at converting net productivity of plants into gross productivity of herbivores. Teal suggested that this resulted from

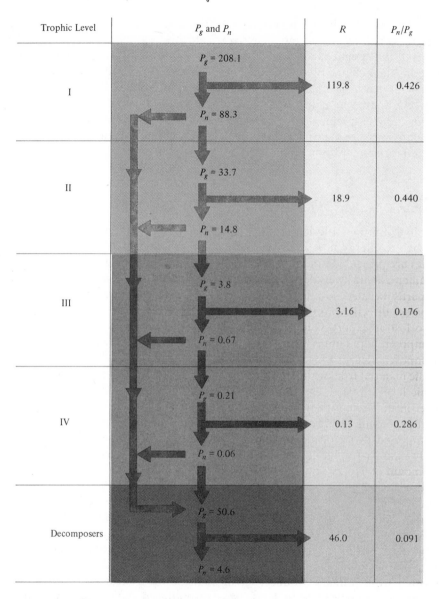

Trophic Level	P_g and P_n	R	P_n/P_g
I	$P_g = 208.1$ $P_n = 88.3$	119.8	0.426
II	$P_g = 33.7$ $P_n = 14.8$	18.9	0.440
III	$P_g = 3.8$ $P_n = 0.67$	3.16	0.176
IV	$P_g = 0.21$ $P_n = 0.06$	0.13	0.286
Decomposers	$P_g = 50.6$ $P_n = 4.6$	46.0	0.091

FIGURE 6-1 Energy flow patterns in Silver Springs, Florida. All values are 10^2 kcal/m²/yr. (Data from H. T. Odum, 1957.)

the "fact that the Root Spring conditions were more favorable to life than those in other environments."

Rapid removal of wastes and sufficient oxygen insured that the activities of the herbivores would not be limited by factors other than the availability of food. Presumably they could operate at nearly peak intake of energy as it appeared in

TABLE 6-2 Efficiencies and assimilations for higher trophic levels of various aquatic ecosystems

	Root Spring		Cedar Bog Lake		Lake Mendota		Minnesota Pond		Silver Springs	
	Teal (1957)		Lindeman (1942)		Lindeman (1942)		Dineen (1953)		Odum (1957)	
	Assim.[a]	Effic.[b] (percent)	Assim.	Effic. (percent)	Assim.	Effic. (percent)	Assim.	Effic. (percent)	Assim.	Effic. (percent)
Primary Consumers (herbivores)	2300	76	148	16.8	416	11.2	92	23	3368	38.1
Secondary Consumers (carnivores)	208	36	31	29.8	23	8.7	34	47	383	25.9
Tertiary Consumers (secondary carnivores)	–	–	–	–	3	23.0	–	–	21	31.3

[a] All assimilations are kcal/m^2/yr.

[b] All efficiencies are $\dfrac{\text{assimilation}}{\text{net production of next lower trophic level}}$

Modified from Teal, 1957

the spring. The only major source of loss was material such as cellulose that was not usable by the herbivores. The high assimilation efficiency of the Root Spring herbivores suggests that they are much closer to exploiting all of the net primary productivity available in the spring. If much of this net primary productivity were essential to producing additional plant biomass, assimilation by herbivores above a certain percentage could reduce net primary productivity during the next growing season and hence reduce the total consumer populations that could be supported. Too high a harvesting rate would lead to extinction of the total system. However, in an ecosystem in which most of the productivity is derived from outside the system, as in Root Springs, the level of herbivory presumably has no impact on what is imported the next year. Hence, selection probably will operate to maximize harvest rate. The assimilation efficiencies for trophic levels above herbivores should be about the same for the cold spring and other ecosystems because all or essentially all energy is derived from herbivores that are part of the system and not from outside. The similarity of assimilation efficiencies of the first carnivore level is evident in Table 6-2.

Food Chain Analysis of Energy Flow

Analysis of energy flow values for a Michigan old field (Golley, 1960) was limited to a single species at each trophic level (Fig. 6-2). The total amount of energy that is used in the chain was small relative to what was available at the next lower level (P_n). In percentage terms 99.7 percent of the plant productivity was not used by the mice and 62.8 percent of the mouse productivity plus immigration of mouse biomass from outside the field was not used by the weasels. Included in these

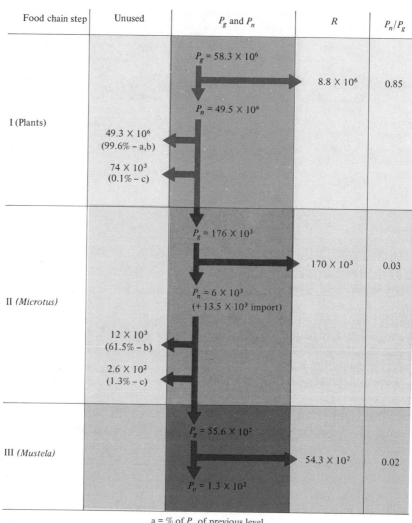

Food chain step	Unused	P_g and P_n	R	P_n/P_g
I (Plants)	49.3×10^6 (99.6% – a,b) 74×10^3 (0.1% – c)	$P_g = 58.3 \times 10^6$ $P_n = 49.5 \times 10^6$	8.8×10^6	0.85
II *(Microtus)*	12×10^3 (61.5% – b) 2.6×10^2 (1.3% – c)	$P_g = 176 \times 10^3$ $P_n = 6 \times 10^3$ (+ 13.5×10^3 import)	170×10^3	0.03
III *(Mustela)*		$P_g = 55.6 \times 10^2$ $P_n = 1.3 \times 10^2$	54.3×10^2	0.02

a = % of P_n of previous level
b = not eaten
c = not assimilated after eaten

FIGURE 6-2 Energy flow in an old field food chain in Michigan. Values are kcal/ha/yr. (Data from Golley, 1960.)

percentages are values for material that was not harvested and energy that was ingested but passed unused through the digestive tract and back into the environment. The latter loss made up a relatively small percentage of the total unused energy at each level (0.1 to 1.3 percent). Of greater importance for the availability of energy for the next step in the chain is the amount of gross productivity that is used for respiration. In the plants this was a relatively low value (15 percent) but in the consumers, both of which are homeotherms and maintain a relatively high constant body temperature as environmental temperature varies, the percentage of their gross productivity that left the chain as respiratory costs was about 97 percent at each level. This means that almost all energy assimilated by the organism was not stored as added biomass and so could not be utilized by the next trophic level.

With this drastic reduction of available energy at each level, a predator on weasels would have to forage over a very large area to acquire enough energy to support a viable population. Golley thought that owls, with foraging ranges far exceeding the area included in the food chain study, were potential predators on the weasels. Owls probably were taking both mice and weasels.

This second major view of energy flow involves following movement of energy between single species. Since the energy being studied moves into only one species at each trophic level, this type of study is referred to as a food chain analysis. Each species forms a single link in the chain and alternative pathways for energy movement are ignored. This technique can produce more detailed information about the partitioning of energy resources by a few species. Much of the data on maintenance is measured in the laboratory; these data usually are collected on single animals and discount all losses or gains associated with species interactions that are expressed in the complete system analysis.

Additional losses of energy from the food chain occur as the result of feeding by organisms outside the food chain. Golley estimated that 24 percent of the plant material produced on the field may have been taken by insects rather than mammalian herbivores. In an ecosystem of this kind energy can easily be imported and exported. At high enough population densities some consumers will leave the field to find other food sources, representing energy loss in this old field but an import term in remote food chains. Golley estimated that 13,500 kcal/hectare/year of *Microtus* entered the field from outside. Food chain analysis provides a clearer picture than the whole system approach of what is happening at a series of specific points within the ecosystem.

Population Analysis of Energy Flow

The analysis of energy movement between single species at each level still does not provide a detailed analysis of many of the important environmental influences on the sources of energy loss in the flow pattern. In particular, it does not investigate in a detailed manner the important parameters that contribute to population energy loss through respiration and the efficiency of energy storage once the energy has been assimilated.

Slobodkin (1960) provided information about energy losses from populations in relation to a number of important environmental parameters, especially the rate of loss of energy to predators and the role of food availability in the channeling of energy into maintenance or growth. To investigate these parameters in sufficient detail requires close control of other variables influencing energy flow. Slobodkin synthesized an ecosystem in the laboratory to solve the problems of control of extraneous variables. His experimental consumer was a species of crustacean, *Daphnia pulex,* commonly called the water flea. He controlled the rate of food input to the *Daphnia* by adding food at different rates. He also served as the predator by harvesting *Daphnia* at different rates. He modified the rate at which *Daphnia* were removed from the population and the characteristics of the animals removed so he could determine what influence old animals versus young animals had on energy storage by the experimental populations.

Slobodkin examined two population energetic properties in particular detail: (a) the ratio of yield to the predator over food ingested (Y/I) and (b) ratio of net to gross productivity (P_n/P_g) of the population in relation to (1) size of organisms harvested and (2) food supplied to the population. The yield to the predator was measured directly from the organisms Slobodkin, as the predator, harvested. Ingestion was equated with the food supplied to the population under the assumption that all food supplied was ingested, an assumption that was verified experimentally.

Unfortunately Slobodkin had no measure of what percentage of the ingested food was actual P_g. He found that Y/I was higher when adults were harvested than when young were harvested. In addition, Y/I increased as predation rate increased, with a maximum of 0.125 when young organisms equal in number to 90 percent of new young added in the preceding four days were being harvested (Fig. 6-3). This means that the maximum efficiency of the conversion of food into biomass

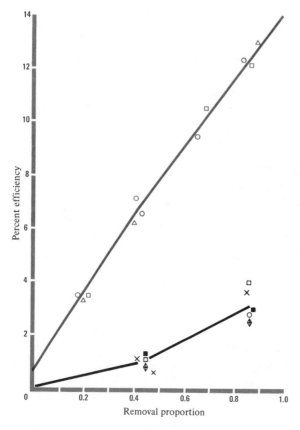

FIGURE 6-3 Relative efficiency of net productivity in *Daphnia pulex* at various rates of removal of individuals from the population. Efficiency is expressed as the ratio of calories removed to calories supplied. The abscissa is the number of animals removed as a proportion of the young produced. The symbols represent different amounts of food added per unit time. In upper line experiments, only young animals were removed. In experiments from which lower line is derived, adults were removed. (After Slobodkin, 1960.)

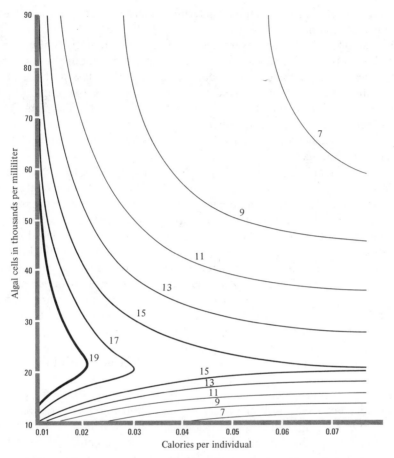

FIGURE 6-4 Isopleths of individual growth efficiencies (in percent) of *Daphnia pulex* as a function of the food available (ordinate) and the size of the animals removed, measured in calories (abscissa). (After Slobodkin, 1960.)

yielded to the next trophic level was 12.5 percent. This maximum occurred when a large proportion of the population was being harvested by the higher trophic level. Size of organisms and food supply were related to isopleths of P_n/P_g for the population (Fig. 6-4). This ratio was lowest when organisms were large and food was abundant and it increased as both decreased. There was, however, an optimum on the food axis at around 20,000 algal food cells per milliliter of medium, and P_n/P_g decreased below this level at all sizes, so that the isopleths are complex in configuration. Maximum net productivity in relation to energy input occurred when organisms were small and food was not too abundant. At this combination of food supply and organism size, 19 percent of the energy input was stored as net productivity. Although the energetic balance for this study is based on some unproven assumptions about the assimilation process, the study does provide additional insight into some of the population interrelationships that may determine energy budgets for total ecosystems or natural food chains.

It is clear from these three examples of energy flow analysis that each specific level of the analysis will answer quite different questions about the movement of energy in ecological systems. The actual causes of energy loss and the pathways of energy movement become blurred as we progress from the single population analysis of Slobodkin to the entire ecosystem approach of H. T. Odum. The final choice of energy flow analysis depends on the type of information desired about the system, but each type of analysis contributes to our final understanding of ecosystem functioning.

In assuming a finite energy supply at the primary productivity stage, all other organisms (consumers and decomposers) must be functionally interrelated by their common dependence on this single energy source. Each consumer organism influences energy availability to higher trophic levels and to decomposers by its own energy requirements and the efficiency with which it uses energy. Until the energy interrelations are known, it is difficult to predict the outcome of changes in the population sizes of individual species on the operation of the whole ecosystem. What is the impact of killing off predators on deer, for instance, or of introducing a parasite on some agricultural pest that is also a component of several surrounding ecosystems? Although we initiated this section with the goal of understanding the functional relationship of ecosystems through energy flow studies, we have only scratched the surface of the important variables. At present the goal is unattainable because there is no general theory in ecology to relate energy efficiencies to other phenomena occurring as part of the energy interdependence of organisms. We must understand predator-prey, producer-herbivore, host-parasite, and many other types of interactions before producing a final picture of the organization of a functioning ecosystem. The energy flow studies just examined have provided some data on one aspect of the interactions: the efficiencies of energy movement under various conditions. In the following section we shall examine in more detail the routes of energy loss in ecosystems. In later chapters we shall relate characteristics of populations and communities to energetic interdependence.

ENERGY LOSSES

An ecosystem leaks energy. Although we have been emphasizing the loss of energy as heat through the respiratory process, there are a variety of pathways through which energy may be lost as it moves from trophic level to trophic level. Whether this energy is lost from the ecosystem depends on whether it is exported from the system (Fig. 6-5). Respiration, of course, does work, but at any trophic level there is much energy wastage which is not realized as P_n or R. Energy that the organism ingests represents less than the total energy available to it. Only part of the energy ingested is actually assimilated by the organism for its own use; the remainder is returned to the environment. Assimilated energy (P_g) is used to repair tissue damage within and otherwise to maintain the organism (R), to add new tissue (P_n), to produce offspring (P_n). In the following discussion we consider in some detail each

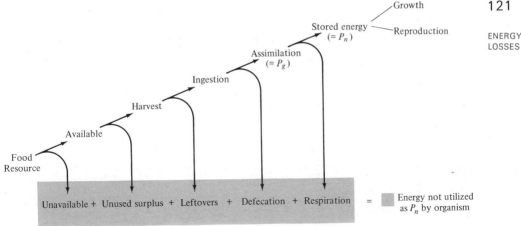

FIGURE 6-5 Flow chart of energy movement and losses between and within trophic levels.
(Courtesy of R. B. Root.)

of the potential routes of energy loss shown in Figure 6-5 as the energy moves into
a new organism and is stored by that organism. In addition, we look at potential
uses for the energy that is stored in consumers as net productivity.

Unavailable Energy

The source of energy for a trophic level above the primary producers is P_n stored
in individuals at the next lower level, that is, if we assume for the moment that
organisms can be placed unequivocally at a single trophic level. Some P_n at the
previous level is stored in a form or position that makes it unavailable for use by
some organisms. Golley (1960) found that about 32 percent of the primary P_n of
an old field in Michigan was in roots of the plants and these were rarely fed on
by the mice. It is probable, however, that most of the plant material is potentially
available to some herbivore in an ecosystem. Nematodes and burrowing insects, for
example, use roots as a food source.

Plants that produce chemical substances that are harmful or repellant to herbivores potentially make unavailable a portion of their P_n (Whittaker and Feeny,
1971). Plants in the milkweed family produce compounds known as cardiac gly-
cosides that harm animals foraging on them. However, some insects (such as the
monarch butterfly) have evolved counter mechanisms and are able to forage on
milkweeds (Brower et al., 1967). Species of herbivores that have not evolved those
counteradaptations cannot make use of energy from the milkweed. Whether energy
of this sort should be classed as unavailable or as unused surplus is debatable. The
energy is unavailable only in the sense that utilization is detrimental to the herbivore.

Another way that P_n at one trophic level is made unavailable to higher levels
is by export from the system. Teal (1962) found that 45 percent of the total net
primary production was carried out of a salt marsh ecosystem in Georgia by the

tides. This export drastically reduced the proportion of the net productivity of the plants that was available to higher trophic levels. Even though primary production in this marsh ecosystem was very high, it could support only three trophic levels, whereas two other aquatic ecosystems, each with less net primary productivity, supported four levels. The absence of the fourth level in the marsh perhaps can be traced to the export of energy from the system, thus leaving insufficient energy at the third trophic level to support a higher carnivore level.

Surplus Energy

The amount of surplus depends on the rate of production of energy at the source trophic level, population sizes of consumers, and feeding rates. If the energy is produced faster than the animals can use it, there will be a surplus; if not, there will be no surplus. If gross productivity at the plant level is greater than the sum of all respiratory costs for the system and there is no net export of plant material, there will be a net increase in ecosystem biomass over the period of measurement. If $P_g = R$ for the ecosystem, then $P_n = 0$ and the ecosystem is at an energetic steady state. If $P_g < R$, the system will lose energy and biomass from one or more trophic levels or may have its gross productivity augmented by the net productivity imported from another ecosystem.

Any litter that is produced represents primary production that is not utilized by consumers and enters the decomposer trophic level. Anyone who has walked through a deciduous forest in the fall will appreciate that at the end of the year many tree leaves, representing plant P_n of that growing season, have not been eaten by herbivores and accumulate as litter on the ground.

Species with marked population fluctuations caused by limiting factors other than food probably often leave food sources unexploited; they are less efficient by our measure at using available P_n than populations that follow fluctuations in their energy resources more closely. On the average, then, trophic levels should become more efficient at using energy from the next lower levels as the ability of the component populations to track fluctuations in environmental resources increases. When resources are increasing in abundance, large amounts are taken if the population can increase its total intake in relation to the increasing abundance.

Predators seldom destroy their prey populations completely, so there is always an unused surplus at the previous trophic level. This unused energy provides the nucleus for regrowth of the prey population. Holling (1959) found that at low and high densities of prey the percentage of prey taken was lower than in moderate density prey populations (Fig. 6-6). There are at least two explanations for decreased predation pressure at low prey densities: either (a) the energy required to find rare items is greater than would be replaced by actually finding and eating those items and/or (b) there is an alternative prey item available that is easier to take. The first explanation assumes that $P_g < R$ for the consumer, which an organism obviously cannot maintain for long periods. At high prey densities the number of predators was insufficient to take a large percentage of the prey population and predation rate declined.

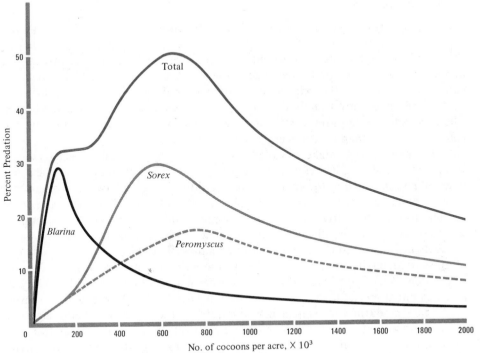

FIGURE 6-6 Percent predation by small mammals on cocoons of sawflies as a function of the density of sawfly cocoons available. (After Holling, 1959.)

If some factor other than food is limiting population size, some of the food will not be used and appear as surplus; then another population may invade the system to utilize it. The success of the second population will depend on whether the first was always kept below the food-limiting level. If not, then the two populations will compete for food and one or the other may be eliminated.

"Leftovers"

Once the energy source has been harvested, some energy often is not eaten by the organism. There are many potential reasons for leaving material after eating. In some cases the size of the energy store represented by a single prey organism may be more than can be consumed by the predator. Some animals, such as rodents, harvest small items but accumulate caches when food is easily available. Depending on the requirements for energy during the periods of low food availability and the size of the cache, there may be some left at the end of the season. Similarly, Steller's Jays (*Cyanocitta stelleri*) collect and bury more seeds than they use later (Grinnell, 1936), and when all of these are not used by other animals, some produce seedlings the next year.

The entire energy packet is rarely in a form that can be entirely assimilated by the consumer. Since the packet comes as a unit, the unused material is discarded as the useful material is ingested. Cellulose is not digested by many herbivores, but accounts for an appreciable percentage of the total energy ingested. Similarly, most carnivores do not eat the skin or fur of their prey despite the energy content of these items; bones are usually discarded as are the exoskeletons of invertebrates. These portions of the prey represent energy sources that are not utilized by the consumer, rather than excess energy that is not consumed.

Consumer energy intake is called *ingestion* (I), but not all ingested food is converted to gross productivity, that is, $P_g/I < 1$. The food remaining may be semi-digested and returned to the environment with little energy removed. In organisms such as mammals that separate metabolic waste and nonassimilated waste, it is possible to measure fairly precisely the amount of ingested energy that is actually assimilated. It is more difficult in animals that combine these two types of waste products before releasing them to the environment (Reichle, 1970) and estimates of gross productivity usually are made by combining the net productivity and the respiration. Welch (1968) found that there were major differences in the ability of aquatic animal species to extract energy from ingested material (Fig. 6-7). However, he did not standardize for similar ingestion rates or quality of food being ingested. He also found that species which assimilate relatively more of the energy from the food they ingest may also channel a smaller proportion of gross into net productivity

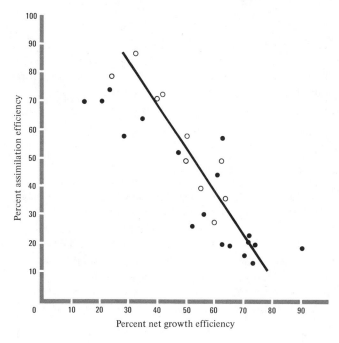

FIGURE 6-7 Relationship between assimilation efficiency and net growth efficiency for a series of animal populations. Open circles are carnivores; solid circles are herbivores. Assimilation efficiency is gross productivity $\times$ 100/ ingestion; net growth efficiency is net productivity $\times$ 100/ gross productivity. (After Welch, 1968.)

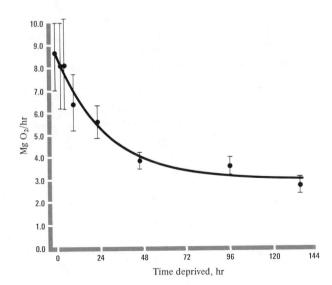

FIGURE 6-8 Effect of food deprivation on oxygen consumption of black bass (*Micropterus salmoides*) acclimated to 20° ± 1°C. Vertical lines are 95 percent confidence intervals. (After Glass, 1968.)

(Fig. 6-7) than do species that assimilate proportionately less of the energy they ingest.

The negative relationship between P_g/I and P_n/P_g suggests that these aquatic species tend to retain a relatively fixed amount of energy as net productivity despite variations of ingestion and gross productivity. Individuals, then, may vary the amount of the total energy intake (P_g) that is spent on respiration (R), depending on the input of energy and the cost of obtaining and processing the energy. As food availability decreases, at least some animals can decrease the amount of energy spent on maintenance. Glass (1968) found that the black bass (*Micropterus salmoides*) decreased its metabolic rate as the length of the food deprivation period increased (Fig. 6-8). After a period of starvation during which previously eaten food was cleared from the gut, there was neither gross nor net productivity left. The fish was utilizing previously stored net productivity (that is, biomass) and operating on a negative energy budget generated by respiration. Reducing respiration would slow the loss of biomass until food was again available.

Maintenance Energy

Assimilated energy (P_g) can be used either for respiration or net productivity. Respiration expenditures include all energy required for the maintenance of the integrity of the organism in the environment such as temperature regulation, replacement of lost tissue (hard to estimate in most energy budgets), and other homeostatic mechanisms, as well as the added costs of assimilation of more energy, interaction with other organisms, and behavior associated with reproduction. All organisms have certain basic respiration costs associated with maintaining their integrity, and those for additional activities depend on the biology of the organism,

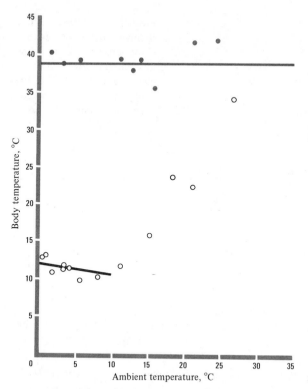

that is, organisms whose body temperature is close to environmental tempera-ture expend a lower proportion of their gross productivity as respiration than do animals that use energy to generate heat internally to maintain a high, relatively constant body temperature (Fig. 6-9). Heterotherms can maintain a high body temperature at low environmental temperatures by absorbing radiant energy (Table 6-3). Homeotherms, on the other hand, normally convert more and more gross productivity to respiration to generate heat to maintain a high body temperature as the temperature of the environment decreases (Fig. 6-10). Thus heat-generating energy expenditures are higher for homeotherms than for heterotherms. In order to reduce respiratory costs, some small homeotherms have evolved means by which their body temperature can fluctuate at times with environmental temperature (Fig. 6-9). The ability to allow body temperature to fluctuate with environmental temper-ature serves as an energy-conserving device because it requires less energy to maintain a body temperature that is nearly identical with environmental temperature than to maintain a body temperature that is high in relation to environmental tempera-ture. The potential energy savings may amount to 90 percent of the energy required to maintain a high body temperature at the same environmental temperature (Hainsworth and Wolf, 1970; Fig. 6-10).

Standard metabolism, that proportion of R used to maintain an individual that is sitting quietly and expending no energy for digestion or other overt activity, is

TABLE 6-3 Cloacal temperatures of three lizards tethered on gravel in sunshine at 7:40 A.M., at 4686 meter elevation, April 18, 1952

		Temperature, C°		
			Cloacal	
Time	Air, in shade	♂, 54	♀, 70	♂, 93
7:40 A.M.	−2	1	1	1
8:15 A.M.	−½	31	27	19
8:35 A.M.	0	28	31	26½
9:20 A.M.	6½	28	29½	31
12:20 P.M.	10½	34	33	–

Measurements, given with sex, are snout-vent, in millimeters.
Pearson, 1954

relatively higher for small than for large homeotherms and heterotherms (Figs. 6-11 and 12). Surface area, the area through which heat is lost to the environment, increases as a square of body size. Volume of the organism, which is a measure of the heat-producing capacity, increases as a cube of the same measure. Small species therefore have larger amounts of surface area relative to body mass than do large species. The relatively larger surface area provides a relatively larger avenue for heat loss and leads to higher respiration per unit body mass for small than for large species.

Bringing together all the potential maintenance costs related to homeothermy

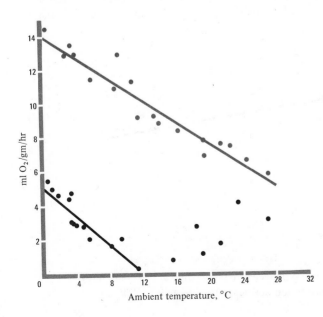

FIGURE 6-10 Relation between oxygen consumption and ambient temperature in the Fiery-throated hummingbird, *Panterpe insignis*. Colored dots are for resting birds; solid dots are for birds in torpor. Each point represents a single determination. (After Wolf and Hainsworth, 1972.)

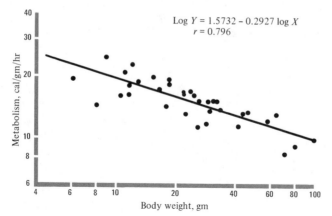

$$\text{Log } Y = 1.5732 - 0.2927 \log X$$
$$r = 0.796$$

FIGURE 6-11 Relation between standard metabolic rate and body weight for a series of passerine (song) birds. (Data from Lasiewski and Dawson, 1967.)

versus heterothermy leads to the conclusion that among animals of similar size homeotherms should expend more of their P_g as R than heterotherms. This relationship is evident in Figure 6-13 which compares R and P_g for populations of both types of animals. Size of animal is not considered in this graph. Although smaller animals should expend relatively more energy for R than larger animals, this generalization probably holds only for equal amounts of P_g for both types. On a population basis, there is a strong correlation between P_g and R, suggesting that animals can compensate for higher R costs by increasing P_g or can increase P_g only by increasing R. On this basis one might expect smaller animals at one trophic level to have relatively higher P_g than larger animals at the same level.

For homeotherms and heterotherms the relation between P_g and R may be a constant for each type of organism, but different between types, at least on a populational basis. At any value of P_g, each type of organism channels a constant,

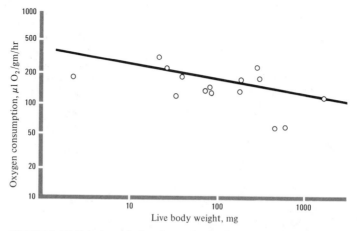

FIGURE 6-12 Relationship between metabolic rate at 20°C and body weight for several arthropod species. (After Reichle, 1968.)

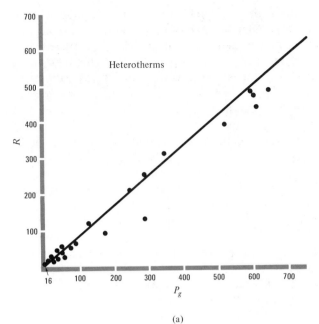

(a)

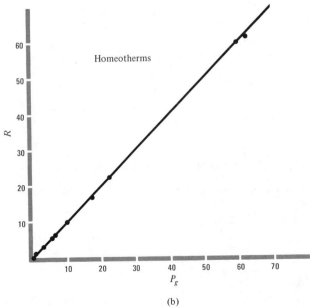

(b)

FIGURE 6-13 Amount of assimilated energy expended for maintenance (R) by populations of heterotherms (a) and homeotherms (b) as a function of gross productivity (= assimilated energy). Units are kcal/m^2/yr. (Data from McNeill and Lawton, 1970.)

but different, percentage of P_g into R (or into P_n). However, this percentage clearly will vary with age structure of the population, time of year, relative availability of food, harshness of physical environment, and many other factors. Averaged over time for populations that are continuously exploiting the ecosystem, the relative P_n and R values may be more or less constant for species with similar adaptations for inhabiting major habitat types.

McNeill and Lawton (1970) suggest that there may be a difference in maintenance costs for long-lived (more than two years) versus short-lived heterotherms (all individuals less than two years; Table 6-4). They postulate that the difference relates to the amount of energy required to maintain the organisms during periods when $P_n = 0$ or when little energy is available for assimilation ($P_g \sim 0$), but when the organisms are still maintaining themselves to operate effectively during the next productive period. For short-lived species more energy could be channeled into production and less into maintenance than for the longer-lived species which are subject to more periods of less than optimum environmental conditions. We encounter a similar hypothesis later to explain a negative relation between generation time and genetic reproductive potential.

Similarly, there probably is a difference between the amount of energy that terrestrial and aquatic heterotherms expend for maintenance. For a series of aquatic animals the conversion constant (or slope of the line relating gross productivity to respiration) is about 82 percent, whereas for a series of terrestrial populations the value is 58 percent (Fig. 6-14). The reasons for this difference between terrestrial and aquatic heterotherms are presently obscure. Costs of osmoregulation and of obtaining respiratory oxygen from the medium may be relatively higher for many aquatic animals than for terrestrial animals.

Species that utilize a relatively abundant or easily handled energy source can convert more of the total energy intake into new standing crop biomass, either by growth of present individuals or by increasing the number of individuals. To some degree this efficiency should be related to the trophic level on which the species operates and the degree of dependence on limited classes of energy sources. As the trophic level position increases, the total availability of energy at the next previous level decreases and, to some extent, also may be found in larger packets.

Primary producers have available a potentially huge energy source in the vast number of photons in sunlight. And, of course, they need not "hunt" for these energy

TABLE 6-4 Equations relating P_n and R for all heterotherms and short-lived heterotherms

---

All heterotherms

$$\log R = 1.0733 \log P_n + 0.3757$$

Short-lived heterotherms

$$\log R = 1.1740 \log P_n + 0.1352$$

McNeill and Lawton, 1970

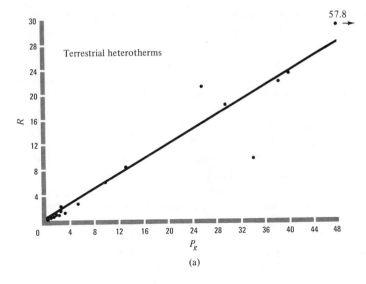

(a)

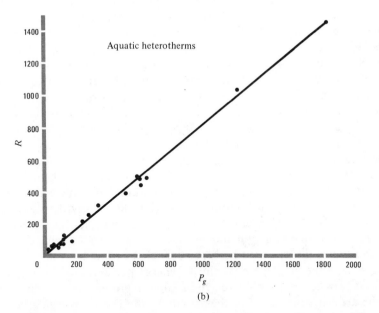

(b)

FIGURE 6-14 Relation between energy expended for maintenance (R) and energy assimilated (P_g) by a series of (a) terrestrial and (b) aquatic heterotherms. Units are kcal/m^2/yr. (Data from McNeill and Lawton, 1970.)

packets as do organisms at higher trophic levels. Photons, however, arrive as tiny packets which must be converted to the larger packets of organic molecules if the energy is to be stabilized in the plants for storage. As we move up the food chain, energy packets tend to become both less frequent and larger. The lower frequency reduces the probability of encounter by a predator, but the larger size increases the energy yield per packet harvested. A successful foraging strategy must balance the energy yield per food item against the cost of harvesting each food item.

Similarly, as species become more and more specialized on particular energy sources, the probability of encounter with an energy packet decreases (MacArthur and Pianka, 1966). Each of these increases in energy cost can be balanced to some extent by other specializations that increase the probability of encountering an energy packet and by the increase in energy per packet. Many herbivorous insects are attracted by chemicals released by plants, and some predators locate prey by learning their behavioral characteristics. A cat that sits beside a mouse runway is more likely to obtain sufficient food even when the mouse population is low than it would by using a random search technique (Pearson, 1964). Similarly, Orians (1969a) postulates that once net primary productivity in a forest reaches a certain minimum threshold level, there are enough energy-rich insects and larvae available to support a class of birds that exploits the environment by sitting for long periods in one position waiting for a sufficiently large prey item to reveal itself and then hovering near the substrate to capture the prey. The bird expends little energy in active search. The proportion of the avifauna of the hovering type declines with the decline in net productivity from tropical to temperate latitudes in the New World (Table 6-5). Following Orians' hypothesis, the critical productivity threshold may rarely be reached in the temperate zone so the system will not support this exploitation type.

Respiration costs also depend to a large degree on the energy expenditures necessary to extract additional energy from the environment (McNab, 1963; Schoener, 1968; Turner et al., 1969; Table 6-6). Species that spend considerable time and energy searching for or consuming prey items (either animals or plants can be prey items) are somewhat less efficient at storing the energy they assimilate than species that can extract the same amount of energy with less energy expenditure. McNab estimated that among mammals, carnivores utilize foraging areas approximately four times the size used by herbivores of the same body size. He suggested that this reflects the greater availability of energy per unit area for the herbivores. For both carnivores and herbivores, respiration is proportional to the home range area—the area in which most of the activity of the organism regularly occurs.

For birds, Armstrong (1965) and Schoener (1968) found that territory size increased faster in relation to body weight than did basal metabolism (Table 6-6). However, the equation relating home range and body weight was particularly influenced by the relatively large numbers of predatory species included in the analysis. Three herbivorous bird species had very similar rates of increase of metabolism and home range size relative to body size and did not differ appreciably from the herbivorous mammals in the relationship between the two variables. Turner et al. (1969) found that among 29 lizard species, home ranges increased faster than

TABLE 6-5 The number of species of birds in different forests that forage primarily by hovering for stationary prey

Forest type	Total number of species	Number of hoverers	Percent of total species
Lowland tropical wet, Panama	40	11	27.5
Lowland tropical wet, Costa Rica	64	20	31.2
Lowland tropical wet, Costa Rica	52	15	28.9
Lowland tropical wet, single species dominant, Costa Rica	54	14	25.9
Lowland tropical moist, Costa Rica	63	16	25.4
Lowland tropical dry, Costa Rica	67	22	32.8
Lowland and tropical wet (disturbed), Panamá	39	9	23.1
Premontane tropical wet, Costa Rica[a]	112	22	19.7
Lower montane tropical wet, Costa Rica	26	7	26.9
Lower montane tropical wet, Costa Rica	32	4	12.5
Premontane subtropical wet, México	41	7	17.1
Lower montane subtropical wet, México	38	5	13.2
Temperate oak-hickory, New York	26	4	15.4
Temperate maple-beech-hemlock, New York	27	3	11.1
Temperate maple-beech, New York	14	0	0.0
Temperate aspen-red maple, New York	14	0	0.0

[a] This census included both undisturbed forest and edge which accounts for both the large number of species and the lower percentage of hovering species.
See Orians, 1969a, for original references.

TABLE 6-6 Regressions of metabolic rate (M) and home range size (A) on body weight (W) in three groups of vertebrates[a]

Group	Relationship	Function	Reference
Mammals	Basal metabolism and body weight	$M = 0.1244\ W^{0.75}$	Kleiber, 1961
	Home range and body weight	$A = 352.3\ W^{0.63}$	McNab, 1963
Birds	Basal metabolism and body weight	$M = 0.1856\ W^{0.668}$	Lasiewski and Dawson, 1967
	Home range and body weight	$A = 11.63\ W^{1.16}$	Schoener, 1968
Lizards	Basal metabolism and body weight	$M = 0.0944\ W^{0.62}$	Bartholomew and Tucker, 1964
	Home range and body weight	$A = 171.4\ W^{0.95}$	Turner et al., 1969

[a] M is in kilocalories per day
A is in square meters
W is in grams
Modified from Turner et al., 1969

metabolism relative to body size. Again predatory species probably had larger home ranges for the same body size than herbivores.

Apparently there are no major differences among mammals, birds, and lizards in the relation of body weight to size of regularly exploited area, at least over the range of body size examined. However, there are major differences within all three classes of vertebrates in the relation of changing size of home range and metabolic rate to body weight. Herbivores tend to change home range size at about the same rate as metabolic rate changes with body weight. This probably indicates the relative lack of excess energy that is expended in either search or prey capture by herbivores. Carnivore home range increases faster relative to body weight than does metabolism. Carnivores operate at a trophic level(s) on which the amount of energy per unit area is less than the availability of plant energy. This means carnivores have to hunt over a larger area to find the same total amount of energy. Because the activity associated with foraging also requires energy at a level above that for resting requirements, the larger the area searched, the more energy spent searching. For carnivores the cost of ingesting the food probably is little different from that for herbivores, but the cost of finding and preparing equal energy units is higher for carnivores than for herbivores because of the increased search and capture energy required.

Another important environmental factor that influences energy expenditures comes from interactions with other organisms. Behavioral interactions require energy and add to total respiratory costs. The courtship displays of some birds are very elaborate and involve considerable activity, but presumably are important in insuring maximal reproductive output by the individuals. Similarly, aggressive encounters

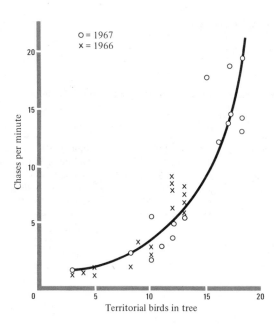

FIGURE 6-15 The number of aggressive chases in a flowering tree (*Genipa americana*) containing territorial hummingbirds as a function of the total number of territorial birds. (After Stiles and Wolf, 1970.)

TABLE 6-7 Summary of energy flow (kcal/day) in a confined house mouse population

	Number of mice	Assimilation	Metabolism	Secondary production	Production Assimilation (percent)
Growing population	26	323.6	312.8	+ 10.8	3.1
Dense population	110	2262.8	2268.3	− 5.5	0.0

Brown, 1963 and Golley, 1968

require energy and can be important components of total respiration expenditures. Stiles and Wolf (1970) have suggested that increasing energetic costs associated with increasing numbers of territorial hummingbirds in a flowering tree (Fig. 6-15) may set an upper limit to the total number of territorial birds present. High density house mouse populations in a confined situation in which relatively more energy should be used in interactions show a negative net productivity (Table 6-7) (Brown, 1963, cited in Golley, 1968). Lower density populations, with relatively less energetic costs for interactions, have a positive net productivity.

Strategy of Partitioning Net Productivity

The final consideration in the flow diagram of energy channeling between and within trophic levels relates to the amounts of energy that are stored as net productivity in a trophic level. We have already seen that the total stored energy varies considerably depending on the species and its mode of exploitation of the environment. In essence, a portion of a limited energy supply that is spent for one activity is no longer available for another. Energy, then, must be rationed for the activities that are most important for the survival and reproduction of the individual. Importance of an activity will not only depend on how much it contributes to survival and reproduction but also by how much it costs in relation to other activities and how much energy spent for one activity reduces the energy available for other essential activities.

Energy channeled into net productivity can (a) make more tissue of the same individual or (b) produce more individuals. The proportion of net productivity put into the two categories will be related to how the organism exploits the environment. Species that breed only once will channel a relatively high amount of energy into reproduction, sometimes to the detriment of the parent if food is limited. Multiple breeders on the other hand sometimes may not breed or may curtail the energy devoted to breeding when energy is limited.

THE ENERGY PYRAMID

At this point it seems reasonable to try to put energy flow back into the context of the entire ecosystem and to summarize the amount of energy lost within the system prior to the final dissipation of the remaining net productivity in the decomposer chain. Each pathway of energy loss must be considered. In each case an efficiency measure may be devised that relates energy available before the loss to energy available after the loss. Virtually every possible efficiency measure has been formulated by ecologists at one time or another. As the energy flow investigations become more specific, the kinds of efficiencies reported increase. Unfortunately, every ecologist does not use the same terminology for the same efficiency and so the whole subject of efficiencies is confused (Kozlovsky, 1968).

The ratio

$$\text{percent efficiency} = P_n \text{ (at level } p + 1)/P_n \text{ (at level } p) \qquad (6\text{-}1)$$

is an important indicator of energy availability at succeeding trophic levels. This ratio is influenced by other efficiencies within and between trophic levels and summarizes the other efficiencies.

Data on total energy fixed at each trophic level are available for four aquatic ecosystems. If we consider total net primary productivity as of unit value (= 1.0), we can plot the values of each higher level as a fraction of 1.0. Combining all four ecosystems in one graph indicates that P_n on each successive level is approximately one tenth that of the preceding level (Fig. 6-16). Since the P_n of each level decreases by an order of magnitude, energy to support a higher level quickly becomes limited and depends on total energy fixed at the first level. Both of the ecosystems that support four trophic levels exceed net primary productivities of 5000 kcal/m²/year. If man is to obtain the most energy from other organisms, he must operate solely as an herbivore; energy utilized as a carnivore is already reduced about ten times from that available to herbivores. Beef may taste good and provide necessary protein, but it is a calorically inefficient food. Some agricultural plant species, such as soybeans, also provide protein and probably will replace animal tissue as our protein source as caloric requirements of the human population increase with the ever-expanding size of the population.

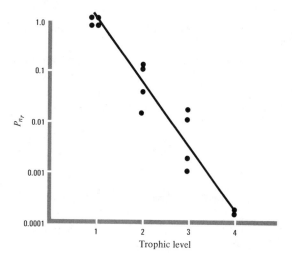

FIGURE 6-16 The relative net productivity of successive trophic levels in four aquatic communities. Values are relative to primary producers which are given a value of 1.0. (Data from Kozlovsky, 1968.)

This index of the efficiency of net productivity is not applicable at the population level except for laboratory cultures in which there is only a single species at each level, and probably for man who expends considerable energy to reduce the number of species receiving energy from his agricultural systems. In nature this measure is relevant as a comparison between trophic levels within an ecosystem. In a food chain analysis there are too many ancillary pathways for energy flow which are not included in the results to make such a measure indicative of energy flow. Movement between trophic levels entails loss via unharvested biomass. Plant material, for instance, is left by the herbivores and enters the decomposer pathway. Dead animals also enter the decomposer food chains if they are not harvested by upper level predators. Unless organic matter is accumulating over the long term, however, the amount of energy fixed by the plants is utilized by either consumers or decomposers. Since man is deriving energy primarily as a consumer and the system as a whole is functioning near the limit of available energy, further additions of consumer biomass are made at the expense of decomposer populations. The total efficiency of a trophic level at storing energy (P_n) finally depends on the relative efficiencies of each organism at that level.

GENERAL CONCLUSIONS

1. Energy flow analyses provide information on
   (a) energetic interdependence of organisms;
   (b) efficiencies of energy transfer.
2. The three major levels of energy flow analyses—ecosystem, food chain, and population—differ in detail of information obtained and sources of energy loss being measured.
3. The major pathways of energy loss between and within trophic levels include
   (a) energy unavailable to a potential user;
   (b) unharvested, but available, energy surpassing the requirements of the organisms;
   (c) "leftovers"—primarily parts of harvested organisms that cannot be used by the harvester or are in excess of the requirements of the harvester;
   (d) energy ingested but not assimilated;
   (e) energy used to maintain the organism, but not contributing to growth.
4. Respiration of an entire population increases linearly with gross productivity, but the slope varies for heterotherms and homeotherms, and for aquatic and terrestrial organisms.
5. Energy is stored by organisms as additional biomass or reproductive output in decreasing amounts at each trophic level. Decrease in P_n is approximately 90 percent of the preceding trophic level at each succeeding level.

SOME HYPOTHESES FOR TESTING

As we move up the trophic levels into consumer populations a number of interesting questions arise from our attempts to provide generalities about energy flow, especially efficiencies. The data presently are at the stage of correlative relations and we have tried to provide some insight into physiological causations. To put the relationships on a firmer base we must understand more about the physiology of energy flow and also the potential genetic bases for underlying physiological variation.

To test the generality that we advanced for terrestrial and aquatic heterotherms, more data are needed on other populations, preferably ones that appear to exploit the environment in obviously different ways and show differences in competitive interactions and the importance of growth and maintenance. Initially, we must clarify the apparent discrepancy produced by good fits of respiration and gross productivity data from all heterotherms, for terrestrial heterotherms, and for aquatic heterotherms to straight lines, each with a different slope. The contradiction is generated by the low values of gross productivity for terrestrial populations, placing all those points at the lower range of the aquatic data. However, using only values of gross productivity less than 70 for aquatic heterotherms still gives significantly different slopes. Can we find terrestrial populations with gross productivity values approaching 1222 kcal/m^2/year as in aquatic systems? If we do, will they be coincident with our expectation that respiration will account for about 58 percent of the gross productivity? Once we can verify (or not) the difference between aquatic and terrestrial populations in the proportion of gross productivity spent on maintenance, we can then ask the physiological question of why this might be so.

As the energy flow relationships are generalized to groups of species and the physiological bases are clarified, the final important question, the evolutionary significance or "why," still remains. If respiration is a constant proportion of gross productivity for all populations, it suggests a strict regulation of energy channeling in relation to energy input or perhaps optimizing the proportion of net productivity in relation to the gross productivity at all of its possible levels. On the other hand, it is likely that the channeling of energy will vary between species depending on techniques of habitat exploitation as well as within species depending on gross productivity. Species with different types of niches might differ in the proportion of gross productivity channeled into respiration and net productivity.

materials flow

7

In the two preceding chapters we considered one of the basic processes in ecosystems: energy flow. In this chapter we consider the general features of the second major process organizing the relationships among ecosystem components: *materials flow.* Movement of energy and materials among the trophic levels and between the biotic and abiotic components integrates the ecosystem into a functional unit. While energy is dissipated as entropy in its passage through the biotic component, chemical substances are returned to the habitat in usable forms. Both materials and energy flow occur through the feeding process. They are intimately coupled since energy is stored in the bonds of organic molecules. As this bond energy is released by respiration to do biological work, the compounds involved are degraded and their chemical constituents released to the habitat. Although materials flow is often referred to as mineral cycling or biogeochemical cycling, we prefer not to use the word "cycle" except when it conveys an accurate impression of how a given chemical substance moves in ecosystems. As we shall see, some substances do not cycle but move primarily unidirectionally through ecosystems.

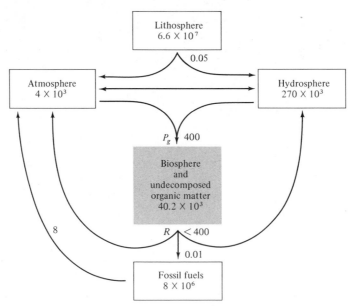

FIGURE 7-1 The carbon cycle estimated as worldwide averages; pool sizes in gm/m² and fluxes in gm/m²/yr. The bulk of the earth's carbon is immobilized in the rocks of the primary lithosphere, with life supported by the air and water pools. (Data from Brock, 1966.)

A consideration of the movement of carbon through the biosphere (Brock, 1966) can demonstrate the general nature of materials flow (Fig. 7-1). By far the bulk of the earth's carbon is immobilized in the *lithosphere* (the earth's rocky crust) as limestone and other rocks. A second large pool is fossil fuel, the excess of past primary production over consumption and decomposition. The two biologically active pools are in the *atmosphere* and the *hydrosphere* (the earth's waters).

The ecosystem balance of many substances is similar to carbon: large and inactive geological pools, smaller and biologically active atmospheric and hydrospheric pools, and a biological pool. The chemical form of the nutrient often varies considerably among these pools. Carbon, for instance, occurs in the lithosphere as carbonates, in the atmosphere as carbon dioxide and carbon monoxide, in the hydrosphere as a variety of forms, in the biological pool as the hundreds of organic molecules synthesized by life, and in fossil fuel deposits as the partial decomposition products of the latter. The chemical form of the nutrient may be regulated by a variety of factors. In the hydrosphere, for instance, carbon occurs primarily as dissolved CO_2 below pH 5, as HCO_3^- between pH 5 and 7, and as $CO_3^=$ above pH 10. The balance among these three forms in aquatic ecosystems depends upon the water's pH.

Consider a hypothetical aquatic ecosystem at pH 4.8. In a community of algae suspended in the water (*phytoplankton*), storage of light energy is accompanied by uptake of carbon dioxide and water and release of the water's oxygen (Fig. 7-2). As the phytoplankton grow and reproduce, they accumulate a variety of inorganic

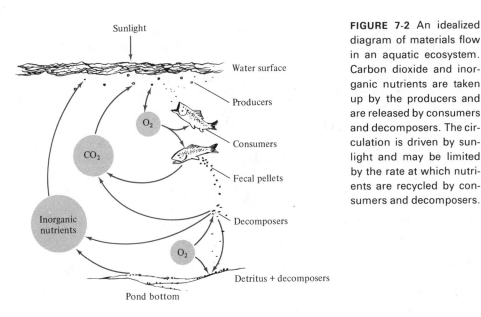

Sunlight

Water surface

Producers

Consumers

Fecal pellets

Decomposers

Detritus + decomposers

Pond bottom

O_2

CO_2

Inorganic nutrients

O_2

FIGURE 7-2 An idealized diagram of materials flow in an aquatic ecosystem. Carbon dioxide and inorganic nutrients are taken up by the producers and are released by consumers and decomposers. The circulation is driven by sunlight and may be limited by the rate at which nutrients are recycled by consumers and decomposers.

chemicals from the water. Animals feeding on the phytoplankton utilize the bond energy of organic molecules stored in the producers and simultaneously consume oxygen and release carbon dioxide. Portions of the molecules will be reassembled into animal tissue, and some will be released to the habitat as fecal pellets. These pellets will be colonized by decomposers which utilize some of the remaining bond energy, releasing carbon dioxide and a variety of other chemicals from the pellet. Carbon dioxide released by respiration can be reutilized by the producers in energy storage. Oxygen produced during photosynthesis can be used in respiration. Chemicals stored by growing phytoplankton become the nutrients of consumers and decomposers.

It is easy to see that a continued flux of energy into this ecosystem depends on cycling of chemicals among the trophic levels. If we eliminated the phytoplankton, there soon would be no energy source for the other trophic levels. Similarly, eliminating the consumers and decomposers would allow the algae to completely deplete the supply of essential inorganic substances. As the growing phytoplankton community depleted some nutrients down to limiting concentrations, growth would cease. In an ecosystem with a supply of nutrients from outside sources, the algae would reach a constant biomass dependent upon rate of nutrient supply. If, however, the ecosystem were closed to outside nutrient supplies, the algae would soon die and settle to the bottom as a layer of stored nutrients. Until these nutrients were released by decomposers, primary production could not continue.

In our hypothetical aquatic ecosystem, a certain proportion of the total organic matter probably would settle into a bottom layer. This layer is called *detritus* in aquatic ecosystems and *litter* in terrestrial ecosystems. Respiration is largely anaerobic in detritus so that decomposition is slow and is accompanied by the release of soluble organic wastes, principally alcohols and organic acids. Since energy, rather than inorganic nutrients, often is the most important limiting factor in detritus, the latter may become a depository of minerals. Certain substances, such as carbon dioxide and oxygen, may cycle much more continuously in the biosphere than others, such as phosphorus, which tend to become immobilized in sediments (Hutchinson, 1957b). Ecosystem support may depend upon small but extremely labile pools of the latter type of chemical.

Chemicals enter the food chain primarily at its base, through producers. In terrestrial ecosystems, consumers and decomposers obtain both their energy and required chemical substances from plant tissues (with the exception of limited direct foraging as, for instance, mammalian "salt licks"). In aquatic ecosystems consumers and decomposers may obtain some of their required nutrients directly from the water. In addition, many aquatic producers take up dissolved organic compounds.

TABLE 7-1 Typical media for a terrestrial plant, an alga, and a protozoan (components in micromoles per liter)

Growth requirement	Terrestrial plant	Alga		Protozoan
KNO_3	4,000	500		–
$Ca(NO_3)_2$	5,000	–		–
KH_2PO_4	1,000	147		3,700
$MgSO_4$	2,000	407		570
Fe	9	36		160
B	23	–		–
Mn	4.6	14.5		6.3
Zn	3.8	12		1.9
Cu	0.3	0.016		12
Mo	0.1	5.2		–
$CaCO_3$	–	1,000		–
$CaCl_2$	–	360		200
EDTA[b]	–	1,000		–
K_2HPO_4	–	–		2,900
Amino acids	–	NaH Glutamate[a] 6,900 Glycine[a] 9,400 Alanine[a] 7,800		Alanine, arginine, aspartic acid, glycine, glutamic acid, histidine, isoleucine, leucine, lysine, methionine, phenylalanine, proline, serine, threonine, tryptophan, valine — approx. 10,000 each
Vitamins	–	–		Pantothenic acid, nicotinamide, pyridoxine, pyridoxal, pyridoxamine, riboflavin, pteroylglutamic acid, thiamine, biotin, choline, lipoic acid — 0.001–50 each
Nucleic acid precursors	–	–		Guanylic, adenylic, cytidylic acids, and uracil — approx. 20 each
Na acetate	–	–		12,000
Tween 80[c]	–	–		24gm
Glucose	–	3,400[a]		17,000

[a] Stimulate growth but not required
[b] A chelating agent
[c] A surfactant
Price, 1970

A comparison of the nutrient media suitable for terrestrial plants, phytoplankton, and zooplankton emphasizes the broad evolutionary differences in their nutritional strategies (Table 7-1). Most terrestrial producers can live solely on inorganic nutrients (plus light, of course) and their growth is not promoted by a supply of organic supplements. Most phytoplankton also are capable of living solely on inorganic media and light, but often grow better if organic compounds are supplemented. Although bacteria and algae may compete for organic substances dissolved in water, the much greater accumulation ability of bacteria forces algae into inorganic nutrition unless there is a huge excess of organic compounds in the water (Hobbie and Wright, 1965).

In contrast to the basically inorganic nutritional requirements of producers, organisms at other trophic levels require complex mixtures of organic and inorganic substances. The protozoan listed in Table 7-1 was the first animal grown in axenic culture (lacking other organisms), and doing so required a variety of organic molecules (Kidder and Dewey, 1951). In addition to requiring organic molecules for their energetic yield, many consumers and decomposers lack the biosynthetic pathways necessary for synthesizing organic molecules and depend upon their food supply for them. Different nutritional requirements will, of course, determine the feeding strategy of organisms and their positions in energy and materials pathways.

CHEMICALS IN THE FOOD WEB

The intimate coupling between energy and materials flow often has been used by ecologists to trace the intricacies of food webs within ecosystems. In an experiment designed to explore food chain relationships in abandoned agricultural fields, radioactive phosphorus (^{32}P) in water was sprayed on grass leaves and the appearance of radioactivity in consumers was followed through time (E. P. Odum, 1962). Radioactivity from the leaves appeared subsequently in both grazing and litter food chains (Fig. 7-3). The most highly labeled organism early in the time period was an ant, which had not been considered an important herbivore prior to the experiment. Since ants sometimes feed on the honeydew produced by plant-feeding aphids, a search was made for aphids but none could be found on the grass. This indicated that the ants fed directly on the plant tissue. Crickets were also important herbivores, and at a later time period grasshoppers became important. Radioactive phosphorus appeared in litter-feeding snails at an early date, but radioactivity levels were never high in the snails, suggesting either that (a) they fed on a number of other materials, diluting the radioisotope or (b) the rate at which the snails excreted phosphorus was greater than for herbivores. Radioactivity appeared in the carnivore populations (spiders) at a later date than in herbivores, reached a high activity level, and then declined rapidly.

Experiments of this type have some pitfalls if care is not taken in the introduction of label into the system. In an experiment in which ^{32}P was introduced into the stem of ragweed (Shure and Pearson, 1969), the radioisotope was preferentially transferred to the flowers. As a result feeding pattern estimates were biased both by the feeding habits of the organisms and by the plant portion used to estimate the initial radioactivity level (Table 7-2). Because much of the ^{32}P was concentrated

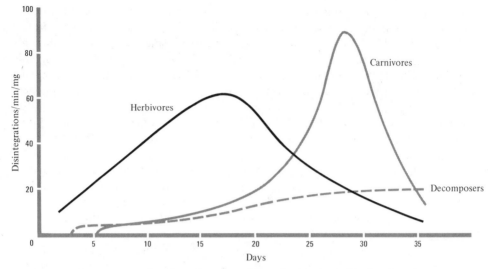

FIGURE 7-3 Appearance of radioactive phosphorus in three trophic levels subsequent to labeling a single plant species at day zero. (After Odum, 1962.)

by the plants in their flowers, the importance of flower feeders in phosphorus turnover was overestimated about threefold if the less radioactive leaves were used as the base level. Leaf feeders, because they consumed tissues with lower radioactivity levels, were underestimated if either the flower or whole plant was used as the base. Although this experiment indicated that care must be taken in introducing radio-isotopes into the system, it also showed that selective labeling can be helpful in determining the specific nature of food chain relationships in ecosystems.

In an experiment designed to determine the relationship between metabolic activity and nutrient turnover, marine zooplankton were labeled with ^{32}P and the subsequent excretion of the radioisotope into the surrounding water was observed (Fig. 7-4). Phosphate excretion rate increased linearly with respiration rate. This suggests that the turnover time of phosphorus is directly related to metabolic rate and emphasizes the important relationship between energetic and materials patterns in ecosystems. Since, as we have seen earlier, metabolic rate is an inverse function

TABLE 7-2 Relative radioactivity of animals feeding on different plant organs. The values given are a trophic transfer index equal to cpm/gm consumer ÷ cpm/gm plant, with the result multiplied by biomass of the consumer population

Feeding habit of animal	Plant organ used as a radioactivity index		
	Leaf	Flower	Whole plant
Flower feeder	143	55	114
Leaf feeder	474	184	379

Shure and Pearson, 1969

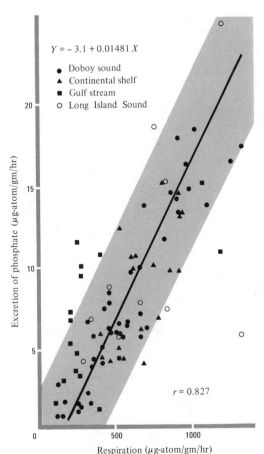

$Y = -3.1 + 0.01481\,X$

● Doboy sound
▲ Continental shelf
■ Gulf stream
○ Long Island Sound

$r = 0.827$

Excretion of phosphate (μg-atom/gm/hr)

Respiration (μg-atom/gm/hr)

FIGURE 7-4 Relationship between the rate at which phosphorus is excreted by zooplankton and their respiration rates. Plankton samples were from different marine locations, as indicated. The shaded area encloses the 95 percent confidence interval of the line. (After Satomi and Pomeroy, 1965.)

of body size, the rate of materials turnover may decrease with increasing size of organism. Microorganisms, such as the principal decomposers, have very high materials turnover rates, whereas larger organisms such as top carnivores have slower rates of materials turnover.

Davis and Foster (1958) examined the distribution of radioisotopes in aquatic organisms in the Columbia River below the Hanford, Washington, plant of the Atomic Energy Commission and found that the concentration of beta-ray emitting compounds, relatively low in the water, was at a high level in plankton and decreased with progressively larger body size (Table 7-3). Whittaker (1961), studying the movement of radioactive phosphorus in aquaria, found that the larger the organism, the smaller the ^{32}P uptake and the slower the rate of phosphorus turnover. Larger organisms generally required a longer time to accumulate radiophosphorus and accumulated it to lower concentrations than small organisms. The rate of disappearance of radiophosphorus as the nutrient was depleted in the ecosystem was slower in the larger organisms. Large organisms are buffered to some extent from

TABLE 7-3 Relative amount of
β-ray emitting radioisotopes in
a flowing water ecosystem

	Percent
Water	<0.1
Plankton	100
Sponge	55
Caddis larvae	40
Snail	25
Shiners	20
Crayfish	10

Davis and Foster, 1958

environmental materials fluctuations by their stable internal materials utilization systems.

PATTERNS OF CHEMICAL FLOW

Chemical flow patterns can be broadly categorized by whether a gas or vapor phase is involved in one of the environmental reservoirs. Materials such as nitrogen and carbon dioxide are characterized by atmospheric reservoirs of various size (Table 7-4). Movement of these materials tends to be cyclic in character, with the material passing through a variety of physical states as it circulates in the biosphere. Other materials, such as calcium, phosphorus, and magnesium, are made available to the

biosphere

TABLE 7-4 Types of materials circulation systems for important elements

Gas/vapor phase present	Gas/vapor phase absent
Water	Calcium
Nitrogen	Phosphorus
Oxygen	Sodium
Carbon dioxide	Potassium
Sulfur	Magnesium
Chlorine	Iron
Bromine	Zinc
Fluorine	Copper
	Manganese
	Nickel
	Cobalt
	Aluminum
	Boron
	Iodine
	Lead
	Selenium
	Molybdenum
	Silicon

ecosystem primarily through weathering of rock or the decomposition of sediments, and do not involve atmospheric reservoirs. These sedimentary systems may tend to be unidirectional in character with the ecosystem functioning as a stage in the flux of material between the lithosphere and accumulative sedimentary deposits in the oceans. Of course, it might be argued that these sediments will be "cycled" in a few millenia, but this seems to be stretching the idea of cycling rather thin.

The nitrogen cycle is typical of materials flow patterns with gaseous reservoirs (Fig. 7-5). The conspicuous properties of cyclic flow patterns are (a) an atmospheric pool, (b) an interlocking network with multiple flow pathways, and (c) a variety of chemical forms in which the elements occur, including a huge array of organic molecules. The nitrogen cycle has a nitrogen gas pool in the atmosphere, which is large and somewhat inactive, and an ammonia pool in the soil that is small and very labile. Most ammonia is converted to ammonium ion on the surfaces of soil particles. Nitrogen occurs in the soil as ammonium, nitrite, and nitrate ions, and as urea from animal wastes. Plants excrete a variety of nitrogenous compounds into the soil profile including phenolics and amino acids. These compounds may be taken up again directly by plant roots, may remain in the soil profile for varying lengths of time, or may serve as substrates for bacterial growth.

The atmosphere is 75 percent nitrogen gas by weight. Since a square centimeter column of air weighs 1.03 kg at sea level, there is about 0.77 kg of free nitrogen over each square centimeter of the earth's surface. The only organisms with direct access to this pool of nitrogen, however, are certain blue-green algae and bacteria. Both groups of organisms are widespread and each converts nitrogen gas into organic molecules, a process called *nitrogen fixation*. Nitrogen gas therefore is incorporated into the food chains through nitrogen fixation and also by uptake of ammonia and nitrate by plants. In the decay process decomposers deaminate organic residues to release ammonia back into the habitat. In addition, other bacteria convert ammonia to nitrite and then to nitrate. Similarly, denitryfying bacteria may convert the same

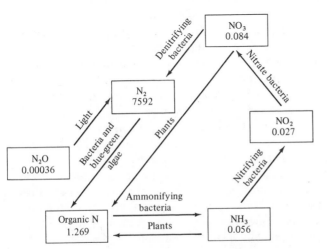

FIGURE 7-5 The nitrogen cycle with pool sizes in kg/m² averaged for the earth's surface. (After Bowen, 1966.)

ions to nitrogen gas. *Denitrification* results in a loss of nitrogen from the ecosystem as the volatile gas escapes to the atmosphere. This process is common under anaerobic conditions and when there is an excess of decaying organic matter in litter. *Nitrification,* in contrast, results in conservation of nitrogen by the ecosystem, since the ammonia may be taken up immediately by microbes or plants. This process is common under aerobic conditions and in soils that are not too acid.

Nitrogen enters ecosystems passively, as a result of rain carrying dissolved nitrate and ammonia, and actively, through nitrogen fixation. In aquatic ecosystems, the major nitrogen fixers are blue-green algae, which convert N_2 into NH_4^+ and then rapidly to amino acids in a series of light-dependent reactions. In temperate terrestrial ecosystems, most nitrogen fixation is accomplished by both aerobic and anaerobic bacteria in the litter and upper soil layers (Brock, 1966). In tropical forests, plants called legumes, which have nitrogen-fixing bacteria of the genus *Rhizobium* associated with their roots, are important. Up to 60 percent of the trees in these forests may be legumes (Richards, 1952).

Bacteria function at almost every step of the nitrogen cycle, accomplishing the release of ammonia from dead organisms (ammonifying bacteria), conversion of ammonia to nitrite (nitrite bacteria), and conversion of nitrite to nitrate (nitrate bacteria). In most other nutrient pathways in ecosystems, one or more key steps involve bacteria.

There are few quantitative data on materials flow. However, Hutchinson (1957b) made estimates of the principal parameters of the hydrologic cycle on a worldwide basis (Fig. 7-6). Most of the world's water is combined chemically with minerals of the primary lithosphere and sedimentary deposits. This water is, of course, unavailable to the ecosystem except through the very slow geological weathering processes that break down rocks. The majority of the available water (98.6 percent) is in the oceans, and another substantial fraction (1.2 percent) is in the ice caps at the poles.

At any one time, less than 0.001 percent of the earth's available water is in the atmosphere. Yet this water supports, through precipitation, the majority of the earth's primary productivity. It was pointed out in the discussion of the world's primary productivity that about 70 percent occurs on land. This is dependent on the tiny pool of atmospheric water from which precipitation originates. This reservoir is renewed by evaporation from open water surfaces and transpiration from plant surfaces. Total annual worldwide precipitation is around 4.46×10^{20} gm/yr, while the atmosphere contains only 0.13×10^{20} gm on an average. Therefore the atmospheric pool is turned over about 34 times a year. In other words, the water vapor in the air is completely replaced every 10.5 days on the average.

Hydrologic cycle data illustrate a fundamental property of materials flow: large unavailable pools are often coupled with small available pools that are extremely labile and highly dependent on flux rates for their maintenance. Atmospheric water content is small relative to the flux through it. Its size at any one time depends on the balance of evapotranspiration and precipitation. Regional precipitation patterns are influenced by altitudinal relief, since rainfall may be caused by the cooling of air rising over mountain ranges. The minor amount of water in the atmosphere as well as its lability is emphasized by the abrupt contacts between high rainfall and

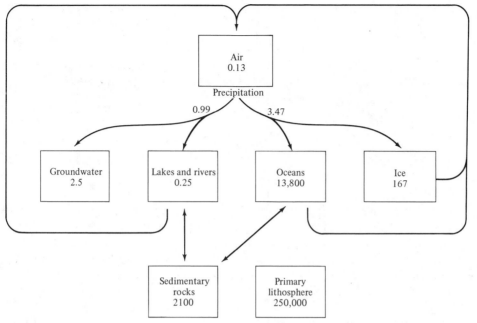

FIGURE 7-6 The world's hydrologic cycle with total worldwide pools in 10^{20} gm and flux rates in 10^{20} gm/yr. (After Hutchinson, 1957b.)

semidesert areas, where air is forced up over mountains by air currents and the water is depleted by precipitation-generating temperature transitions.

The fragility of the water circulation pattern on which terrestrial life depends suggests that climate modification schemes are risky (Whittaker, 1967). Plans to modify regional precipitation patterns through wholesale weather modification techniques are questionable at best and potentially disasterous at worst. The earth's climate has changed constantly and dramatically through its history. But rapid changes in climate, of the type that might result from the application of sophisticated weather modification technology, have in the past generated wholesale species extinctions and rapid evolution of surviving species. Given the lability of the atmospheric water pool, substantial modification of fluxes seems particularly unwise since there is a high probability of upsetting a delicately balanced ecological cycle that is essential for terrestrial life, including man.

Materials like water and nitrogen, with a gaseous phase, cycle continuously in the ecosystem. Other chemicals, such as phosphorus, calcium, and potassium, lack a volatile phase. Movement of these substances often begins in weathering of the lithosphere and ends in sediments, with cycling occurring primarily between litter and organisms within a given ecosystem. These materials are often lost from the ecosystem almost as rapidly as they become available through weathering. Studies of calcium balance in a deciduous forest in New Hampshire (Bormann and Likens, 1967) indicate that the organisms are a detour on an import-export flux (Fig. 7-7). Of the calcium lost from the ecosystem, 98 percent was dissolved in runoff water.

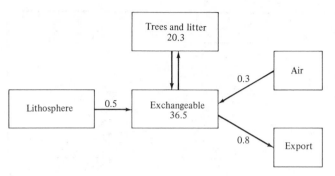

FIGURE 7-7 Calcium flow in a New Hampshire forest with pool sizes in gm/m² and fluxes in gm/m²/yr. (After Bormann and Likens, 1967.)

Two other minerals also showed a net loss from the ecosystem over the study period. Sodium had an input of 1 kg/ha and an output of 6.3 kg/ha for a net loss of 5.3 kg/ha. Magnesium had an input of 0.7 kg/ha and an output of 2.5 kg/ha for a net loss of 1.8 kg/ha. Although the estimates are crude at this stage of the study, we assume that the export quantities, plus the materials that are immobilized as a result of the ecosystem's net productivity, must be derived from weathering of the rock. The minor cycling of these elements that occurs involves atmospheric "dust."

To examine the circulation of materials in white oak forests, Witherspoon (1964) used a radioisotope (^{134}cesium) that mimics calcium as a tracer. It was injected into tree trunks, and its appearance in the ecosystem was monitored (Fig. 7-8). Of the portion that moved out of the labeling site, the bulk was returned to the wood of the trunk and the other major portion went into the litter layer from leaf fall in the autumn. The large pool of calcium in the trees of the New Hampshire forest results, as in this oak forest, from the immobilization of materials in wood.

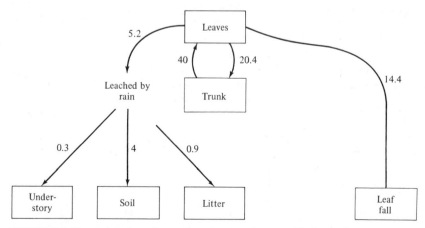

FIGURE 7-8 Flow of radioactive cesium in an oak tree with fluxes in percent per year based on 100 percent introduced into the tree trunk in April. (After Witherspoon, 1964.)

Sediments of dead organic material deposited on the ecosystem substrate form important reservoirs in materials flow. MacFadyen's (1963) studies of energetics showed that decomposers can contribute up to 72 percent of total nonplant respiration in grasslands. Utilizing the energy of organic molecules in litter, decomposers are instrumental in the release of litter minerals.

In an attempt to determine the relative importance of large and small decomposers in leaf decay, Edwards and Heath (1963) enclosed fragments of oak leaves in mesh bags with openings of 7 mm and 0.5 mm to exclude different size classes of decomposer populations. Bags were placed in the litter in July and decomposition was followed through the following April. At the end of this period, about 90 percent of the leaf mass in the 7 mm mesh bags had disappeared, but only 30 percent had been lost from the 0.5 mm mesh bags. This experiment suggests that there is a size hierarchy in litter feeding, with large arthropods and snails eating large litter fragments, fungi and microarthropods utilizing the fecal material of larger decomposers as well as small litter fragments, and bacteria colonizing the most finely divided sedimentary materials.

Witkamp (1966) found that decomposition of litter in forests was a function of the density of fungal and bacterial populations in the litter, and that decomposer abundance varied conspicuously on leaves from different tree species (Fig. 7-9). Mulberry leaves support very high decomposer densities, and they decayed almost completely over a one year period. Pine needles, in contrast, supported low densities of decomposers, and were not even half consumed at the end of a year. Different decay rates of different litter material indicate that ecosystems containing a number of different species will have litter pools with different decay rates. Litter from certain species may be so readily decomposed that substantial accumulations never occur. Litter of other species may be so resistant to decay that large quantities of relatively differentiated organic matter may accumulate with time.

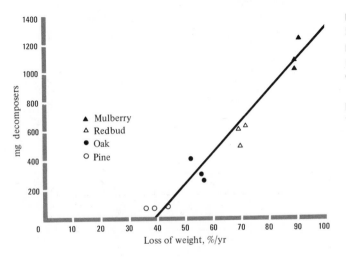

FIGURE 7-9 Relationship between size of decomposer communities supported by different types of leaves and the rate of leaf decomposition, in percent of weight loss per year. (After Witkamp, 1966.)

▲ Mulberry
△ Redbud
● Oak
○ Pine

mg decomposers

Loss of weight, %/yr

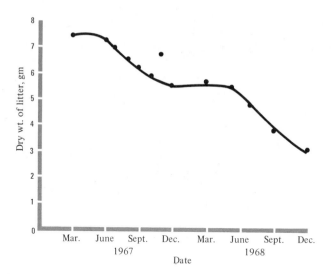

FIGURE 7-10 Pattern of litter weight loss over a two-year period in an Illinois abandoned field. (Data from Old, 1969.)

Old (1969) examined litter decomposition patterns in an Illinois prairie and found that the rate of decomposition varies seasonally (Fig. 7-10). During the first year of decomposition, rate of litter decrease was 0.173 gm/month. Decomposition fell to zero during the winter as cold temperatures interrupted metabolic activity. During the second year, the rate of decomposition was 0.293 gm/month. As we saw earlier (Chapter 5), soil organic matter declines from south to north. This undoubtedly results from climatic limitation of decomposer activity.

Old also found that, because of different rates of elemental loss, litter may become enriched in certain elements during the decomposition process (Table 7-5). Although the amount of nitrogen declined more rapidly than total weight during the 21 month period, phosphorus content of the litter increased fourfold, and potassium increased sixfold. Enrichment of litter concentration of certain minerals through the decomposition process causes litter to become a preferential depository of certain chemicals. If this pattern of decomposition continues over an extended period, the organic sedimentary layer will have higher concentrations of phosphorus and potassium, and a lower concentration of nitrogen than the plants and animals from which it is derived.

The dynamics of a sedimentary layer, litter or detritus, may be described (Jenny et al., 1949; Olson, 1963) by

$$dL/dt = I - kL \qquad (7\text{-}1)$$

TABLE 7-5 Change in elemental composition of litter during decomposition

	%N	%P	%K	%Ca	%Mg
New litter	0.96	0.04	0.04	0.35	0.04
Old litter—70% decomposed	0.79	0.14	0.25	0.43	0.06

Old, 1969.

where dL/dt is the change in amount of litter per unit time, I is the rate of input

to the litter layer from leaf fall or detritus settling, k is a constant describing the decomposition of litter, and L is the amount of litter present (in grams per square meter or some similar measurement). There are three possible consequences of this relationship. If input is greater than the decomposition rate, litter will accumulate in the system. If decomposition rate is greater than input, there will be a loss of litter from the system; if $I = kL$, then $dL/dt = 0$, and the amount of litter is constant.

Olson argued that the amount of litter present, if the ecosystem is given enough time to develop, will become a constant; that is, that dL/dt will become zero and $L = c$, where c is some amount of litter that does not vary with time. A balance will be reached between the litter production rate of the ecosystem and the amount of litter that is accumulated in the ecosystem. The determining factor in this balance is k, the rate parameter. This constant is a measure of the decomposer activity of the ecosystem. The larger it is, the greater the decomposer activity and the more rapidly litter will be decomposed. The balance between litter accumulation and litter production is determined by k (Fig. 7-11). For moist tropical forests, k is large and litter accumulation is minor compared with litter production. In temperate pine forests, litter accumulation is large in spite of a small litter production, because k is small.

The data from the Illinois prairie indicated that season was important in determining the rate of litter decomposition, with the rate being zero in winter. The warm tropical climates of the Congo, Ghana, and Colombia allow rapid litter decomposition, so these forests fall near the line where $k = 4$, whereas the pine forests fall near the line where $k = 0.0156$. In addition, pine needles, as we saw previously, are relatively unfavorable substrates for decomposer populations, which places an additional restriction on k. In fact, Figure 7-9 indicates that k will vary for leaves

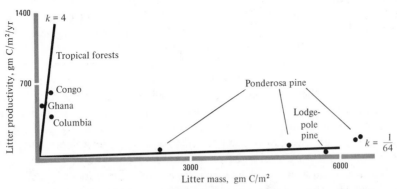

FIGURE 7-11 Relationship between the litter input rate and the mass of litter accumulated for different values of the litter decomposition rate constant (k). If k is high, very little litter is accumulated in spite of high input; if k is low, large litter amounts may accumulate even with relatively small inputs. (After Olson, 1963.)

TABLE 7-6 Nutrient composition in parts per million of soils supporting different types of vegetation in Wisconsin

Vegetation	Ammonium ion	Phosphorus	Potassium
Mesic forest	9	34	96
Savanna	12	22	124
Mesic prairie	29	38	129

Curtis, 1959

of different sources in the same ecosystem. The Illinois prairie studies indicate that k could be further subdivided to describe different nutrients within the litter layer. Although the accumulation of litter mass may reach some point where $dL/dt = 0$, there still may be enrichment phenomena producing gradual changes in the chemical composition of the litter.

The energy analyses of several ecosystems by Ovington and Lawrence (1967) discussed earlier indicated that although proportionately more of the ecosystem energy occurred in the litter layer in the prairie than in the forest, the amount of litter was much greater in the forest than in the prairie. Furthermore, the examination of mineral flow in the New Hampshire forest (Likens et al., 1967) showed that minerals were lost from the system as rapidly as they were released by weathering of the primary lithosphere. One of the primary differences between woody and herbaceous ecosystems is the distribution of stored materials between various ecosystem compartments. In the forest almost all of the materials are stored in either dead or living organic matter and the soil is notoriously poor in nutrients (Table 7-6). In herbaceous systems the soil acts as an important storehouse of nutrients. The conspicuously different agricultural productivities of prairie and forest regions are well known.

The high agricultural productivities of regions such as the North American corn and wheat belts, developed on prairie soils, has been achieved at the expense of nutrients stored prior to cultivation (Fig. 7-12). Global fertilizer use from 1938 to 1966 skyrocketed from 9.31×10^6 to 46.7×10^6 metric tons, a 500 percent increase, with over two-thirds of the world's consumption occurring in Western Europe and North America (Borgstrom, 1969). The increase in agricultural productivity during this period has been about 200 percent. Although the agricultural increase has been widely publicized, the massive increases in fertilizer consumption have been less widely remarked. In fact, about 60 percent of the fertilizer usage increase was required just to offset depletion of the original stored nutrients.

The litter input term (I) is partially related to the growth form of the plants, that is, whether they are evergreen or deciduous, woody or herbaceous. And it is also, of course, a rate term with the same dimensions as productivity and represents a certain fraction of net productivity. The difference between temperate pine forest and tropical broadleaf forest litter storage in Figure 7-11 represents the balance between producer activity (the ordinate) and decomposer activity (k). The larger litter accumulation in the pine forests as compared with that of the tropical forests is a consequence of the fact that decomposer activity is much greater, relative to

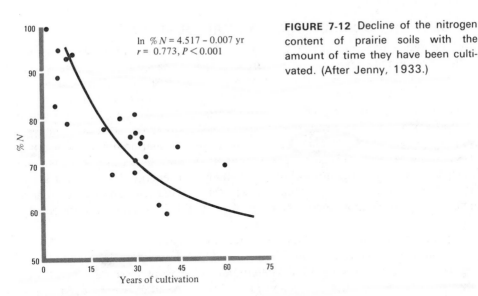

ln %N = 4.517 − 0.007 yr
r = 0.773, P < 0.001

FIGURE 7-12 Decline of the nitrogen content of prairie soils with the amount of time they have been cultivated. (After Jenny, 1933.)

producer activity, in the tropical forest than in the temperate forest. A similar pattern can be discovered by examining the nutrient storage compartment in prairie soils. Jenny (1930) found that soil nitrogen content was a decreasing function of temperature throughout a constant rainfall region of central North America (Fig. 7-13). The increase in stored nitrogen from south to north resulted from decomposer activity being more severely limited by low temperature than producer activity: k is a positive function of ecosystem temperature.

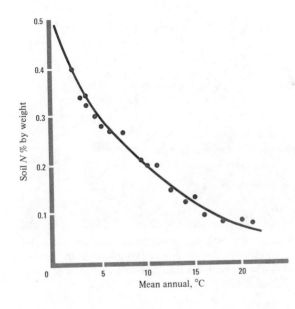

FIGURE 7-13 Decline of soil nitrogen with increasing mean annual air temperature when going from North to South in the Central United States. (After Jenny, 1930.)

The importance of organisms in an ecological system cannot always be measured by their function in energy flow. There are a large number of functional roles within an ecosystem, some of which may not be measurable solely in energetic units.

Some species can play a role in nutrient flow that far exceeds their impact on the system in terms of energy flow. Their importance to the community as a whole cannot be measured by energy flow, but must take into account the nutrient cycle pathways in the system. *Modiolus*, a mollusc that lives in estuarine situations, removes about one-third of the particulate phosphorus from the water per day, thus making it available to other organisms (Fig. 7-14). Its role in energy flow, in contrast, is almost inconsequential on a percentage basis. Removal of such a species would have minor initial effects upon energy flow but substantial effects upon nutrient circulation.

The importance of dogwood and pine trees in the circulation of calcium in pine plantations has been examined by Thomas (1969). As in the *Modiolus* example, the dogwood, which was less important in determining biomass of the stand, accumulated disproportionate quantities of calcium (Table 7-7). The litter was nearly twice as

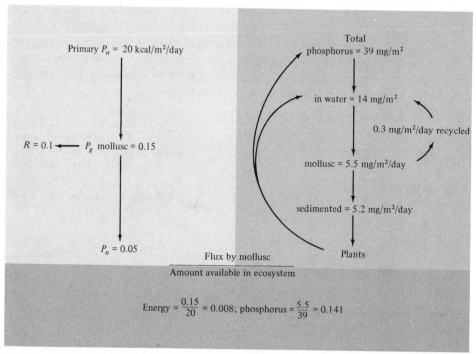

FIGURE 7-14 Comparison of energy flow (on the left) and phosphorus flow (on the right) in a marine mollusc population. At the bottom are the relative contributions of the animal to energy flow (about 0.8 percent of total) and phosphorus flow (about 14 percent of total). (After Kuenzler, 1961a, 1961b, and Odum, 1963.)

TABLE 7-7 Comparative importance of dogwood, pine, and litter in the energy and calcium budgets of pine forests

Component	Percent of total		Ratio Ca/*B*
	Biomass	Calcium	
Dogwood	0.2	1.8	9.00
Litter	10.5	24.1	2.30
Pine	89.3	74.1	0.83

Thomas, 1969

important in the calcium reserves as in biomass, whereas dogwood was nine times more important in calcium balance than in weight. The pine, which was by far the most important species in determining biomass, was relatively less important in the calcium pool. Thomas found that calcium content of dogwood leaves was constant at around 3 percent of dry weight, so leaf mass was particularly important in determining the total amount of calcium incorporated into the dogwood. Because of the pronounced decay resistance of pine needles in comparison to dogwood leaves, calcium was cycled more rapidly through the dogwood than through the pine. Seemingly unimportant species, such as *Modiolus* and dogwood, may perform important community functions in materials flow although they are insignificant components of biomass or energy flow.

Materials flow is both (a) buffered by a variety of alternative pathways and diverse organisms performing similar functions and (b) a fragile pattern that may be catastrophically disrupted if interrupted at a flux "bottleneck." The hydrologic cycle, we have seen, might be subject to severe disruption through modification of the flux of water into or out of the atmospheric pool. We also have seen that agricultural exploitation of the Central United States depleted ecosystems there of a substantial portion of their materials capital. Equally significant in modifying ecosystem function may be the introduction of unusual biological poisons into materials flow. We consider this phenomenon in detail in the next chapter.

SYSTEMS ANALYSIS OF ECOSYSTEM PROCESSES I: A SPECIFIC MODEL

Our examination of litter decomposition is an initial approach to systems analysis of ecological phenomena. A *system* may be defined as a group of interacting elements forming a functional unit. *Systems analysis* is an analytical approach employing differential equations to describe interactions among system elements. One of the fundamental powers of the scientific method has rested in *reductionism,* the simplification of complex phenomena into a series of unit components that are examined one at a time. Although systems analysis dissects complex systems into a series of unit components, it attempts to consider each of these as part of the whole.

Let us examine an organism and a single environmental component as an isolated example of the systems approach. There are two important things that we need to know if we are to accurately describe the organism-environment system: (a) flux rates and (b) pool sizes. Taken together these describe the partitioning of materials (or energy) between the organism and its environment. In an experiment designed to determine the pattern of iodine circulation in a lake (Short et al., 1969), a radioisotope of iodine (^{131}I) was introduced into water. Its subsequent appearance was monitored in amphipods and bottom sediments in separate perforated Plexiglas chambers suspended in the water (Fig. 7-15). Separate components of this experiment can be regarded as models of ecosystem compartments or pools. The appearance

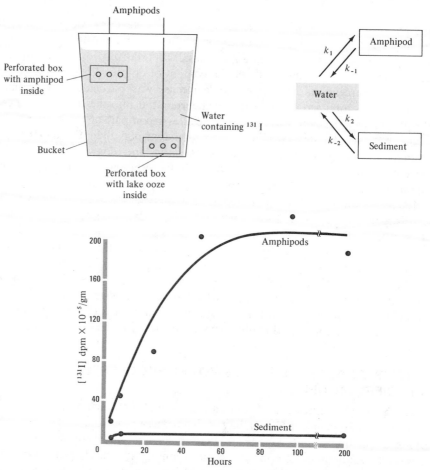

FIGURE 7-15 Summary of an experiment in which radioactive iodine was introduced into an experimental aquatic ecosystem containing a consumer population (amphipods) and a decomposer community with associated sediments (lake ooze). A compartmental model at the upper right shows iodine exchanges described by rate constants (k). At the bottom is shown the pattern of iodine accumulation in amphipod and sediment compartments. (Data from Short et al., 1969.)

of ^{131}I in each compartment was monitored over a period of time. Iodine concentration in the water was relatively constant, with a slight decline as time passed. In the amphipods there was a rapid increase in label early in the experimental period, with a subsequent leveling off after a few days; in the bottom sediment from the lake the radioisotope accumulated slowly and to a much lower concentration than in the amphipods. Since the volume of the habitat pool (water) was very large relative to other pools, large changes in concentration of ^{131}I in the latter had little effect upon habitat concentration and it could be assumed to be constant.

Because of the way the experiment was designed, with all of the compartments isolated, there were four exchange processes occurring: (a) uptake by the amphipods, (b) excretion by the amphipods, (c) uptake by the sediment, and (d) loss from the sediment. Uptake by eating would complicate the organismal compartments slightly. We are not interested in the nature of these exchange processes at the present time, since it is clear that uptake by a compartment may be a combination of passive absorption as well as physiologically based accumulation of the element. Much of the accumulation by the sediment compartment, for instance, may be the result of absorption of iodine onto soil particles and detritus in the ooze, whereas most of the amphipod uptake was probably active, thus requiring energy. We are primarily interested, at this time, in the rate constants rather than the mechanisms determining the magnitude of the constants. Physiological ecology is concerned primarily with these mechanisms. Changes in these constants may be important mechanisms of both adaptation and acclimation.

For the amphipod compartment, we can write

$$dI/dt = k_1 E - k_{-1} O \tag{7-2}$$

where dI/dt is the change in radioiodine concentration with time, k_1 is the rate constant describing the accumulative capacity of the amphipods, E is the concentration of radioiodine in the water, k_{-1} is the rate constant describing the iodine excretion capacity of the amphipods, and O is the concentration of radioiodine in the amphipods. This, of course, is almost identical to the litter equation and is the equation for a straight line when the Y-axis is dI/dt and the X-axis is O. The Y-intercept will be $k_1 E$ and the slope is k_{-1}. A plot of this type for the amphipod population (Fig. 7-16) is a reasonable fit to a straight line:

$$dI/dt = 6.740 - 0.029\ O \tag{7-3}$$

So, for the amphipod, k_{-1} is 0.029 hr and $k_1 E$ is 6.740×10^5 dpm/hr. The amount of iodine is measured as radioactive disintegrations per minute (dpm). Since we know from monitoring the water that E is 0.78×10^5 dpm of ^{131}I per milligram of water,

$$k_1 = 6.740 \times 10^5 \text{ dpm/hr} \div 0.78 \times 10^5 \text{ dpm} = 8.64/\text{hr} \tag{7-4}$$

This gives us some interesting insight into the operation of iodine pools and the accumulative function of the amphipods. The uptake rate constant (8.64/hr) is about 300 times greater than the excretion rate constant (0.029/hr). This means that the amphipods are capable of substantial accumulation of iodine under the experimental conditions and that the flux through the amphipods is fairly rapid. The X-intercept

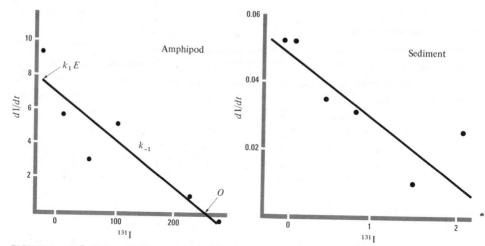

FIGURE 7-16 Relationship between concentrations of radioactive iodine in amphipod (*left*) and sediment (*right*) compartments, and the rate at which radioactive iodine count changed. As the pool size increased, the rate of uptake decreased. The maximum uptake rate occurs at the *Y*-intercept, and the maximum pool size occurs at the *X*-intercept. Note the strikingly different scales for axes of amphipod and sediment graphs. (Data from Short et al., 1969.)

is 232×10^5 dpm of ^{131}I/gm of body weight; this is an estimate of the equilibrium concentration of iodine in the amphipod tissues. At this concentration, excretion and uptake are just balanced.

For the sediment compartment,

$$dI/dt = 0.045 \times 10^5 \text{ dpm/gm/hr} - 0.016 \; S \qquad (7-5)$$

where S is the concentration of ^{131}I in 10^5 dpm/gm. The uptake constant k_2 is 0.058/hr, and the loss constant, k_{-2}, is, of course, 0.016/hr. A striking difference between the sediment and the amphipods is the much larger uptake constant for the amphipods. The amphipod term is 150 times greater than the sediment term. This is not unexpected since the amphipods undoubtedly have a much larger active transport component than the sediments taken as a whole. The active component of sediment uptake results from the presence of decomposer populations. A more realistic approach would necessitate further subdivision of the sedimentary fraction into living and dead fractions. At $dI/dt = 0$, the concentration of ^{131}I in the sediment is 2.64×10^5 dpm/gm, compared with 232×10^5 dpm/gm for the amphipod. So the iodine accumulating ability of the amphipod is about 80 times the accumulating ability of the sediment.

Uptake rate of the radioisotope is given by the *Y*-intercept, that is, when internal concentration is zero. When $dI/dt = 0$, $k_1E = k_{-1}O$. The uptake rate is probably constant, but the excretion rate increases as a function of internal concentration until the two processes are equal.

An important property of a materials pool is *turnover time*, which is pool size divided by flux rate. This gives us an estimate of stability of the pool and the degree to which it will be affected by environmental perturbations. As we saw for the hydrologic cycle, the atmospheric water pool is completely replenished by evaporation and precipitation every 10.5 days. From the iodine data we can determine the turnover time T by dividing the pool size by the Y-intercept. For the amphipods,

$$T = 232 \times 10^5 \text{ dpm/gm} \div 6.74 \times 10^5 \text{ dpm/gm/hr} = 34.4 \text{ hr} \qquad (7\text{-}6)$$

where T is turnover time in hours. For the sediment,

$$T = 2.647/0.045 = 58.8 \text{ hr} \qquad (7\text{-}7)$$

So the amphipod iodine pool is replenished about twice as rapidly as the sediment pool. In other words, the flux of iodine is faster through the amphipods than through the sediment.

This approach to studying materials flow provides us with considerable insight into the coupling of three important ecosystem pools. It allows us to estimate the rate constants governing exchanges among pools and to gain some insight into the potential effects that disruption might have on the system. To extend this to a natural ecosystem, we would have to have information about the relative sizes of the pools in addition to the concentrations of the element within the pool. As a population grows, the amount of materials or energy in the population will increase even though the amount per individual may be constant.

It would be relatively straightforward though laborious to obtain such data through surveys of natural ecosystems. We would expect, for instance, that the total mass of the sediments would be huge compared to the mass of the amphipods. Since the amphipods must compete with the sediment for iodine in the water, the large accumulation constant is necessary to keep all of the iodine from becoming immobilized in the more inactive sediment pool. By knowing the specific activity of the radioisotope in the system (dpm per gram or mole of iodine), we could determine the amount of iodine present in each of the pools. It seems likely that the amphipods are accumulating iodine much in excess of their growth needs under the conditions of the experiment. Given the magnitude of their accumulation constant, if there were no other limiting factors in the bucket, the amphipod population would grow, reducing E and secondarily O and S until O reached the minimum level that would support a living amphipod. At this point the concentration of iodine in all of the compartments would be minimal, but the absolute amount would be maximal in the amphipods and minimal in the water and sediment compartments. The net effect of growth by the amphipods on the lines of Figure 7-16 would be to slide them toward the origins. For a series of organismal compartments that were utilizing the same material, populations would adjust their relative population size according to the interactions among ks for all of the growth requirements.

SYSTEMS ANALYSIS II: GENERAL MODELS

By using a more general form, we can extend the three compartment iodine model to the soil, plant, and litter nutrient exchange system considered earlier. This will allow the development of more precise predictions regarding nutrient circulation in herbaceous and forest ecosystems, and the governing factors that lead to different pool sizes. The model must be modified to include unidirectional transfers and import-export pathways (Fig. 7-17). In this example there are six rate constants: k_1 describes the movement of materials from the plant compartment to the litter compartment, k_2 describes the movement from the litter to the soil compartment, k_3 and k_{-3} describe the movement from soil to plant and vice versa, and k_4 and k_{-4} are weathering-import and leaching-export coefficients, respectively. We can now write a series of equations to describe the dynamics of materials circulation within the system. For the plant compartment,

$$dP/dt = k_3S - k_1P \tag{7-8}$$

where dP/dt is change in size of the plant materials storage compartment through time, S is the size of the soil compartment, P is the size of the plant compartment, and the rate constants are as previously defined. If the size of the plant compartment is constant over a given time period ($dP/dt = 0$), it is because uptake rate, k_3S, is equaled by the loss rate, k_1P, over the time period.

Similarly, for the litter compartment,

$$dL/dt = k_1P - k_2L \tag{7-9}$$

where L is the size of the litter compartment and the other terms are as defined previously. This, it should be noted, is exactly the same as the litter equation (7-1) that we examined earlier except that the input term, called I previously, is replaced by the composite k_1P term.

For the soil compartment, a somewhat more complicated expression is required since this is a more "leaky" compartment:

$$dS/dt = k_2L + k_{-3}P + k_4E - (k_3 + k_{-4})S \tag{7-10}$$

where E is one or more pools external to the ecosystem under examination and S is the soil compartment.

If we now return to our consideration of the fundamentally different patterns of materials circulation in different types of ecosystems, we can consider the mechanisms generating the differences somewhat more precisely. From Table 7-6 we know that forest soils are poorer in nutrients than prairie soils. In other words, the soil compartment of Fig. 7-17 is smaller in forests than in prairies. From the study of Ovington and Lawrence (Table 5-5), we know that the litter compartment is larger in forests than in prairies. For forests, then, we can deduce that k_2 of Figure 7-17 is much smaller than k_3; that is, it seems likely that trees have a much larger materials accumulation constant than herbs, which allows them to keep the soil nutrient level very low. Much of the litter from trees is likely to be wood from twigs and limbs, which is more resistant to decomposer attack than leaves and herbaceous tissues; therefore k_2 is smaller in woody systems. The different materials circulation systems of forests and grasslands can be summed up by pointing out that in forests, P, L, k_3, and k_{-4} will be large and S, k_1, and k_2 will be small, whereas in herbaceous communities, S, k_1, and k_2 will be large and P, L, k_3, and k_{-4} will be small.

The k_{-4} term, which represents export from the system, must be large in the

TABLE 7-8 Percentage soil nutrient contents of a temperate soil from the Eastern United States and a tropical soil from Western India

Nutrient	Temperate	Tropical
K_2O	1.95	0.75
P_2O_5	0.10	0.038
N	0.16	0.10
Organic matter	5.30	1.33

Data from Curtis, 1959 and Singh, 1968

New Hampshire forest discussed earlier. This undoubtedly is a rainfall term, in large part reflecting the degree to which runoff and soil leaching carry nutrients out of the system. The New Hampshire forest is delicately poised on the k_4 and k_{-4} terms, with the plants being a minor accumulation system draining small quantities of nutrients off of a comparatively massive import-export flow.

Nutrient content of the soil declines toward the wet tropics (Table 7-8). This is in spite of the fact (Fig. 7-11) that litter production is often high in these ecosystems. We saw from Figure 7-11 that tropical forests have a small litter compartment. High decomposer activity associated with warm temperatures results in a large k_2 so that litter is rapidly decomposed. In addition, high rainfall assures a large k_{-4} as a result of leaching. In Figure 7-17 we included a k_{-3} rate constant regulating the movement of materials from the plants to the soil. This constant represents the leaching of available plant nutrients directly out of the foliage in rainfall. Tukey et al. (1958) have shown that the plant compartment is rather leaky and that substantial quantities of readily available plant nutrients may be leached directly out of the foliage by rainfall. A 24-hour mist spray leached about 35 percent of the sodium and manganese and 5 percent of the calcium, magnesium, and sulfur out of leaves. Foliar leaching may be a particularly important factor in nutrient circulation in tropical forests. The morphological organization of many tropical plants suggests that there has been strong selection for minimizing or interrupting the $k_{-3}P$ flux before it reaches the soil where it subsequently may be lost. The development of "drip tips" on the leaves of tropical trees suggests a device for concentrating rain flow in limited areas to

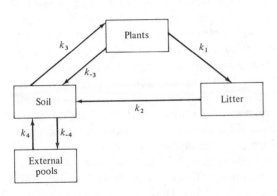

FIGURE 7-17 A simple generalized compartmental diagram for nutrient flow between plants and habitat pools. Exchanges with only a single arrow and rate constant are essentially unidirectional.

keep it from spreading over the leaf surface in a broad nutrient-leaching sheet. The abundance of epiphytes in tropical forests suggests that they function as traps for falling leaves and nutrient-containing rainfall from upper stories.

The materials circulation scheme of Figure 7-17 is, of course, simplified by the deletion of consumer populations which may be quite important in maintaining materials fluxes. And although decomposer populations have been represented only implicitly by their function in determining the magnitude of k_2, it should be recognized that they represent another compartment between litter and soil compartments.

The systems approach to ecosystem processes provides a powerful tool for generalizing the organizing processes in ecosystems. The basic arguments we have developed are (a) the ecosystem may be regarded realistically as a group of compartments, (b) these compartments are functionally integrated by exchanges of energy and materials, (c) flux rate in an exchange can be described by the product of the size of the donor compartment and a rate constant, (d) change in the size of a compartment is the difference between incoming and outgoing fluxes, (e) size of a compartment at any point in time will be a consequence of the historical balance between incoming and outgoing fluxes, and (f) there will tend to be a balancing of fluxes with time, so that compartment size approximates a steady-state. Although we have been considering the application of these principles to the analysis of entire ecosystems, they can be extended to a consideration of single species patterns, food chains, or other levels of biological organization.

Our discussion has centered on linear compartmental models in which it is assumed that the coefficients are constant over the time period studied. The dynamics of variable coefficient models are extremely complex by comparison with our consideration here. At the present time, so little is known about flux rates and pool sizes in real ecosystems that it is impossible to speculate about the applicability of constant and variable coefficient models in ecology.

ECOSYSTEM MODELING

Our consideration of the application of systems analysis to iodine exchange in a simple aquatic ecosystem provides us with an introduction to the concept of modeling. A *model* is a tentative conceptual structure used as a testing device. We can consider a model as a series of hypotheses believed to describe some phenomenon—in our case, an ecosystem. The model provides an integrative framework relating the separate hypotheses into a total viewpoint over the ecosystem. A model, however, is not merely a string of hypotheses. Those hypotheses must be related to one another in a logical, sequential fashion. This organization is imposed on the model, as on the ecosystem, by the driving forces. *Driving forces* are those environmental inputs which, interacting with the biological component, generate the flow of materials and energy in the ecosystem. The ultimate driving force in every ecosystem is energy, particularly sunlight. This driving force, of course, is modified by other environmental variables which regulate energy flux into the ecosystem.

In a systems approach to grassland ecosystems, a model has been developed (Bledsoe and Jameson, 1969) which contains four basic driving forces: sunlight, air temperature, wind speed, and precipitation (Fig. 7-18). These also represent *extrinsic*

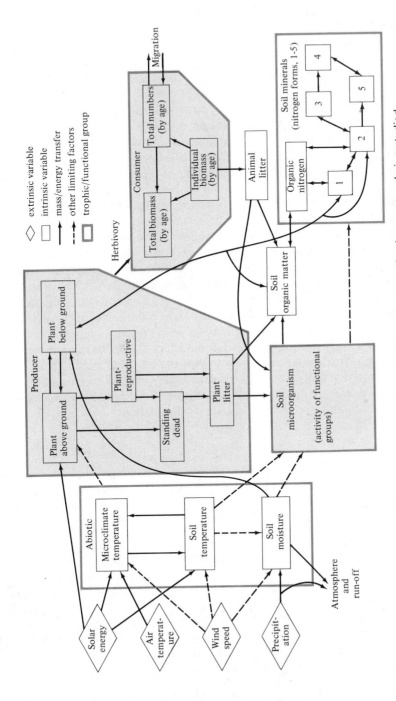

FIGURE 7-18 Diagram for energy and materials flows in a grassland ecosystem being studied in the International Biological Program. Each arrow is represented by at least one equation in the ecosystem model. Extrinsic variables are more or less independent of the state of the ecosystem, while intrinsic variables are affected by the state of the ecosystem. (After Bledsoe and Jameson, 1969.)

variables, factors with values generally independent of the state of the ecosystem itself. That is, sunlight is a function of time of year and degree of cloudiness, rather than a function of, for instance, plant biomass. The *intrinsic variables,* in contrast, take values which depend upon the ecosystem itself. Soil moisture, for instance, is influenced by the rate of transpiration as well as by precipitation. Although the driving forces in the grassland model are extrinsic variables, this need not always be the case.

The diagram is not a model in itself. The model consists of a series of differential equations like those we applied to iodine exchange, albeit more complex. The equations may be organized into a *transfer matrix* which describes the flow of energy and materials among compartments (Fig. 7-19). Each of the nonzero intersections in this matrix represents at least one equation in the ecosystem model (Van Dyne, 1969).

Van Dyne points out that there are two particularly difficult problems in applying such models to real ecosystems, apart from the logistics of collecting the massive amount of data required to evaluate exchange coefficients and pool sizes. These are: (a) many of the inputs to ecosystems are *stochastic,* that is, they vary in a non-systematic fashion with time, and (b) it is possible to reproduce observed patterns artificially without really describing ecosystem function.

In the first case, certain ecosystem variables, such as many weather patterns, may be extremely variable with time and the ecosystem produced by these driving forces will vary also. Evaluating the importance of such stochastic inputs is difficult. In the second case, it is possible to do what modelers call "massaging the model" to generate output predictions very close to real patterns, but produced by different forces. An important problem in ecological modeling is to make appropriate decisions about what are relevant driving forces and transfer functions so that results describe ecological reality and not spurious secondary correlations.

From \ To	Atmosphere	Live plants	Standing dead	Plant litter	Microflora	Soil	Herbivores	Omnivores	Carnivores	Soil fauna	Animal litter
Atmosphere	−	X	O	O	X	O	O	O	O	O	O
Live plants	X	+	X	X	X	X	X	X	O	O	O
Standing dead	O	O	−	X	O	O	X	X	O	O	O
Plant litter	O	O	O	−	X	O	O	O	O	X	O
Microflora	X	X	O	X	+	X	O	O	O	X	O
Soil	X	X	O	O	X	−	O	O	O	X	O
Herbivores	X	O	O	O	O	X	+	X	X	O	X
Omnivores	X	O	O	O	O	X	O	+	X	O	X
Carnivores	X	O	O	O	O	X	O	O	+	O	X
Soil fauna	X	O	O	O	X	X	O	X	O	+	X
Animal litter	O	O	O	O	X	O	O	O	O	X	−

FIGURE 7-19 A matrix of transfer functions between ecosystem components. The *X*'s represent major transfers, the O's are minor transfers, the +'s represent competition or transfers within a population, and −'s represent lack of transfer. In an ecosystem model, each compartmental interaction must be represented by at least one equation. (After VanDyne, 1969.)

Although we have treated systems analysis and ecosystem modeling under the topic of materials flow, it should be apparent that they may be applied to many other ecological phenomena. We have covered them in this context merely because some of the best data for demonstrating the basic approach occur here.

GENERAL CONCLUSIONS

1. Materials flow
   (a) The movement of chemicals in the ecosystem is coupled to energy flow by the feeding process. Whereas energy is dissipated in its passage through organisms, chemicals are returned to the habitat in usable forms.
   (b) Chemical turnover by organisms is a positive function of metabolic rate per unit biomass.
   (c) Chemical flow may be cyclic or unidirectional.
      (1) In cyclic movement the ecosystem depends on internal circulation among multiple pools.
      (2) In unidirectional movement the ecosystem may be a minor detour on a large import-export flux.
   (d) The importance of different populations in energy and materials flow may be quite different, depending on their nutritional strategies.
2. The "system" in ecosystem
   (a) An ecosystem may be regarded as a group of compartments.
   (b) These compartments are connected by exchanges of energy and chemicals.
   (c) Exchanges between compartments A and B can be described by

$$\boxed{A} \underset{k_{-1}}{\overset{k_1}{\rightleftarrows}} \boxed{B}$$

$$dA/dt = k_{-1}B - k_1A$$
$$dB/dt = k_1A - k_{-1}B$$

4. For many exchange processes, particularly those involving energy, k_{-1} will be zero, that is, the exchange will be unidirectional.
5. Partitioning between the compartments eventually may approach a steady-state where $k_{-1}B = k_1A$. When this occurs, dA/dt and dB/dt both equal zero. This will occur only if both the absolute sizes and concentrations within the compartments either (a) become constant or (b) change in counterbalancing fashion—when one goes up the other goes down proportionately.
6. Ecosystem processes may be modeled through systems of differential equations organized in a transfer matrix. Realistic driving forces are an essential component of any such model.

SOME HYPOTHESES FOR TESTING

There is something of a contradiction in the materials flow data which would be interesting to examine. Early in the chapter, we considered data on zooplankton populations, demonstrating that the rate of chemical turnover was a positive function of respiration rate per unit biomass. Then, in a later section, we observed that the rate of nutrient cycling by marine molluscs and by dogwood trees was somewhat greater than the rate of energy flow through the populations. The latter data suggest that energy flow per population is not an accurate index of chemical flow through that population. It would be interesting to examine the relative importance of organism size (as an index of energy flow per unit biomass) and population size in defining a species' relative importance in energy and materials flow.

The mollusc and dogwood data also suggest that relatively insignificant populations may perform important functions in maintaining ecosystem integrity. Certainly we may expect that the structure of estuaries would be modified if the mollusc were eliminated. It would be valuable to have a relatively complete materials budget for an ecosystem to examine the relationship between population abundances and functional roles.

A massive research area in ecology may develop out of the systems approach to ecological phenomena. This appears to represent a powerful tool for organizing our approach to the complex interrelationships of ecosystems. However, only well-designed tests will tell whether this approach can live up to its promise.

biocides in ecosystems

8

From our consideration of materials flow we have seen that as energy is processed by organisms, they also process a huge variety of chemical substances. Many of these will be accumulated to some equilibrium level dependent upon environmental concentration, the uptake process, internal concentration, and the excretion process. We generally focused on chemicals that performed important functions in biological processes. Other chemicals, of course, may have ecological effects because they are toxic. These move through ecosystems according to the same general principles developed earlier. If, however, the equilibrium concentration of a toxic substance in an organism exceeds the tolerance level, that organism will suffer death or other, less total, detrimental effects. In Chapter 2 we briefly considered the effects of one such substance, carbon monoxide, upon humans. One of the consequences of man's technological innovation has been an increase in the variety, uniqueness, and pervasiveness of such toxic chemicals in the biosphere.

Among man's earliest technological innovations was the discovery of smelting, the process by which a metal is separated from its ore through heating. By 2500

B.C., copper was being combined with tin to form the alloy, bronze. One of the most common uses of bronze, as cooking and food storage vessels, led to chronic copper poisoning. Since copper has a distinctively bitter taste, Greeks and Romans used lead to line household vessels. This technological solution to the bitter taste of copper resulted in widespread lead poisoning because food and wine acids combine with lead to form soluble salts that are readily absorbed during digestion. Bones of Romans of the classical period contain lead concentrations sufficient to indicate substantial poisoning of population members (Ehrlich and Ehrlich, 1970).

Among the most controversial (Carter, 1967; Jukes, 1968; Johnson, 1970; Lykken, 1970; White-Stevens, 1970) of more modern technological innovations is the widespread use of chlorinated hydrocarbons, organophosphates, and other biocides to poison insects. The most widely studied as well as the most widely used of these biocides is DDT. It is one of a class of compounds called *chlorinated hydrocarbons,* compounds with hydrocarbon chains to which chlorine atoms have been appended; DDT, for instance, is the acronym for dichloro-diphenyl-trichloroethane. It is only one of a huge variety of similar compounds including its degradation products, DDD and DDE, and such widely used biocides as dieldrin, chlordane, and methoxychlor.

The first and most famous of these compounds, DDT, was synthesized in 1873 by a chemist interested in hydrocarbon addition compounds. There was no known practical use for the compound until 1939. Its insecticidal properties were discovered by Paul Mueller and in 1943 it was used to quell a typhus epidemic in Naples. Typhus is a rickettsial fever which is transmitted by body lice and results in cerebral disorder. DDT was used to delouse the citizens of Naples, thereby stopping transmission of the disease and saving the city from a severe epidemic. For his discovery Mueller received the Nobel Prize in Medicine in 1948.

The chlorinated hydrocarbons are presumed to act as poisons of the central nervous system, but after 25 years of use, with usage of biocides in the United States running over 175 million kg annually (Warner et al., 1966), the specific mode of action of the most popular biocide remained unknown (O'Brien and Matsumura, 1964).

Production of DDT in the United States increased sevenfold in the decade following 1948 and the production of six other chlorinated hydrocarbons developed commercially during this period was five times the total DDT production at the beginning of the decade (Ehrlich and Ehrlich, 1970). These massive increases in production were employed primarily in agricultural poisoning of insect pests infesting crop plants and animals, in public health mosquito and fly poisoning programs, and in household poisoning of domestic pests.

The devastating impact of chlorinated hydrocarbons on susceptible insect populations arises in part out of the fact that the hydrocarbon molecule, although soluble only to a concentration of one to two parts per billion (ppb) in water, is highly soluble in fatty substances, including the waxy cuticle of arthropods. It rapidly penetrates the cuticle so that it need not be ingested by these organisms, but will kill them on contact. In addition, acute toxicity to human beings is extremely low. Studies of people working in the pesticide industry who are subject to massive exposure reported no authenticated cases of acute DDT poisoning (Laws et al., 1967) and experiments in which convict volunteers ate large amounts of DDT did not produce any pathological symptoms (Hayes et al., 1956).

ORGANIC BIOCIDES: EXTENT OF OCCURRENCES

In 1945 DDT was released for civilian use, and in subsequent decades millions of kilograms of it and similar compounds have been applied to ecosystems. In 1963, less than two decades after DDT became widely available, J. L. George (1963) estimated that about 5 percent of the conterminous United States was treated in any one year and that 25 percent had been treated at least once since 1945. Although only a fourth of the United States had suffered a direct biocide application by 1963, "It was becoming increasingly difficult to find uncontaminated animals in any part of the United States, the adjacent Canadian provinces, or the neighboring oceans. It seemed possible that the entire globe was contaminated, at least in minute degree" (George and Frear, 1966). Studies of Antarctic animals discovered DDT residues in penguins, seals, fish, and skuas (George and Frear, 1966). Maximum concentrations in parts per million (ppm) wet weight were 0.15 in seals, 0.18 in penguins, 0.44 in fish, and 2.8 in skuas. The penguins, fish, and seals are all species that do not venture far from the ice pack, so contamination of these species suggests contamination of organisms thousands of kilometers from the nearest known site of DDT application.

Peterle (1969) has subsequently shown that the Antarctic snowfall contains DDT at concentrations of 0.04 ppb, which indicates that DDT is introduced into the Antarctic ecosystem partly through precipitation. Since it can be vaporized directly from the soil into the atmosphere, where it may serve as part of a moisture condensation nucleus or become absorbed onto dust particles, it is likely that DDT and other chlorinated hydrocarbons occur in every ecosystem in the world.

One of the properties of chlorinated hydrocarbons that make them especially efficacious compounds as insecticides is their persistence. Fourteen to 17 years following the application of chlorinated hydrocarbons to fields in the United States (Nash and Woolson, 1967), an average of 30 percent of the compound remained in the soil. Kokke (1970) has shown that this persistence of chlorinated hydrocarbons in the habitat is having a substantial effect on the properties of soil bacterial communities. He streaked water or soil suspensions on nutrient agar containing 0.8 ppm ^{14}C labeled DDT and used X-ray film to detect the presences of DDT accumulating colonies in the samples. In colonies from tap water, less than 1 percent of the colonies accumulated DDT; in polluted surface water, from 2 to 7 percent; in garden soil, 60 to 70 percent; and in nursery soil subject to recent DDT treatment, 80 to 95 percent of all colonies were DDT accumulators. In soil samples that were believed never to have been exposed to direct DDT treatment, 11 to 36 percent of the colonies accumulated DDT.

The somewhat lower level of DDT accumulators in the polluted surface water ecosystem than in soil communities probably arises out of the low solubility of DDT in water, with less stringent natural selection for DDT-resistant bacteria in such aquatic ecosystems. In nutrient agar with 25 ppm DDT, bacterial growth was only

0.6 percent of control values when the surface water was used as inoculum; when the nursery soil was used as an inoculum, growth was 16 percent of controls. Before the soil bacteria showed a growth inhibition equivalent to that occurring in aquatic bacteria at 25 ppm, DDT concentrations of one part per thousand had to be reached.

Another class of biocides that has been widely utilized to poison arthropods is the organophosphates, including such compounds as malathion, parathion, and Azodrin. One of the nerve gases developed by chemists in Nazi Germany was diisopropyl-flurophosphate, which disrupts the nervous system by inactivating the enzyme responsible for the transmission of impulses across nerve junctions. The organophosphates are derived from this nerve gas. These compounds are rapidly decomposed in comparison with the chlorinated hydrocarbons, with 95 percent disappearing from the soil 6 weeks to 6 months following application (Edwards, 1965). They are, however, highly toxic to vertebrates, including man, in comparison with the chlorinated hydrocarbons, and deaths are known in persons working with organophosphates. Restrictions on the use of chlorinated hydrocarbons have caused a switch to organophosphates in agriculture and this switch is likely to be accompanied by an increased human death rate as a result of careless handling.

Whereas pesticides have been widely dispersed directly into the habitat, other compounds that are not directly applied to the environment have begun to appear in organisms. One class of these compounds is the polychlorinated biphenyls (PCB), which are used primarily as coolants and "plasticizers" in manufacturing. These compounds are even more resistant to decomposition than DDT. Samples of a wide variety of organisms from Swedish coastal areas contained PCB concentrations ranging from 0.017 to 190 ppm in fresh tissue and 0.75 to 14,000 ppm in fat (Jensen et al., 1969). Like other hydrocarbons, PCB is accumulated to much higher concentration in fatty tissues because of its high fat solubility and low water solubility. The highest concentrations were reported in the fat around the flight muscles of white-tailed eagles, a declining species. Although little is known about the toxicity of PCB to humans, pathological changes in laboratory animals are known and it has been reported to be one of the ten most potent poisons of a hundred tested by injection into fertile eggs (Jenson et al., 1969).

Jensen et al. (1969) found that the PCB concentration generally decreased from south to north in the Baltic, suggesting that the compounds may be entering the ecosystem from industrial discharges in southern Sweden. In mussels, the only sedentary species sampled, the concentrations of PCB in fat were 2 ppm in samples from the Atlantic, but 5.2 ppm in samples from the Archipelago of Stockholm. Two seal pups from the Gulf of Finland which were only one week old contained 42 ppm of chlorinated hydrocarbon pesticides and 6.5 ppm of PCB in fat. Seal milk from the same area contained 36 ppm of the pesticides and 4.5 ppm of PCB, indicating that even very young organisms have substantial poison loads in their bodies. Because hydrocarbons are stored in fats, they are subject to alternate storage and mobilization as energy requirements and dietary properties change. The fat content in herring muscles varied from 1 percent in the spring to 10 percent in fall samples. Organisms therefore may be subject to periodic "flushes" of fat-soluble poisons as fat residues are mobilized.

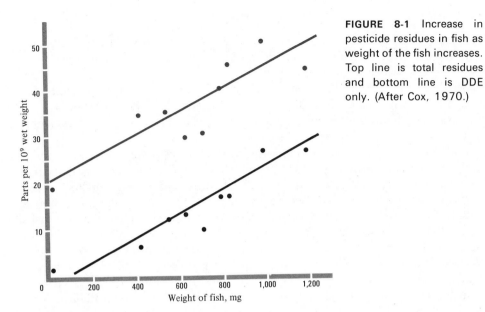

FIGURE 8-1 Increase in 173
pesticide residues in fish as
weight of the fish increases.   ORGANIC
Top line is total residues   BIOCIDES:
and bottom line is DDE   EFFECTS IN
only. (After Cox, 1970.)   FOOD WEBS

Most studies of pesticide residues in organisms consisted of pooled samples of individuals varying considerably in size and developmental state. Cox (1970), however, separated samples of a fish from the Gulf of California, *Triphoturus mexicanus,* according to size and found that the concentration of DDT increased with size of individual (Fig. 8-1). These fish came from water that is not subject to direct chlorinated hydrocarbon input from industrial or agricultural sources, but the total level of these pesticides ranged from 19 to around 50 ppb wet weight of total tissue. Over this range, there was a linear relationship between pesticide accumulation level and individual biomass. These data contrast markedly with the observations cited in the previous chapter that the concentrations of phosphorus and β-ray emitting radio-isotopes declined as size increased. Their only likely explanation is that fat content increased with size in these fish.

ORGANIC BIOCIDES: EFFECTS IN FOOD WEBS

Shortly after DDT and its family of compounds were first used, evidence began to appear that it had effects beyond those for which it was intended. Two years after release of the compound for civilian use Bishop (1947) reported that mosquito control spraying had detrimental effects on plankton communities in treated waters. A gnat control program at Clear Lake, California, in the 1950s, which had been preceded by extensive pilot testing for effectiveness, was reported to have resulted in unexpected deaths of Western grebes and other fish-eating birds and in the accumulation of chlorinated hydrocarbons to 25 ppm in largemouth bass (Hunt and Bischoff, 1960). The appearance of DDT in sport fish alerted many of the scientists working in areas of applied ecology, like game management and forestry, to the potential dangers of pesticide use. Natural resource management journals, with increasing frequency, began to carry reports of unintended effects of pesticide use. This evidence was marshaled in 1962 by Rachel Carson in her monumentally important book, *Silent Spring*, which argued from massive evidence that the compounds initially called insecticides, and later pesticides, are, in fact, biocides, killing unintended organisms, often very far from the site of application. This book created a storm of controversy in both the scientific and popular press that continues to rage (Johnson, 1970; Lykken, 1970; White-Stevens, 1970).

One of the first proposed civilian uses of DDT was poisoning of insects living on trees (Whitten and Parker, 1945). Within a year there were reports that DDT spraying of forests resulted in bird mortality (Hotchkiss and Pough, 1946). Wallace (1959) reported that repeated treatment of the Michigan State University campus with DDT was accompanied by virtual extinction of robins on the campus. Wurster et al. (1965) examined the effect of spraying elms with DDT on bird populations in Hanover, New Hampshire. Just 1 mile west of Hanover is the town of Norwich, Vermont, which was not sprayed and was therefore utilized as a control. A 6 hectare (ha) area of each town, representing 2.25 percent of Hanover's area, was surveyed daily from early April to mid-May by pairs of observers who counted all birds heard or seen from 7 to 8 A.M. There were four observers who were rotated between the two towns to eliminate personal sampling differences. Data were presented as 5-day moving averages with May 1 set equal to 100 percent, when an average of 12 robins were counted in the 6 hectares in Hanover and 16 in Norwich. In addition, dead or dying birds were collected by residents of the two towns who responded to radio and newspaper requests. DDT was applied to Hanover elms on the nights of April 15 to 18, at which time the robin populations were near their peaks in each town (Fig. 8-2). The populations subsequently declined in both towns, presumably as migrants moved farther north, but a relatively stable size was reached in Norwich by May 1. In Hanover, however, the population continued to decline until its size ultimately dropped to 30 percent of the May 1 level. During the precipitous population crash in the DDT-sprayed town, the number of dead robins in the same town increased steadily. The chlorinated hydrocarbon residues in 43 of these dead birds was compared with 5 control birds collected outside of Hanover. The average whole body level of chlorinated hydrocarbons was 2.79 ppm in the control birds. The average chlorinated hydrocarbon load in the dead birds from Hanover was over 103 ppm. A total of 151 dead birds were recovered in Hanover, including robins, chipping sparrows, myrtle warblers, song sparrows, catbirds, and white-breasted nuthatches. Only 10 dead birds were found in Norwich.

In 1967 Wurster et al's. study of these two towns was entered as testimony in a court case in Wisconsin (which was won) requesting that DDT use be banned in that state. The Environmental Defense Fund attorney arguing the case, Vincent Yannacone, was quoted (Carter, 1967) as saying that the scientists called to testify were "virgin witnesses." It was subsequently (Jukes, 1968) argued that, "The virginity extolled by Yannacone as an attribute of scientists would be helpful to those who wish to accept the extrapolation, made by Wurster, based on a sample of 12 robins. Worldly readers, however, might wonder if some of the Hanover robins had flown

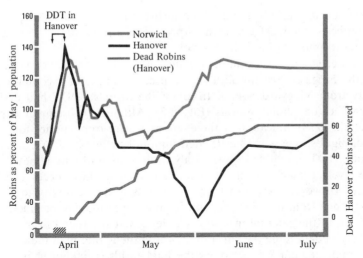

FIGURE 8-2 Robin density in a town sprayed with DDT (Hanover) for insect control and an adjacent unsprayed town (Norwich). Also included are the number of dead robins found in Hanover during the study period. (After Wurster et al., 1965.)

across the Connecticut River to Norwich." This suggestion that the differences between the two towns represented differential migration certainly must be counted among the most disingenuous ever adduced to explain population phenomena. Since the data were presented as five-day moving averages, 60 robin records are responsible for the number estimated in the sampling area. And, as already mentioned, there were 15 times as many dead birds found in Hanover as in Norwich and the average chlorinated hydrocarbon load in dead Hanover robins was over 30 times the level found in birds collected outside Hanover. Accepting the differential migration hypothesis, we would have to believe that whereas most of the healthy birds in the area sought out Norwich, most of the comatose birds (which only incidentally had high body biocide levels) sought out Hanover.

Among the first evidence that "insecticides" could have detrimental effects on organisms other than the target organisms was Bishop's (1947) report that DDT affected plankton in mosquito control areas. Wurster (1968) performed experiments designed to determine the effects of DDT on photosynthesis of marine phytoplankton. Cultures were grown for 24 hours in seawater with added nutrients at 17°C, a 14-hour photoperiod, and a light intensity of 5000 lux. After 24 hours, cultures were divided into control and experimental sets. To the controls, a small volume of ethanol was added. To the experimentals, a small volume of ethanol containing 1 to 2 μl of a commercial DDT preparation was added. The cultures were incubated for another 24 hours and then $NaH^{14}CO_3$ was added to measure photosynthesis as ^{14}C uptake. Cultures were again subdivided and both experimentals and controls were incubated for a further 5 hours in either light or darkness, at the end of which time the amount of radioactivity incorporated into the algae was determined. The counts incorporated in light minus those incorporated in darkness was a measure of photosynthesis; the DDT effect was expressed as a ratio: photosynthesis in DDT plus ethanol to photosynthesis in ethanol only.

Wurster used single species cultures of algae native to both the Long Island Sound area and the Sargasso Sea. In addition he used a sample of a complex plankton community from Vineyard Sound. In all of the samples DDT had a significant inhibitory effect on photosynthesis (Fig. 8-3). Although subsequent experiments (Menzel et al., 1970) indicate that the effect may vary from species to species, the majority of algal species shows a decided inhibitory effect by DDT. As Menzel et al. point out, DDT is soluble in water only to the extent of only about 1 ppb, well below the level at which major alterations of photosynthesis occur in laboratory experiments. However, they also note that plants have the ability to concentrate most compounds from the environment, including presumably biocides, to a level far above that in the external medium. The level actually found in the algal tissue is more likely to determine the effect on photosynthetic rate than the concentration in the external medium. So, although the final result is unknown, it can be said with certainty that pesticides currently in use can have a detrimental effect on primary productivity of some systems, and perhaps of all systems. The mechanisms of the effect and the relative importance of concentrations in the external medium still are not sufficiently well known to make any definitive conclusion.

Wurster (1968) attempted to determine the relationship between exposure level and effect by varying the number of algal cells in a medium with a constant concentration of DDT (Fig. 8-4). He concluded that the typical sigmoid dose-response curve "suggests the absence of a threshold concentration of DDT below which no effects occur." These experiments demonstrate conclusively that pesticides are biocides, as Rachel Carson called them, and often have detrimental effects on physiological processes in organisms remote from the intended target.

In addition to the effects of *insect*icides on plants, *herb*icides may have detrimental effects on animals. In experiments in which the toxicity measure was the amount of herbicide required to kill half of the rainbow trout tested after 48 hours of exposure, trifluralin was effective at 0.011 ppm, 2,4-D (perhaps the most widely used

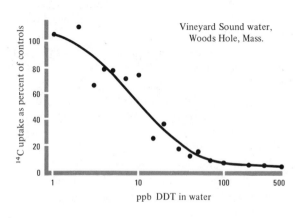

FIGURE 8-3 Relationship between relative net productivity of a marine phytoplankton community sample and concentration of DDT in the water. (After Wurster, 1968.)

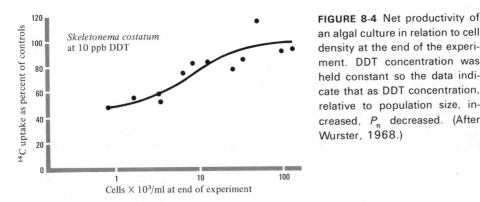

FIGURE 8-4 Net productivity of an algal culture in relation to cell density at the end of the experiment. DDT concentration was held constant so the data indicate that as DDT concentration, relative to population size, increased, P_n decreased. (After Wurster, 1968.)

herbicide) killed half the fish at 0.29 ppm, and hydram was effective at 0.29 ppm (Cope, 1966). All of these compounds, which were designed to be herbicides, were capable of killing trout, bluegill, stonefly nymphs, and water fleas.

Another property of biocides that makes their presence in the habitat ominous is their tendency to accumulate at increasingly higher concentrations in increasingly higher trophic levels. Since DDT is only slightly soluble in water, the huge volume of the habitat relative even to the millions of kilograms of DDT used has often been cited as a reason that concern for DDT use is misplaced. It has been repeatedly demonstrated, however, that chlorinated hydrocarbon concentrations increase along food chains. In a study of a detritus-based food chain on the Great Lakes (Hickey et al., 1966), the concentration of these compounds increased an order of magnitude with each feeding step (Fig. 8-5). The tendency of these compounds to accumulate

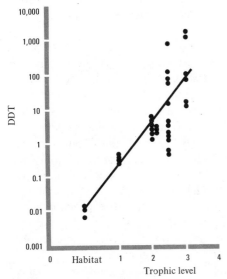

FIGURE 8-5 Increase of average bodily DDT concentration (ppm) at progressively higher trophic levels in a detritus-based Great Lakes food chain. (Data from Hickey et al., 1966.)

to much higher concentrations in fatty tissues than in the body as a whole may result in deposition in the nervous system. In a study of a food chain on Long Island Sound (Woodwell et al., 1967), concentrations of chlorinated hydrocarbons in the brain of a dead ring-billed gull were 1,500,000 times the concentration in Sound water.

The impact of pesticides on the reproductive capacity of organisms other than the target species has only recently begun to be fully appreciated. Many species of birds show a correlation of increasing pesticide residues and decreasing thickness of the egg shell (Peakall, 1970; Cade et al., 1971). Thinner shells are more fragile and reduce the chances that the egg will remain intact until hatching. Residues of DDT have an impact on steroid metabolism in some birds (Peakall, 1970) and influence calcium deposition in the shell of the egg. High DDT concentrations are also correlated with high levels of infertility. Presumably this is related to the increased susceptibility of embryos to pesticide toxicity, but the actual cause of embryo mortality or lack of fertilization is not known (Ames, 1966). *Pelecanus occidentalis,* the brown pelican, is declining in numbers in California as their fertility declines in relation to increasing pesticide residues in the ocean (Peakall, 1970).

In soil, DDT has a half-time of about 10 years. Its principal breakdown product is DDE. This compound appears to be virtually indestructible, with only negligible degradation detectable after 10 years (Ehrlich and Ehrlich, 1970). Studies of the effect of feeding chlorinated hydrocarbons to mallard ducks (Heath et al., 1969) indicate that this decay-resistant product of DDT metabolism is fully as disruptive of bird reproduction as DDT itself. Cracking of egg shells reached 25 percent in birds fed on a DDE-containing diet, compared with about 5 percent in control birds. Combining this loss with subsequent low hatchability resulted in a 30 to 50 percent reduction in reproductive success. Similarly, DDT in the diet reduced reproduction by 35 to 50 percent. This suggests that the effects of DDT use are going to persist for a long time into the future as DDE remains in materials circulation systems. The effects on bird reproduction need not be accompanied by death of the birds ingesting the biocide; there is often no direct mortality, only a disruption of reproduction.

It has been shown that sublethal amounts of DDT have a disruptive effect on behavioral responses of brook trout (Anderson and Prins, 1970). The response, a waving of the fish's tail, was elicited using mild electric shock in the throat region as the unconditioned stimulus. The conditioned stimulus was a light flashed prior to the electric shock. Five "correct" responses, tail waving when the light was flashed but before the shock, was defined as appropriate learning. Experimental fish were exposed to 20 ppb DDT added in 0.45 ml acetone to 9 liters of continuously aerated water 24 hours prior to testing. Controls consisted of the addition of acetone only or of no additions. There were no differences between the two control groups. The presence of DDT in the water had a decidedly detrimental effect on the ability of the trout to learn the conditioned response (Fig. 8-6). Of the 16 experimental fish exposed to DDT, 10 failed to respond at all after 100 trials, and the remainder required over 60 trials to become conditioned. Anderson and Prins concluded that "Insecticide pollution may be even more harmful to fish than the well-publicized lethal effects that have been reported. Learning, for example, is probably a critical

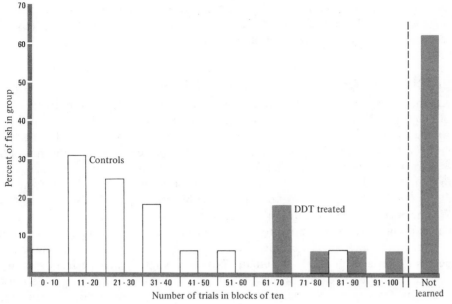

FIGURE 8-6 Experiment demonstrating the disruptive effect of subacute levels of DDT on ability of rainbow trout to learn a conditioned response. The number of trials needed to reach the defined response are grouped in blocks of ten. (After Anderson and Prins, 1970.)

element in such behavioral activities as territorial defense and migration. Any pollutant that might affect such behavior patterns is obviously of potential ecological significance. To the extent that insecticide-induced changes in behavior may adversely influence the survival of the species, the term 'sublethal' applying to the concentrations used, is clearly a misnomer. No-effect levels rather than no-kill levels should serve as the basis for establishing water quality criteria."

Probably the most bizzare of the biocide effects in ecosystems began in 1957 when a mysterious disease struck the chicken-raising industry, causing the loss of millions of dollars (Abelson, 1970). It took 5 years of persistent research to trace the "disease" to poisoning by dioxin, the most toxic chlorine-containing compound known. In addition to causing death, the compound has caused birth defects and neurological disorders when fed to laboratory animals in subacute amounts. The dioxin in the chicken diet was traced to vegetable oils processed at high temperatures to liberate fatty acids. Some of these vegetable oils contained 2,4,5-T, a herbicide applied to the oil-producing plants; the 2,4,5-T was apparently taken up through plant roots and appeared in the oils. During the fatty acid isolating treatment, part of the herbicide was apparently converted to dioxin. It has been shown, however, that dioxin is present as a contaminant in 2,4,5-T manufacture, and an alternative explanation is that contaminating dioxin was concentrated by the plants in their fatty deposits. As Abelson points out, "We are manufacturing thousands of chemicals. In their preparation, side reactions are producing many thousands of unwanted and even unidentified substances. Companies producing fat-soluble, nonbiodegradable, organic chemicals should give careful attention to the question of what they may responsibly set loose on the environment."

BIOCIDES AND THE HUMAN POPULATION

Kraybill (1969) estimates that humans in the United States take in about 35 mg of DDT + DDE annually, the bulk of it coming through the food chain (Fig. 8-7). Another important source is cosmetics which, of course, contain lanolin, the wool fat from sheep. Since sheep are exposed to direct biocide application for the control of ectoparasites as well as in their diet, chlorinated hydrocarbons may be expected in their fat. Chlorinated hydrocarbons have been found in hair dressing, lipstick, hair lotions and sprays, and eye shadow. Within the dietary intake (Duggan and Weatherwax, 1967), the DDT + DDE load is distributed as follows: meats 41 percent, fruits 23 percent, dairy products 14 percent, grains 8 percent, leafy vegetables 3 percent, oils and fats 3 percent, and other miscellaneous foods the remaining 8 percent. The level of DDT in human body fat is highly variable from sample to sample but appears to have leveled off at about 12 ppm since the mid-1950s after increasing from zero in 1942 (Ehrlich and Ehrlich, 1970). A sample of 254 individuals from Israel taken in 1963 and 1964 recorded an average of 19.2 ppm, and 65 individuals from Delhi, India, had an average DDT load of 26 ppm in body fat, indicating that DDT occurs in widely scattered human populations at varying levels of economic development.

The significance of such DDT levels in the human population is unknown. As we saw earlier, people involved in the pesticide industry and people exposed to high dietary DDT intake (Laws et al., 1967; Hayes et al., 1956) reported no detectable adverse effects resulting from massive DDT exposure. We have seen repeated

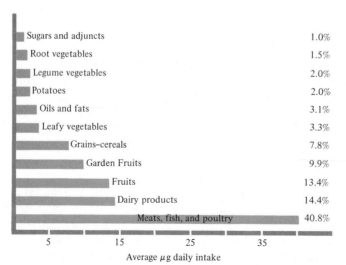

Sugars and adjuncts	1.0%
Root vegetables	1.5%
Legume vegetables	2.0%
Potatoes	2.0%
Oils and fats	3.1%
Leafy vegetables	3.3%
Grains–cereals	7.8%
Garden Fruits	9.9%
Fruits	13.4%
Dairy products	14.4%
Meats, fish, and poultry	40.8%

5   15   25   35

Average μg daily intake

FIGURE 8-7 Human intake of pesticide residues in average total daily diet. (After Duggan and Weatherwax, 1967.)

evidence, however, indicating that biocides have widespread nontarget impact, and that low-level exposure may disrupt reproduction and learning behavior without being lethal. Radomski et al. (1968) examined the levels of chlorinated hydrocarbons in the liver, brain, and fatty tissue of humans during autopsies. They found that the concentrations of these biocides were higher in patients suffering from liver degeneration, hypertension, and nervous system degeneration than in patients dying from infectious diseases. There are, however, at least two alternate interpretations of such correlative evidence: (a) chlorinated hydrocarbons were responsible for the fatal conditions or (b) such medical problems predispose a person to higher chlorinated hydrocarbon accumulation.

In addition to the chlorinated hydrocarbon biocides, man is exposed to a wide variety of biocides containing heavy metals. Among the earliest known biocides, as we saw, was lead. Until the last half century, lead came into humans primarily through the diet and as a result of children eating flakes of lead-base paint. More recently huge quantities of lead have been released to the atmosphere as a result of the utilization of lead in automobile gasolines to reduce engine "knock." About 1 kg per person per year is utilized in the United States as a gasoline additive (*Chemistry,* 1968). From half to two thirds of this is discharged as particles and the remainder is released as a gas which can be transported for considerable distances by wind.

Patterson (1968) reported that the amount of lead in the Greenland ice sheet has increased dramatically since 1950, soaring up to 200 μg/ton of snow in recent years (Fig. 8-8). In contrast, relatively small amounts of lead were recorded in snow samples from the southern hemisphere. The minor exchange of air between the two hemispheres results in the confinement of most of the lead to the heavily industrialized northern hemisphere. Lead poisoning through the diet was largely eliminated centuries ago; more recently it has been regarded as an occupational disease

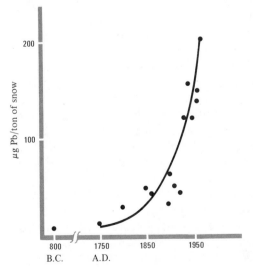

FIGURE 8-8 Accumulation of lead in the Greenland ice pack from 800 B.C. to the present. (After *Chemistry,* 1968.)

(often called painter's colic or plumbism) that can be controlled by hygienic working conditions. It is well known from previous experience in industry that lead may be absorbed through the respiratory tract. It is hard to imagine that continued volatilization of huge quantities of lead will not have a detrimental effect much more widespread than previous occupational hazards.

Another previous occupational disease that may become increasingly widespread in the general population is mercury poisoning. The expression "mad as a hatter" came to be applied to hatmakers, beginning with London's Roger Crab in the seventeenth century, and was popularized by Lewis Carroll in *Alice in Wonderland.* A mercury compound was used in hatmaking and continued exposure to it led to brain damage and increasingly eccentric behavior. The United Nations Food and Agriculture Organization (1967) gives "recommended tolerances for organo-mercurials as none."

The first evidence that mercury poisoning was becoming more widespread than a mere occupational hazard was the poisoning of over 100 Japanese in 1953 and 1961 from eating fish contaminated by the mercury discharge from a plastics factory. Then in 1969 seven members of a New Mexico family were poisoned by pork that had been fattened on grain treated with an organo-mercurial fungicide (*Newsweek,* April 20, 1970). Three of the seven family members may have suffered permanent brain damage. In the fall of 1969 officials in Alberta, Canada, canceled the pheasant and Hungarian partridge hunting season because of high levels of mercury in tissues of these birds (*New York Times,* November 9, 1969). Montana is just across the international border from Alberta and tests were immediately run on 20 birds drawn from different areas in the state. Mercury contents ranged from 0.05 to 0.47 ppm. Although the World Health Organization's suggested maximum level in the human diet is 0.05 ppm, the Montana hunting season was not called off. Hunters were merely warned not to eat more than two birds, and not to eat *any* livers. One daily newspaper in the state, *The Missoulian,* called this policy "weird and fantastic" and continued, "Montana's game birds are, by scientific measurement, poisonous or potentially poisonous to human beings—they should *not* be eaten."

Subsequent tests for mercury in aquatic ecosystems showed toxic levels in at least 33 states and massive amounts were discovered in parts of Lake Erie, San Francisco Bay, Lake Calcasieu in Louisiana, the Delaware River, Brunswick Bay, Georgia, Lake Champlain, and the Tennessee River (Lyons, 1970). As a result, commercial fishing was banned in parts of California, Louisiana, Michigan, and Ontario. Then G. Y. Harry, a Federal marine biologist in Seattle, found toxic levels of mercury in fur seals from the Pribilof Islands south of Alaska. This led to the recall of 25,000 liver supplement pills by the Food and Drug Administration when it was found that pills made from the livers of fur seals killed in 1964 contained mercury levels 60 times the permissible limit (Associated Press, October 1970). The discovery of mercury in a fourth of the canned tuna sampled by B. McDuffie, an analytical chemist at the Binghamton campus of the State University of New York, led the Food and Drug Administration into a widespread testing program that confirmed McDuffie's findings (*Newsweek,* December 28, 1970). Said Ralph Nadar, "If they have to wait for some professor to launch them into action, it is clear these people should quit their jobs. What has the FDA been doing with its authority? Why weren't these tests first made in FDA laboratories?"

It should be pointed out, however, that studies such as this lack an essential scientific component, a control. An implicit part of the statements that decry the high mercury content of fish used for food and place the blame on pollution sources is the assumption that mercury content of fish has been increasing recently. The appropriate control in such an investigation would be determinations of the mercury content in fish caught in the past. New York State's Department of Environmental Conservation (Bird, 1971) has analyzed tissues from fish caught as long ago as 1927, utilizing fish preserved at the state museum in Albany. These analyses revealed mercury concentrations as high as 1.03 ppm in 1939 samples from a bay near Rochester. Fish from isolated waters where there had never been any manufacturing operations had mercury levels as high as 0.5 ppm. In the case of naturally occurring biocides such as mercury, we can be less sure that past levels were zero than we can be with such synthetic biocides as DDT and PCB. It is quite likely that food chains may always have carried loads of certain naturally occurring biocides.

Another ecosystem contamination problem that has direct impact on humans is radioactive fallout resulting from nuclear explosions in the atmosphere. Here the problem again involves chemical substances that have an impact on ecosystems far from the site of initial occurrence. Radioactive strontium and cesium, important fallout constituents, have contaminated the lichen-based food chain on which Eskimos in North America and reindeer herders in northern Eurasia depend (Schulert, 1962; Hanson, 1967). Lichens are extremely effective concentrators of these substances, which are then passed on to the caribou or reindeer and then to man with about a twofold increase in concentration at each step. The ^{132}cesium body burden of Eskimos was about 50 to 100 times that experienced by man in temperate latitudes where lichens are insignificant at the primary production level of the food chain. The biological half-life for cesium in lichens is of the order of 10 to 15 years; so even when input of radioactive cesium into these tundra ecosystems is stopped, it will be some time before the inherent danger of cesium radioactivity is eliminated from the system. This is perhaps one of the most subtle examples of man's inhumanity to man. So long as man thoughtlessly continues to disregard the integrated functioning of all of the earth's ecosystems, there is little hope for a rational environmental policy.

There are, of course, additional chemical pollutants that function in materials circulation systems. We have already considered some of those related to gaseous cycles; some of those important in the degradation of aquatic systems are considered later.

ORGANIC BIOCIDES: EFFECTIVENESS AND ALTERNATIVES

The impact of pesticides on a natural system will depend on the species that are most affected. The loss of a rare species may have less impact on the organization of the system than the loss of a very common species. Uncommon species, however, may be extremely important in the nutrient cycling portion of the system. Not all species are equally important in both energy and materials flow in the system; we commented earlier on the importance of *Modiolus,* an estuarine mollusc, in phosphorus cycling in certain salt marsh systems. Reduction or elimination of populations of *Modiolus* could influence the functioning of the system far in excess of its role as measured merely in terms of abundance. Microbial components of ecosystems are often quite sensitive to environmental pollutants, and one of the dangers of man's practice of filling the environment with toxic materials is that the toxins may interrupt essential materials cycles.

Another potentially dangerous outcome of broadcasting biocides in the habitat is that progressive extinctions of species in food chains will channel more and more of the total biocide amount into the remaining species (Harrison et al., 1970). Repeated extinctions may have a "domino" effect as the biocide load on the remaining species becomes increasingly large. The first wave of extinctions that is now appearing, such as the brown pelican and raptors, may be just a harbinger of things to come. The fact that man is in the same position on the trophic system as many of these species is not reassuring.

One of the really difficult questions about biocides, and particularly those designed to increase agricultural yield and control disease by insect poisoning, is whether they really accomplish their intended purpose. As Strickland (1966) points out, "There are few data on the benefits accruing from field use of pesticides." Most of the evidence cited in support of pesticides is anecdotal in nature and consists of nonquantitative correlative evidence to the effect that "use of DDT in the malaria control program in Ceylon [provided] some 15 years of virtual freedom from this major killer" (White-Stevens, 1970). Certainly a decline in malarial frequency accompanied widespread biocide use, but whether the biocides were more important than better sanitation and new medical technologies is largely conjectural.

In an experimental series in Georgia, millet was planted in a field tightly fenced against mammals. The field was divided into two sections, and one section was sprayed with an insecticide while the other section was sprayed with water (Barrett, 1968). The insecticide was Sevin, which inhibits an enzyme (cholinesterase) important in transmission of nerve stimuli; its inhibition results in disruption of the nervous system. After application on July 28, the concentration of Sevin on plant tissues declined exponentially with time (Fig. 8-9). Arthropod populations were devastated by the spray (Fig. 8-10) and required about 6 weeks to recover. Decrease in arthropods included destruction of spider populations. There was a highly significant decrease in litter decomposition after spraying as a result of the destruction of decomposer populations along with the "target" organisms, herbivorous arthropods. The insecticide disrupted reproduction in cotton rats and this population declined in the sprayed area. Its decline, however, was offset by a substantial increase in the house mouse population.

The question remains, How effective was the biocide in the desired effect of increasing the yield of millet? In fact, it had exactly the opposite effect of the objective: seed yield was 0.85 $gm/m^2/day$ from the unsprayed plot and 0.66 $gm/m^2/day$ from the sprayed one. Similarly detrimental agricultural effects have been documented (Vanden Bosch, 1968; Shea, 1968) as a result of application of

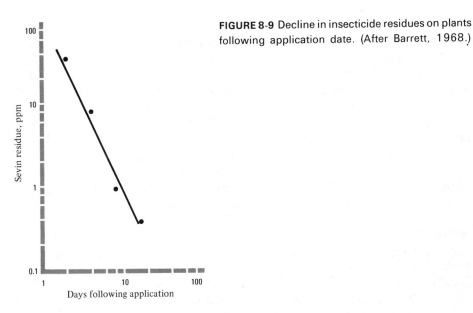

FIGURE 8-9 Decline in insecticide residues on plants following application date. (After Barrett, 1968.)

185

ORGANIC
BIOCIDES:
EFFECTIVENESS
AND
ALTERNATIVES

Azodrin to cotton in California. And estimates of the percentage of the total agricultural yield lost to insects were 10 percent in 1948 and 10 percent in 1969 (Ehrlich and Ehrlich, 1970).

Whether biocides have had a significant promotive effect upon agricultural yields or have been more important than improved sanitation and widespread availability of sulfa drugs and antibiotics in reducing disease is not clear. But there is extensive evidence that the duration of biocide effectiveness is strictly limited for many target organisms (A. Brown, 1969). We already have examined an experimental treatment

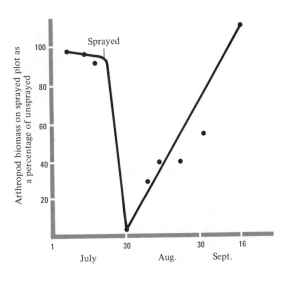

FIGURE 8-10 Effect of Sevin-spraying on comparative size of arthropod community biomass over the period before and after spraying. Biomass declined precipitously upon spraying but then increased to a size exceeding that of unsprayed plots. (After Barrett, 1968.)

that resulted in DDT resistance in fruitflies in a few generations (Chapter 3). Insect resistance to poisons in the field has been known since 1908 when the San José scale became resistant to lime sulfur treatment of apple orchards. By 1967 over 225 species of pests had developed resistance to one or more biocides (A. Brown, 1969). The first serious cases of biocide ineffectiveness developed in public health applications, particularly housefly and mosquito control programs. As Brown points out, "Among the anopheline mosquitoes, vectors of malaria which since 1955 have been attacked by the vast and intensive Global Eradication Campaign, no less than 38 species have developed resistance to chlorinated hydrocarbons."

One important side effect that was not anticipated in using nonspecific pesticides on insects was the immediate impact of the pesticides on the natural predator controls of some insect pests. With application of pesticides, predators often were virtually eliminated from the system, thus releasing the prey (and pest) species from effective biological control (Muir, 1965; Fig. 8-11). Unless the pesticide was repeatedly applied, pest populations increased to levels beyond those attained before application of the pesticide. As our understanding of environmental factors controlling insect populations increases, our ability to deal effectively with insect pests also will increase. Indiscriminate use of pesticides for insect control assumes primarily that the populations of the pest are not controlled by predators but by physical factors or the availability of food. However, many studies (for example, Huffaker and Kennett, 1956) and theoretical arguments (Hairston et al., 1960) lead to the conclusion that many insect populations may be predator controlled to an important degree. For

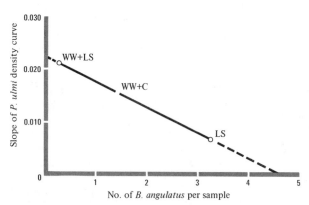

FIGURE 8-11 The influence of predator (*Blepharidopterus angulatus*) density on population growth rate for its prey, an apple orchard mite (*Panonychus ulmi*). The data are from plots sprayed with winter wash (WW = 0.1 percent DNOC and 3 percent petroleum oil), lime sulfur (LS), and captan (C). Note that as the treatments reduce the number of predators the growth rate of the prey population increases. Each treatment was to some extent toxic to both predator and prey populations. (After Muir, 1965.)

these populations, removing the predator by inappropriate use of pesticides may cause tremendous pest outbreaks.

Alternatives to the use of biocides are mainly of four types: (a) enhancement of populations of predator (including parasitic) species so that the pest is kept at a low level by predation, (b) use of reproductive disruption such as sterilization, (c) utilization of specific chemical attractants in combination with either of the above or with localized biocide application, and (d) utilization of a mixed species agricultural system so that food species are harder to find. For instance, it has been shown that capric acid (a fatty acid), which is less toxic than most pesticides, is a potent mosquito larvacide when applied at concentrations of 150 to 300 ppm to larvae-containing pools (Maw, 1970). Following a three-day period in which the larvae died, the pools became extremely attractive as oviposition sites for mosquitoes. During the two months following treatment, an average of 287 egg rafts per pool were deposited in treated pools while untreated pools averaged less than one raft per pool. The ability of capric acid treated pools to act as powerful oviposition attractants suggests that, in addition to the initial larvacidal effect, a few treated pools might act as a congregating mechanism useful in combination with other control mechanisms.

One control mechanism that has been spectacularly successful is the sterile male program suggested by E. P. Knipling to control screw worms, an animal ectoparasite, in the southern United States. Since screw worm male flies mate several times and females only once, a technique was developed in which male flies were sterilized by high energy irradiation during pupation (Bushland and Hopkins, 1953). Subsequently, large numbers of sterile males released into wild populations resulted in almost total reproductive failure in these populations (Baumhover et al., 1955). The release of 435 sterile males per square mile per week resulted in 100 percent sterile egg masses on experimental goats and a population crash of 99 percent in the screw worm population. Subsequent application of this technique to the continental United States has resulted in almost complete eradication of a pest previously causing extensive livestock losses.

Another alternative to chemical controls that has been employed for almost 100 years in California's citrus industry is the use of predators to maintain pest species at noninjurious population sizes (Stern et al., 1959). In 1868 the cushiony cotton scale was introduced into California on imported Australian acacias and threatened to destroy the citrus industry; two predators were introduced 20 years later and the scale was kept at low levels for over 50 years. However, a new discovery, DDT, was applied to orchards in the San Joaquin Valley, resulting in poisoning of the predators and a resurgence of the pest. Spraying was subsequently discontinued.

Successful biological control of both animal and plant pests is conspicuous in its persistence in comparison to the transitory nature of most biocidal control. Huffaker (1970) points out that "On a worldwide basis, about 250 cases of partial to complete success (through biological control) are on record, with about one third of these being so complete that the species now presents no problem at all." Biological control has been characterized by careful laboratory experimentation and extensive pilot tests in the field to demonstrate effective economic benefit prior to large-scale implementation. Biocidal control, in contrast, seems to have been predicated on the question: Will it kill insects without immediately killing humans? Control of organisms noxious to humans seems an extension of the "magic bullet" myth from medicine to the world as a whole. This is the myth that chemicals are magic bullets that are shot at a detrimental target. It is clear in both medicine and ecology that there are no magic bullets, but that chemical bullets strike multiple targets, either by ricocheting or directly on their way to the intended target.

We have concentrated on pesticides in this chapter because they exemplify many of the ecological problems associated with man's technological innovation. These innovations usually are widely employed without meaningful evaluations of cost effectiveness, even to human society, much less to the biosphere as a whole. Often, as with pesticides, certain unexpected effects subsequently become apparent. Interpretation of these effects and their significance lags far behind diffusion of the innovation. There is much more social inertia to overcome in evaluating technologies than in implementing them. Finally, as with pesticides, many innovations have no obvious detrimental effect upon man himself and there is no public consensus about the importance of other detrimental effects. Clearly, to most of us the demise of a few wild animal species is of no significance. We do not yet know enough about the functioning of species in ecosystems to state categorically that such extinctions have any significance whatsoever, other than the esthetic loss to those who enjoy seeing an occasional eagle or pelican.

From the economic standpoint, of course, the long-term ineffectiveness of biocides makes developing them a profitable enterprise. Because they have to be reapplied continually it is possible to generate substantial economic activity around an array of products. Most alternative mechanisms of pest control, in contrast, are ultimately unprofitable from the producer's standpoint since, if they do their job appropriately, implementation is a limited enterprise and there is no need for recurrent purchase of a "product." This makes it extremely unlikely that nonchemical methods of control will be developed by profit making organizations.

GENERAL CONCLUSIONS

1. Human technological innovation has released into the habitat a large variety of compounds with unusual biological properties. Their toxicity to organisms varies widely according both to the nature of the compound and the type of organism.
2. Many of these compounds have become widespread in the biosphere and are found in ecosystems remote from their origin.
3. Biocides introduced into certain ecosystems to control the abundances of populations affecting man's well-being have had unexpected detrimental effects on nontarget species and ecosystems. Many target species, in contrast, have rapidly evolved biocide resistance.
4. In addition to death, many biocides have disruptive effects upon such phenomena as reproduction and behavior when present at subacute levels.
5. A variety of substitutes for biocides is available. Implementing them is hindered by insufficient ecological knowledge and the fact that their potential long-term effectiveness makes development essentially a nonprofit enterprise.

SOME HYPOTHESES FOR TESTING

Although the problem of biocides has developed primarily around applied rather than basic ecological phenomena, many contributions to basic knowledge can come through biocide research. One of the problems that needs to be solved, both in basic and applied ecology, is the importance of species variety in maintaining ecosystems. In later chapters we consider some fundamental problems associated with species diversity. Research on the effects of biocides, however, could serve as an efficient vehicle for examining the functional roles of different species in ecosystems. The tendency for chlorinated hydrocarbons to accumulate in food chains, with resulting devastating effects upon predator populations, has already taught us that predators often are important regulators of their prey. Some of our knowledge of predator-prey phenomena, as we see in a later chapter, has arisen from attempts to define alternatives to biocidal control. Comparisons of forests in gypsy moth free areas, in gypsy moth outbreak areas, and in sprayed outbreak areas might provide considerable information on the effects of herbivore density upon producer function (P. Skaller, personal communication).

Perhaps the most pressing applied information that is required about biocides is their effects upon humans. Almost all data are correlative, anecdotal, or poorly controlled. Abundant evidence that subacute amounts may disrupt behavior and reproduction of other species suggests an imperative need for appropriate data on humans. This, however, is a medical problem—and a very intractable one from the standpoint of designing meaningful experiments. Clearly, few of us would be willing to sacrifice ourselves to obtaining such information. Like most aspects of applied ecology, the biocide problem is ultimately a socio-moral phenomenon.

PART 3

POPULATION AND COMMUNITY PROCESSES

This part consists of a discussion of the processes of energy and materials flow as they are expressed in changes in numbers of organisms. In Chapter 1 we considered a violet population growing on a forest floor. We now have some realization of how population members are coupled with one another and with other organisms through transfers of energy and chemicals. This part will consider how these transfers are manifested in the population itself and in co-existing populations of associated organisms. Chapter 9 discusses the general process of population growth. Chapter 10 considers the energy flow process in detail, as energy flow is manifested in the feeding process. Chapter 11 examines the effects of demand for shared resources upon co-occurring organisms. Chapter 12 concentrates on the mechanisms regulating population size. Chapter 13 deals with the manifestations of resource demand and changing environmental organization on communities through extended time periods. Finally, Chapter 14 is an extension of these principles to a special case: the dynamics of aquatic populations in lakes.

population growth

9

A population was defined in Chapter 1 as one or more individuals of a single species that co-occur in time and space. A single colonizing individual of a vegetatively reproducing species is a population; two or more individuals are required for obligately sexual reproducers. The total size of a population ranges from these low values to a maximum of all living individuals of a species. Usually a population is composed of fewer than all individuals in a species since groups of individuals tend to be spatially separated from other such groups.

In evolutionary terms the important component of separation is not spatial but genetic—individuals in different populations interbreed less freely than members of the same population. The degree of gene exchange necessary to define members of the same and different populations is subjective and cannot be rigidly established. Geographically isolated groups of individuals achieve reduced gene exchange by distance and the inability of individuals carrying "foreign" genotypes to contact other populations. However, apparently continuously distributed clusters of individuals also can be relatively distinct genetically, thus producing local populations within a larger array of individuals. In Chapter 3 we showed how strong selective pressures can lead to local ecotypes on mine spoils and even may lead to divergent breeding systems (self- and cross-fertilization) in an array of individuals of plant species continuously distributed from mine spoils onto pasture areas with low heavy metal content. Kerster and Levin (1968) showed how local differentiation could occur because of limited movement of genes by pollinators. In animals gene flow among local populations also can be restricted by aggressive behavior (Selander, 1970). In humans similar restrictions of gene flow derive from ethnic and religious affiliations as well as living patterns.

Most populations in a local area start with a relatively small number of individuals, increase to a larger number, and then eventually may become extinct, perhaps to be reestablished later by other colonists. Between initiation and demise of a local population the numbers of individuals change through time, often in predictable ways. Simultaneously the individual members of the population interact with one another and with members of other species in the same locality. The types of interactions may vary considerably depending on the species of organisms and the niche characteristics of the individuals. Some interactions are overt and easy to observe, and some may require long-term study or experimentation.

Individuals progress through a characteristic series of developmental stages from birth, growth to a mature individual capable of reproduction, and finally to death. Changes in population size through time and the types and outcomes of interactions are closely linked with developmental changes at the individual level. In the following four chapters we examine characteristics of changes in population numbers through time, the evolution of reproductive rates, the types of interactions among individuals of the same and different species, and, finally, the mechanisms of population regulation.

GENERAL MODEL

The ability of population numbers to change through time is determined by the difference between births and deaths in the time interval,

$$dN/dt = b - d \tag{9-1}$$

where dN/dt is the change in numbers of the population over time t, b the number of births, and d the number of deaths over the interval (this assumes no migration). If the number of births exceeds the number of deaths, the population increases; if the number of deaths exceeds the births, the population declines. In the unlikely event that the number of births exactly balances the number of deaths, the population will remain stable.

Population growth essentially involves partitioning finite resources between two compartments, one for organisms and one for environment (Fig. 9-1). We saw earlier (Chapter 7) that the flow between compartments is regulated by the amounts of resources in the two compartments and by the uptake and the return coefficients characteristic of the organisms. If we denote the amount of resource in the environment by R, the amount of resource in the population by N, the uptake coefficient k^+, and the return coefficient k^-, we have

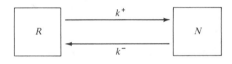

FIGURE 9-1 Interrelationship between resource pool and numbers of individuals in a population. k^+ is the uptake coefficient and k^- is the loss coefficient. This model allows for negative feedback from a required resource to population growth.

195

GENERAL MODEL

$$dN/dt = k^+RN - k^-N \qquad (9\text{-}2)$$

which indicates that the rate of the resource flow into the population depends on the uptake and the return coefficients and amounts of resource in the two compartments. Rearranging,

$$dN/dt = (k^+R - k^-)N \qquad (9\text{-}3)$$

This suggests that population growth rate is a function of population size, N, times the term $k^+R - k^-$, which describes the ability of individuals to accumulate resources. The latter term is commonly called r, the realized reproductive output per individual member of the population. The term r is also equal to the difference between average birth and death rates per individual. It is apparent that r must include individual growth rates since individuals in a population will not all be of identical size from birth to death. Assuming that they are interchangeable in the population, however, we can call r the population size increase per individual per time period.

Realized growth rate may be positive (an increasing population), negative (a decreasing population), or zero (a stable population). The actual change in numbers of the population over time, then, can be related to the initial numbers in the population by the following equation:

$$dN/dt = rN_0 \qquad (9\text{-}4)$$

where N_0 is the initial number. This equation does not indicate whether the numbers are increasing, decreasing, or stable. These trends will depend on the value and sign (positive or negative) of r.

Organisms generally have the ability to more than replace themselves with offspring. Organisms not having this capability gradually would become extinct. This reproductive ability was evident to some early scientists and philosophers; it led Thomas Malthus in 1798 to his significant consideration that the population of man could increase faster than man could increase his food resources. If the population of organisms is more than replacing itself over a short-time period and the value of r remains essentially constant, the population will increase exponentially (Fig. 9-2a). Transformed to semilogarithmic axes (base e) the figure appears as a straight line, the slope of which is equal to r (Fig. 9-2b). In arithmetic terms the growth of the population can be related to initial numbers by integrating equation (9-4) to

$$N_t = N_0 e^{rt} \qquad (9\text{-}5)$$

where N_t is the number after a time interval t and e is the base of the natural

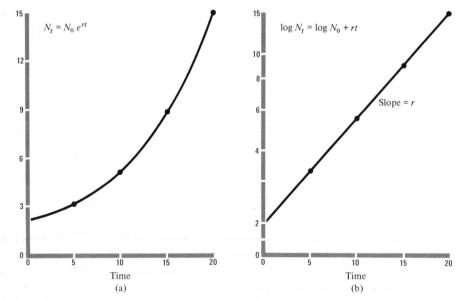

FIGURE 9-2 (a) Arithmetic plot of the exponential growth curve for a population based on equations (9-4) and (9-5). (b) The exponential growth curve of (a) transformed to semilogarithmic axes.

logarithms. The capacity for exponential growth depends on the excess of births over deaths. Presumably each population will reach a size at which deaths begin to exceed births and the population begins to decline. Malthus argued that the point at which the human population began to decline could be established by the inability of the exponentially expanding human population to increase food available at the same rate.

There are numerous studies that document the genetic basis of the reproductive output of organisms (for example, Robertson, 1957). Each individual has a genetic capability to reproduce that has evolved in relation to environmental conditions. The actual reproductive output of any individual is the combined action of this genetic information and modifications brought about by environmental conditions. In a hypothetically optimal environment, however, in which the environment (for example, predators, competitors, and physical factors) has no modifying impact on the genetic capacity of the individual to reproduce, r should equal the *innate,* or *intrinsic, reproductive capacity* of individuals.

There is but a single set of theoretical environmental conditions that is optimal and hence that allows the population to grow at its intrinsic rate of increase. Because the environment is so complex, the probability of being able to define this value for any population is remote. However, under each specified set of environmental conditions there is a maximal rate, r_m, at which the population is able to grow, based on its genetic capabilities. If these environmental characteristics do not change, presumably the population will continue to grow at the maximal rate and exhibit exponential growth:

$$dN/dt = r_mN_0 \qquad (9\text{-}6)$$

The actual growth rate in any environment depends on how much r_m is modified by environmental changes. Remember that r_m is an average of the genotypically controlled rates of increase for individual members of the population. The value for the individual genotypes might be called the *Malthusian parameter, m* (Fisher, 1958). Rates of increase are estimated from the composite reproductive output of a group of individuals. However, since the optimum density of most sexually reproducing forms is greater than one individual, it is impossible to measure m, the value for an individual genotype (Hairston et al., 1970).

The maximum r for an environment, r_m, occurs when the difference between birth and death rates per individual is at a maximum. Many sexually reproducing organisms have an optimal density greater than one female and one male for maximum offspring production per female (Fig. 9-3). It is possible that there is some sort of facilitation of reproduction through environmental modification at optimal densities that makes output higher than at densities both slightly above and below the optimal. On the other hand it is possible that sexual reproducers require a minimal population size to insure that each female is fertilized. Because offspring output per female must be an average value in populations of more than one female, any unfertilized females will cause the average output to drop. At the optimal density

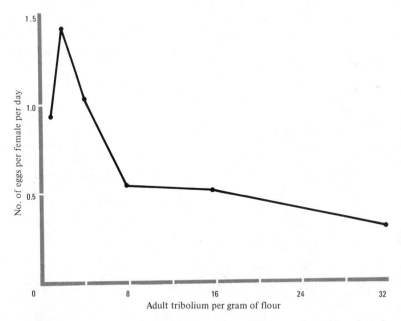

FIGURE 9-3 Relationship between the average output of eggs per female and the density of adult *Tribolium confusum* in the experimental culture. Note the peak in egg output which defines the optimum adult density for per individual egg output under these culture conditions. (After Andrewartha and Birch, 1954.)

for reproduction the death rate could be somewhat higher than at lower densities. In this case r would be maximized at some density slightly different from the optimum for reproduction.

Clearly the environment includes other individuals of the same species as well as potential mortality hazards and specified resource levels. Each change in environmental conditions produces a new environment and potentially a new r value. Hence a growing population must be in an expanding environment including space to maintain its r_m. Very few populations occur in their optimal environment for a long period of time. The environmental conditions change and reduce the capacity of the environment to support the same rate of population growth. Either mortality factors increase or the population begins to deplete resources to limiting levels. Einarsen (1945) showed that pheasants introduced to an island off the coast of the state of Washington initially grew at an exponential rate, with slight annual declines after the breeding season, for about five years; the growth rate then began to decrease as the birds colonized the entire pheasant environment on the island (Fig. 9-4). Many laboratory experiments (Fig. 9-5) also show that small populations initially introduced to a large volume of culture medium will grow at a decreasing rate per individual as size increases. The actual growth rate of a population depends to a large degree on how the individuals respond to environmental stimuli acting to modify r.

The following discussion will focus on populations that decrease in relative growth rate as the size of the population increases and as the modifying impact of the environment on r increases. Other populations (to be discussed later) show little or no decrease in growth rate as the size of the population increases to near maximum population size.

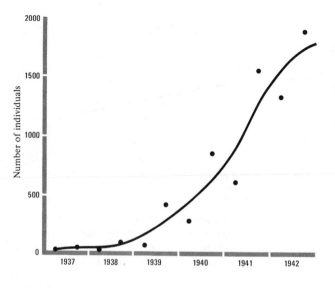

FIGURE 9-4 Growth of a pheasant (*Phasianus colchicus*) population after being introduced into an uninhabited site, Protection Island, Washington. (Data from Einarsen, 1945; After Odum, 1959.)

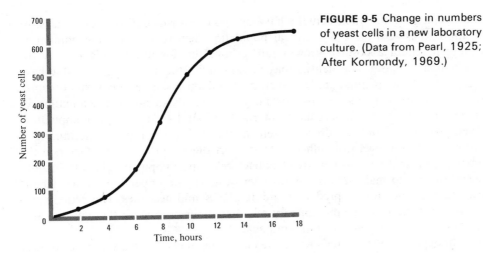

FIGURE 9-5 Change in numbers of yeast cells in a new laboratory culture. (Data from Pearl, 1925; After Kormondy, 1969.)

K-RESPONSIVE POPULATIONS

The modifying impact of the environment on r as population size increases eventually sets an upper limit (K, or *carrying capacity*) on the population size that the environment can maintain; when this size is reached, $r = 0$ (Fig. 9-6). For organisms that respond rapidly enough and with sufficient magnitude to environmental influences, the numbers in the population can stabilize around the carrying capacity, K, of the environment. If, however, the population exceeds this carrying capacity, it will tend to decrease in size. If it drops below the carrying capacity, the environmental influences that reduce the growth rate decrease and then the population tends to increase again. If the population is stable, then dN/dt and r are zero. Figure 9-6

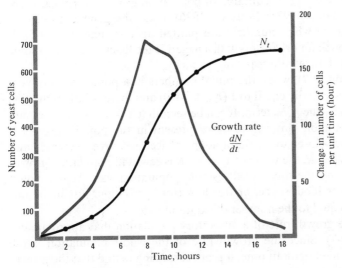

FIGURE 9-6 Rate of change of yeast population in the laboratory culture shown in Figure 9-5, which also is included here. This growth rate curve is based on the entire population. (After Kormondy, 1969.)

density dep

includes pop. dens factors, & outside factors?

shows that all populations actually have two points at which dN/dt equals zero. As N approaches zero there is a critical population density at which the population cannot maintain a positive growth rate. Above this lower threshold density the population increases. This density may be zero, indicating the obvious need to have a founder for any population. However, many organisms, especially sexually reproducing species, may require two or more individuals to insure insemination of sufficient females to counterbalance mortality of adults. At the upper population level, growth rate (dN/dt) drops to zero as the critical resource(s) determining K are reduced to a level just sufficient to support the population. Any factor in the environment that has this negative feedback relation to population growth in sufficient strength to make dN/dt equal to zero can set the upper limit. Such factors can include food supply, predators, hiding places, and nest sites. The actual regulating factors need not be the same for all populations or even for those of the same species at different points in time and space.

Although some characteristics of the environment can force dN/dt to zero, they do not do so in a continuous negative feedback loop with population size. They only set the carrying capacity of responsive populations for short periods, and the impact may not always depend on population size. For instance, weather may adversely affect reproduction of a population, but the effect may occur regardless of the population size. Weather could force dN/dt to zero, but might not set an upper limit to population size through a continuously present negative feedback mechanism involving density.

Since carrying capacity influences the maximal exponential growth rate of the population, eventually driving it to zero, equation (9-4) can be rewritten to reflect the negative feedback of carrying capacity on an r_m, assumed to be constant:

$$dN/dt = r_m N_0 (1 - N_0/K) \qquad (9\text{-}7)$$

This equation, known as the logistic equation of population growth (Verhulst, 1838; Pearl and Reed, 1920; Lotka, 1925; Volterra, 1926), takes the general form of a sigmoid, or S-shaped curve when numbers are plotted against time (Fig. 9-5). It describes population growth for organisms with a negative feedback response to some environmental factor(s) that sets the carrying capacity.

According to equation (9-7), when the initial numbers in a population are very near zero, $1 - N_0/K$ is essentially equal to 1 ($N_0/K \rightarrow 0$) and there is little reduction in growth rate described by the $r_m N_0$ term. It can be seen that $1 - N_0/K$ is a measure of the proportion of available limiting resource present in the population. As N increases and approaches closer to K, the amount of free resource decreases and N_0/K increases. When $N_0 = K$, the value of $1 - N_0/K$ is zero and population growth rate is zero, thus achieving the theoretically stable population size. At this point, the free resource, whether food, space, or freedom from predation, just maintains the population at a size equal to the number of extant individuals.

Figure 9-6 shows the growth rate of a laboratory population that approximates the logistic equation (9-7). Since the growth rate of the whole population is the number of individuals added per unit time, it is not surprising to find that the growth rate increases for a time as more and more reproductives are added, even though

the impact of the negative feedback from the environment is increasing. At the peak
of the dN/dt curve the modifying influence of the environment becomes sufficiently
strong so that even the large numbers of reproductives cannot maintain the increase
in population growth rate. The growth rate begins to decline, eventually reaching
zero as the population stabilizes at the carrying capacity.

The effect of environmental feedback on population growth would be perhaps
easier to visualize if equation (9-7) were modified to reflect the growth rate of the
population per individual rather than the actual increase in numbers through time.
We do this by dividing through equation (9-7) by N_0

$$dN/N_0 \, dt = r_m(1 - N_0/K) \tag{9-8}$$

Rearranging slightly this equation becomes

$$dN/N_0 \, dt = r_m - r_m/K(N_0) \tag{9-9}$$

which is the equation for a straight line with r_m, the maximal capacity to increase,
being the Y-intercept and K the X-intercept (Fig. 9-7). The slope of the line is $-r_m/K$,
which indicates that the growth rate of a population per individual decreases as
the number of individuals increases. If the equation could be shown to hold for
any population, it would be relatively easy to calculate the maximal capacity to
increase by establishing the Y-intercept and to calculate the carrying capacity from
the X-intercept (Smith, 1952). Unfortunately most populations do not grow according
to the logistic equation.

K-UNRESPONSIVE POPULATIONS

Many populations do not respond to an environmental carrying capacity with a
decreasing rate of population growth as the population density approaches K. This
unresponsiveness to increasing environmental pressure might not be evident in
populations that are not dense enough for any decreased growth rate to be noticed.
However, there are populations that do not show a decreased growth rate at densities

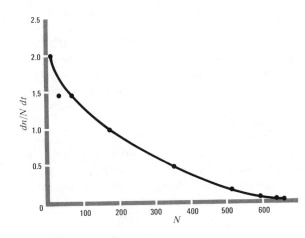

FIGURE 9-7 Relation between per individual population growth rate and population size for the yeast population shown in Figure 9-5. Note the slightly concave shape of the curve rather than the straight line predicted by the logistic equation.

near the carrying capacity. In some, r may remain high or even increase as the population size increases. Since some factor, available energy if nothing else, will become limiting in the environment, the population eventually will cease to grow and often abruptly decline in numbers.

A laboratory example is the blowfly populations studied by Nicholson (1955, 1957). He raised the flies in closed containers and carefully regulated the amount of food given at daily intervals. The regular addition of food maintained the energy-determined K value at a fairly constant level. Nicholson ran three sets of experiments in which he rationed food to the larvae, to the adults, and to both life stages. The unrationed life stage received *ad libitum* food. Population sizes were measured as adult density.

The blowfly larvae need a certain level of food intake to grow large enough to pupate and to emerge as adults. Above this threshold food level, the size at which the larvae pupate increases somewhat to an upper size limit with increasing food per larva. Larger larvae produce larger adults that can lay more eggs. Adult females require protein to lay eggs. As the rations of protein per adult were decreased the numbers of eggs laid per female also decreased. Thus in conditions of high larval density and low larval food supply both the number of larvae pupating and the number of adults produced would be small. As this level of food supply was approached the egg output of the smaller adults would decrease, as it would with limited food supply to the adult.

At the start of an experiment Nicholson introduced a few adults. The larvae from the eggs laid by these adults would pupate if they reached the critical size. Even under conditions of limited larval food the number of adults produced in the first generation was high. This population was supplied with *ad libitum* food and laid large numbers of eggs. If larval growth and survival were reduced by food limitation, the next generation would have fewer adults. Nicholson found that when one or the other life stage was rationed the population fluctuated widely through time (Fig. 9-8), with the peak population sizes positively related to the amount of food in the daily ration. The decline in number of adult flies produced under conditions of food limitation for either the adult or the larvae was fairly rapid and, coupled with overlapping generations, the low density in the oscillating population was close to zero adults. These oscillations were continued in the laboratory for over a year. Nicholson concluded that neither the adults nor the larvae responded to the approach to K (food-determined) by reducing their per individual acquisition of the limited resource. The population of adults diminished rapidly each time after the food was in such short supply that very few adults emerged. The independent action of limiting factors on adults and larvae allowed the growth rate of the population per individual to remain high and actually to increase somewhat as the population approached and surpassed K (Fig. 9-9). In this situation the population did not respond to an approach toward K density by decreasing per individual population growth rate.

Apparently the independence of the limitations on each life stage produced the oscillations, since when both life stages were rationed simultaneously, the population cycles were damped and showed an approach to what might have been the K density.

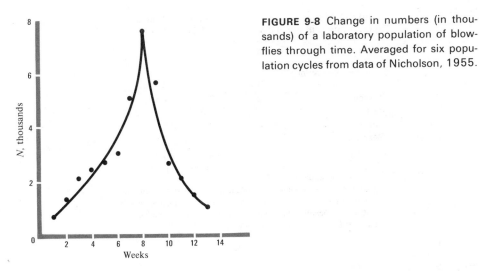

FIGURE 9-8 Change in numbers (in thousands) of a laboratory population of blowflies through time. Averaged for six population cycles from data of Nicholson, 1955.

203

K-UNRESPONSIVE POPULATIONS

The time lags inherent in rationing only one stage are removed and the limitation is quickly translated throughout the population.

Blowflies might exhibit a variety of responses in either life stage to insure continued, but reduced, output of adults as *K* is approached, thus becoming *K*-responsive. These responses include intrinsic behavioral and chemical means of distributing the resources such that the *K* individuals are maintained and replace themselves. (These mechanisms are more fully discussed in Chapter 11.) The responses do not appear in either life stage of the blowfly, suggesting that the blowflies have not been selected for precise regulatory mechanisms relative to *K*. As will be argued later it is very likely that these individuals have evolved in situations in which a response to *K* is disadvantageous because the population rarely or unpredictably approaches *K*. The apparent regulation of the population when both life stages are rationed results from a chance summing of two *K*-unresponsive stages that offset each other and reduce the severity of the oscillations.

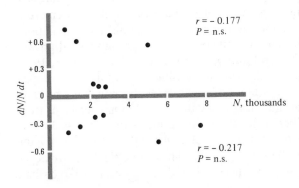

FIGURE 9-9 The per individual growth rate for the blowfly population shown in Figure 9-8. Note that the growth rate becomes negative as the population declines and that there is a rapid shift from a high positive growth rate at the apex of the curve to a high negative growth rate at the next sampling time. (Data from Nicholson, 1955.)

TABLE 9-1 Summary of correlations of per individual growth rate with population size for 71 populations

Category	A	B	C	D
Invertebrates other than insects	1	–	4	–
Insects	2	3	12	–
Fish	–	3	4	–
Birds	4	5	14	–
Mammals	–	5	13	1[a]
Totals	7	16	47	1

[a] The human population of the world.
Column A: species eliminated because census data were not different from series of random numbers; B: correlation coefficient not significantly different from zero; C: correlation significantly negative; D: correlation significantly positive
Tanner, 1966

NATURAL POPULATIONS

Tanner (1966) examined the possibility that natural populations show the negative feedback essential to population stability. He surveyed 71 populations (Table 9-1) using published data from censuses taken at fairly regular, short intervals. Initially he tested the census data to make sure that the sequential population estimates differed from a series of random numbers; those that did not (seven populations) were not considered further. Using the correlation coefficient as an index of the relation of per individual growth rate and population density, Tanner found that for 16 populations the correlation coefficient did not differ significantly from zero, and hence there appeared to be no trend in per individual growth rate as population size changed. For 47 populations the correlation coefficient was significantly different from zero and also negative, indicating that as the populations increased the per individual growth rate decreased, as expected in the logistic model of population growth. For a single population, man, the correlation coefficient was also significantly different from zero. Here, however, it was positive, indicating that as the population of man increases the per individual growth rate also increases (Fig. 9-10), making man much more like the blowfly populations than like the K-responsive populations.

Tanner's conclusions about the importance of negative feedback from environmental factors in causing a decline in the per individual growth rate of 46 populations depended on the value of the correlation coefficient relating per individual growth rate to population size. Recently several authors (Eberhardt, 1970; Maelzer, 1970; St. Amant, 1970) have shown that, in the various possible forms this relation may take, the conclusion regarding the importance of negative feedback may not be valid because of the high probability of obtaining a negative correlation coefficient from random censuses of this type. Eberhardt (1970), using much the same technique of comparison as did Tanner, found that the correlation for the two-point moving average of a series of random numbers was -0.705. Thus to be significantly negative, the value of a negative correlation coefficient should be significantly larger than -0.705.

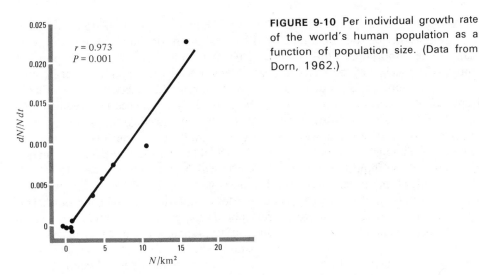

FIGURE 9-10 Per individual growth rate of the world's human population as a function of population size. (Data from Dorn, 1962.)

St. Amant and Maelzer both concluded that another technique of assessing the importance of negative feedback in control of population growth rate, that is, the relation between the logarithm of numbers after a time interval and at the start of a time interval, is also subject to inherent sampling errors. St. Amant finally concluded that the best tests for negative feedback should include data other than from censuses and that populations should be altered to sizes above and below the size at which there is apparent regulation. A regulated population, that is, one that follows the negative feedback assumptions of the logistic equation, should return to the same total population size set by the environmental carrying capacity regardless of whether the population initially started above or below that level (Murdoch, 1970). All four authors conclude that there are few unequivocal data to support the idea of a strong negative feedback component in population growth.

THE HUMAN POPULATION

There are several possible explanations for the positive relation between per individual growth rate and human population size. First, man may be a species that does not respond to negative feedback from the environment by decreasing per individual growth rate as population size increases. Judging from other populations that grow similarly, it is likely that if man is a nonresponsive population, he will overshoot the carrying capacity of his environment and crash to a small population size. It will not be possible to predict the crash, either in terms of human numbers or time, since we do not recognize the impact of the carrying capacity and presumably do not recognize the factor(s) that determine it. This suggested growth pattern for the human population differs quite significantly from predictions of many national and international agencies. The usual prediction (Ternes, 1970) is that the population of man will eventually level off and be maintained at some upper limit. At this population size dN/dt will for some reason, currently unknown, become zero.

A second possible explanation of why the human population does not grow according to the logistic equation is that cultural input, or the reasoning ability of man, is the important factor influencing changes in the growth rate. A possible cultural feedback may be the conflicting demands placed on time, energy, and monetary resources of an individual in a highly developed society. The individual's response often seems to become more self-oriented with a consequent reduction in reproduction. For this self-orientation to be effective in affecting population growth, large-scale cooperation of governmental groups and medical personnel would be required to enable individuals to control birth rate. It would also require major rethinking of the importance of population growth to the national welfare.

The cultural input also may reflect intolerable living conditions; high per individual growth rates often are associated with low standards of living conditions. The countries that most closely approach zero population growth are among the highly developed and technological nations. One country, Japan, which had reached a stable population size by appropriate planning and medical facilities, has now altered its national goal to increased population size so as to maintain a base of young people who serve as factory workers and care for older people unable to work (Boffey, 1970). Japan's population policy, however, is unique only in its reversal to a growth syndrome; most developed nations view continued growth of the population and technological capabilities as desirable and probably as inevitable. The lesson from other populations, and from logic, is that growth cannot be forever positive in any ecosystem. We return to the human population in detail in Chapter 20.

MODIFICATION OF THE LOGISTIC EQUATION

The discussion of the logistic equation of population growth was based on a series of implied and very essential assumptions.

1. Each individual is equivalent to all others in the population.
2. Times lags (delay in effect of environmental change) are unimportant in influencing growth characteristics of the population at varying densities.
3. K, the carrying capacity, is constant.
4. There are no threshold phenomena associated with response to K.

Although these assumptions are probably never met in an actual population, in some cases (see, for example, Figs. 9-6 and 9-7) in which the departures are of sufficiently small magnitude, the growth rate of the population shows little variation from the predictions of the logistic equation. In view of this similarity and the inherent mathematical simplicity of the logistic equation, the equation still has a major place in theoretical models of population growth (and population interactions). However, for most populations the logistic equation is a poor or nonfunctional description of the actual growth rate.

Under controlled laboratory conditions the assumptions of the logistic equation can sometimes be approximated. For instance, K can often be held nearly constant. Similarly the effect of time lags in producing differential population growth can be minimized by continually expanding the culture conditions as the population increases so that each individual throughout the history of the population is maintained at the same density and hence under the same density-related lag conditions (Smith, 1963). Time lags are still important, but they are equal for equal-aged individuals as the population grows and they can be assumed constant for the experiment. Assumption 4 can be met by carefully choosing the factor that will be important in setting K. Under laboratory conditions the major assumption that is not met, or at least not controlled, is the equality of individuals. This inequality extends from individuals of different age groups (Slobodkin, 1960) to the same individuals at different population sizes relative to K (King and Anderson, 1971). That is, density-dependent acclimation and adaptation both violate essential assumptions of the logistic.

Populations that do not meet the assumptions of the logistic equation require either a new model or a modification of the logistic equation to describe population growth. The two models discussed in the following paragraphs represent modifications of the logistic equation; the first is a general modification based on inequality of individuals, and the second is an attempt to modify all four assumptions to produce a specific equation for the growth of a single species under controlled laboratory conditions.

From the earlier discussion of Slobodkin's predation experiment (Chapter 6) it was evident in *Daphnia,* and by analogy in all organisms that survive after terminating growth, that the greatest efficiency of energy conversion, measured in terms of energy storage, is probably in the growing stages. Adults do not accumulate energy although they continue to utilize energy to maintain themselves and reproduce. The utilization efficiency of the K-determining resource probably also varies with the density of individuals in many cases. If K is food-determined, then as the population approaches K the food per individual will gradually diminish to an amount just sufficient to maintain each individual. If search time is an important part of the energy expended in food uptake, the efficiency of energy utilization will decrease as food items become more rare. On the other hand if the major inefficiency occurs at the stage of food use once it has been obtained, then over a wide range of food densities the consumer will not show a decrease in feeding efficiency.

General Modification

Smith (1963) used the expanding culture technique to maintain populations of *Daphnia* at the same relative density as the total population grew through time. In this way he maintained K constant, held time lags more or less constant, and varied only the efficiency at which the limiting resource, food, was being used. He then made four assumptions that allowed him to modify the logistic equation to take into account the inefficiencies of food exploitation.

1. The law of mass action, inherent in the logistic equation, also applies to a system in which rate of food use is the measure of population size.
2. The relation between the rate of food use and biomass can be described with a simple linear regression involving the specific growth rate.

3. The efficiencies of conversion of food into new mass and into replacement mass are similar.

4. The replacement of dead individuals at saturation density can be used to estimate the replacement rate at all population densities.

On the basis of these assumptions Smith introduced a modified logistic equation:

$$\frac{dM}{dt} = r_m M \left[\frac{K - M}{K + (r_m/c)M} \right] \qquad (9\text{-}10)$$

where M is biomass of the population, r_m the maximal rate of biomass increase with unlimited food (mg/mg/day), K the value of M at saturation density, t is time, and c is the rate of replacement of biomass in the population at saturation density (mg/mg/day). Smith then estimated the three parameters, K, r_m, and c, independently of the population growth curves for his experimental populations. With these constants he compared the expected growth curves [calculated from equation (9-10)] with the observed curves and found very close fits (Fig. 9-11). In each population the per individual growth rate declined very rapidly through initial increases in density and then declined much more slowly at higher densities. This concave relationship between population size and growth rate held for per individual growth rates that were expressed either as numbers or, more appropriately for a food-limited population, as the increase in biomass of the population with time. In these experiments inefficiency of resource exploitation was manifested as a decline in per individual growth rate principally at early stages of population growth.

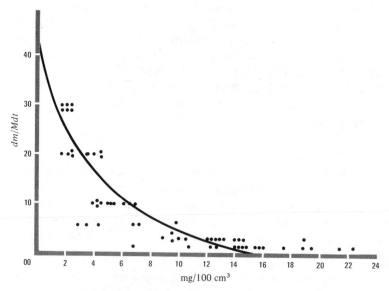

FIGURE 9-11 Comparison of actual (dots) and predicted (line) rates of biomass change in relation to biomass densities of *Daphnia magna* in laboratory cultures. Predicted values were determined independently of the experimental cultures using equation (9-10). (After Smith, 1963.)

Smith also suggested that under certain conditions, such as a positive relation between aggressive activity and population density, the curve could become a convex hyperbola. In this case the environmental limiting factor would have little influence on per individual population growth until the density approached K. The limiting factor then approaches a sort of threshold influence; that is, below the threshold there is little impact, whereas above the threshold the impact is nearly maximal. Caperon (1967, 1968) suggested that this could be the case in a nutrient-limited population below K when the limitation of growth rate was based on how fast the individual could use the nutrient and not how fast it could obtain the nutrient from the environment. At very high population densities, as the amount of nutrients available per individual declined, a point would be reached at which the rate of uptake from the environment would be more limiting than the rate of assimilation once the food had been obtained.

Specific Modification

One of the few attempts to modify the logistic equation in a precise manner for a single species was summarized by Frank (1960) for laboratory populations of the crustacean, *Daphnia pulex*. Frank and his co-workers (1957) had previously calculated, under specified conditions, such constants as r. Frank then examined additional responses of individual *Daphnia* to changing density. He measured the change in rate of growth of small individuals in low- and high-density cultures, the change in survival rate, and the change in reproductive output per female at varying densities. Finally, he attempted to ascertain the frequency with which he should take a census of the population so he could account for changes in the relative numbers of old and young individuals as population size changed. More young (small) individuals mean a large proportion of reproductive individuals at some later time, a time that depends on the rate of attaining maturity. A high proportion of old (large) individuals means that the proportion of reproductives at some slightly later time will probably decline, but that there will be a large proportion of young. To insure an adequate measurement of the impact of age structure of the population on later growth characteristics, Frank sampled every two days.

On the basis of his experimental populations, Frank generated theoretical curves describing population growth. The equations were all expressed as numbers changing through time (Fig. 9-12), but curves b and c were calculated from expected changes in biomass units and then converted back into numbers. Curve a was calculated with numbers of individuals regardless of their size. The logistic curve (a) predicts that the population will eventually reach a stable level of about 600 individuals under the experimental conditions. The shape of the curve indicates that the population should reach the carrying capacity fairly quickly. Curve b in Figure 9-12 is the logistic equation modified to include discrete population growth intervals and the effects of initial age distributions. Curve c assumes, in addition, that the important lags to consider in modifying the logistic equation are the time from the increase in population size (by adding newborns) to the time that this increased size will have an influence on offspring production by females. Frank estimated that the time lag should be about five days, the approximate time for a young to reach maturity.

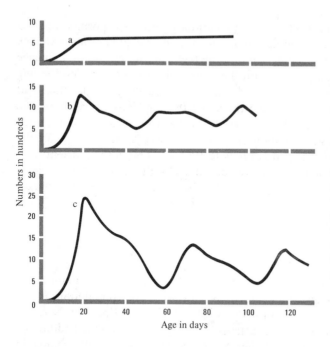

FIGURE 9-12 Estimates of population trends in *Daphnia* cultures started from 25 individuals. Curve a is based on numbers of individuals; curves b and c are based on biomass which has been reconverted to numbers. Curve a is the logistic estimate; curve b assumes time lags associated with growth rates and mortality rates as a function of changing density; curve c is like curve b with the added time lag associated with changes in natality as a function of changing density. (After Frank, 1960.)

The critical test of these modifications is to see if any that are based on known population characteristics sufficiently alter the logistic equation to make the resultant description of population growth conform more exactly to the realized pattern from the experimental cultures. Figure 9-13 illustrates the similarity of a series of replicate populations. The fit of one experimental population to curve c in Figure 9-12 is illustrated in Figure 9-14. Both Figure 9-12c and the experimental populations tend to overshoot the K value, then decline below it, increase again, and finally oscillate with decreasing amplitude about K. Apparently time lags were important in generating these overshoots and resultant declines; but as the population continued in time the heterogeneity of age structure tended to damp these oscillations.

The preceding experiments indicate that the logistic equation has to be modified for each population in relation to deviations of the populations from each of the assumptions. The necessity of studying each population in some detail before describing the growth rate mathematically probably accounts for two current conditions in ecology: (a) the continued use of the logistic equation despite its limited applicability; and (b) the few attempts that ecologists have made to modify the logistic equation to describe the growth characteristics of a population, especially in natural conditions.

POPULATION VITAL STATISTICS

Two factors determine the growth rate of an isolated population: the birth rate and the death rate. Each is influenced by all the other factors that comprise the environment and each also has genetic components that have evolved in relation to environmental conditions. Growth rate depends on the reproductive output of each female (r if over a measured time interval) times the number of reproductive females. To predict the growth rate accurately, then, it is necessary to

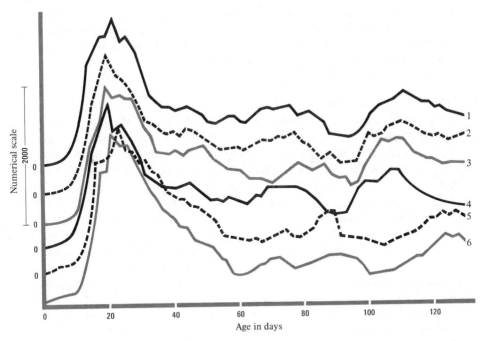

FIGURE 9-13 Six replicate populations of *Daphnia* started with 25 individuals. The ordinate has been shifted so that each population starts at the same value. Notice the close similarity of the six populations. (After Frank, 1960.)

know the distribution of individuals in the different age classes, that is, the age structure of the population. The age structure is produced by the birth rate at various times in the past and by the different mortality rates of the various age groups and their variation through time. High mortality of young for a short period will produce a population with relatively few reproductives since not many young survive and

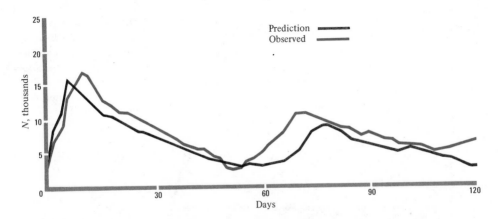

FIGURE 9-14 Changes in numbers in *Daphnia* populations through time according to prediction c and actual laboratory results. (After Frank, 1960.)

reach maturity. Although both birth and death rates contribute to the actual rate of population growth and the age structure, they are, in fact, separate entities and will be thus treated in the following discussions. This is not to imply, however, that there is no evolutionary relation between the two.

Survivorship (l_x) Curves

It is true, of course, that every organism has a finite lifespan. However, individuals usually vary in longevity. The ages at which they die is often a population characteristic that provides important clues to the age at which the strongest selection pressures occur (Emlen, 1970; Williams, 1966), and especially to the parent's degree of effort to insure the survival of the offspring. For more practical purposes, knowing the probability of an individual surviving to a particular age is the basis of the life insurance industry. Insurance companies use actuarial tables to predict how much they have to charge to make a profit insuring people against the probability of death before a certain age. For both man and other organisms, what is needed to construct a table of survival probabilities is an accurate method of estimating the death rate in a population from a sample. Deevey (1947) summarized much of the actuarial information on animals that was available at the time of his review. He also provided a discussion of the theoretical framework for survivorship investigations.

Survivorship data are usually presented as a graph (Fig. 9-15) showing the number of survivors at a given age (l_x) from a known initial group of individuals (called a *cohort*). Most life tables are not constructed from age zero, the moment of conception. Mortality in some portion of the lifespan is usually impossible to follow

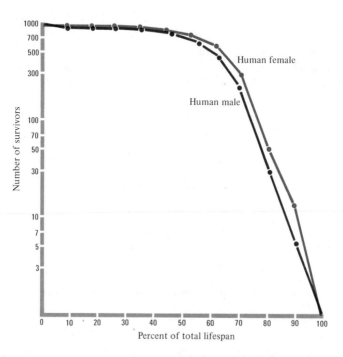

FIGURE 9-15 Survivorship curve for human, white male and female. The abscissa is percent of the total lifespan. Lifespan for males is 108 years and for females is 109 years. From these data the mean lifespans are 68.5 years (male) and 73.2 years (female). (Data are for 1939–1941; from Dublin et al., 1949. After Kormondy, 1969.)

accurately because of inaccessibility of the individuals. The construction of a survivorship curve generally depends on one of three methods of obtaining data, each providing the information *a posteriori* rather than *a priori* so that the predictive power of the data depends on the similarity of mortality factors through time. One method records the age of death of all the individuals in a cohort from age "zero" to the death of the last individual. In assuming that these individuals are representative of the population as a whole, this method provides a probability estimate that any individual will still be alive at age x. Alternatively, in natural populations the number of known survivors can be recorded as a function of age. Assuming that all individuals missing since the last census have died, the data are comparable to the death record. For sedentary organisms any accurate mapping technique will allow repetitive sampling of the fate of individuals (see Connell, 1961, 1970, for examples of this method). If immigration can be controlled or eliminated from the data and the appearance of the individuals in the cohort is nearly simultaneous, this procedure provides a workable survivorship curve for a single population through time.

A somewhat less accurate technique for constructing survivorship curves depends on being able to determine the age of dead organisms or parts thereof. With a large enough sample the proportion of individuals dying at a particular age can be ascertained. This estimate of the probability of living to age x assumes a constant probability throughout the time the dead organisms accumulated and a constant and equal probability of finding the remains of any age group. This technique is appropriate for many natural populations of animals because of the difficulty of maintaining contact with sufficient numbers of individuals throughout their entire lifespan, especially long-lived organisms. Deevey (1947) was able to construct an approximate survivorship curve for mountain sheep in Mt. McKinley Park, Alaska, using data collected by Murie (1944). Murie saved all the mountain sheep skulls with horns that he found in the field and then determined the age at death of each individual by the growth rings on the horns.

Types of Survivorship Curves

An infinite number of survivorship curves are possible; however, they can be categorized into three types that reflect major differences in the timing and severity of mortality relative to the length of the life cycle (Fig. 9-16). An unlikely survivorship schedule (curve a) is for each individual to live to an almost identical age. Presumably this schedule reflects either complete, catastrophic disaster or the inherent longevity of individuals in a population influenced by no outside mortality factors. Pearl (1928) was able to produce this type of survivorship by placing newly emerged adult *Drosophila* (emergence is then time zero) in vials containing no food. The ability of each adult to withstand starvation was about equal and the population died nearly synchronously.

The inverse of curve a is curve c in which nearly the entire population dies at a very early age. This is probably a very common occurrence in nature, especially among organisms that produce a large number of young and provide them with essentially no protection from the environment. The curve might reflect the mortality of the offspring of a colonizing species in which the probability of finding a suitable

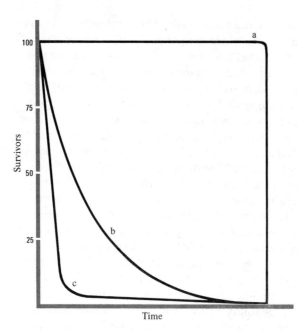

habitat is very low. For example, many marine invertebrates produce numerous small larvae that form part of the floating plankton. The larvae eventually settle on a solid substrate to mature. Mortality by predators that feed on plankton coupled with the low probability of finding a suitable substrate for the adult yield a very high mortality in the early stages of the life cycle and a very short life expectancy for a newly hatched young.

Survivorship curve b represents a constant *proportion* of the living individuals dying during each time interval. Death rate is independent of age because the probability of dying during any time interval is the same for all individuals alive at the beginning of each interval. Curve b can be transformed into a linear function by changing the Y-axis to logarithms. Many bird populations approximate type b curves for adult mortality (Fig. 9-17).

As will be demonstrated later the survivorship curve is one factor fixing the population r. The other factor is the age specific birth rates discussed in the next section. Selection pressures that influence the relative fitness of an individual through changing r values potentially will have an important impact on the survivorship curve. The coordinated relationship between the survivorship curve and the birth schedule curve in determining r also tends to balance, evolutionarily, one age-dependent function against the other. The survivorship curve then is an evolutionary product for each species that reflects the composite pressures on r and the age specific birth and death rates. Colonizing species will be forced into extensive birth efforts to insure that some offspring reach newly opened habitats. Species faced with strong pressures from other individuals to occupy habitat space as it is vacated may encounter strong selective pressures to retain their position in the community (that

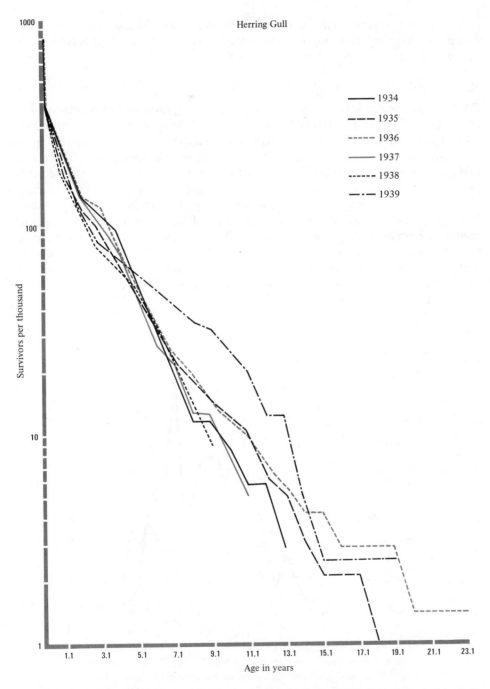

FIGURE 9-17 Survivorship curves for Kent Island, Maine, herring gulls hatched in 1934–1939. Data obtained from returns of numbered aluminum bands attached to gull's legs as chicks. (After Paynter, 1966.)

is, to reduce the probability of mortality of individuals already present) at the expense perhaps of producing new individuals. We shall return to this contrast of selective situations later.

Birth Schedules

Data for age specific birth rates (usually designated m_x) similar to the survivorship curves can be obtained by recording the reproductive output of known age individuals, females in bisexual species. For many species the curve will closely approximate that shown for *Oncopeltus fasciatus,* the milkweed bug, in Figure 9-18. There is a variable prereproductive period followed by a fairly abrupt increase in the birth rate per individual. The peak is then followed by a more or less gradual decline in the reproductive output to the age at which reproduction is no longer possible. For species that reproduce only once, the shape of the curve will depend on synchrony of reproduction among individuals. The offspring output often will increase with increasing adult size for species, such as plants and many heterothermic animals, that may continue growing for much of their reproductive lifespan. Thus the age specific birth rates may be a continually increasing function of age rather than a peaked curve, as in the *Oncopeltus* example.

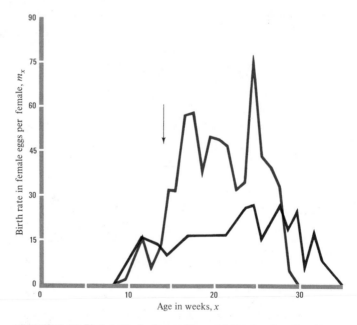

FIGURE 9-18 Birth rates in two cultures of *Oncopeltus fasciatus,* the milkweed bug, maintained at 23°C, with 16L-8D photoperiod. *Dark line:* population initially contained 20 adult pairs; these were not replaced as mortality occurred. *Colored line:* population initially contained 20 adult pairs and then the adult population was reduced by one half at age 14 weeks (arrow). Note higher birth rate indicated by colored line is accompanied by a shorter reproductive span. (After Dingle, 1968.)

Maintaining a constant mortality schedule and a constant birth rate ($r \geq 0$) eventually leads to a fixed proportional distribution of individuals among the various age classes. The distribution for such a population that is increasing ($r > 0$) eventually approaches what is called the *stable* age distribution. For populations that maintain a constant size ($r = 0$) the proportional age distribution is usually called a *stationary* age distribution. Each distribution will depend on the mortality schedule and the age specific birth rate. Since $r = 0$ for a population with a stationary age distribution, the population will not be increasing and the percent of individuals in each age class should be indicative of the mortality rate associated with each age class.

Significant changes in either schedule will produce a marked change in the age distribution. For example, a baby boom in the United States, such as at the end of World War II, will provide an exceptionally large base for the age distribution that will move up the distribution through time as a proportional bulge. When these babies reach maturity, they probably will generate another bulge at the bottom of the distribution unless some other factor reduces their reproductive output.

Age distribution graphs can be used to estimate population growth characteristics, assuming that the mortality and reproductive characteristics of the population remain essentially unchanged, at least over the study period. For expanding populations the relative number of young should be high, whereas declining populations should have a relatively large proportion of older individuals (Fig. 9-19). A stable population will have some intermediate shape that depends on the length of pre- and post-reproductive lifespans and the time units for the distribution (Fig. 9-19).

CALCULATING r_m

Under specified environmental conditions that remain constant long enough for a stable age distribution to develop, age specific mortality and age specific birth rates can be defined for a single population. With these data it is then possible to calculate r_m (Birch, 1948). Specifically once the l_x and m_x values are known for a population,

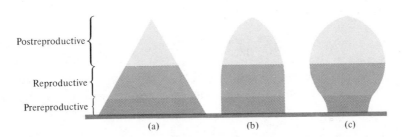

FIGURE 9-19 Comparisons of age structure in populations showing three different growth patterns. Population a with many prereproductives and reproductives should continue to expand in size; population b should remain stable in size; population c, with mostly individuals past reproductive age, should continue to decline in size. (After Kormondy, 1969.)

the equation for calculating r_m can be obtained in the following manner. Assume that the population is growing in an unlimited environment. Under these conditions the offspring of any particular generation will be estimated from equation (9-6)

$$N_t = N_0 e^{r_m t}$$

where N_t is the total population of offspring after time t. For one generation, the females of which all produce offspring at the same instant, the total number of offspring will depend on the total number of surviving females (l_x) and their reproductive potential (m_x). To estimate the number of adults it is necessary to know how many were in that generation when they were born. The adults alive at some later time (that is, the time of the present reproduction) can be estimated from the age specific mortality data. From a given number of newborn, N_0, where N_0 is the number of females born into a *single* generation at some earlier time, the total that survive to age x will be $N_0 l_x$. And the number of offspring produced, N_t, is

$$N_t = N_0 l_x m_x \qquad (9\text{-}11)$$

Similar calculations can be made for each generation represented in the population by changing the value of x in the equation. The total number of offspring produced at a particular breeding period will be the sum of the contributions by each generation represented at the time:

$$N_t = \sum_{x=1}^{\infty} N_0 l_x m_x \qquad (9\text{-}12)$$

The equation can then be divided on both sides by N_t to give

$$1 = \sum_{x=1}^{\infty} (N_0/N_t) l_x m_x \qquad (9\text{-}13)$$

The N_0/N_t term is the same as N_0/N_t from equation (9-5) if each generation is growing in an unlimited environment. By assuming that the time interval, t, is equal to age x and slightly rearranging equation (9-5) to

$$N_t/N_0 = e^{r_m x}$$

and finally to

$$N_0/N_t = 1/e^{r_m x}$$
$$= e^{-r_m x}$$

we can substitute back into equation (9-13) to arrive at an equation predicting the total reproductive output of a seasonally breeding species with overlapping generations,

$$1 = \sum_{x=1}^{\infty} e^{-r_m x} l_x m_x \qquad (9\text{-}14)$$

The usual calculating procedure is to try successive values of r_m in the equation until arriving at the closest approximation possible from the data.

From this formula for calculating r_m it is easy to see that both age specific mortality rates and age specific birth rates play important roles in determining the rate at which a population is increasing (or decreasing) under specified (and constant) environmental conditions. Since the two factors are in part related to each other, it is evident that the resultant r_m value for an exponentially growing population will reflect an evolutionary balance between the two schedules.

219

ENVIRONMENTAL
MODIFICATION
OF THE
EXPRESSION
OF r_m

ENVIRONMENTAL MODIFICATION OF THE EXPRESSION OF r_m

There are three nonexclusive modifications of reproductive output of a population on a per individual per time basis (r) that can be produced by the environment. The r value can be changed by changing (a) the number of offspring per time interval, (b) the mortality rate of the offspring or the adults (or both), or (c) the developmental rate of the offspring to produce a change in the age at which reproduction starts in the next generation. Each type of modification occurs in some natural populations. Any environmental factor (see Chapter 2) may influence the r value of a population, and probably most factors have some influence on one population or another. However, certain aspects of the environment are often implicated as affecting r. These include the availability of energy for the production and care of offspring, the role of timing cues in developmental processes, and the effectiveness of antimortality adaptations.

Physical factors encompass a wide variety of environmental conditions with diverse effects on the reproductive output of an organism, primarily by influencing the rates of energy uptake and paths of energy channeling. Temperature, light, moisture, and other climatic factors are of major significance in the biology of most organisms, but the degree of importance depends to a large extent on other properties of the environment. Thus water is probably a major limiting factor for plant growth in a desert, whereas in high mountain areas temperature is probably of critical importance. The importance of the various factors can also vary to some extent depending on how they influence the biology of the organisms. For instance, temperature has a major effect on the energy expenditure and general activity levels of most organisms. A relatively few species of organisms, the birds and mammals, have evolved homeothermy which helps compensate for temperature fluctuations in the environment. However, this evolutionary adjustment is expensive energetically, at least when the ambient temperature drops low enough to require an energy increase above minimum levels (see Chapter 6). For heterothermic organisms temperature variations in the environment potentially change the internal temperature causing changes in rates of enzymatic reactions, general activity levels, and presumably influencing the rate of uptake of additional energy.

Temperature is an environmental variable that is not precisely predictable. Light, on the other hand, except for short-term changes produced by cloud cover and taller organisms, is remarkably predictable in its occurrence from year to year. There are no out-of-season long days as there may be out-of-season high temperatures. The inexorable course of the changing photoperiod in the temperate zones is an important proximate factor in timing reproductive seasons for many species (Marshall, 1961). Light is essential to many organisms, both diurnal and nocturnal, in determining the length of the activity periods. Lack (1968) suggested that birds in the northern temperate regions can raise more young than similar tropical forms because they have more daylight hours to feed the young. For growing plants the longer availability of incident radiant energy provides a potentially higher daily gross productivity than in the tropical regions. At higher elevations less of the radiant energy is filtered out in the atmosphere before it reaches the plants and animals, thus providing a potentially higher gross productivity per hour than at the same latitude at lower altitudes.

Assuming that a limited amount of energy is available, the distribution of energy into various reproductive and maintenance activities will have significant effects on the timing and extent of the reproductive output per individual. The amount of energy finally channeled into actual reproduction can be divided in a variety of ways among the offspring. For some organisms the energy will be divided among a few relatively large offspring and for others among a very large number of small offspring. Also if the individual practices parental care, energy expenditure is required.

The first critical factor in reproductive ability is the total amount of energy that can be channeled into offspring. In years when the supply of the small rodents used for food by the tawny owl (*Strix aluco*) is low, the owl produces fewer eggs than in years of high populations of rodents (Table 9-2). Similarly, Perrins (1965) concluded that availability of food for the female great tit (*Parus major*) was an important factor in determining the number of eggs she would lay.

For many organisms parental input to insure a genetic contribution to the next

TABLE 9-2 Breeding of Tawny Owl, *Strix aluco,* in Wytham woods, Oxford, England

Year	Number of pairs	Percent pairs breeding	Mean clutch	Number of young fledged	Young fledged per pair	Abundance of rodents on arbitrary scale
1947	16	69	(2.5)	20	1.3	?
1948	20	65	(2.0)	19	1.0	?
1949	20	90	2.8	25	1.3	5
1950	21	81	2.7	27	1.3	6
1951	22	50	2.0	6	0.3	3
1952	24	70	2.6	20	0.8	5
1953	25	60	2.1	19	0.8	4
1954	26	69	2.4	16	0.6	5
1955	27	15	(2.0)	4	0.1	2
1956	29	79	2.2	23	0.8	6
1957	30	60	3.0	20	0.7	7
1958	30	0	–	0	0	–
1959	30	–	–	29	1.0	8

Based on Southern (1959). The mean clutch was based on only a sample of the nests (at least 11 each year, except for only two in each year where the average has been placed in parentheses). The density of rodents, *Apodemus sylvaticus* and *Clethrionomys glareolus,* is scored on a relative scale from eight (extremely abundant) to one (extremely scarce), but the intervals between successive numbers do not necessarily correspond to equal differences in density. (Lack, 1966.)

breeding generation ends with production of offspring. For some organisms, however, the adults provide further care of the offspring, either by feeding or by protecting (or both), to increase the young's probability of survival. This further energy expenditure must be included in calculating the total energy that is channeled into reproduction; in many cases, however, it can be acquired while the young are present, thus potentially reducing the energy required for actual production of young. For example, human parents continually acquire energy while raising their children. In periods of limited energy availability this places the parent in a situation in which the probability of survival of one or both generations is decreased.

221

ENVIRONMENTAL
MODIFICATION
OF THE
EXPRESSION
OF r_m

David Lack (1954, 1968) has generalized this reasoning to the hypothesis that the average clutch size of most birds with parental feeding responsibilities probably is related to the ability of the adults to feed the young or to the availability of energy for egg production (or to both). In birds there are at least two ways to adjust the number of young that are being fed in the nest relative to the available supply of food. First, the number can be adjusted at the time of egg production—either through limitations on energy available to the female to produce eggs or by genetic limitations on egg number resulting from long-term trends in ability to raise young (see Chapter 3). The second time of adjustment can come after the young hatch. At this stage the number of young that can be successfully reared by birds practicing parental care will depend on the availability of food. Brood reduction is a term given to the phenomenon of starvation of younger members of a brood due to the inability of the parents to feed all young in the nest (Ricklefs, 1965). Generally this reduction occurs among young that hatch asynchronously such that the first hatched, and hence bigger, young get the food first and if there is sufficient food, then the smaller young are fed. Another possible method of "brood reduction" is the cannibalism among nest mates practiced in some predatory birds. In either case the number of young produced is relatively large and then is adjusted to the available food supply through selective mortality. In years of high food supply the reproduction of the adults is not fixed by a genetic limitation on egg output and in bad years some young survive at the expense of the smaller ones. Not all birds have evolved brood reduction and it seems likely that the method of adjustment of offspring output depends in part on how predictable environmental conditions will be at various stages in the nesting cycle. If the usual limitation on energy is early in the cycle, the major impact may be felt at the time of egg production. If food for the young is usually unpredictable, the major adjustment may be during the nestling stage. As will be discussed later it is also critical to keep in mind the life expectancy of the adult as an important component in the evolution of reproductive rates. Reproduction by an adult in future years at the expense of high current reproduction may be advantageous, but this partly will depend on the probability of survival of the adult to breed again.

Energy is also required to develop and maintain the adult organism, at least through the time of reproduction. Ultimately the amount of energy that can be channeled into maintenance for the adult, into offspring production, and into parental care is competing within the individual for the finite energy supply accumulated by the individual. The division of this energy supply will determine the probability of survival of both the adult and the offspring and consequently will determine the reproductive effort of the adult. Selection presumably will optimize energy partitioning among the categories in such a way as to maximize, evolutionarily, the genetic contribution of individuals to the next generation.

Dingle (1968) reported a series of experiments designed to describe the impact of three environmental features on the ability of populations of the milkweed bug to reproduce. He tried to make all other conditions optimum, including providing more than sufficient food. The three factors were (a) temperature, (b) photoperiod, and (c) density of adults. Each of these factors has some influence on the life cycle of the milkweed bug and produces predictable changes in the r value of the population. An interesting outcome of the study, and one that will be discussed later in the chapter, is that the populations with the most offspring per adult female were not always those with the largest rate of growth—that is, the largest r value.

Table 9-3 presents the results in the several sets of environmental conditions. The highest values of r for this species are achieved under relatively high temperatures, relatively long photoperiods, and at the lower of the two experimental densities of adults. Of the three experiments run at this density the one yielding the highest r value also was the one that had the fewest number of daughter eggs produced per female egg in the previous generation (R_0). The low R_0 value could reflect either low egg output per female or relatively high mortality of females in the adult generation from the egg stage to maturity. Figure 9-20 shows the smoothed survival (l_x) curves for various experimental conditions; clearly the survival curve is steepest (earliest average mortality) for the experimental group that produces the largest r value. There obviously must be a compensating mechanism in the output of eggs per living female. The population with the highest r value has the highest peak output of eggs of any of the populations, but only slightly higher than the one with the next highest r value (Fig. 9-21). This peak occurs also early in the potential lifespan of the population. The high mortality of this population seems to be associated with the peak of reproduction.

Dingle thought that each population showed a positive relationship between increased reproduction and increased mortality, which might be related to the increased energy expenditures required by the females to reproduce. Thus increased reproduction for these bugs at any point in time is accompanied by a decreased probability of survival through the next unit of time compared to the same time

TABLE 9-3 Rates of increase and doubling time for populations of *Oncopeltus fasciatus* under several sets of environmental conditions

Photoperiod[a]	Temperature (°C)	Initial density	Increase per day (r)	Doubling time (days)
16L-8D	27	10 pairs/box	0.0861	8.42
16L-8D	27	20 pairs/box	0.0810	8.90
16L-8D	23	10 pairs/box	0.0736	9.75
12L-12D	27	20 pairs/box	0.0593	12.03
16L-8D	23	20 pairs/box	0.0499	14.24
12L-12D	23	20 pairs/box	0.0369	19.13

[a]L = hours of light; D = hours of dark
Data from Dingle, 1968

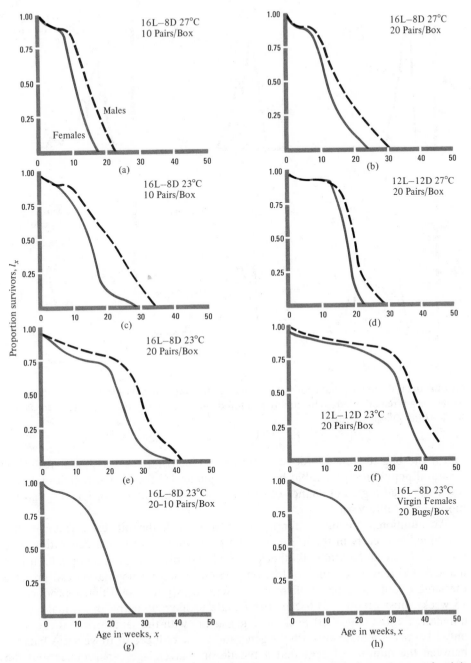

FIGURE 9-20 Survivorship curves (male, dotted; female, color) for cultures of adult *Oncopeltus* under a variety of experimental conditions. Age zero indicates date of hatching. Mortality of adults caused decline in population sizes. Eggs were removed at two-to-three-day intervals. Survivorship prior to adult life stage was determined in separate experiments. In *g* the density was halved at 14 weeks; in *h* only virgin females were introduced. (After Dingle, 1968.)

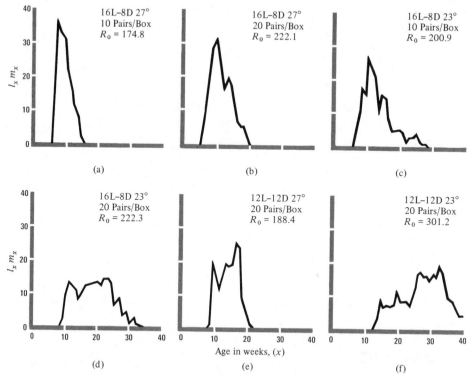

FIGURE 9-21 Combined data for probablility of survival and average egg production per female as a function of age. The cultures pictured here are the same as those used for Figure 9-20. (After Dingle, 1968.)

unit if the bug did not reproduce. Murdoch (1966*a*) reported a similar result for a natural population of carabid beetles in England. He found that females that lay fewer eggs had a greater chance of surviving over the winter than females that lay more eggs (Table 9-4).

An additional advantage for the population with the highest r value is that reproduction is early in the lifespan of the population; that is, the generation time has been reduced. Holding all other potential variables constant, a population that matures and reproduces in two-thirds the time of a second population can produce the same r value with fewer offspring per female. A population that reproduces in 15 weeks must produce 31.6 eggs per female to have as many individuals in the population at the end of 30 weeks as one starting with the same size that produces only 10 eggs per female but with the generation time reduced to 10 weeks. A balance between the number of eggs that a female of a given age can produce and the probability of survival to that age presumably will be reflected in the timing of reproduction of a population.

Since evolution is dependent on the differential contribution of genotypes to the next breeding generation, the maximal reproductive output should reflect some

TABLE 9-4 Relation between reproductive output (or breeding status) and the probability of survival over the next winter for several species of carabid beetles in England.

Type of population	Species	Population number	Breeding status or mean mature egg number	Estimated survival (%) May-winter
Experimental	A. fuliginosum	–	Breeding	36
		–	Nonbreeding	72
Natural	A. fuliginosum	1	3.2 ± 0.6 (May–June)	25
		2	6.0 ± 0.6 (May–June)	6
	A. fuliginosum	2	6.2 ± 0.5 (May–July)	6
		3	3.7 ± 0.6 (May–July)	48
	A. thoreyi	2	4.7 ± 1.3	14
		3	0.9 ± 0.5	30–60

Murdoch, 1966

compromise with the environmental conditions the population normally encounters. In the milkweed bugs Dingle noted that this migratory species moved into the temperate zone in the spring when the temperatures are fairly high, photoperiod is long and still increasing, and the density of insects, especially other milkweed bugs, is lower. Under these conditions a female presumably should reproduce as fast as possible so that her progeny will help fill the nearly empty ecological space. However, as fall approaches with decreasing temperatures, shorter and decreasing photoperiods, and higher densities of milkweed bugs, it is advantageous for an adult to increase its chances of survival for a longer period to enable it to migrate to a more equitable climate. In the spring the females expend energy for eggs to insure maximal reproductive rates, whereas in the fall the energy output for reproduction is decreased; this increases the probability of an adult surviving and migrating to where it can reproduce young that will also survive. Dingle found that the reproductive responses of the bugs to his experimental conditions were generally an attempt to maximize reproductive output in view of the normal seasonal events but that this maximization need not always derive from laying the most eggs.

The conclusion is that an individual's reproductive output is an evolutionary balance between effort expended to maintain the organism and effort to reproduce. The optimal strategy maximizes the reproductive potential of the individual in relation to environmental changes.

EVOLUTIONARY FACTORS INFLUENCING r_m, AVERAGE MAXIMUM POTENTIAL INDIVIDUAL REPRODUCTIVE OUTPUT

Cole (1954a) has examined from a theoretical view the question of how changes in the life history of individuals may affect their genetic capability to reproduce. He first simplified the problem by assuming that many of the variables we have just discussed are, in fact, constants. He held offspring production per unit time constant and also assumed that mortality was not important. These two assumptions then allowed Cole to examine the effects of other changes in the biology of an organism on its maximum potential reproductive output r_m.

Cole first asked why some species are single brooded and some reproduce more than once. Although reproducing more than once is equivalent to reproducing for infinite time according to Cole's assumption regarding mortality, it will become clear that a very few repeated reproductions approximate to infinitely prolonged reproduction. Cole calculated the gain in r_m that an individual might acquire by becoming *iteroparous,* or multiple brooded, as opposed to *semelparous,* or single brooded. He concluded that the gain for semelparous species that breed annually would be equivalent to adding a single individual to the brood of the semelparous female. In other words, according to Cole an annual plant that suddenly becomes a perennial does no more to its r_m value than it could by adding a single offspring to its total output as an annual. In most climates it would probably be easier to increase the annual reproductive output by one than to make physiological adjustments of sufficient magnitude to change from an annual to a perennial. This would depend in part, of course, on the number of young that the annual produced. A change from 1000 to 1001 offspring is proportionately much less and presumably would require fewer total adjustments than a change from one to two offspring. Thus the relative efficiency of the two possible changes would depend on the physiological adjustments required for surviving the nonreproductive periods or for producing an extra offspring. A further consideration that must always be included is the ability of selection to accomplish specified changes of current genetic properties.

Gadgil and Bossert (1970) modified Cole's assumption to include some mortality during the first year. They then found that the effect on the reproductive output of changing from an annual to a perennial was not equivalent to adding a single individual to the brood. On the assumption that each reproductive individual exactly replaces itself from generation to generation (that is, the population is stable), changing to a perennial would approximate to doubling the size of the brood. Thus for a female producing relatively large numbers of offspring, with only one surviving to reproduce the next year, the fact that the female has survived is more valuable by her addition to the next breeding generation than producing a single additional offspring that probably will not survive. Although the relative addition remains about two times for increasing offspring output, the value of a reproducing female relative to one of her offspring obviously increases substantially as the average number of offspring she produces as an annual increases. Bryant (1971) noted that a variety of effects of iteroparity are possible depending on assumptions related to mortality rates and constancy of brood sizes. Thus if the female more than replaces herself (as undoubtedly happens in some insects in the north temperate zone), her value relative to an additional offspring declines from the peak of two times the brood size. (It is possible in a declining population that the female is actually worth more than two times the brood size.) The range of relative values for iteroparity as opposed to semelparity is limited, but within these limits virtually any value is possible depending on the assumptions about dN/dt, mortality rates and timing, costs of physiological adjustments versus brood size adjustment, and so on.

Environmental conditions that produce a moderately high probability that the offspring produced in any one year will not survive to maturity should produce

selection pressures to increase the chances that a female will reproduce more than
once. This is the extreme condition of the increasing mortality investigated by Gadgil
and Bossert. In this case the potential increase in r_m is infinite if no young survive
and the female becomes iteroparous. The actual selective value will decrease as the
probability of successful reproduction in a year increases. This probably has been
an important selective pressure for iteroparity in many organisms (Murphy, 1968).

The gain in r_m achieved in going from an annual breeder to an iteroparous
individual was predicted on the assumption that all individuals started breeding at
the end of one time unit and then the iteroparous individual continued to reproduce
once every subsequent time unit. Thus the generation time for an annual and the
interval between breedings for an iteroparous individual were identical. A further
factor in the evolution of reproductive rates is that some species have a much longer
generation time than the time between repeated reproductions, even if it is assumed
that subsequent reproductions occur at regular, unitary intervals. For example, a
human female could reproduce once a year after reaching maturity at ages 10 to
13. Figure 9-22 shows percent gain in r_m from changing to an iteroparous habit from
semelparous as a function of the age of initial reproduction (called alpha or α) and
brood size. The percent gain in r_m is defined as

$$(r_{m_o} - r_1) \times 100/r_1 \tag{9-15}$$

where r_{m_o} is the r_m value of an interoparous individual and r_1 is the value for a
semelparous one.

The relative gain by becoming iteroparous is highest for small brood sizes and
for individuals that otherwise begin to breed at a relatively old age. It is evident
from the kind of gain to be expected from Cole's data comparing annual to perennial
individuals that the relative gain declines as the brood size of the semelparous
individual increases. The age-related gain reflects the total number of reproductive
efforts that can be achieved by a single individual in the time required for the
offspring of a semelparous individual to achieve sexual maturity. The relationship
defined by Cole appears to be more or less asymptotic so that if initial reproduction
is delayed to an old enough age, there is little further gain to be achieved by
becoming iteroparous as compared to individuals that begin reproducing at a slightly
younger age. Hence an individual with a constant brood size and a high age of first
reproduction that became iteroparous would gain no more than a semelparous
individual that started breeding somewhat earlier. However, it also means that an
individual that becomes iteroparous and has a high age of first reproduction (that
is, well out on the asymptotic portion of the curve) would not benefit much from
decreasing the age at which it reproduces the first time.

A salmon that produces a relatively large number of eggs, at the expense of
continued survival, at age four years would gain relatively little by becoming iteroparous, according to Cole's graph. A major gain might be made by decreasing the
age of reproduction to three years, but this must be balanced against the decreased
numbers of eggs that a slightly smaller-sized individual can produce as the result
of having one less year to grow to maturity. Since Cole's assumption of no mortality

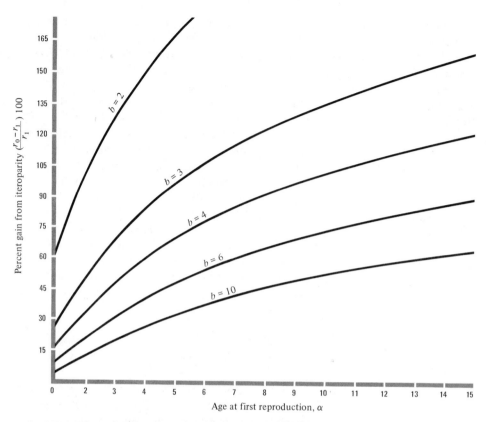

Percent gain from iteroparity $\left(\dfrac{r_0 - r_1}{r_1}\right) 100$

Age at first reproduction, α

FIGURE 9-22 The percent gain in intrinsic rate of increase attainable by an individual changing from a single-brooded habit to reproducing indefinitely. The gain is plotted as a function of brood size and the age of first reproduction. Brood size is the number of female offspring per brood (usually one-half of brood size). (After Cole, 1954a.)

is obviously false for salmon, the selective pressures on timing and extent of reproduction also include the probability that an individual will be able to survive to the following year if it delays reproducing.

This suggests an important relation between brood size and the age of first reproduction (alpha or α) in determining r_m (Fig. 9-23). The disparity between the r_m values of individuals producing large and small broods declines as the age of first reproduction increases. The delay in the onset of reproduction tends to reduce the importance of brood size in determining the reproductive contribution of a female. At high alpha the gain achieved by decreasing alpha one unit is minimal over a wide range of brood sizes, at least to an order of magnitude change in brood sizes. When alpha is low, however, the relative impact of decreasing this age and increasing brood size is again about equal, but the total impact is much greater than for high alpha.

Another way to view the relative advantages of increasing the age of first repro-

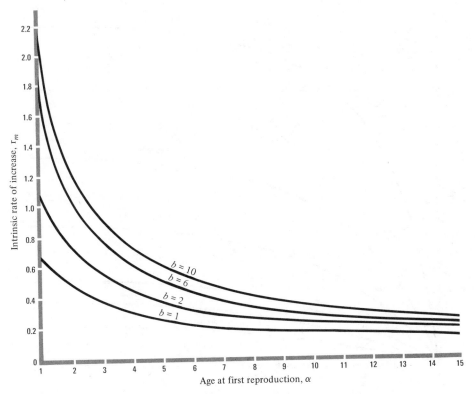

FIGURE 9-23 The relation between intrinsic rate of increase and the age of first reproduction for several brood sizes among iteroparous breeders. (After Cole, 1954a.)

duction is by examining the impact of alpha on r_m among individuals in a series of species that share all other traits for reproductive output. Cole chose species that have a single young at each reproduction, produced once a year at a more or less synchronous annual breeding period. He then could relate the r_m value for an individual female to the total number of young produced. He equated this with the number of annual births to establish the relation between total reproductive effort and the maximal capacity to increase.

Species that begin reproducing earlier have higher r_m values for increasing number of young produced (Fig. 9-24). The r_m of each species rises to a plateau at about seven to ten annual births. The initial rise in r_m for species with higher plateaus is much faster, thus making the first few annual births extremely important. For a species such as the passenger pigeon that may have an average reproductive span of about four to five years and had a relatively stable population, any major environmental factor that reduces the average annual births by one to two per female would have a marked effect on the ability of the population to maintain its numbers. It is theoretically possible that the extinction of the passenger pigeon was caused by the impact of man on the average reproductive lifespan of female passenger

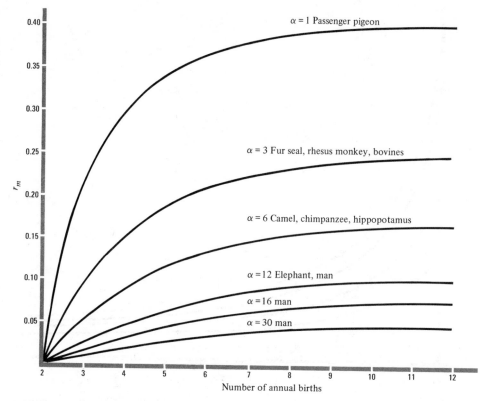

FIGURE 9-24 The relation between intrinsic rate of increase and the number of births for species that have a single young (effective brood size = one-half) each year after reproduction starts. The lines indicate the effect of the age of first reproduction. Notice that each line approaches an asymptote after about seven annual births. (After Cole, 1954a.)

pigeons, thus sharply reducing r_m and hence the growth potential of the population.

In order to ask similar questions for a single species Cole calculated theoretical values for humans at varying alphas and varying numbers of annual births and plotted these along isopleths of r_m (Fig. 9-25). This graph illustrates the conclusion that two mechanisms for reducing the maximal capacity for increase of the human female are by (a) decreasing the number of annual births (that is, the total number of young a female produces) and (b) increasing the age of first reproduction. The latter mechanism presumably lacks the overtones of governmental control that might be associated with the first alternative.

REPRODUCTIVE STRATEGIES

These theoretical arguments by Cole and others bring the discussion of reproductive rates full circle to the importance of energy, mortality, and developmental rate on the capacity of a female to contribute to population growth. The age of first repro-

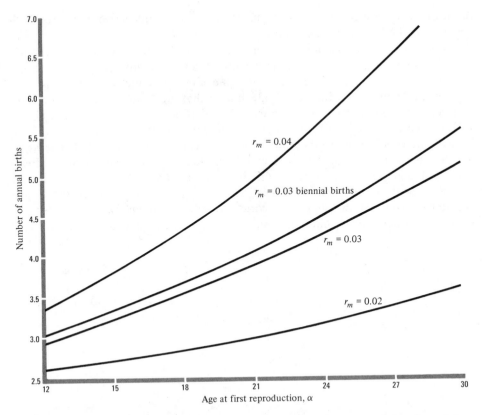

FIGURE 9-25 Relation between the number of annual births for a human female and the age at which she must start reproducing to achieve a specified intrinsic rate of increase. Notice that delayed onset of reproduction reduces the affect of large numbers of children on the population growth rate. (After Cole, 1954a.)

duction, brood size, and the total number of reproductive efforts are major contributors to the r_m value. If the assumption of a limited energy supply is made, it is reasonable to conclude that the various methods of influencing r_m are directly affected by the relative amounts of the limited energy that are channeled to each characteristic influencing changes in r_m. To decrease alpha requires earlier maturation, which normally reflects an increased growth rate and hence more of the total energy put into an increased biomass. To decrease mortality requires energy expenditures for antipredator devices or for structural and physiological adaptations designed to combat other causes of mortality. Increasing brood size requires either a decreased total amount of energy per young or an increase in the total energy per reproduction. In the latter case the energy is derived at the expense of other possible uses, and the former distribution probably reduces the probability of survival of each offspring. Thus changes in r_m are a good example of what Cody (1966) refers to as the *principle of allocation;* that is, energy division among the various activities of an organism are reflections of balances between advantages and costs of each activity for producing changes in r_m.

Larger organisms are more protected from climatic fluctuations than smaller ones. However, these larger organisms have usually achieved this advantage by increasing the total energy cost and time required to achieve maturity and hence by decreasing the potential value of r_m. Bonner (1965) showed that the relation between body size and generation time was a positive exponential (Fig. 9-26). Earlier Smith (1954) had pointed out that the relation between generation time and r_m was negative on logarithmic axes. Taken together the relationships confirm the expectation that larger size has been accompanied by decreased reproductive capability. Pianka (1970) suggested that body sizes tend to fall into two broad categories approximately coincident with the division of animals between invertebrates and vertebrates; he concluded that these two major categories probably represent two mechanisms of dealing with a relatively unpredictable environment. In small organisms reduced

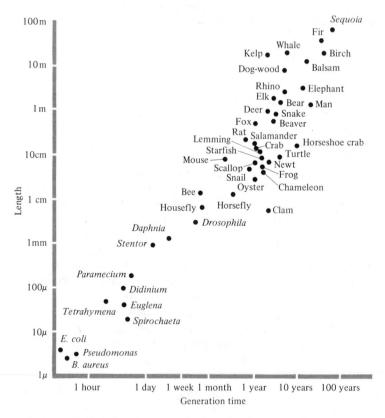

FIGURE 9-26 Relation between the length of an organism (estimate of total size) and the generation time for that organism. Data are on log-log axes. (After Bonner, 1965.)

probability of adult survival results in selection to produce large numbers of off-spring. In larger vertebrates reduced offspring production may be counteracted by increased probability of adult survival.

As the severity or unpredictability of the environment increases, more mainte-nance energy is required to compensate for environmental uncertainties; hence only those individuals that have made this energy demand will be able to maintain their populations at or near the carrying capacity of the environment over long periods of time. Species that use little energy for maintenance generally fluctuate in popula-tion size with environmental changes and will, on the average, occur in numbers below the carrying capacity. The average level of the population will depend largely on the balance between individual adjustment capabilities and the range of environ-mental fluctuations. For populations that regularly occur in or enter unfilled envi-ronments, such that the population is usually below the carrying capacity, consider-able energy usually is available to support new offspring. Additionally, colonizing species usually have a low probability of an individual reaching a suitable habi-tat. Under both conditions large numbers of offspring may be selectively advan-tageous. On the other hand species that are close to or at the carrying capacity for long periods of time probably will produce offspring that enter habitats in which there is relatively little energy not being exploited by the adults, provided the carrying capacity is determined by energy limits. In this situation the most successful offspring will be those that can effectively compete with other offspring and with adults for the available energy. To facilitate early growth and development of the young, which helps to insure high competitive ability, the parents may channel more energy into fewer offspring.

In this case the population regularly occurs at or near K and selection is strong for a relatively few, well-developed offspring that stand a good chance of surviving. For species that regularly occur at population levels below K, selection often will operate to increase reproductive rates. Hairston et al. (1970) called these two types of selection death and birth selection, respectively. Birth selection increases the reproductive output to fill normally empty ecological space. Death selection operates to reduce the probability of mortality of the offspring produced. The dichotomy is not perfect, but the two types represent the ends of a continuum of reproductive strategies. Species that expend much energy to reduce probability of individuals dying generally fall closer to the death selection extreme, and species that channel proportionately more energy into reproduction fall closer to the birth selection extreme (Gadgil and Solbrig, 1972).

REPRODUCTIVE STRATEGIES IN BIRDS

Cody (1966) attempted to explain the reproductive output of various bird species by employing the time and the energy compromise as a basis for the evolutionary differences in clutch sizes of tropical and temperate species and continental and island species. Although it is likely that a large number of possible factors influenced the spending of time and energy, Cody chose to concentrate on only three. His

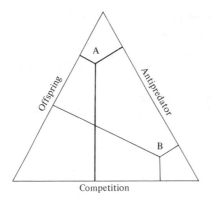

Competition

FIGURE 9-27 Two potential energy channeling strategies associated with reproduction. The length of a line from the central point is proportional to the energy spent in that activity. The sum of the three lines for an individual equals 100 percent. Individual A puts most of its energy into competitive efficiency and has little remaining for antipredator activities or for offspring production. Individual B uses most of its energy for offspring, sacrificing energy from competitive and antipredator activities.

arguments then could be visually represented on a three-dimensional graph (Fig. 9-27). He let one dimension be energy channeled into competitive interactions by the adults, the second was energy channeled into predator avoidance for adults and dependent young, and the third axis was energy channeled into egg production. The interior of the equilateral triangle in Figure 9-27 represents all possible combinations of energy expenditure for these three activities. Cody concluded that on islands the predator factor would be relatively unimportant because of a general reduction in the predators that reach islands compared to predators on the mainland. Moving values along the predator line to zero permits more energy expenditures for egg production and competitive interactions. On the average, islands also have fewer species than a similar-sized area on the mainland (MacArthur and Wilson, 1963, 1967), thus potentially reducing the variety (and presumably energetic requirements) of interspecific competitive interactions (although intraspecific might increase proportionately). These arguments led Cody to the conclusion that island birds should have higher clutch sizes than similar mainland forms. Unfortunately at present there are not sufficient data to test this conclusion adequately.

It is well known that clutch sizes in tropical birds tend to be somewhat smaller than the clutch sizes of closely related temperate zone forms (Cody, 1966). Even in tropical latitudes, however, species that occur in highly seasonal areas such as savannas tend to have larger clutches than species in evergreen forests, which are probably less seasonal (Snow and Snow, 1964; Lack and Moreau, 1965). From Cody's analysis the argument would follow that more seasonal tropical areas and the seasonal temperate zone tend to keep populations somewhat below the carrying capacity of the environment, so that the energy required for interspecific competition is somewhat reduced in comparison to less seasonal areas. At the same time many naturalists (Skutch, 1967) feel that predator pressure is somewhat higher in the tropics than the temperate zone, making the expenditure of energy for predator avoidance more costly in the less seasonal tropics. The increased costs of predator avoidance and interspecific competition associated with the less seasonal tropics then reduce the amount of energy available for egg production leading to the smaller clutches of tropical birds.

The theoretical occurrence of death and birth selection suggests that certain of the factors determining r_m might occur together more frequently than others. In other words there may be a finite series of reproductive strategies associated with certain techniques of habitat exploitation. Tinkle et al. (1970) examined this possibility for a series of lizard species for which reliable reproductive data were available. They found that there tended to be clusters of species within four distinct reproductive strategies (Fig. 9-28). Viviparity, the production of free-living offspring, was not a common strategy among the species nor was the strategy of large clutches associated with early reproduction and multiple broods. Both strategies require considerable expenditures of energy by the adult lizard and presumably reduce the probability that the adult will reproduce over a relatively long period. However, viviparity requires relatively large energy expenditures for few young. It is presumed that this strategy is effective principally in nonsevere climates and areas of low predator pressure. On the other hand the few species that reproduce early with large clutches

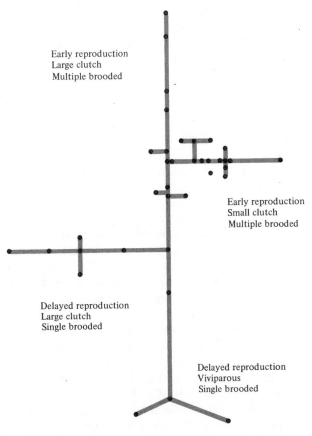

Early reproduction
Large clutch
Multiple brooded

Early reproduction
Small clutch
Multiple brooded

Delayed reproduction
Large clutch
Single brooded

Delayed reproduction
Viviparous
Single brooded

FIGURE 9-28 Clustering of reproductive strategies of 37 species of lizards as determined by an analysis technique known as Prim network analysis. Each dot represents a single species. (After Tinkle et al., 1970.)

and are multiple brooded must operate in areas where the probability of survival of adults and young is very low. Hence the optimal strategy for an individual that has reached reproductive maturity is to produce as many offspring as possible in the short time available.

Lizards differ in this regard from birds in that the lizard species in more equitable, tropical habitats generally reproduce early with relatively large clutches or are multiple brooded. In the few tropical bird species that have been studied in detail (for example, Snow, 1962) high adult survival is associated with delayed maturity in offspring and small clutches. In large part the difference probably reflects the environmental pressures that influence longevity in the two vertebrate classes. Adult survival is high in the birds and probably is very limited for the lizards. The mortality curves for the two classes in tropical environments probably approximate to a type c (Fig. 9-16) for the birds with high mortality of young and high survival of adults, whereas the lizards approximate to a type a or b (Fig. 9-16)—probably somewhere between the two—but with a short lifespan. The lizards would have reasonably similar survival of young in comparison to adult survival but continuing mortality pressure throughout the lifespan. The differences may reflect predator pressures. For birds the relatively helpless eggs and young may be much easier to find and "capture" than the adults, whereas adult lizards may be almost equally "catchable" as young and yet provide more energy per capture. The increased efficiency for a predator taking adult lizards as compared to young might produce selective predator pressure on the adults. So far there are too few studies of tropical populations of either birds or lizards to allow any clear-cut test of these ideas.

REPRODUCTIVE STRATEGIES IN HUMANS

The production of live young that have been nourished in the female's body by movement of food from her circulation is almost universal in mammals. This viviparous habit affords constant protection by the adult to the developing embryo. Through evolution adults that provide parental care can reduce the large, short-term input of energy required to produce an egg with a large yolk by nourishing the developing embryo throughout the prenatal period. However, viviparity reduces the number of young that can be produced simultaneously, with numerical limits set by the body size of the female and the need to retain a degree of mobility. This mobility requirement presumably has been an important reason that no birds have developed viviparity or ovoviviparity, although bats are viviparous.

Human infants have a relatively long-term dependence on the adult for food and protection. In part this reduces the short-term energy demands on the adult by spacing the demands over the seven or so years that are required before the relative independence of offspring. This dependence period may be important in assuring information transfer between generations by learning, thus reducing the amount of information that must be genetically coded for the offspring to survive.

Man differs from many other mammals only in the number of young produced at birth and the length of the dependent period. Whereas there are undoubtedly some size-related constraints on the number of young, these differences among mammal species may be partly a function of the distribution and nature of the food supply exploited by the species. A seasonal food supply, as might be expected for herbivores on the African savanna where man presumably evolved (see Chapters 1 and 20), would produce either strongly cycling population sizes or migrations to areas where the seasons are sufficiently different from the abandoned areas to assure the presence of food. The mass migrations of the grazing herbivores of the Serengeti Plain, Tanzania, is a clear example of these food-induced movements.

Animals that may have been dependent on these herbivores for food, such as early man, would be forced to move with the animals or switch to available stable food supplies, such as roots supplemented with fruits and whatever meat happened to be obtainable. This movement requirement coupled with the long period of dependency of the young reduce the number of equal-aged young that can be cared for simultaneously. The care requirement and the inability of a female to have many young simultaneously mean that one adult could care for the offspring of several adults, freeing the others for the work of providing food and shelter for the entire group. The gains from such cooperation may have led to establishing social units that are larger than the family unit.

REPRODUCTIVE STRATEGIES IN SEED PLANTS

Similar strategic analyses can be made for flowering plant reproduction (Salisbury, 1942; Harper et al., 1970; Baker, 1972) with the important proviso that most plant species will tend to be like planktonic animals in that dispersal and subsequent growth of the offspring in unpredictable conditions are important considerations in the balance of energy channeling. Baker (1972) has analyzed some reproductive characteristics of the California flora by measuring seed weights of taxa from distinct habitat types and with distinct life forms. He found that there was a significant increase in seed weight from herbs and grasses through shrubs and trees (Fig. 9-29). Similarly, there tended to be an increase in seed weight in plants from drier areas. In general, the average seed weight of plants decreased with increasing altitude. This trend was found not only in a range of species from different elevations but was also present in altitudinal populations of the genus *Penstemon* (Fig. 9-30). The trend within *Penstemon* represents a quite different strategy from that reported by Johnson and Cook (1968) for *Ranunculus* in Oregon. These authors found that the average seed weight of *Ranunculus* remained constant with increasing altitude but that the number of seeds decreased. In *Ranunculus* and many other species of plants there seems to be a strong correlation between total reproductive output and the amount of energy that is stored by the plant. Thus for *Ranunculus* there was a good correlation between number of seeds (of equal weight per seed at all elevations) and the length of the growing season (Fig. 9-31).

The total amount of energy that can be transformed into reproductive output is partly a function of the total photosynthetic capability of the plants, which in turn is related to a number of environmental factors, including length of growing season, light input, and average temperatures. The final distribution of energy in the plant will reflect the energy requirements of the plant to maintain itself to reach

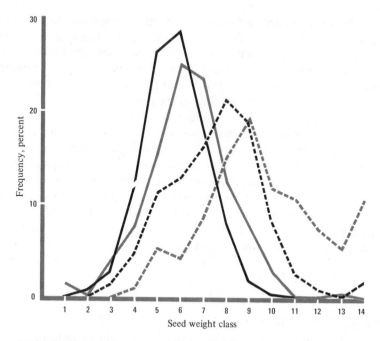

FIGURE 9-29 Frequency distribution of seed weights in the California flora. Data on actual weights are transformed to categories of increasing weight. Each category has a range of weights that is three times the range of the next previous category (for example, category 2 ranges from 0.003 to 0.009 mg and category 3 ranges from 0.010 to 0.031 mg). ———, annual herbs; ▬▬▬, perennial herbs; ▬ ▬ ▬, shrubs; ▭ ▭ ▭, trees. (After Baker, 1972.)

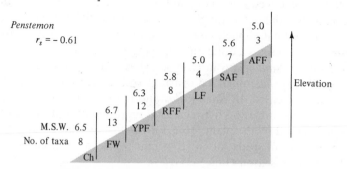

FIGURE 9-30 Changing seed size (M.S.W.) in the genus *Penstemon* with increasing elevation in California. Ch = Chaparral; FW = Foothill Woodland; YPF = Yellow-pine forest; RFF = Redfir forest; LF = Lodgepole-pine forest; SAF = Subalpine forest; AFF = Alpine fields. (After Baker, 1972.)

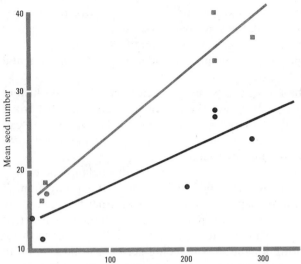

Maximum number of days between frost (growing season)

FIGURE 9-31 Seed number as a function of maximum number of days between frosts. Squares are means for cultivated plants; circles are means of field measurements. (After Johnson and Cook, 1968.)

reproductive maturity and the amount of energy that can be fixed by a mature plant to be stored in seeds and associated structures (Harper and Ogden, 1970). For a plant, as for most animals, the channeling of energy in the young plant probably has a profound effect on the amount of energy that can be fixed during the later reproductive stage.

Energy that becomes available for storage in seeds can be distributed into one or more of a number of compartments: seed coat, endosperm, dispersal adaptations such as a fleshy coat, sticky surfaces, and various substances that will reduce the probability of mortality of the seed or seedling (antipredator chemicals) or plant toxins that reduce competitive interactions with nearby plants (Whittaker and Feeny, 1971; Harper et al., 1970). The actual distribution of the energy among these potential storage compartments will reflect the selective pressures placed on the seeds and seedlings in the environment. This conclusion leads back to recognizing seed strategies for plants that are based on the floristic life form (tree, shrub, and herb) or the particular community type. Baker's analysis of these strategies in seed weights for the California flora represents an important step in understanding strategies of reproduction in flowering plants. It will be interesting to compare seed plant strategies over elevational and latitudinal gradients. However, in all organisms the analysis of reproductive strategies is really just beginning and presents a newly emerging area of ecology that deserves much attention.

GENERAL CONCLUSIONS

1. dN/dt = births − deaths.
2. K-responsive populations recognize K, the carrying capacity of the environment, and the dN/Ndt value in that the growth rate decreases as the population approaches K. The decrease may be fairly regular as in the logistic equation or there may be more or less threshold responses.

3. K-unresponsive populations usually show no relation or a positive relation between dN/Ndt and N. These populations are ones that rarely if ever encounter K, and probably have been selected against responding to K.

4. r_m is the maximum rate of increase of a population under specified environmental conditions.

5. Since r_m is genetic, it can be influenced by natural selection. Major life history aspects that influence the value of r_m are the age of first reproduction, brood size, and the number of broods produced in a time interval. The value of r_m can be calculated if age-specific natality and mortality rates can be measured for a population growing in an optimum environment.

6. The reproductive strategy of an organism depends on the balance of energy channeled for growth and maintenance of the individual and production of offspring. Energy channeled into offspring can be divided further between many or few offspring, and some can be used for parental care, including devices that help insure survival of offspring in the absence of the adult. More successful organisms leave more offspring for the next breeding generation. Success of an individual will depend on the amount of energy channeled into each use, relative to the particular set of environmental conditions the individual is exploiting.

SOME HYPOTHESES FOR TESTING

Much of the discussion of reproductive rates in this chapter depended on the value of K, the carrying capacity of the environment. Although this can be carefully controlled in laboratory situations, the value of K in any natural system is extremely difficult to measure, especially since it is unlikely to be constant for extended periods of time. Measuring may be difficult in nature, but it is nonetheless important that ecologists identify those parameters that are important in setting K (that is, limiting factors) and begin to estimate how they vary through time.

Once the important parameter(s) determining K are recognized, it is possible to ask questions about the role of K—and the probability of encountering K—in the evolution of reproductive rates and other biological properties of individuals and populations. For example, we suggested that threshold effects in the relation between dN/dt and N could occur if the limitation that a resource sets on population growth was manifested at certain stages in utilization of the resource. Furthermore we can begin to recognize the role that depletion of a resource has on future availability of that resource.

Finally, much of this chapter was devoted to recognizing and understanding various types of reproductive strategies. The possible divisions of energy among the various requirements for successful reproduction are nearly limitless and yet there appears to be a finite collection of actual strategies employed, at least among a series of lizard populations and some flowering plants. An extremely fruitful area for further work will be in describing and comparing reproductive strategies in various organisms, within and between habitat types, in an effort to understand the constraints placed on reproduction by certain classes of environments and to understand the constraints on energy channeling that are physiologically set within the organism.

predator-prey interactions

10

WHAT IS A PREDATOR?

In the broadest sense of the word predators are organisms that eat all or parts of other live organisms as an energy source. By this definition any consumer organism is a predator, and herbivores as well as carnivores are included. Here we will restrict the definition of *predator* to an animal whose feeding removes the prey individual from the population. Because of this removal, the prey taken no longer will deplete further the resources used by the prey population or contribute to their growth in numbers. Although grazing herbivores that do not destroy their "prey" require different theoretical treatment than is given in this chapter, they will be included in the discussion of predator strategies where useful. However, consumer organisms that move plant reproductive parts to new areas are not included. A fruit-eating bird, for instance, that disperses the seed of a plant is not a predator in the same sense as a beetle larva that destructively eats a seed. Parasites are not included because they usually have evolved a sufficiently close relation with their normal host to allow it to remain alive and continue to provide them with energy. *Parasitoids,* organisms that are parasitic but kill the host, albeit gradually, are, however, usually included in predator-prey discussions. The adult parasitoid locates the host or prey and lays an egg in or on the host. The larva then parasitizes and eventually kills the host. Thus different stages of the life cycle of the parasitoid are responsible for finding and killing the prey and the time lag from attack to death is usually considered as minimal in classical predator-prey interactions.

GENERAL MODELS

One model of the influence of predators on the growth of prey populations modifies the logistic equation (see Chapter 9) to account for losses of prey individuals relative to the population sizes of predator and prey:

$$dH/dt = r_mH(1 - H/K) - cPH \qquad (10\text{-}1)$$

where H is the number of prey individuals, K the carrying capacity for the prey population, P the number of predators, and c is the constant predation rate per predator. Similarly, the rate of change of the predator population is a function of its ability to capture prey and of its rate of increase per individual.

$$dP/dt = k_2PH - k_3P \qquad (10\text{-}2)$$

where k_2 is the ability of a single predator to transform prey into more predator individuals and k_3 is the death rate of predators in the absence of prey. According to equations (10-1) and (10-2) prey populations will increase fastest, per individual already present, when predator populations are low. Thus the growth rate per individual predator will be highest at peak prey density.

There always will be some value of $H > 0$, which depends on k_2 and k_3, needed to make $dP/dt > 0$. In other words predator populations cannot start growing until after the prey populations have grown to some threshold density. The peak growth rate per individual of predators occurs when H is greatest, but it can still be greater than zero when H begins to decline. Therefore predator populations will lag behind prey populations and eventually will decline drastically, generating inherent oscillations of predator and prey populations as in Figure 10-1. If the oscillations become sufficiently violent, which may occur if the predator increases at very low prey densities, then it is always true from the model that the predator will become extinct and that sometimes the prey will also become extinct. The only situation that could prevent the extinction of predator populations would be for some other factor(s) in the environment to damp the fluctuations, thus keeping the populations from crossing either axis (Fig. 10-1). However, the general predator-prey equations (10-1) and (10-2) contain only terms for predator and prey populations and their impact on each other.

In early experiments attempting to show the oscillatory nature of the predator-prey interaction Gause (1934) used protozoans in the hope that such simple organisms would reduce sufficiently the inherent time lags caused by the complexity of the organisms to permit continued oscillations (see Chapter 9 and discussion of modifications of the logistic equation for further information about possible time

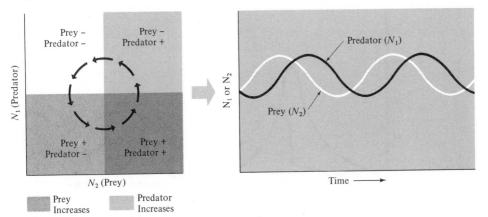

FIGURE 10-1 Relation between prey and predator density plotted against each other and through time. Based on a predicted result of equation (10-1). (After Wilson and Bossert, 1971.)

lags). When *Paramecium* was used as the prey and *Didinium* as the predators, and the relationship was allowed to develop unhindered, predator extinction always occurred (Fig. 10-2). Either the predators drove the prey to extinction or enough prey escaped from the predators to leave a reservoir of prey individuals after the extinction of the predators; the prey population then grew to its monoculture equilibrium level. Gause was only able to generate oscillations by periodically introducing new individuals into the experimental cultures to maintain both popula-

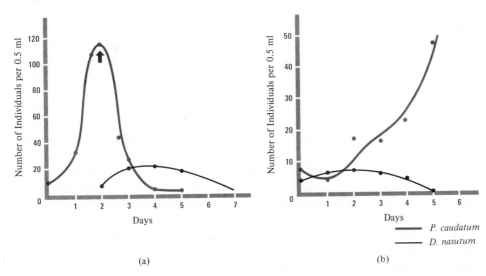

FIGURE 10-2 Two examples of the outcome of species interactions using *Paramecium caudatum* as the prey and *Didinium nasutum* as the predator. (After Gause, 1934.)

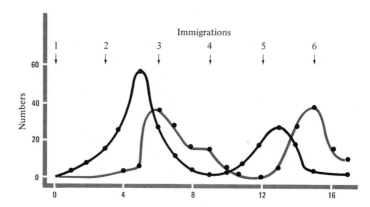

FIGURE 10-3 Changes in numbers of *Paramecium* (prey) and *Didinium* (predator) through time. Predator and prey individuals added at times indicated by arrows. Only by these repeated introductions (immigrations) was Gause able to establish the oscillations predicted by the simple predator-prey equations. Black line is the prey; colored line is the predator. (After Gause, 1934.)

tions (Fig. 10-3). Appropriate numbers and timing of the introductions would presumably allow the system to oscillate indefinitely.

Huffaker (1958) devised laboratory studies of predator-prey interactions using two species of mites, an herbivorous species preyed on by a carnivore. He found that in a simple experimental design the oscillations of predator and prey populations quickly caused dual extinction (Fig. 10-4). However, by increasing habitat complexity the probability of prey escape was sufficiently high to maintain three oscillations of numbers before extinction of the prey (Fig. 10-5). In a simple habitat, feeding areas for herbivorous mites were the surfaces of oranges placed uniformly on a board divided by a grid system. By covering various portions of the oranges, Huffaker regulated the total feeding area. He also made sure there were few or no barriers, other than space, to movement of the mites from orange to orange. The single oscillation of predator and prey populations produced in a simple habitat resulted from a delay in predator increase until after prey populations had reached a certain size. Subsequent destruction of the bulk of the prey population by the predator followed.

In the most complex habitat design, Huffaker introduced petroleum jelly barriers to dispersal and also severely restricted the available surface area per orange. In this case the prey population could escape to oranges not inhabited by predators and its population could increase somewhat before predators arrived and wiped them out. Differential dispersal abilities of the predator and prey populations allowed the prey to maintain itself by colonizing new feeding sites before the predators destroyed a local population. At the same time the predators did not become extinct, because they could find the prey populations sufficiently fast. The single oscillation to extinction was repeated on each individual orange.

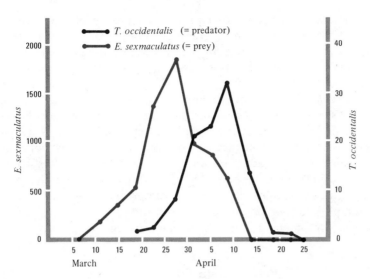

FIGURE 10-4 Changes in numbers of predator and prey mites through time. The experimental universe was portions of the surface of 20 oranges alternating with 20 surfaces with no food. See text for more details (After Huffaker, 1958.)

Predators specific for certain prey species often may be faced with prey inaccessibility, either on a seasonal basis or because of the inherent oscillations of the interaction. Salt (1967) found that the protozoan *Woodruffia*, a predator specific to *Paramecium*, had evolved the ability to encyst at low prey densities and to become active again as the prey population increased. In laboratory experiments *Woodruffia*

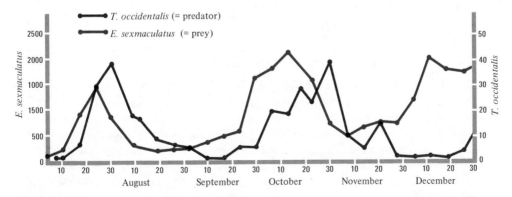

FIGURE 10-5 Oscillations of prey and predator numbers through time in a relatively complex experimental universe. The feeding surface per orange was less than the entire orange and there were vaseline barriers between oranges to increase the difficulty of dispersal for the mites. This was the most complex universe Huffaker used in this series of experiments and was the only universe in which he obtained recurrent oscillations as predicted by the simple predator-prey equations (After Huffaker, 1958.)

encysted at *Paramecia* densities that permitted regrowth of the prey population following predator encystment. In this case oscillations predicted by the mathematical model were generated by a refuge for the prey (nonfeeding by encysted predators at low prey density) and the ability of the predators to maintain a viable population at low prey densities.

The *Woodruffia-Paramecium* example of predator escape from extinction is essentially the same as that produced by the complex environments Huffaker used in his mite experiments; but it represents a different evolutionary adaptation for escape and hence for maintenance of both predator and prey populations. For Huffaker's mites selection acted on dispersal rates. The *Woodruffia-Paramecium* interaction selected for predator individuals that were food specific and that encysted at low prey densities, temporarily removing the predators from the prey niche. In the mite example, selection produced principally an escape mechanism for the prey, whereas in the protozoans the specific predator evolved an escape response.

GRAPHICAL REPRESENTATIONS OF PREDATOR-PREY INTERACTIONS

If equation (10-1) for prey is set equal to zero (that is, the rate of change of the prey population is zero) signifying a stable prey population, it is possible to identify all prey population densities (excluding a density of zero) that are stable at specified predator population densities. It must be remembered that all the assumptions inherent in the logistic equation of population growth, plus the assumption of a constant rate of predator attack, are contained in this formulation.

Under the conditions producing a stable prey population, equation (10-1) simplifies to

$$P = \frac{r_m(K - H)}{cK} \tag{10-3}$$

which can also be expressed as

$$P = r_m/c - (r_m/cK)H \tag{10-4}$$

Since r_m, c, and K are all constants, this equation represents a straight line where the Y-axis is predator density (P) and the X-axis is prey density (H). This line, called the *prey isocline,* represents stable prey densities at varying predator densities.

The relationship predicted by this general model is always negative ($-r_m/cK$); that is, decreasing predator population sizes are required to maintain a stable prey population at increasing prey densities. This is simply a result of the fact that reproductive output per individual in a population growing according to the logistic equation is decreasing because of K-responsive internal regulatory mechanisms. Therefore as density increases less and less predator impact is required to drive prey growth rate to zero. According to this model, small prey populations should always increase, since very large predator populations would be required to drive the growth rate of the prey population to zero and large predator populations cannot be supported by low prey densities. Plotted on a graph in which the X-axis is prey density

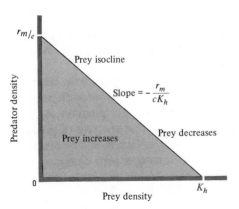

Prey isocline

$r_{m/c}$

Slope $= -\dfrac{r_m}{cK_h}$

Prey increases

Prey decreases

Predator density

0

Prey density

K_h

FIGURE 10-6 Equilibrium line predicted by equation (10-1) for prey populations as a function of predator density. Below the line toward the origin the prey population can increase; above the line the prey population decreases.

and the Y-axis is predator density, equation (10-1) for prey populations gives the relationship shown in Figure 10-6. It has already been argued that predators will increase in population size above some threshold prey density and decrease below it. Thus the predator isocline will be a vertical line through the X-axis at the threshold prey density. Predators presumably will encounter some maximum upper limit to density, so the predator isocline bends at 90 degrees at this threshold predator density.

Rosenzweig (1969) has presented arguments to support the idea that the prey isocline should have a positive slope at low prey densities and a negative slope at high prey densities (Fig. 10-7). The argument follows from more reasonable considerations of predator behavior and from the shape of the logistic growth curve for the prey population. Rosenzweig contends that the rate of predator attack required to drive dH/dt to zero at low prey densities decreases with increasing and decreasing prey density on either side of the peak predator density for stabilizing the prey population (see Fig. 9-5). Predators are not likely to attack rare prey; they are probably hunting for other prey species or not hunting at all. Rare prey also have low population growth rates.

As density of rare prey increases, the likelihood of an attack, k_1 in equation (10-1), increases until the population reaches a sufficiently high level to satiate the predators effectively. At this prey density any further increases in density decrease the proba-

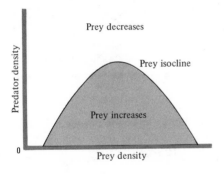

Prey decreases

Prey isocline

Prey increases

Predator density

0

Prey density

FIGURE 10-7 Prey isocline as predicted by Rosenzweig and MacArthur (1963) on the assumption of changing predation rates with changing prey density.

bility of an attack on an individual. The shape of the curve relating prey density to predator density (or attack rate) required to make the prey population stable (the prey isocline in Fig. 10-6) depends on the reciprocal responses of predators and prey to changing densities of each population.

The changing total predator pressure, $-k_1 PH$ of equation (10-1), with changing prey density can be derived from (a) changing the number of predators with constant rate of attack per predator, or (b) changing the rate of attack per predator, k_1, with a constant number of predators. Equation (10-1) assumes that the rate of predator attack per predator is constant over all prey densities, but, in fact, as will be discussed later in this chapter, the rate of attack per predator changes with changing prey density. If the rate of predator attack increases as prey density increases, predation impact on the prey population growth rate will be higher than predicted by equation (10-1) and a smaller predator population will be required to drive dH/dt to zero. This could yield a positive slope to the prey isocline as initially suggested by Rosenzweig and MacArthur (1963), provided the predator attack rate increases fast enough relative to the growth rate of the prey population.

Although the shape and position of the prey isocline may vary, a general model may be used to predict the outcome of different kinds of predator-prey interactions (Fig. 10-8). The prey isocline will be humped (Rosenzweig, 1969) and pass through the origin. Any time the predator density (graphed on the vertical axis) is above the prey isocline, the population of prey will decline; if the predator density falls below this prey isocline, the prey population will increase. A similar isocline can be made for the predator population (Fig. 10-8a). In general it is expected that predator populations normally increase when the prey population exceeds a threshold density and decrease below that threshold. For simplicity it is assumed here that the predator isocline is a vertical line. In reality the predator isocline probably occurs at higher prey densities as the predator density increases. There is an upper limit of predator density above which additional prey will not support additional pred-

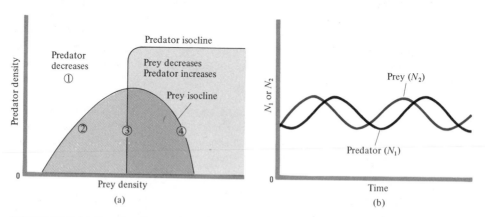

FIGURE 10-8 (a) Prey isocline and predator isocline plotted as functions of prey and predator densities. See text for further explanation. In (b) the changing population sizes are plotted through time. (After Rosenzweig and MacArthur, 1963.)

ators. At this upper threshold of predator density, growth rate of the predator population also decreases to zero. The actual position and height of the isocline can vary depending on environmental conditions and the efficiency with which the predators use the prey. Starting from any predator and prey relationship, graphs relating predator and prey isoclines to densities of each species can now be used to predict their pattern of population growth.

249

GRAPHICAL
REPRESENTA-
TIONS OF
PREDATOR-
INTERACTIONS

An example is illustrated in Figure 10-8, where the predator isocline passes through the peak of the prey isocline. Point 1, representing population densities of predator and prey that fall to the left of the predator isocline and above the prey isocline, indicates that the prey population is sufficiently uncommon and the predator sufficiently common that both populations are declining. The population declines continue until the predator density drops below a level, point 2, at which the prey can begin to increase in numbers (drops below the prey isocline). As prey density increases, the predator population continues to decrease until the prey population reaches a sufficient density for it to cross the predator isocline, point 3. At this point the predator population also begins to increase. As the predator population increases it eventually crosses the prey isocline (point 4) and the prey population starts to decrease until it is no longer large enough to support increasing predator populations and the two populations arrive back at or near the starting point. Figure 10-8b illustrates how Figure 10-8a would appear if translated onto a numbers-versus-time graph with both populations on the same graph.

By shifting the relative position of the two isoclines it is possible to predict population cycles that are gradually damped (Fig. 10-9) and cycles that eventually lead to the extinction of one population or the other (Fig. 10-10). In either case in the latter cycles the predators become extinct. Extinction of prey leads to extinction of predators without alternate prey, as in the example from Gause (1934) and in

Damped cycle

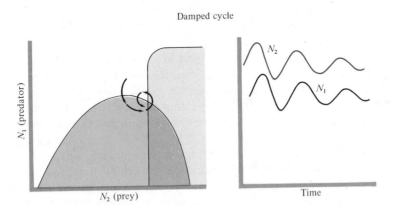

FIGURE 10-9 The damping effect on predator-prey interactions of having the predator isocline at a prey density beyond that at which a maximum number of predators are required to stop growth of the prey population. Note that the magnitude of population fluctuations decreases through time. (After Rosenzweig and MacArthur, 1963.)

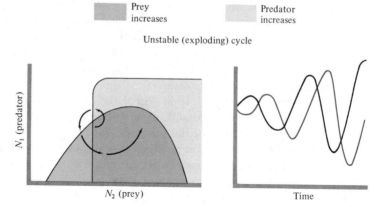

FIGURE 10-10 Moving the predator isocline to the other side of the "hump" in the prey isocline (relative to Fig. 10-9) generates oscillations in numbers of predator and prey that increase with time. This interaction leads to the extinction of one or both populations. (After Rosenzweig and MacArthur, 1963.)

equation (10-2). Extinction of the predator releases the prey population from predatory attack and their population grows until stopped by another environmental factor.

Rosenzweig (1969) applied this graphical method of predator-prey analysis to the data accumulated by Huffaker (1958) for his complex environment allowing three oscillations of the predator and prey populations before extinction (Fig. 10-11). As we noted earlier the dispersal ability of the prey gave them a temporal refuge by permitting small isolated populations to expand before the predators' arrival. Rosenzweig included this in the graph by showing a density of prey below which predators could not expand.

If, for instance, a local rabbit and coyote interaction were similar to that shown in the graph of Huffaker's mite populations, the two populations would cycle for a period of time and then at least the coyotes would go extinct because the rabbit density was so low over so long a time that the coyotes starved. At low prey densities the coyotes, if they are dense enough, can find and take sufficient numbers to keep rabbits from increasing. However, if the coyotes are also taking more rabbits than are being produced, then the size of the rabbit population will gradually decline so that the coyotes are no longer able to find enough rabbits to reproduce themselves at a rate that offsets mortality losses. As coyotes die off their pressure on the rabbits is reduced and the rabbit populations can begin to expand. After the rabbits reach a critical density the coyotes again can harvest enough prey to reproduce at a rate that causes the coyote population to increase. However, the rabbits will continue to increase until the predator population gets large enough or some other factor in the environment becomes limiting to slow growth of the prey population. Then the coyote population will continue to increase since rabbit density is high. The high

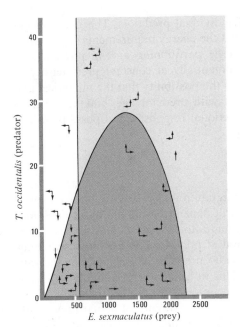

E. sexmaculatus (prey)

251

PREDATOR
RESPONSES TO
CHANGING
PREY
DENSITIES

FIGURE 10-11 Graphical representation of changing prey and predator densities in the most complex habitat used by Huffaker (1958) in his mite experiments. Arrows represent direction of change of prey and predator densities starting at the intersection. From the trajectories of the two populations, the predator-prey isoclines can be drawn. (After Rosenzweig, 1969).

coyote density then causes a decline of the rabbit population. Eventually the coyote pressure on the rabbits causes a decline in the rabbit population to a level below which the coyotes are no longer able to harvest enough rabbits to maintain a reproductive level sufficient to replace mortality losses of the coyotes, and the cycle would start again. While this example presents a hypothetical case for the oscillatory population densities generated by the predator-prey model of Rosenzweig and MacArthur, it is not meant to explain how predators influence natural populations. The subject of predator impact on prey populations will be covered in detail in Chapter 12.

PREDATOR RESPONSES TO CHANGING PREY DENSITIES

Solomon (1949; 1964) distinguished two major predator responses to changing prey density: numerical and functional responses. In the *numerical response* the predator species becomes more or less abundant as the prey density changes while attack rate per predator is constant. In the *functional response* an individual predator changes its attack rate as prey density changes but predator density is constant. These two types of response can occur separately or simultaneously. The effect on an increasing prey population would be identical—predator pressure increasing with prey density. In considering the predator population, however, equivalent changes in energy flow from prey to predator are achieved in two different ways, one essentially on an individual predator basis and the other based on population responses. Within each response type there are several ways of producing the expected result.

To these predator responses to prey density Murdoch (1971) added developmental responses associated with the ontogeny of individual predators. The new concept associated with *developmental responses* is that the energy requirements of predators change with age due to differences in energy partitioning among physiological functions. Developmental responses occur potentially at constant prey density as the individual predators mature. There also is the possibility that the numerical and functional responses of a predator may vary with predator age, making developmental responses an age-specific case of functional responses, and possibly also of numerical responses.

Numerical Responses

A principal numerical response is predator migration into or out of an area as prey densities change. Rapidity of this response will depend on the ability of predators to find locally increasing prey populations. In addition predators must have mechanisms to assess predator density as a function of prey density. For most searching predators an estimate of prey density relative to predator density may be the rate at which new prey items are encountered. This information presumably will have to be integrated with the probability of encounter with another predator as an estimate of the predator population to arrive at an estimate of actual prey availability. Prey availability to a single predator over a time period depends on total prey in the area and the number of predators exploiting the prey population.

In Nicholson's and Bailey's (1935) original formulation of parasite-host interactions they assumed that the area searched by a parasite (predator) is constant at all parasite population densities. More recently, however, Hassell and Varley (1969) obtained data showing that search area decreased as predator density increased (Fig. 10-12). This reduction could result from physical interference between two individuals, or behavioral interactions tending to separate activity areas of different individuals. Reduced search area generated by predator competition might reduce predation rate sufficiently to enable alternate predator species to exploit the host population effectively. Hassell and Varley suggest that if such interference is a common phenomenon, the potential inability of a single parasite species to keep the host population within economically permissible size limits may require the introduction of more than one parasite species. Rather than searching for the single "best" parasite, it may be more appropriate to introduce several species to achieve lower host populations.

Pitelka et al. (1955) found that the number of predators feeding on lemmings and breeding around Point Barrow, Alaska, increased as lemming density increased. Average territory size of breeding pairs of snowy owls (*Nyctea scandiaca*) remained more or less constant (2 to 4 square miles) over the range of lemming densities; however, the total number of resident owls, including nonbreeding individuals not holding territories, was much higher in high lemming densities than at low or moderate densities. Furthermore, owls were uncommon at low lemming densities. Size of the pomarine jaeger (*Stercorarius pomarinus*) population, also a lemming predator, increased and breeding territory size decreased as lemming density increased. At low lemming densities in 1951 pomarine jaegers did not breed. Thus

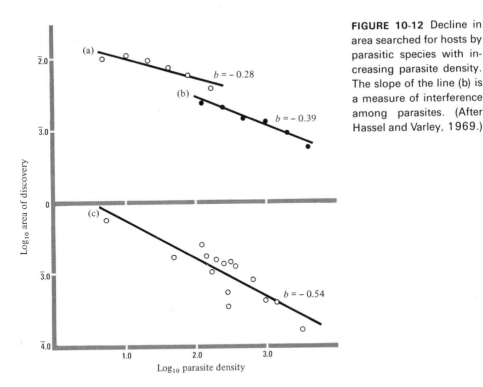

253

PREDATOR
RESPONSES TO
CHANGING
PREY
DENSITIES

FIGURE 10-12 Decline in area searched for hosts by parasitic species with increasing parasite density. The slope of the line (b) is a measure of interference among parasites. (After Hassel and Varley, 1969.)

both predator species moved into the Point Barrow area to exploit a locally abundant food supply, but the interactions among individuals of the predator species were somewhat different in response to increased density.

Another potential numerical response of predators is change in reproductive output associated with differences in prey density. Tawny owl (*Strix aluco*) egg clutches were larger and more young were produced (Southern, 1959) in years of high vole (a small rodent) numbers than in years of low vole densities (Fig. 10-13). Changes in the number of potential offspring from changes in the number of eggs produced are probably a common response to changing food availability. Predator reproductive output also may be regulated by food quality. Nicholson (1957) reported that the number of eggs laid by sheep blowflies (*Lucilia*) in the laboratory was related to the availability of protein to females. When protein intake was restricted, the number of eggs laid declined.

Both food quantity and quality therefore may influence the total energy that can be channeled into the reproductive effort by the female (and male if his activities are important to the postmating reproductive success of a pair) and hence can influence the potential reproductive output. The magnitude of the numerical response then depends on the relative availability of energy and the quality of that energy.

For many species probability of survival of offspring may be dependent on energy stored with the offspring at the time of reproduction (see Chapter 9). In plants, for example, food stored for the growing embryo, seed chemicals

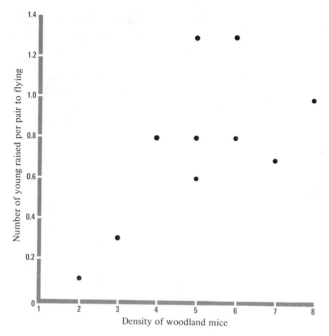

FIGURE 10-13 Relation between an index of reproductive success of the Tawny owl (*Strix aluco*) and the density of primary prey in Wytham Woods, near Oxford, England, from 1949 to 1959. Density of mice on an arbitrary scale from 1 (uncommon) to 8 (abundant). (Data from Southern, 1959; figure after Lack, 1966.)

serving to reduce seed predation (Janzen, 1969b), or seed properties affecting dispersal are all produced at some energetic cost to the adult. Reproductive output also can be influenced directly at a later stage of the reproductive cycle in those species that provide parental care. For many species of birds for which the clutch size is more or less constant, and apparently genetically fixed, the major reproductive numerical response is to raise more of this fixed number of young to maturity in years of high food availability than in poor years. Other species of birds have evolved brood reduction responses (Ricklefs, 1965) in which the young usually hatch at somewhat different times, so, initially, they are of slightly different sizes. The largest young tend to be fed first and smallest young last. Thus in years of low food availability the smallest young starve to death, but the adult has expended little energy to feed the smallest young at the expense of the largest young.

Developmental Responses

The total energy requirements of smaller, young organisms are less than for larger, older organisms of the same species, but the efficiency of young individuals in converting energy into new biomass is greater than for mature individuals, except possibly for reproducing adults (see Chapters 6 and 9). Therefore the rate of utilization of the prey population at a constant predator density and constant prey density will vary with the age structure of the predator population. Furthermore, the predatory efficiency may change with age, especially in predators capable of learning. These ontogenetic changes in predator traits and energy requirements obviously influence the rate of prey use. However, at this time little research has been done 255

PREDATOR
RESPONSES TO
CHANGING
PREY
DENSITIES

to provide empirical examples of this influence.

Functional Responses

The third type of predator response to changing prey density is the functional response: change in the number of prey taken per unit time *per predator* as prey density varies. The major series of experiments associated with functional responses was performed by Holling (1965, 1966) using praying mantises and mice as predators. Holling characterized three types of functional response (Fig. 10-14). The *linear,* or Type I, *response* (Fig. 10-14a) is characterized by a constant rate of increase to a plateau level in prey taken as prey density increases; at the plateau level the predator presumably is satiated and can take no increased number of prey per unit time even though prey density increases. Holling suggested this type of response would be characteristic of filter feeders or sedentary organisms that encounter prey items in direct proportion to their relative density. Filter feeders must process volumes of water to extract food items. If the food is randomly distributed throughout the water volume, the amount of food extracted will be proportional to the volume of water moved and the abundance of food per volume. At low food densities and for short periods of time predators will presumably move the maximum amount of water, thus extracting food in proportion to its density. At very high food levels the amount of water moved will decrease because the food requirements can be met with less than the maximum water volume filtered.

The maximum amount of prey taken per unit time could be a function either of (a) maximum filtering rate of the individual or (b) total energy required to sustain the individual (or a combination of the two). If the asymptote indicates the maximum filtering rate, as the maximum is approached, presumably the filtering mechanism will gradually become saturated with material; the filtering rate will then no longer be a function of the prey density, but will depend on the rate at which filters can be cleaned. This should produce a curvilinear response to increasing prey density as the asymptote is approached.

If the asymptote is determined by the total energy requirements of the predator, then the rate of prey uptake will depend on hunger responses and rate at which the predator digests its prey. Again a gradual diminution of the increase of rate of uptake would seem more likely than a sharp break in the functional response curve. These conditions are very similar to some of the determinants of the invertebrate type of functional response defined by Holling. It seems improbable that any organism would show the Type I response, since most will have a decline in relative efficiency of intake as the plateau is approached. Evidence adduced in favor of the Type I response (Holling, 1965) lacks sufficient data to differentiate between a Type I and a Type II response.

In Holling's experiments (1966) on the *invertebrate type* of functional response (Type II) only a single prey species was available. The predator, the praying mantis (*Hierodula crassa*) in Holling's experiments, was capable of active search and there was a very rapid initial increase in the number of prey taken at low prey densities

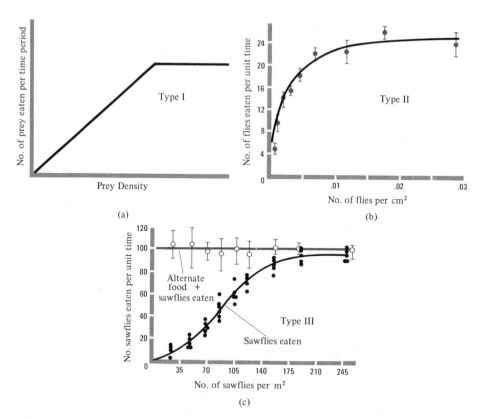

FIGURE 10-14 The three functional response types of predators to changing prey density. In (a) is the Type I response in which the number of prey a predator takes remains a constant proportion with increases in prey density, at least to an asymptote that probably reflects the satiation level for the predator. In (b) the predator is a praying mantis exposed to varying densities of houseflies. This situation provides a good example of a Type II response in which there is a very rapid initial increase in number of prey taken with increasing density of prey followed by a slow increase to the satiation level. In (c) the deermice that are serving as predators are offered a choice between sawfly cocoons and some alternate prey type. The relatively sharp increase in number taken over small density changes at moderate prey densities characterizes the Type III functional response. [(b) and (c) after Holling, 1965.]

(Fig. 10-14b). As the plateau of predation rate was approached the predator captured prey sufficiently fast that the handling time of prey items and the time lag to the next period of high prey capture "motivation" caused a decrease in the slope of the prey intake curve. This response occurs throughout the measured range of prey densities for the Type II experimental situations used by Holling. The importance of handling time and the "digestive pause" produced a negative acceleration to a plateau generating the curve in Figure 10-14b.

The third functional response is what Holling called the *vertebrate type* (Type

III). This primarily takes into account an additional factor: that a predator with alternative prey available will increase its predation rate faster over moderate densities as compared to low and high prey densities (Fig. 10-14c). In other words the third functional response curve suggests that predators without strong food preferences will shift their feeding habits in relation to prey availability. The shift may come about as a result of learning of prey characteristics as the prey become more dense or simply by rejecting unrewarding prey items when the prey reaches a sufficiently high level.

Holling assumed that the Type II (invertebrate) curve was typical of many predators that cannot learn prey characteristics and hence must respond solely to changing densities of all the potential prey organisms. Recall that Holling's experiments producing the Type II functional response never provided more than one prey species. Murdoch (1969), however, found that when alternate prey items were available, the functional response of several species of marine snails depended on relative preference for alternate prey. If the snail had a strong preference for one particular food type, the functional response approached Type II even though alternate prey were available. When the preference was weaker, the response could be shifted to a Type III (vertebrate) response. A Type II response would be expected in those cases in which the predator could discriminate clearly between alternative prey species but had a marked preference for one or more species, or where the predator could not discriminate between the available prey species. In the latter case the individuals of each species would appear similar and prey density would be the sum of the densities of each species.

An additional possibility is that at (a) very low prey densities, lower than those examined by Holling using the praying mantis as a predator, and (b) with alternative prey available, the predator with a strong food preference may take more than expected of the less preferred prey, because the energy requirements of the predator could not be met with the preferred food. In this case the functional response curve would be similar to the Type II curve with a short tail at low prey densities, thus producing a curve rather similar to the Type III response. A shift to an alternate but less preferred prey when the principal or preferred prey is at very low density suggests that the Type III curve may be the general one for all predators. In this case the Type II curve becomes a limited case when (a) the predator has extremely strong preferences for one of the prey items; (b) the predator cannot discriminate prey species; or (c) search capabilities of the predator are such that encounter rate is not really a factor limiting predation rate even at very low prey densities.

The ability of a predator to take proportionately more prey of one species than would be expected on the basis of random sampling of alternative prey has led to the concepts of (a) specific search images and (b) switching. Tinbergen (1960) was among the first to suggest from field data that the number of prey taken is not a linear function of prey density, but rather may be a sigmoidal function (that is, a Type III response). If predators were capable of learning characteristics of their prey, especially those that would facilitate prey location or increase prey capture efficiency, there would be a dramatic increase in the rate of capture over a small change in prey density.

Tinbergen's hypothesis, since used in part by Holling to explain the Type III response, was that when the reward rate to the predator from capturing a particular prey species reached a critical threshold, the predator would "switch" its attention to that particular prey item in a disproportionately higher amount than the prey item represented in the total spectrum of available prey species. All that is required is that the predator disproportionately increase its rate of attack on one prey species compared to alternative prey species present. This can be accomplished in other ways than acquiring a search image for that prey species. For example, if prey are sufficiently abundant to allow the predator the time and energy to be selective, it is likely that the degree of specialization will increase (Schoener, 1969a, c; 1971); the more efficient prey items will then be taken to the exclusion of less efficient items. Efficiency is measured here as the cost of uptake divided into the energy acquired. Although this appears to be a specific search image, it is in fact not the learning of prey traits that facilitates search, but rather the ability to learn the characteristics necessary to make discriminatory rejections of less suitable prey.

Murdoch (1969) in his study of predatory behavior of marine snails showed that the preference of some predators for a particular prey species can be learned. He found that with two prey equally available, a significant selection is made, especially at relatively high prey densities. However, other species of snails apparently have such an innate preference for certain prey species that even after a period of exposure to prey of another species their original preference remains.

As many authors have noted prey characteristics that might be learned are numerous and to some extent influence the type of searching and the nature of effective predatory behavior. If prey traits that aid in search are learned, the length of time required to find a prey item may decrease abruptly with a small change of prey numbers at a threshold density. However, if two prey items are equally dense and equally easy to find and capture, the response could be one of preference. In fact the results Holling obtained when he compared the mouse feeding rates on sawfly cocoons when dog biscuits or sunflower seeds were the alternative food suggested that the mice responded primarily to some characteristic of the food other than to an effective search technique (Fig. 10-15).

Individual predation rate relative to changing prey densities also must reflect the predator's evolutionary background, which may dictate the degree of behavioral adjustment that is possible in selecting prey species. There may be no possibility of adjustment when the predator has an innate preference for one prey item that is not susceptible to modifying influences; or the predator may have virtually catholic tastes that are then subject to environmental modifiers to determine its actual diet. Predators with the ability to learn may change their diet to reflect the relative energy reward values associated with different prey species. This could be learning of prey characteristics that facilitate capture or the efficiency of uptake of one prey item relative to others available at the same time and place. Various combinations of preferences and modification of predatory behavior relative to varying prey densities will produce the Type II and Type III functional responses; it is unlikely that the Type I response occurs in nature. Type II and Type III responses are the ends of a behavioral and genetic continuum.

259

PREDATOR
RESPONSES TO
CHANGING
PREY
DENSITIES

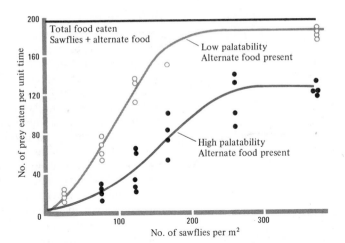

FIGURE 10-15 The influence of variable palatability of alternative prey on the functional response of a Type III predator. Sawfly cocoons are the primary prey with dog biscuits as the low palatability alternate prey and sunflower seeds as the high palatability alternate prey. The predator is the deermouse. Note that a high palatability prey reduces the peak intake and lowers the rate of approach to the peak intake of the primary prey. (After Holling, 1965.)

    Another type of functional response was suggested by Gibb (1962), who noted that for hidden prey (larvae of moths buried in conifer cones), the predators (titmice) seemed to respond as if they had learned the average number of prey at each subsite within a large foraging area. In areas with relatively low density, predators might take a higher percentage of prey than would be expected from predation in areas with average density. However, at very low densities the predation rate might be low because of the difficulty of locating prey. Further, in high density areas the predator would leave when an average number of prey had been taken and would consequently take a smaller percentage of prey than at moderate or low prey densities. The predation rate in which the predator "expects" certain densities might produce the curve illustrated in Figure 10-16 as compared to an inverted U-shaped curve for a species with a specific search image. In this case, however, the predator moves from population to population of prey so it is continually sampling hidden prey of unknown densities.

    In summary, the entire functional response system probably can be viewed as a differential reward system in which the quality and quantity of the prey provide the reward and the predator response depends on the balance of possible rewards from alternative prey coupled with the degree to which predator behavior may be nongenetically modified. The different functional response types were used by Rosenzweig (1969) as the basis for his explanation of a positive slope on the prey

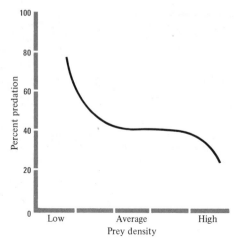

FIGURE 10-16 A possible relation between percent predation and prey density based on the "predation by expectation" hypothesis of Gibb (1962). The predators expect average densities and may take relatively more prey from low density populations and relatively less from high density populations than from populations with the average density. Alternatively, at low densities, the predation percent may also be lower than at average densities making the figure an inverted U. The major result is at high densities where the percent predation should decrease under this hypothesis. Note that this result would also be obtained if the predators were satiated at high densities and maintained food intake at levels associated with average densities.

isocline since that explanation required changing predator pressure with changing prey density.

FACTORS INFLUENCING PREDATOR PREFERENCE

The preceding discussion was based on the assumption that predators can evolve specializations in diet that will be reflected in their numerical and functional responses to prey density. A basic conclusion was that preference has an important impact on the functional response of a predator. If the preference is strong and not easily modified by prey availability, a Type II functional response is most likely. If the preference is weak, the functional response is more likely to be Type III with major predation on the more abundant or more efficiently exploited prey species. These results lead directly to a consideration of the factors involved in the specialization of prey types among predators. At this point in the discussion it is not important whether the preference is learned or genetic, although this will have major effects on the functional response type.

Energetic Efficiency

Effective foraging behavior ultimately requires that total energy intake exceeds the energy output required to obtain food, so that some energy remains for other maintenance and reproductive costs. In the long run the cost of a particular foraging strategy cannot exceed the potential rewards. The ratio of energy intake to energy expenditure will depend on the total amount of time spent in foraging behavior, the energetic cost of that behavior, and the energy assimilated from the food. The first two variables reflect the probability of encountering and capturing prey as a function of time spent and foraging behavior. These variables also reflect the fact that the predator normally must spend more time locating a rare prey than a common one; however, in the same time interval he will expend less energy walking than running. Similarly, the time and energy required to capture prey once they have been observed will depend on their escape abilities. Thus energy return to a predator will be optimal at a certain balance between the costs of different foraging techniques and the total amount of time required by each technique to find, capture, and ingest a prey item of a certain energy content.

With increasing density of all potential prey, a predator can be more specific in his choice without any marked change in the time and the energy spent in foraging; however, decreased time and energy might be spent if the predator is nonselective. If the predator specialized on prey that provided a high energy return for total energy expended (high efficiency), it presumably would be as efficient at extracting energy from the environment at high food densities as would a more generalized predator at lower food densities (MacArthur and Pianka, 1966; Emlen, 1966, 1968; Schoener, 1969a, c; 1971). A general predator, with a wide variety of alternate prey, will eat more of the prey encountered and so reduce the total foraging time.

A specialized predator can reduce its foraging time at high food densities by eating only energy-rich food. With increasing specialization the predator also can evolve morphological and behavioral specializations that will increase the efficiency of its search for and capture of prey. Such specializations could include both the type of habitat searched and the type of search employed (MacArthur and Pianka, 1966). Increased prey density also might allow a predator to take prey of lesser quality if this would optimize his time budget (Emlen, 1966). With high enough prey densities it may make little difference whether the predator processes more easily obtainable, albeit low quality, food or lesser amounts of high quality food, if it extracts the same energy. This could be especially important for a predator that normally spends considerable time and energy in search.

The energy taken by predators occurs as packets—the bodies of prey organisms. Thus distribution of energy in the environment depends on prey size distribution through time and space. Efficiency of uptake for the predator is dependent on energy distribution and costs of acquiring the energy. Large packets of food, such as the large grazing herbivores on the East African savanna, often may be most efficiently exploited by large predators. Small packets of energy may be inefficient for large predators but highly efficient for small ones. Clearly, however, feeding strategy is primarily dependent on the balance between energy yield and expenditure. For instance, the specialization of whalebone, or baleen, whales on small planktonic crustaceans (Russell-Hunter, 1970) demonstrates that the size of the packet of energy may be relatively unimportant if the total energy available is sufficiently high and predictable for efficient foraging techniques to evolve that will compensate for small prey size.

Large predators have greater total energy requirements per individual than smaller predators and therefore require, on the average, more food to support a viable population. If net primary productivity of an area is low, small predators could maintain a larger population than larger predators. Since there is a lower size limit for a viable population, especially for sexually reproducing forms, it would seem that under conditions of low net primary productivity, and hence less energy availability to higher trophic levels, small predators would be at an advantage. Alternatively, the support of larger, especially homeothermic predators, requires relatively high net primary productivity. However, it must be remembered that these are not the only selective pressures on body size.

The primary productivity required to support a third or fourth level consumer population would be higher than that needed to support an herbivore population of about the same size. Within an area an individual herbivore presumably would have to search, on the average, over a smaller area to find sufficient food as compared to a third or fourth level consumer. This results from the inefficiencies of energy transfer up the food chains as discussed in Chapter 6.

McNab (1963) has shown that among mammals with a given body size, herbivores generally exploit a smaller area than carnivores. In general the area is about one-quarter that exploited by the same-sized carnivores. But as expected the larger predators within a trophic level use more foraging space than the smaller species (see Chapter 6 and Fig. 10-17).

Available energy, however, is only a relative measure since each organism must maintain a positive energy balance over the long period. Thus if the distribution of energy or prey escape mechanisms are such that even abundant food cannot be

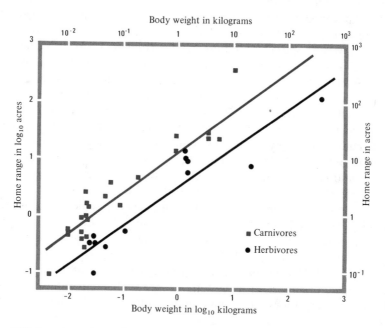

FIGURE 10-17 Relationship of size of regularly utilized foraging area (home range) and body weight in some mammals. Colored line is for carnivores and the black line is for herbivores. (After McNab, 1963.)

exploited efficiently enough to maintain a positive energy balance, the predator must either shift to another prey, move to another area with even higher prey abundance, or starve. The requirement of efficient energy uptake is a strong selective force on predator-foraging strategies. In an ecological sense predator-prey relationships can be looked at as an evolutionary game. There is strong selection on predators for maximizing the ratio of predation yield to predation cost. At the same time there will be strong selection on prey for reducing yield to predators.

Predators must locate, attack, and, finally, ingest their prey. For herbivores the last two items may be relatively inexpensive in terms of time and energy requirements. On the other hand extraction of energy from the plant material may require a considerable fraction of the total time and energy cost of feeding. For example, digestion in ruminants, such as cattle and antelope, may require considerable time (Bell, 1971).

Locating suitable energy sources often may be a major time and energy cost for all predators, including herbivores. Grazing herbivores on the African savanna are relatively "prey" specific, eating only certain plants or plant parts. Since availability of these foods varies with season and geography, the animals may make relatively lengthy and appropriately timed migrations to locate the specific food items taken (Bell, 1971).

A small predator that searches throughout its general foraging area for prey may require relatively large amounts of search time to locate enough prey to replenish energy spent in foraging and still leave enough for other activities. Long-billed marsh wrens (*Telmatodytes palustris*) in the state of Washington spent all their daylight hours during the breeding season either foraging or in behavior associated with reproduction or maintenance (Verner, 1965). Foraging by the wrens takes, on the average, about 50 percent of their active hours. Surprisingly they allocated no time during the day to sitting quietly in the marsh.

The wrens were foraging at times of maximum insect production in the marsh. Gibb (1954, 1960) found that in England some titmice and kinglets, small insectivorous birds, may spend as much as 100 percent of their active time in winter foraging. The larger species in the area spent a smaller percentage of foraging time to maintain themselves. Thus species that depend on small energy packets that are relatively hard to locate and capture may spend long periods of time foraging. These individuals will probably be selected to minimize costs associated with foraging in an effort to maximize the foraging efficiency.

At the other end of the time spectrum are species that utilize food sources that are very dependable, predictable, and rich. Hummingbirds forage on flower nectar. Although hummingbirds do not destroy the prey individual (and hence are really not predators), they nonetheless provide a good example for this extreme case of foraging behavior. The flowers provide an easily visible cue to presence of potential food. Through repeated visits to the source the birds can assess its quality and quantity and forage only at flowers that provide maximum efficiency. Since flowers are stationary and readily visited, this foraging strategy should reduce the total time required for foraging. Wolf and Hainsworth (1971) reported that foraging time, including 1 to 2 percent spent catching flying insects, may be as little as 5 to 10 percent of the total time available to a hummingbird. If the reward in energy provided by the flowers is sufficient, then this foraging system allows for developing energy-foraging techniques that are relatively expensive; the hovering flight employed by many hummingbirds is an example. This foraging technique is advantageous also to the flowers since it provides a selective value for floral traits that reduce the possibility of other nectar feeders visiting the flower without effective pollination. This increases the probability of visitation and, hence, pollination, for an outcrossing flower that must receive pollen from another individual to set seed (Grant and Grant, 1968; Wolf et al., 1972).

Food Quality

Food quality can be measured in terms of the total energy available to the predator, in terms of the difficulty of obtaining energy, or in terms of the types of materials obtained from the food. In the previous section we examined the influence of the first two factors. In this section we examine the last.

Among the vast herds of ungulates that graze on the Seregenti Plains in Tanzania, Africa, are several species that are especially important numerically and show characteristic foraging patterns relative to each other (Gwynne and Bell, 1968; Bell, 1970). At certain times of the year when food is scarce, a major problem for each ungulate species is extraction of sufficient protein from available plant food. Ruminant ungulates, such as wildebeest (*Connochaetes taurinus*), Thomson's gazelle (*Gazella thomsoni*), and topi (*Damaliscus korrigum*), have a digestive system that can extract a high proportion of the protein from plant tissues. But a fairly long time is required to process the food. A nonruminant, the zebra (*Equus burchelli*), extracts proportionately less of the protein, but can process more total food per unit time than the ruminants. Given the appropriate ratio between efficiency of extracting protein and the amount of protein passed through the animal per unit time, the nonruminant could obtain more total protein per unit time than could the ruminant. When protein and calorie availability are low, the ruminant would presumably be better off if the total number of calories was still sufficient to support the population.

The characteristic differences in digestive processes, coupled with the differences in protein content of plant parts, have resulted in the evolution of distinct foraging behavior for each species (Table 10-1). Differences in foraging are especially evident when the availability of ground layer vegetation is low as in East Africa during the dry season. Zebras tend to graze the tops of grasses which are poor in protein but contain enough calories to make energy extraction relatively easy. Much of the energy is passed through the zebra unused, for the animal processes large amounts of material to obtain adequate protein. The ruminants take more leaves and sheath material which is relatively high in protein, but from which longer and more complete digestion also yields sufficient caloric intake.

In general, ungulate migrations through the Serengeti area can be predicted from the relations between food type and animal species. The zebra usually moves into an area first and eats the tops of the grass while at the same time trampling the vegetation. The grasses then put out new growth that can be used by the ruminant

TABLE 10-1 Mean percentage of different parts of plants occurring in the vegetation intake of three ungulate species in the dry season, September, 1967

Ungulate species	Total dicotyledon	Grass leaf	Grass sheath	Grass stem
Wildebeest	0	17.2	52.7	30.1
Topi	0	9.4	53.4	37.2
Zebra	0.1	0.2	48.7	57.0

Bell, 1970

ungulates. At the same time younger grasses that would normally be hidden under the taller vegetation are exposed for ruminant harvest. According to Bell and his co-workers the passage of the first herbivore species prepares the area for use by later species, the general movement pattern being generated by the quality of food available and the ability of the various ungulates to extract the necessary material from the food.

This choice of food items by a predator may be, in part, a reflection of preferentially different qualities of prey available. Organisms also may store chemical compounds that influence the probability of attack by predators (Whittaker and Feeny, 1971). These antipredator substances will be considered in the section dealing with prey escape.

Other Predators

Density of available food is dependent not only on the action of a single predator species, as was implied in earlier sections, but also on the requirements of coexisting predator species that may be taking or attempting to take the same prey. In other words food competition among predators may influence the yield-expenditure relationship.

A high-density food supply together with large numbers of other predator species may represent lower relative food availability. The total predator strategy must take into account possible losses to other predators and also possible restrictions of food availability following behavioral interactions with other predators. Schoener (1969b, c, 1970) and Soule (1966) have both shown that the average size of adult individuals of certain lizard species on islands was negatively correlated with the number of co-occurring lizard species (Fig. 10-18). Each thought that larger numbers of species reduced the total available food and hence that selection was for smaller individuals, as previously indicated. However, it is also possible that when size is especially important in determining access to the food supply, as in dominance relationships of hummingbird species (Wolf, 1970), there could be strong selection for larger individuals.

We have restricted this discussion primarily to evolutionary aspects of predator characteristics, including coevolutionary relationships influencing food preference. Although each factor was discussed individually, we did attempt to interrelate the various influences. The ultimate predator strategy for optimizing foraging efficiency

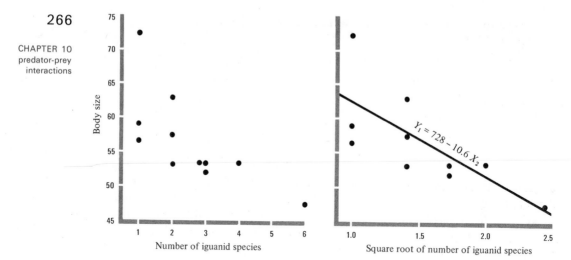

FIGURE 10-18 The relationship of body size in the lizard *Uta stansburiana* and the number of lizard species of the family Iguanidae coexisting on deep-water islands in the Gulf of California, Mexico. The right-hand plot is the least squares regression line calculated after the square root transformation of the number of species. (After Soule, 1966.)

relative to other necessary requirements on its time and energy budget will represent an evolutionary compromise among the many factors discussed. An important problem not yet considered is that although predators optimize their foraging strategies, prey are continuously evolving escape mechanisms. These mechanisms require further adaptations from the predator and lead to a continually coevolving relationship between predator attack and prey escape.

TECHNIQUES OF PREY ESCAPE

We have been discussing predator strategies for optimizing time or energy budgets; at the same time, however, prey species evolve characteristics to reduce predation pressures. These escape techniques presumably involve two primary considerations: (a) escape in time or space and (b) escape *in situ* by various defensive mechanisms. The two strategies are not mutually exclusive, but represent ends of a continuum of escape techniques.

Escape in Time

Prey populations may reduce predator pressure by mechanisms that make it improbable that predator densities can respond efficiently to periodic high prey density. An example of this type of escape occurs in the periodic cicadas (Lloyd and Dybas, 1966). In a local geographic area these cicadas emerge as adults almost synchronously once every 13 or 17 years. A predator must have the same reproductive periodicity, annual reproduction, or rapid reproductive response to increased prey density in order to coincide exactly with emergence of adult cicadas. Predators on the adult cicadas, such as birds, presumably feed principally on other species between cicada emergences, and then switch to cicadas because of their high availability. Synchronous emergence of adult cicadas not only insures that mating occurs, it also effectively floods the environment with many more cicadas than can be eaten by predators with population sizes established relative to other prey species. Presumably no predator can maintain a sufficiently high population to remove a large proportion of the adults. Adults that emerge in off years do not have the advantage of the swamping effect and stand a greater chance of being captured, whereas synchronous emergence produces the positive selection associated with swamping of the predator population. Similar explanations probably apply to the large, synchronous emergences of many insect species on one or a few days in the year.

Annual species invite counteradaptations by predators, however, and thus help time the increase of predator populations to the availability of emerging insects. Many birds breeding in the Arctic take advantage of large populations of chironomid or tendipedid (midge) larvae by timing the production of offspring to coincide with high insect availability (Holmes, 1966, 1970). Similarly, large numbers of tropical birds may congregate in areas where winged termites are emerging (Eisenmann, 1961). This short-term food source abundance represents more biomass than can be exploited by long-lived predators that must depend on other prey species for the remainder of the year.

A similar escape in time, but one determined by prey size rather than local habitat conditions, was described by Jackson (1961) for the tigerfish in Africa. This predator feeds by ingesting whole prey individuals. Predator size determines the largest size prey that can be eaten. Jackson concluded that prey individuals larger than 20 cm (that is, ones with sufficient time to grow to 20 cm) were exempt from predator pressure in the areas he studied. An important question in this system is why there is not a sufficient size increase in the predators to make all prey subject to exploitation.

Escape in time also can be attained if the prey population reproduces under existing environmental conditions at a rate sufficiently high to keep a predator population from limiting the increase. Most prey species have higher reproductive potentials than their predators, probably partly as a response to predator pressure. The prey simply outbreeds the predator, especially when the season for population growth of both is somewhat limited. Burnett (1970a, b) found that the population size of an herbivorous mite in Ontario, Canada, that eats alfalfa plants and in turn is preyed upon by other species of mite was not controlled in the natural situation by predation. Using a combination of field studies and greenhouse experiments, he showed that reproductive rates of prey and predator were strongly temperature dependent. However, the predator also was dependent on the prey to provide sufficient energy to reproduce. At relatively low temperatures, that is, 58°F, the prey population increased very slowly and predators increased even less. At slightly higher temperatures the rate of prey increase exceeded predator increase by amounts sufficient to allow the prey population to expand. As temperature reached a critical threshold the predator population grew fast enough to control the prey provided the time period was sufficient. At the latitude of these experiments the predator was never able to control the prey in alfalfa fields; the growing season was too short to permit the more slowly growing predator population to increase to a level at which it could make an appreciable impact on its prey. In warm years the alfalfa mite becomes a pest because it has time to reach large population sizes. A long warm summer might give the predatory mite sufficient time to control the pest mite, but Burnett did not observe this result during his field studies.

Escape in Space

Prey escape in space involves (a) evolution of dispersal mechanisms that are more efficient than those of the predators and (b) running and hiding abilities that reduce the risk of predation for at least a portion of the prey population. Errington (1946) thought that a portion of the muskrat population in Iowa marshes was fairly well insulated from predation by each individual muskrat restricting its activities to one area of the marsh that it knew sufficiently to facilitate escape. To maintain this advantage other muskrats were kept out of the restricted areas through aggressive behavior. In general the major impact of predators was on dispersing, nonterritorial individuals. By removing only nonterritorial individuals, predators also could help reduce fluctuations in numbers of prey individuals from year to year. If only dispersing individuals were taken and each territory were filled, then the population would be relatively stable at a size equal to the number of territories that could support muskrats (see also Carl, 1971).

Dispersal also provides for a spatial escape mechanism for several possible reasons. Huffaker (1958) showed that as the complexity of his laboratory habitats—and hence the difficulty of dispersal to unoccupied but suitable sites—increased, the probability of maintaining a prey population in the presence of an efficient predator was increased (compare Figs. 10-4 and 10-5). If the prey can disperse better than the predator, prey colonists may build up a population in a refuge prior to predator arrival. Continued existence of the prey in this situation requires that its population expand rapidly enough to colonize other such predator-free areas before the arrival of the more slowly dispersing predator and the destruction of the local population.

A natural example of the continued presence of a "prey" population in the face of a very efficient "predator" is provided by the interaction of *Opuntia* cactus and the *Cactoblastis* moth in Australia. The moth was originally introduced into Australia to control the cactus population that had covered, by 1925, about 60 million ha of usable range land (Andrewartha and Birch, 1954). Moth larvae burrow into cactus pads and stems, opening the way for disease organisms that destroy the cactus. The moth can effectively destroy a local population of cactus, but apparently is more restricted in dispersal ability than the cactus. Thus although there are no longer large expanses of cactus growing in Australia and the total land area covered has been markedly reduced, established local populations continue to flourish until moth arrival. Cactus abundance apparently is being maintained at "acceptable" levels by the moth.

The *Opuntia-Cactoblastis* interaction is an example of biological control techniques—use of organisms to control populations of other organisms—that are being tried as alternatives to chemical control of pest organisms. The effectiveness of biological control depends on relative ability of control agents, usually a predator, to control or regulate pest populations. An important consideration that increases the effectiveness of biological control measures is the narrow specificity of prey shown by many insect predators. Thus introducing a new insect, the control agent, into the biota of a region should have relatively little harmful impact on the native biota since it is specific for the pest, which often is also an introduced species. Predator effectiveness in regulating a prey population depends on factors that have been discussed in this chapter. Coevolution between predator and prey populations, which is so important in biological control, also has many implications for the regulation of diversity in natural communities. As will be discussed more fully in Chapters 11 and 19, by controlling prey populations below the level of the carrying capacity of the environment as set by other conditions, predators may free resources that can then be exploited by additional species at the same trophic level as the prey and hence can increase prey community diversity.

Another important selective advantage for dispersal results from the concentration of predators at high densities of prey populations—the classical numerical response of Solomon and Holling. In this case dispersal evolves as a mechanism to reduce local density of prey items by increasing uniformity of distribution (Lidicker, 1962; Janzen, 1970) and by decreasing the probability of predator-prey encounter. Janzen suggested that in tropical areas in which climatic conditions allow predator populations to remain at relatively high levels, wide spacing of prey individuals reduces the likelihood that a prey (in this case a tree or its offspring) will be found by a predator. Clumped prey are more likely to be driven to extinction by locally dense predator populations. Only widely scattered individuals would survive over time, thus generating the low population densities characteristic of many tropical tree species.

Escape in space also is possible when portions of the habitat are not effectively exploited by the predator population. In a long-term study of the interaction of snails of the genus *Thais* and several of their prey, principally barnacles of the genus *Balanus,* Connell (1970) found that barnacle larvae (*Balanus glandula*) settled throughout the intertidal region on San Juan Island, Washington. However, in areas inhabited by snails, adult barnacles survived only in a "refuge" at the top of the shoreline. From experiments on feeding rates and measurements of prey density Connell concluded that the snails were probably responsible for most or all of the mortality of barnacles at the intertidal level below the top of the shoreline. He thought that the upper limit of predation by the snails was determined by the ability of the predator to move up, feed, and move back down to a safe site during a tidal cycle. Thus the barnacles that can survive physiologically above the upper limit of snail predation are effectively removed from predator pressure. The width of this safe refuge is determined by the foraging capacities of the predator and the ability of the barnacles to survive in the desiccating environment of the upper intertidal region.

Another strategy of prey escape is the presence of noxious substances that reduce the palatability of, or actively ward off, predators. The spray mechanism of the skunk is an obvious example. Thomas Eisner and his associates (Eisner, 1970) have been working for some time on the defensive secretions of arthropods, such as the cyanide produced by millipedes (Eisner et al., 1967) and the blast of quinones sprayed at potential predators by some species of tenebrionid beetles (Eisner and Meinwald, 1966).

Prey species that can produce or incorporate some noxious element may also derive some protection from predation by keeping predator populations at sufficiently low levels in areas of high prey density for some individuals to escape, or attacks may be reduced if predators are capable of learning to avoid individuals with noxious properties.

Some plants produce or incorporate a variety of chemical substances deleterious to predators. Many legume species store alkaloids and free amino acids in their seeds (Janzen, 1969b) that serve as potent defenses against insects, especially bruchid beetles that attack the seed crop. Janzen points out the apparent relationship between the presence of these chemical defenses and the relatively larger size of seeds thus protected as compared to other legumes in the same areas without these chemicals. His conclusion is that those species that have not evolved chemical defenses are forced to increase seed output partly by channeling more energy into reproduction and by making individual seeds smaller to satiate the predator population. Additionally the plant still must produce enough viable seeds to insure reproduction. Janzen assumes that the benefit of a novel biochemical defense mechanism is energetically less expensive than a similar benefit derived by increasing seed output and so would allow the plant to use more of its energy for maintaining its position in the community. These chemicals apparently are produced by the plant as direct antipredator devices and are not of direct metabolic importance to the plant (Janzen, 1969b). To be effective these compounds must be stored in all the seeds prior to the arrival of the potential predators.

A potentially more efficient mechanism would be to mobilize chemical defenses at the time of attack. Green and Ryan (1972) found that natural or artificial wounding of tomato or potato leaves induced an accumulation at the wound site and throughout the leafy part of the plant of a chemical substance that they identified as an inhibitor of proteinase activity in the insect digestive system. The substance is present normally in the plant, but accumulates in the leaves after they have been damaged by insects (or mechanically) (Table 10-2). The proteinase inhibitor reduces plant digestibility and could cause death of the insect if sufficiently active. Thus the plant is either newly synthesizing or transporting the inhibitor to sites damaged by herbivores. This chemical defense mechanism probably is less expensive energetically than would be maintaining high levels throughout the plant.

Many insects have evolved the capacity to store noxious compounds produced by food plants as antiherbivore devices. Such compounds provide a sufficiently bad experience for a learning predator so as to reduce the probability of its attacking

TABLE 10-2 Evidence of accumulation of proteinase inhibitor in leaves of tomato plants damaged by Colorado potato beetles. The beetles were allowed to feed for 24 hours and the plants were assayed for inhibitor activity (immunological assay) 24 hours later. Leaf damage by the beetles was variable and inhibitor concentration varied directly with damage. Values are means of 11 leaves; ranges are given in parentheses.

Leaf condition	Average inhibitor concentration (μg/ml)		
	Leaves	Main stem	Roots
Damaged	202 (77–235)	52 (0–73)	<15
Undamaged	47 (0–120)	<15	<15

Green and Ryan, 1972

another individual of the same prey species. The prey individual may also be rejected before being destroyed by the predator. The Browers have been especially active in investigating plant poisons that are incorporated in animals, such as the monarch butterfly, that consume the plants (Brower et al., 1967).

An organism that can signal its noxious properties to a potential predator before attack can presumably reduce the possibility of mortality from predation. This leads to the evolution of warning or surprise coloration, often bright and bizarre patterns, and warning behavior (Cott, 1940). At the same time it opens the possibility of nonnoxious species evolving similar coloration to the noxious species. The palatable forms present the same warning colors to potential predators as a means of reducing predator pressure on themselves. Similarity of two potential prey species resulting from convergence to reduce predation is a common form of mimicry. Mimicry of color patterns is especially common in the tropical regions of the world, suggesting that predator pressure there may be a much more important selective agent than in the temperate regions.

Two classical kinds of mimicry are often referred to as Batesian and Mullerian mimicry after early naturalists who recognized these types. *Batesian mimicry* involves a noxious species, called the model, and at least one similarly colored nonnoxious species, called the mimic (Fig. 10-19). In this case the mimic depends on the noxious qualities of the model for its protection from learning predators. The Browers have provided much of the available experimentation on this subject (for example, Brower, 1960), especially experiments dealing with the question of how relatively common the mimics can be and still derive protection from looking like the models. An important complicating factor in understanding optimum relative abundance of the mimic is that the degree of the model's noxiousness varies as does the similarity of the model to the mimic. Each of these factors will influence the optimum ratio of models to mimics that still maintain an advantage for the mimic.

Mullerian mimicry involves a series of models that look alike so that each provides a noxious experience to the shared predator (Fig. 10-20). The feedback for the predator from the experiment of capturing a prey item is always negative, thus enhancing the advantage achieved by any of the potential prey. The possible number of negative rewards increases as more models appear in the Mullerian complex. In

FIGURE 10-19 An example of a Batesian mimicry complex existing over a broad geographic area. At the left are palatable mimics of distasteful danaid butterflies on the right. At the top are the palatable southern viceroy and the distasteful queen butterfly. At the bottom are the palatable northern viceroy and its distasteful model, the monarch. (Photo by John G. Franclemont.)

view of the obvious advantages of Mullerian mimicry an important question is why Mullerian mimicry complexes are not more common with even more species in each complex. Evolutionary bottlenecks probably occur at two points in the interaction. The first, and probably the more important, is the ability of the model to store or produce a noxious substance. This ability requires genetic adjustments that are probably not simple to make. The second, and of unknown importance, is the ability of a predator to achieve counteradaptations that make all the models palatable and thus open the Mullerian complex as a whole new set of potential prey—each species very clearly marked.

Two other kinds of mimicry have been suggested recently. Brower and his colleagues (1970) found that some butterfly species that derive their noxious material from milkweed plants (family Asclepiadaceae) may also feed on species in that family that are less noxious. If larvae of the same butterfly species feed on different plant species in the same geographic area, it is possible that only a portion of the butterfly population will be noxious. In this case the predator obviously cannot tell the difference between noxious and palatable individuals; here again the advantage of this mimicry situation, called *automimicry,* will depend on the ratio of noxious to palatable individuals as well as on the predator's learning capacity.

FIGURE 10-20 Combination Mullerian and Batesian mimicry. The two upper species are Mullerian mimics (both are distasteful), while the lower species is a palatable Batesian mimic of the other two species. (Photo by John G. Franclemont.)

Holling (1966) has proposed that in certain situations in which the total number of predators is limited by some factor other than prey availability (for example, territoriality) two nonnoxious prey species may evolve as mimics to supersaturate the environment with that prey pattern. The efficiency of this *palatable mimicry* depends on the functional response of the predator, the learning ability of the predator, and the probability that the prey population will normally occur in high enough densities for the functional response of the entire predator population always to be on the right-hand portion of the functional response curve (Fig. 10-21). This mimicry depends on the fact that percent predation decreases as prey density increases above the level at which the predation rate reaches its asymptote. If a prey item were taken to the complete satiation of the predator, it might be an advantage to mimic another prey population. However, it could also be an advantage to insure that the apparent prey density is very low on the Type III functional response curve. This sort of apparent density could be achieved by hiding or by behavioral interactions to promote spreading of the available prey over a larger area.

Another way to achieve a small population size of prey is to evolve a series of morphological types so that the predator perceives the prey population as a number of distinct types to each of which it responds, both functionally and numerically, as a distinct prey population. The small populations then might be subject to a lesser total predation pressure than if the population were composed of morphologically

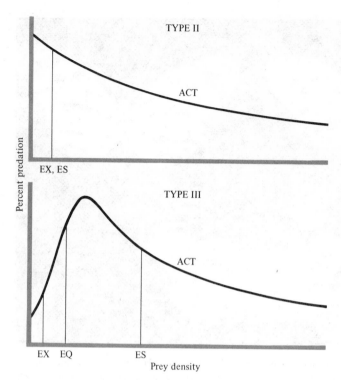

FIGURE 10-21 Predicted effects of Type II and Type III functional responses on population growth of prey. The ACT curve is the predicted actual percent predation of each functional response type; below density EX the population should go to extinction; above density ES the prey should numerically escape from the predator and at density EQ an equilibrium between predator and prey populations is established. Note the lack of an EQ point and the low density for ES in a Type II functional response. (After Holling, 1965.)

identical individuals. This selection pressure would produce a polymorphic prey population. A similar argument could be used in reverse to suggest that if prey can learn characteristics of their predators, it would be advantageous for a predator population to evolve polymorphs. D. R. Paulson (personal communication) has suggested that this is a selective basis for the polymorphisms in many birds of prey, including neotropical hawks.

One important type of protection from predators derived from color patterns is camouflage. Organisms that evolve coloration similar to their background (Fig. 10-22) or morphology that disrupts the shape of the prey (Fig. 10-23) may reduce the possibility that predators recognize them during active search for prey.

A final predator strategy that is very similar to the prey strategies discussed earlier is *aggressive mimicry*. In this case the predator evolves characteristics that give it the appearance of a nonpredator, which decreases the escape response by the prey. A beautiful example comes from the work of Lloyd (1965) on fireflies. Male fireflies of the genus *Photurus* climb to the top of blades of grass at night and signal in a specific pattern with the luminescent organ. Females of the same species flying by perceive the light signal and, if they are ready to mate, answer with flashes of a certain pattern. The female then homes in on the male and mating occurs. Fireflies of the genus *Photinus,* which are predators on other fireflies, also fly at night as a search technique. The flash of a male *Photurus* elicits a flashing response from

FIGURE 10-22 An example of how morphology can camouflage a potential prey individual. The katydid is brown and approximately the same color as the dead leaves on the forest floor. The wings fold over the back and are patterned to resemble a dead leaf. (Photo by Larry L. Wolf.)

FIGURE 10-23 The morphology of this walking-stick tends to disrupt the general body outline of the insect making it more difficult for a predator to recognize the organism as a potential prey item. (Photo by Larry L. Wolf.)

the predatory *Photinus* that mimics the response of a female *Photurus*. The male *Photurus* responds as if the approaching firefly is a potential mate, allowing the predator to home in for an attack. Other examples of aggressive mimicry from insects are well known (Eisner et al., 1962). Willis (1963) postulated that aggressive mimicry was the selective force producing a color pattern and flying behavior in the predatory zone-tailed hawk that is very similar to that shown by the non-predatory turkey vulture.

GENERAL CONCLUSIONS

1. Predators ingest prey as an energy and nutrient source. Predator-prey theory is based on complete removal of a prey individual from the prey population. With appropriate modifications these theories also may be applied to herbivores and other organisms that crop only part of an individual, leaving the remnant as a viable, but somewhat reduced, entity in the population. The boundary between the two types breaks down when we consider that some parasites kill the host and others do not. Also a seed predator can be considered as taking part of the energy of the parent plant or as removing a potential individual from the next generation.
2. Predators can show numerical and functional responses to changing prey density. The general form of the functional, or per individual predator, response is probably sigmoid, but can be modified by preference and availability of alternative prey.

3. It is probable that most predators show developmental changes in their requirements for prey individuals. These developmental responses will modify numerical and functional population responses, if age structure of the predator population changes.

4. Predator preference is influenced very strongly by energetic foraging efficiency associated with potential prey items. Efficiency is determined by energy spent in foraging and energy uptake. Energy spent will be a function of foraging costs and time spent. Energy uptake will depend on the amount of energy per prey individual, and the number of prey captured per unit time, clearly a function of the efficiency of predation and prey density. Predator preference also is dependent on prey quality. Nutrients and energy that can be extracted from the prey depend in part on what has been stored by the prey and the ability of the predator to use the material in the form in which it is stored.

5. Predators presumably optimize their foraging strategy to achieve a maximum reproduction rate under prevailing environmental conditions. This means that time for activities other than foraging must be considered important in determining the optimal time costs allowable for foraging and the prey types taken.

6. Prey must be continually evolving mechanisms of escape from predators. Generally these mechanisms can be grouped under two categories, escape in time and escape in space. Escape in time involves being at the same location as the predator, but at a time when the predator is not active or when the predator population is sufficiently low to allow a high probability of escape for a prey individual. Escape in space is being at a spot where the predator does not occur or being at the same spot as the predator but for various reasons being relatively unavailable to the predator. These reasons might include camouflage and other protective coloration or the evolution of characteristics that make the prey distasteful or harmful to the predator.

SOME HYPOTHESES FOR TESTING

Although the theory of predator impact on prey populations continues to expand into various hypothetical situations, little progress has been made in understanding predation phenomena under natural situations. This reflects the difficulty of measuring the availability of prey relative to the foraging strategies that might be employed by the predator. A lesser problem that is being overcome to some extent through the examination of stomach content of predators is actually measuring the uptake of prey items by predators. Analyses of stomach contents leave much to be desired as estimates of prey selection since they represent the activities of the predator over an unknown length of time after exposure to unknown prey populations. An important area of future research involves measuring the impact of predators on natural, multispecific prey communities. Some data are available from the biological control literature, but the predators in these observations generally are very specific for a certain prey species.

Another important area of investigation for understanding predators is the conditions under which specific predator preferences might evolve and the role of genetic information and juvenile learning periods in these preferences. Food plant preferences for some insects are known to depend in large part on the species of plant on which the young were raised (Jermy et al., 1968). But Brower and his co-workers found that monarch butterfly larvae were genetically conditioned to grow on milkweed plants; it required an apparent genetic change to establish a population of larvae on cabbage. Similarly, Murdoch found that several species of marine snails differed markedly in the influence that training would have on their choice of food. One species was labile in its preference, whereas the preference of the other seemed to be relatively specific and unmodifiable.

The preferences exhibited by a predator depend in part on the predictability of quality and quantity of available prey, and also on the ability of the prey individuals to evolve antipredator devices. This coevolutionary relationship between predator and prey populations opens many new pathways of increasing diversity in ecological communities as the prey reach new adaptive zones that allow them to radiate with little predator impact. As the predators make the appropriate evolutionary changes to exploit these prey, a new adaptive zone is opened for the predator population, leading to a series of specialists such as are presently associated with seemingly noxious plants, for example, milkweed.

competition

11

In the previous chapter we considered individual interaction in which one individual served as an energy source for the second. Although our discussion was limited to individuals of different species, individuals of the same species may interact as predator and prey (that is, cannibalism). Another type of interaction between individuals involves common requirements for a resource in limited supply. If a resource is limited relative to the requirements of the individuals or if the resource quality varies, individuals may *compete* for that resource. Whereas predator-prey interactions are primarily among individuals at different trophic levels, competition is primarily among individuals at the same trophic level.

COMPETITION DEFINED

As with many ecological terms, there is no unique definition of competition currently recognized and used by all ecologists (Harper, 1961; Milne, 1961). We define *competition* as the biological interaction between two or more individuals that occurs when

(a) a necessary resource is in limited supply or (b) resource quality varies and demand is quality dependent. This definition expands the common concept that competition is for a resource in short supply to include attempts to utilize different qualities among a resource type. In other words organisms may compete if each attempts to obtain the "best" resource along a quality gradient or if two individuals try to occupy the same space simultaneously. Competition is an active process that has major influences on the ability of competing individuals to survive and reproduce. The type of influence can vary from (a) direct interference with access to a resource to (b) reducing the efficiency of exploitation of a resource (Wolf and Hainsworth, 1971).

Every organism has requirements that are shared to some extent with other organisms of the same or different species. Competition between members of the same species is *intraspecific* competition. Competition between members of different species is *interspecific* competition. Potential outcomes of competition depend in part on the genetic relation of the competitors. Although intraspecific and interspecific competition sometimes are treated separately, our discussion will combine the two classes when appropriate. Similarity of ecological requirements is more likely among genetically similar individuals from the same gene pool than among individuals from different gene pools. The likelihood that gene exchange between competitors will influence competitive interactions resides, in general, only in intraspecific competitive situations.

Competition requires that niches of competing individuals overlap sufficiently that they share utilization of some resources. The entire niche of one organism need not be within the niche of the other, but there must be some overlap of fundamental niches. The probability that fundamental and realized niches will overlap is much greater for members of the same species than for members of different species. At the same time there will be different genetic constraints on the evolutionary adaptations to reduce competition among individuals that share a common gene pool and members of populations with different genetic bases.

In general ecologists assume that competition may play a major role in the functional organization of coexisting species, especially in determining the number of species that can coexist. Some authors (Andrewartha and Birch, 1954; Bowman, 1961) argue that present lack of good evidence for competition makes it difficult to be certain that competition in the past was important in generating the present ecological relationships of species and individuals.

GENERAL MODEL OF COMPETITIVE INTERACTIONS

The potential outcome of an interspecific competitive interaction is either (a) the *extinction* of one or both forms or (b) the continued *coexistence* of each form. In the first case the better competitor wins before the lesser competitor can make ecological adjustments allowing continued coexistence. Coexistence can be maintained if the organisms are sufficiently different to provide an exclusive refuge for the less effective competitor or if relative competitive abilities change sufficiently rapidly for neither competitor to eliminate the other.

Competitive interactions, in general, will tend to work to the detriment of both competitors and will be expressed as a decreased rate of growth in both populations. General competition equations indicate the effect that populations have on each other's abilities to grow. We saw in Chapter 9 that a single population influences its own ability to continue growing, at least if it is K-responsive. A first approximation of the impact of a second species can be expressed as a modification of the logistic equation (Lotka, 1925; Volterra, 1926; Gause, 1934; Slobodkin, 1961):

$$dN_i/dt = r_{mi}N_i \frac{K_i - N_i - \alpha_{ij}N_j}{K_i} \tag{11-1}$$

where dN_i signifies the change in numbers of species i, r_{mi} is the maximal rate of increase of species i, N_j is the number of individuals of species j which is the competitor with species i, and α_{ij} is the competition term. Actually α_{ij} is the relative impact of one individual of species j on the growth rate of the population of species i compared to the impact of one individual of species i on the growth rate of species i.

When α_{ij} is greater than 1.0, the impact of an individual of species j has more influence on the growth rate of species i than does a single individual of species i. If an individual of species j required twice as much food per unit time as an individual of species i, its impact on the food-determined K of species i would be twice as great as the impact of one individual of species i. The conditions under which two species will coexist, or one will displace the other, are related to the potential carrying capacity of the environment for each species and the relative effect of each species on the other.

This theoretical formulation neglects the second possible type of competition, which is competition for resource quality classes. In general this is a behavioral relationship that depends on the ability of individuals to win in "aggressive" interactions that follow the simultaneous attempt to use a resource unit. For example, the more aggressive of two hummingbirds will be able to control the most energy-rich patch of flowers to the exclusion of the other hummingbird. Aggression must be defined broadly here to include interactions such as occur between plant seedlings that are trying to grow at the same spot. The limitation is resources in that particular space, not space in general, and the interaction leads to the exclusion of one or both individuals.

THEORETICAL OUTCOMES OF COMPETITION

Two Species Systems

Assume that the outcome of a competitive interaction between two species is not completed until each population reaches a stable size (even if this is extinction for one population). Then the outcome of competition may be estimated from relationships inherent in equation (11-1). At equilibrium, population growth rate for both species equals zero. Species 1 is equivalent to i and species 2 is equivalent to j in equation (11-1). Setting equation (11-1) equal to zero and rearranging yields

$$N_1 = K_1 - \alpha_{12}N_2 \qquad (11\text{-}2)$$

that is, density of species 1 at equilibrium equals the carrying capacity of the environment for species 1 minus the impact of species 2 on the carrying capacity of species 1. If species 1 has driven species 2 to extinction, then

$$N_1 = K_1 \qquad (11\text{-}3)$$

If species 2 has driven species 1 to extinction, then

$$N_2 = K_1/\alpha_{12} \qquad (11\text{-}4)$$

which means that the density of species 2 in the absence of species 1 depends on the carrying capacity of the environment defined in terms of species 1 modified by the competition coefficient. Similar equations hold for the impact of species 1 on 2. When species 2 replaces species 1,

$$N_2 = K_2 \qquad (11\text{-}5)$$

or if 1 replaces 2,

$$N_1 = K_2/\alpha_{21} \qquad (11\text{-}6)$$

If the axes of a two-dimensional graph are the densities of species 1, N_1, and species 2, N_2, a straight line with slope $-1/\alpha_{12}$ describes the relative proportions of N_1 and N_2 at equilibrium of N_1 (Fig. 11-1a). The N_1 intercept for this line will be K_1 and the N_2 intercept will be K_1/α_{12}. Similar calculations lead to a straight line (Fig. 11-1b) with slope $-\alpha_{21}$ representing the relative numbers of N_1 and N_2 when N_2 is at equilibrium. In this case the N_2 intercept is K_2 and the N_1 intercept is K_2/α_{21}.

Both population isoclines can be drawn on the same axes in four different positions relative to each other to represent the four possible equilibrium outcomes

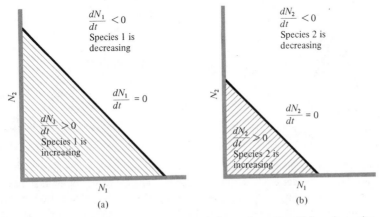

FIGURE 11-1 Equilibrium lines for populations of species 1(a) and species 2(b) plotted as a function of the relative abundances of the two species. Equilibrium values were calculated from equation (11-1).

of competitive interaction for populations that fit the Gause competition equations: (a) extinction of species 2; (b) extinction of species 1; (c) unstable equilibrium with the eventual winner dependent on relative population sizes; (d) stable equilibrium and coexistence (Fig. 11-2). Although it is unlikely that many populations fit these equations exactly (see Chapter 9), they do provide a theoretical framework within which we can explore possible competitive relations between two species.

If individuals vary genetically in a single population, competition can be represented by the same set of equations with 1 and 2 representing different genotypes rather than different species. In the logistic equation we were forced to assume that all individuals were identical and had the same impact on K, thus eliminating the $\alpha_{12}N_2$ term from the competition equation. Since it is unlikely that two individuals

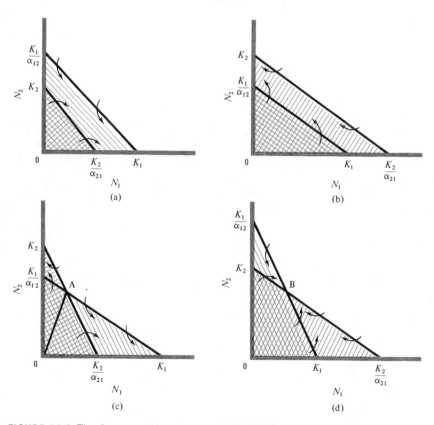

FIGURE 11-2 The four possible outcomes of interspecific competition for species that are not ecological equivalents. (a) species i always wins; (b) species j always wins; (c) an unstable equilibrium at the intersection of the two isoclines; any deflection from this point leads to one or the other species winning depending on the direction of the deflection; (d) a stable equilibrium at the intersection of the equilibrium lines; a deflection from this point is damped back toward the intersection.

of a sexually reproducing species are genetically identical, the more complex competition equations are more appropriate.

The same resource may set different carrying capacities for different species cultured separately (K_1, K_2, with K now defined in units of individuals of each species). If an individual of species 2 has a different impact on K relative to an individual of species 1 than it has on itself, the ratio of $K_1/K_2 \neq 1$. If $K_1/K_2 = 2$, this means that at equilibrium in monoculture, species 1 has twice as many individuals as species 2. Since the resource determining K is the same for each species (for example, 10,000 calories, 60 gm of nitrogen, and so on), this must mean that an individual of species 1 has only one-half the impact on K as an individual of species 2. In other words an individual of species 1 would have to have twice as much impact on the ability of species 2 to increase in order to be the ecological equivalent of an individual of species 2. If this were true, then

$$\alpha_{12} = K_1/K_2 = 1/\alpha_{21}$$

The competitive interaction in this situation would be unstable, clearly in violation of our assumptions, because neither species would have a clear advantage in terms of its impact on the growth rate of the other species and eventually, by chance fluctuations of populations, one species or the other would go extinct.

When $\alpha_{12} = 1/\alpha_{21}$, individuals of the two species are not identical, but their niche characteristics are identical; this situation is very unlikely. An equally unlikely situation exists when $\alpha_{12} = \alpha_{21}$, meaning that individuals of the two species are competitively identical.

If $\alpha_{12} < K_1/K_2$, species 2 has proportionately more impact on itself than it does on species 1. This could occur in a situation in which a species at or near K is intraspecifically territorial but not interspecifically territorial. The population size of the territorial form would be limited by the behavioral interaction but not the population growth of the other species. A similar disproportionate impact on conspecifics could also occur when injurious metabolites are intraspecifically active but have reduced or no effect interspecifically (Slobodkin, 1961).

If $\alpha_{12} > K_1/K_2$, an individual of species 2 has more impact on the growth rate of the population of species 1 than on species 2. In this case the activities of species 2, such as metabolic products that are released to the environment or behavioral interactions, have more impact on the ability of the competitor 1 to exploit the environment than on members of its own population.

When $\qquad\qquad \alpha_{12} < K_1/K_2$

and $\qquad\qquad \alpha_{21} > K_2/K_1$

then the isocline of zero population growth for species 1 will lie outside that of species 2 at all points on the graph (Fig. 11-2a). This means that after species 2 has reached its equilibrium density, species 1 still will be able to increase in density and eventually will win the competitive interaction regardless of the initial population sizes or the r values for each species. To individuals of species 1, when species 2 is at equilibrium, the environment is not filled to K and some resource is still available for continued population growth. To individuals of species 2 the environment is filled, and as species 1 continues to increase, the environment appears supersaturated, driving dN/dt negative until species 2 becomes extinct.

A similar explanation holds for Figure 11-2b, except that the conditions have been reversed and species 2 will always win. In either case the losing species harms itself more in competitive situations than it harms its competitor. Similarly, the winning species harms the loser more than it harms itself. The species affected most detrimentally will go to extinction.

The remaining two cases produce equilibrium populations containing both species, the relative abundance of each species depending on where the isoclines cross (Figs. 11-2c, d). The important difference in the two cases is that the equilibrium in Figure 11-2c is unstable, whereas that in Figure 11-2d is stable. Figure 11-2c provides a third type of isocline, a line that designates equal probabilities of winning the competition. To the left or above this line species 2 will eventually win and species 1 will win to the right or below this line. Thus the competitive outcome depends on the initial starting concentrations since each harms the other more than itself in competitive interactions. Hence the winner depends on which species has the largest population relative to K and the respective alphas as competition intensifies. If the populations happen to start on or to reach the competitive isocline or the unstable equilibrium point, any chance fluctuation in population sizes will displace abundances from the competitive isocline and one or the other species will win, depending on the direction of the displacement.

Finally, Figure 11-2d represents a situation in which each population harms itself more in competitive situations than it harms the competitor; that is,

$$\alpha_{12} < K_1/K_2$$

and
$$\alpha_{21} < K_2/K_1$$

As expected this allows the two populations to coexist and strongly suggests that each has some critical limiting factor that is not shared with the other species. An important part of the ecological niche of each species then is exclusive or at least important to one and not to the other. At other points, however, the niches overlap sufficiently for each species to influence the total population size of the other.

For predicting the outcomes from these equations it is essential to know the values of alpha (α) and K for each species and to know that the K resource is somehow shared between the species. The simplest situation in which these data are available is the laboratory where K and r can be estimated under controlled conditions in monospecific culture. Once K and r are known it is possible to place mixtures of the two species together and to estimate the impact of the second species on the rate of growth of the population of the first species. By comparing the rate of growth of a population of size N under monoculture and competitive conditions, it should be possible to estimate the $\alpha_{12}N_2$ term of equation (11-1). Theoretically the equation depends on constancy of α_{12}, a very unlikely situation (Slobodkin, 1961). However, mean values of α_{12} with estimates of variance could also be used.

Despite the potential importance of competition in the organization and func-

tioning of ecological communities (see Chapters 13 and 19), there have been few
detailed studies of competition, especially in natural systems. Laboratory systems
can be initiated and controlled to produce competitive results; these generally lead
to the extinction of one species. However, an expectation for coexisting species in
natural systems is that competition is either not important or has not produced
competitive elimination of any of the other species. A further problem in natural
systems is that noncompetitive limiting factors may keep populations below the size
at which competition plays an important role. In fact, we might expect that birth-
selected species often would not be subject to strong competitive interactions.

TWO EXAMPLES OF COMPETITION

To establish the possible importance of interspecific competition in nature, experiments initially were performed in the laboratory with conditions controlled so that
populations were forced to compete for a known common resource.

Gause (1934) studied populations of two species of *Paramecium* (*P. aurelia* and
P. caudatum) in a medium with food input controlled. The populations quickly
expanded in size, leading to competitive extinction of *P. caudatum* (Fig. 11-3). Under
the experimental conditions, *P. aurelia* always won in competition and it was clear
that the equilibrium isocline of *P. aurelia* was above the isocline of *P. caudatum*.
Gause later (1935) was able to modify the outcome of competition by using *P.
bursaria* rather than *P. caudatum* as a competitor against *P. aurelia*. *Paramecium
bursaria* could survive in the lower part of the culture medium, a habitat not exploited
by *P. aurelia*. The difference may have reflected tolerance to the buildup of excreted
metabolites or the advantage to *P. bursaria* contributed by its symbiotic algae. There
was always a lower spatial position in the medium where *P. bursaria* could survive,
even though it was outcompeted in the upper level by *P. aurelia*. These two species
competed for the limited bacterial food supply, but coexistence was allowed by a

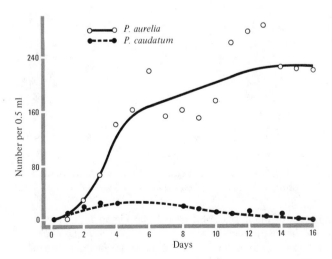

FIGURE 11-3 Population
changes through time for a
mixed population of *Paramecium aurelia* and *P.
caudatum*. (After Gause,
1934.)

spatial refuge for each species. Coexistence depended on relative independence of the limiting resource in the two sections of the medium.

Competition in natural situations is much harder to discover and document. On rocky intertidal shores, however, it can be documented. There is a relatively limited amount of space that can be exploited by sessile animals. Many of these have motile larval populations that can overcolonize an area so that resulting adults could potentially occupy more than the available space. Therefore there will be competition for that space on a fairly regular, recurring basis.

It is well known that many invertebrates tend to occur as horizontal bands on marine shores. As physical characteristics of the environment change from aquatic to terrestrial along the shoreline, the break between bands of closely related species seems to be rather sharp. This suggests that something besides the physical gradient might regulate the distribution of these invertebrates. We already have seen (Chapter 10) that some of the horizontal banding may be produced by predator action (Connell, 1970; Paine, 1971), but earlier studies showed that competition between species was also important in maintaining the distribution patterns.

Connell (1961) found that larvae of two barnacle species, *Balanus* and *Chthamalus*, settled throughout much of the vertical intertidal range (Fig. 11-4). Adult

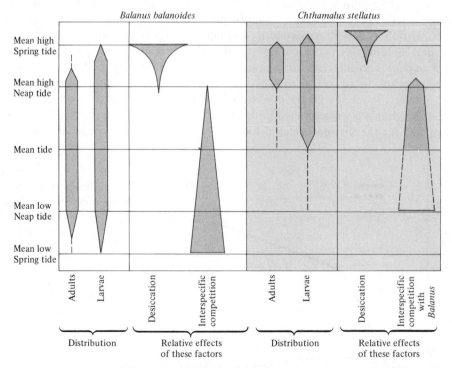

FIGURE 11-4 Pictorial illustration of the impact of physical and biological factors in the intertidal environment on the distribution of adults of two species of barnacles in Scotland. The width of the bars indicates the severity of the factor or the numbers of individuals surviving at that tidal height. (After Connell, 1961.)

TABLE 11-1 Comparison of the mortality rates of young and older *Chthamalus stellatus* on transplanted stones; MTL is mean tide level

Stone No.	Shore level	Treatment	Number of *Chthamalus* present in June 1954		Percent mortality over one year (or for six months for 14a) of *Chthamalus*	
			1953 year group	1952 or older year groups	1953 year group	1952 or older year groups
13b	1.0 ft below MTL	*Balanus* removed	51	3	35	0
		Undisturbed	69	16	90	31
12a	MTL, in a tide pool, caged	*Balanus* removed	50	41	44	37
		Undisturbed	60	31	95	71
14a	2.5 ft. below MTL, caged	*Balanus* removed	25	45	40	36
		Undisturbed	22	8	86	75

Connell, 1961

barnacles finally developing from these settling larvae tended to be restricted to different levels (Fig. 11-4). Connell discovered that one species, *Balanus,* could not survive the desiccating environment of the high intertidal area. *Chthamalus,* however, was able to survive in the low levels of the intertidal shore in the absence of *Balanus* (Table 11-1).

Connell measured growth rates of the two species in several places and found that *Balanus* grew faster than *Chthamalus* above the mean tide level (Table 11-2). He also noted that *Balanus* could overgrow or crowd out *Chthamalus* in the lower intertidal area. In a series of experiments he selectively removed *Balanus* from rocks placed at various levels in the intertidal area. The results compared survival of *Chthamalus* in the presence and the absence of *Balanus.* Its survival at positions normally inhabited by *Balanus* tended to be much better in the absence of *Balanus* than when present (Table 11-1). He concluded that distribution of adults on the

TABLE 11-2 Growth rates of *Chthamalus stellatus* and *Balanus balanoides*

	Chthamalus		*Balanus*	
	No. measured	Average size, mm	No. measured	Average size, mm
June 11, 1954	25	2.49	39	1.87
November 3, 1954	25	4.24	27	4.83
Average size in the interval	3.36		3.35	
Absolute growth rate per day × 100	1.21		2.04	

Measurements of *Chthamalus* were made on the same individuals on both dates; of *Balanus* representative samples were chosen.
Connell, 1961

intertidal shoreline represented an interplay of biological and physical limits. *Balanus* was more susceptible to desiccation, leaving an area at the top of the intertidal region that served as a refuge for the more desiccation-resistant *Chthamalus*. When both species could survive, the faster growth rate of *Balanus* allowed that species to win competition for space. Connell also discovered that preferential predation on *Balanus* by marine snails of the genus *Thais* tended to reduce the competitive interaction and increased the chances of survival of *Chthamalus* at lower tide levels.

MULTISPECIFIC ($>$ 2 SPECIES) COMPETITIVE SYSTEMS

The general form of equation (11-1) for a community of m species would be (MacArthur and Levins, 1967)

$$\frac{dN_i}{dt} = r_{mi}N_i \left(\frac{K_i - N_i - \sum_{j}^{m} \alpha_{ij}N_j}{K_i} \right) \tag{11-7}$$

on the assumption that the impact of m other species on species i is merely the sum of the individual impacts of each of the m species on the growth rate of species i. Therefore the combination of species k and j (a potential coalition against species i and hence a higher order interaction) has no different impact on the growth rate of species i than does the sum of the two two-species interactions alone. The assumption that there are no higher order competitive interactions in a community reduces the complexity of mathematical (and empirical) treatments of community organization; but the assumption is not very realistic (Vandermeer, 1969). Higher order interactions assume that the basic interaction in equation (11-7) must be further modified to include the effect of various species combinations on the growth rate of species i.

To test for the possibility of higher order interaction, Vandermeer (1969) initiated experiments in a very simple system on the assumption that this system would minimize the probability of detecting higher order interactions. He used four species of protozoans: *Paramecium aurelia, P. bursaria, P. caudatum,* and *Blepharisma* sp. The experiments were run at 25°C in the dark (partly to reduce the effect of the symbiotic algae in *P. bursaria*). He grew each species initially in monospecific cultures and estimated r_m and K for each species by fitting the observed values from a series of replicate populations to the logistic equation for population growth (see Chapter 9). Two of the species, *P. aurelia* and *P. caudatum,* gave very good fits with little scatter of observed values about the fitted logistic. The other two species fit somewhat less well, but the logistic still provided a reasonable description of the dynamics of population growth.

Vandermeer then ran a series of experiments in which replicates of all possible two species combinations were tested to estimate competitive abilities of the species relative to each other. Values of the various competition coefficients (α) were calculated from equation (11-1) using the values of r_m and K previously calculated for monospecific cultures. Finally all four species were grown in the same culture chamber to see if the population dynamics of each could be predicted from the values of the two-species interaction coefficients or whether there were more complex interactions among species as the number of species in the "community" increased. He reasoned that if he could effectively predict the outcome of the four-species interaction from the values obtained in monocultures and two-species interactions, he could conclude that higher order interactions were not important.

Vandermeer was able to predict most outcomes of the four-species system from data obtained in monoculture and the series of two-species competitive interactions. In only one species, *Blepharisma,* was growth rate reduced more than expected, suggesting higher order interactions among its competitors. In general, though, for the community as a whole (that is, the four-species system) there was no need to invoke higher order interactions to explain the outcome. As Vandermeer noted this result merely means that experiments with more complex systems are required to elucidate the importance, if any, of higher order interactions in community dynamics. In this type of experimental system only positive results (that is, demonstrating higher order interactions) permit clear-cut statements about the importance of these interactions in the ecological organization of communities.

STOCHASTIC PROCESSES

If species conform exactly to the above competition equations, the outcome of most two-species interactions can be predicted if the K and α are known for each species. It is clear that in these equations N at any given time [except where initial numbers are important (Fig. 11-2c)] and intrinsic rate of increase are unimportant in the competitive outcome. However, in Figure 11-2c it is necessary to know the relative starting numbers of each species.

In all cases we have assumed that competition coefficients do not change throughout the experiment. If α does change, the outcome becomes less predictable. Additionally as α_{ij} approaches K_i/K_j for both species, the outcome presumably will become completely unpredictable when $\alpha_{ij} = 1/\alpha_{ji}$, which is inherently unstable with chance factors providing the key to the eventual outcome. The only equilibrium would be after the elimination of one of the species. Presumably, as α_{ij} approaches $1/\alpha_{ji}$ chance fluctuations of environmental conditions play more of a role in the outcome.

The stochastic nature of competitive interactions was examined by Thomas Park (1954) with flour beetles (*Tribolium*). He initiated competitive situations under controlled environmental conditions and measured outcomes in large numbers of replicate experiments. From earlier discussions about the influence of climatic conditions on the rate of population growth and efficiency of energy utilization, it is not surprising that outcomes of competition in Park's cultures varied depending on environmental conditions. Table 11-3 presents data on the outcome of competition between *T. castaneum* and *T. confusum* in a series of experiments in which temperature and relative humidity were varied to approximate conditions ranging from hot and wet to cool and dry.

TABLE 11-3 Competition between two species populations of flour beetles

Climate	Temperature (°C)	Relative humidity, %	Percentage of replicate experiments in which only one species survived	
			Tribolium castaneum	*Tribolium confusum*
Hot-wet	34	70	100	0
Hot-dry	34	30	10	90
Warm-wet	29	70	86	14
Warm-dry	29	30	13	87
Cool-wet	24	70	31	69
Cool-dry	24	30	0	100

Park, 1954

Each species was favored at one environmental extreme, but they tended to respond differently to variations along the two environmental gradients. *Tribolium castaneum* eliminated *T. confusum* in all experiments in hot-wet conditions (34°C, 70 percent RH), but in only 10 percent of the experiments when the conditions were equally hot but drier (34°C, 30 percent RH). Conversely, in the cool conditions *T. confusum* tended to win whether the humidity was high or low. At intermediate conditions of temperature the success of each species was about equal, but *T. confusum* won most often in the dry environment and *T. castaneum* in the wetter environment.

In four out of the six situations the outcome was a probability function, and there was always a tendency for one species or the other to win under each set of conditions. This reflects the possibility contained in Figure 11-2c of the competition equations, that is, that interaction may depend on which species achieves numerical predominance first even though the cultures for the experiments discussed so far were all started at equal densities of both species. Park and his co-workers (Neyman et al., 1958) did further experiments that showed that the outcome under certain environmental conditions (24°C, 70 percent RH) could be influenced by the initial densities of each species (Fig. 11-5). When the initial densities were nearly equal, the outcome was uncertain, whereas when one species had a clear numerical dominance at the start it was assured of winning.

In equation (11-2), r_m was assumed to play no role in the competitive outcome (it was canceled out when the equation was set equal to zero). However, results similar to those described above where chance fluctuations in numbers influenced the outcome might occur with species having different rates of increase at low densities or different types of reproductive response to changing density. A population that could grow extremely rapidly at low densities might be able to reach a critical numerical predominance before a competitor population that grew less well at low densities. So far there has been little work on the importance of r or r_m in determining competitive outcomes.

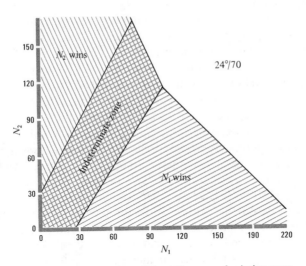

FIGURE 11-5 Illustration of the influence of relative numbers on the outcome of competitive interactions between *Tribolium confusum* (=N_1) and *T. castaneum* (=N_2) in mixed culture at 23°C and 70 percent relative humidity. At high relative abundances the outcome is predictable, but becomes less so as the relative abundances become more similar. Near equality of abundances the outcome is unpredictable. See Table 11-3 for the probability of each species winning when equal numbers are used to start the cultures under a variety of experimental conditions. From Table 11-3 it is clear that the indeterminancy varies with environmental conditions. (After Neyman, et al, 1958.)

COMPETITIVE EXCLUSION

An important conclusion from theoretical considerations of simple competitive systems is that species can coexist only when they are sufficiently different that each harms itself more in competitive circumstances than it harms the other species. In another view the species must be ecologically distinct enough for each to have some refuge in ecological space (the *n*-dimensional hypervolume) such that strong competitive interactions for a particular resource lead to exploitation of another resource on which the species is clearly able to win. The ability to win might be due either to the nonshared character of the refuge resource or to the presence of two shared resources, on each of which a different competing species is superior. Thus if two species are competing for a single prey species which becomes limiting, the two competitors could continue to coexist spatially if each uses another prey species to the exclusion of the second competitor.

The principal conclusion from these theoretical considerations is that species that are complete competitors cannot coexist indefinitely (Hardin, 1960). Gause was among the first to formalize this result and it has come to be called the *competitive exclusion* (or Gause) *principle*. Another statement of the principle is that similar species coexisting for extensive time periods must have some differences in their ecological niches that allow coexistence. This statement points up a major problem with the competitive exclusion principle—it is probably a truism since not even two individuals of the same sexually reproducing species (neglecting identical twins and clones) are likely to be genetically identical. But it can lead to extremely interesting studies of how two species can be sufficiently different to coexist.

From a series of experiments on two species of fruit flies (*Drosophila pseudoobcura* and *D. serrata*) grown together in small culture bottles, Ayala (1969b, 1970) concluded that the competitive exclusion principle did not operate and was invalid for these two species. The two species were very similar, at least morphologically, and they continued to coexist in his culture bottles throughout the experiment (Fig. 11-6) even though there was evidence of strong interspecific competition. Ayala obtained evidence for and an estimate of the competitive interaction between the species by calculating the K values for each species in monospecific cultures and then comparing the K values to the competition coefficients calculated from the two species interactions. The method is very similar to that used by Vandermeer in the experiment on higher order interactions. However, Ayala was dealing with more complex organisms that had both a larval and an adult stage that were quite different in form and ecological requirements. He counted only adult flies in making his population estimates.

Two problems arise from Ayala's interpretation. First, his method of calculating the competition coefficients was on the basis of the outcome of his experimental

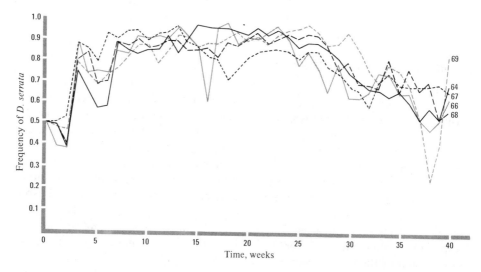

FIGURE 11-6 Competitive population changes in mixed cultures of *Drosophila serrata* and *D. pseudoobscura*. Five different experimental cultures (64, 66, 67, 68, 69) are each represented by a separate line. Note the similarity of results. (After Ayala, 1969b.)

manipulations relative to single species cultures. Unfortunately, this method yields but a single value for the competition coefficient when in fact the value may not be constant. The value probably varies throughout the growth of the population toward equilibrium (Vandermeer, 1969) and also varies between larval and adult stages (Gause, 1970). The calculation also requires equilibrium populations, which is an invalid assumption in these experiments (see Fig. 11-6). The other major problem involves the complex life cycle of the flies and the possibility that competition may be somewhat different at the several stages to the point of favoring one species during one portion of the cycle and the other species during the other portion. In fact, Ayala (1969b) remarked that *D. serrata* was favored in larval competition and that *D. pseudoobscura* was favored in adult competition, hence the niches contain "refuges."

Experiments by Miller (1964a, b) indicate that larval competition in *Drosophila* is probably for enough food to reach a sufficient size to pupate and then emerge as an adult. One clear result of competitive interactions of many *Drosophila* species when food is limiting is a dramatic decrease in adult size (Bakker, 1961). Adult *Drosophila* may compete for egg-laying space in crowded cultures or the larvae may reduce the quality of the medium for oviposition sites (Miller, 1967). Thus discrete stages of the life cycle probably interact in interspecific competition for two somewhat different resources. Adults of one species may be at an advantage in adult competition, whereas larvae of the other species may win in larval competition. If the interactions were sufficiently balanced in magnitude and timing, it would be possible to maintain a two-species population through time, but likely not an equilibrium population.

TYPES OF COMPETITIVE INTERACTION

Competition can take major forms defined by the types of interactions between competing individuals. In *exploitation* competition (Elton and Miller, 1954; Park, 1954; Miller, 1967) an individual depletes the resource without reducing the probability that another individual can still exploit the same resource pool. The only reduction in availability to a competing individual would be by removal of the resource. There are no behavioral interactions restricting use of the resource by other individuals, and the only determinant of use is the presence or the absence of the resource.

In *interference* competition, in contrast, access to a resource is influenced by presence of a competitor (Elton and Miller, 1954; Park, 1954; Miller, 1967). Interference usually involves some sort of behavioral or chemical interaction between individuals prior to actual use of the resource. The most obvious behavioral restriction of access is by defense of the resource, which, if spatially fixed, may lead to territorial behavior (Brown, 1964).

The two competitive interaction types are ends of a behavioral continuum associated with utilization of different types of resources. In general, exploitation or scramble (Nicholson, 1954) competition is characteristic of species that initially are rare relative to the carrying capacity of the resource, for example, the number of fly larvae initially exploiting a beef carcass. The carcass represents a resource that is relatively superabundant when it is present, but one that is unpredictable in time or space and that will disappear regardless of the activities of the flies. Such a resource appears as a short-term pulse of new resource. Optimal reproductive strategy of an adult fly is to find the carcass first and to leave a maximum number of eggs that will yield larvae to scavenge it.

Regular superabundance of a resource would select against genotypes that leave fewer offspring, each having greater competitive ability, since this competitive ability would be important only as the population became food limited. This limitation, of course, cannot be predicted by an egg-laying female. Although food will eventually limit larval growth, the inability of the female to correctly "predict" the approach of the larval population to K reduces natural selection for more competitive larvae rather than gross fecundity. Where "K" can be "predicted" accurately, as in parasitoid wasps where only a single adult can emerge from a host, there will be strong selection for behavioral responses reducing the probability of oviposition in an already occupied larval site (Price, 1971).

As the predictability of the resource increases, there will be selection for interference if energy spent on reducing the probability of use of the resource by other individuals enhances the relative reproductive output of the genotype. This need not occur only at or near K, although an individual in a population at or near K will tend to be selected for interference competition when the probability of successful reproduction depends on access to the resource at some later time. To select for interference behavior, availability of the resource must be reduced by use of the resource by the second individual. Generally the resource must not be "instantly" renewable, but must be removed from the resource pool at the time of use. There are probably relatively few resources of the instantly renewable type. There also may be natural selection for interference if the resource is superabundant, but resource quality differs. In the fiery-throated hummingbird (*Panterpe insignis*) the quality of the male territory as measured by the energy in nectar produced by flowers on the territory is important in determining the number of females that nest in or near that territory (Wolf and Stiles, 1970). These females normally mate with the male holding the territory. Thus if the quality of the territories varies as a result of clumped flower distributions, there will be some in which males mate with more females and can leave more offspring than can males in territories of poorer quality. Although there may be sufficient numbers of territories for all males, there still will be strong competition for the better territories (Orians, 1969b; Verner, 1964; Verner and Willson, 1966; Wolf and Stiles, 1970).

We have been arguing that exploitation competition has evolved because females producing offspring "perceive" the environment as unfilled. This is very similar to our argument that maximal reproductive output will be selected (that is, birth selection) in species that regularly find themselves in populations at some distance from K. The expectation, then, is that most species exhibiting exploitation competition will be selected for high birth rates. This is not to say that all species selected for high birth rates are exploitation competitors, because that depends on how resource availability is perceived by reproductive individuals or whether the resource being competed for regulates reproductive success. Male dragonflies often hold mating territories at ponds, but increased reproductive output of females might still be selected. Males compete for available females; the females lay eggs according to abundance of some other resource.

Far from being primitive, as suggested by Miller (1967), exploitation competition probably is a general phenomenon among species that regularly exploit initially superabundant resources that only become limiting after a period of use. Also important to exploitation competition is the fact that availability of further resource to the user at some later time is independent of the activities of the users. Exploitation competitors would be expected among species that tend to utilize resources that potentially are unpredictable in time and space. This unpredictability makes the value of behavioral interactions less certain; reproductive success from searching for more resources or using located resources as fast as possible may exceed the reproductive gains from the time and energy expended in interference behavior.

MECHANISMS OF COMPETITION

An organism in a competitive situation either (a) competes effectively; (b) is eliminated; or (c) moves elsewhere. Terrestrial plants are limited to the first two responses once the seed or other reproductive structure has reached the growing site. Motile organisms, on the other hand, can migrate to sites where competitive pressures are reduced or lacking. Sessile animals, such as many marine intertidal invertebrates, of course, also lack migratory ability as adults. According to the theoretical model, competitive devices ultimately come down to the question of which individuals reproduce more under those conditions. Mechanisms for maximizing reproductive output under competitive conditions form the subject of this section. The major focus will be on mechanisms allowing individuals to persist in the environment until they are able to reproduce more than their competitors. To be more successful at reproducing obviously means that the individual must survive to reproductive maturity and, finally, must channel more energy into reproduction. The outcome is increased abundance of the more successful genotypes in the next generation.

The mechanisms of individual competition can be considered in two categories: *intrinsic mechanisms,* those that act within the organism to make it more likely to survive and reproduce in competition, and *extrinsic mechanisms,* those that result from the activity of the individual but operate by reducing the competitive ability of other individuals. The dichotomy is not perfect but serves to distinguish between two major methods of winning at competition.

Intrinsic Mechanisms

An important ability in competition among sessile or relatively nonmobile organisms is the ability to survive under conditions of reduced resource availability, as, for instance, energy or mineral nutrients are depleted. Individuals that can slow their rate of growth or survive for periods with little additional resource may stand a greater chance of survival. Depending on how long the limitation continues this reduced growth may lead to decreased adult size. A common response in dense *Drosophila* cultures is for adult flies to be much smaller because the larvae pupate at smaller sizes (Bakker, 1961). If there is also a major change in body form required to achieve the adult stage (for example, butterflies) and there is a size threshold required for the transformation, reduced growth rate also could limit the production of adults in a competitive situation (Bakker, 1961).

Plants also are capable of size plasticity under competition. Brome grass (*Bromus*) grown in pots at varying densities shows a decreased weight per individual as density increases and an increasing weight per individual plant as the time from initial planting increases (Fig. 11-7). Obeid et al. (1967) were able to show that plants at low density had normally distributed weight frequencies but gradually developed log-normal distributions of weight through time; at higher densities the shift occurred faster. That is, the frequency of individuals in smaller weight classes increased relatively more than in larger weight classes. Even at densities in which the weight distribution at early growth stages was normal, the energy and nutrients were being preferentially taken up by fewer and fewer individuals. If the density becomes great enough or the competitive interaction stringent enough, the smaller individuals die first, thus gradually shifting the average per individual plant weight to higher values as the smaller individuals are lost from the population by differential mortality.

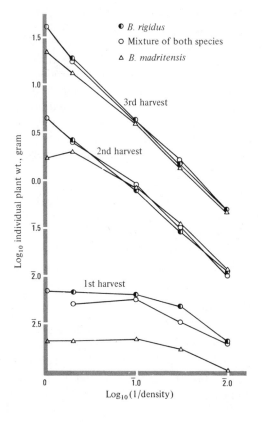

FIGURE 11-7 The relation between weight of individual plants and planting density (plants per pot) for *Bromus rigidus, B. madritensis,* and an average for a mixture of equal numbers of both species. The plants were harvested about 2 months, 2 3/4 months, and 10 months after planting. (After Harper, 1961.)

White and Harper (1970) have tested the hypothesis that intraspecific competition among plants in crowded conditions produces a higher mortality of smaller individuals in the population. They examined data from a variety of plant species and suggested that the "$3/2$ law" of Yoda et al. (1963) was a good description of competition effects on plant biomass. The law states that

$$W = Kp^{-3/2} \qquad (11\text{-}8)$$

where W is the weight of an average individual, K is maximum size of an individual, and p is the density of the surviving plants. Transformed to log-log axes this equation becomes

$$\log W = \log K - \tfrac{3}{2}\log p \qquad (11\text{-}9)$$

which is the equation for a straight line with slope of -1.5.

For a series of plant species, White and Harper found that the slope could not be statistically differentiated from -1.5 and therefore accepted the idea that this relation is somehow fundamental to intraspecific competitive interactions of plants. The Y-intercept ($\log K$) estimates the size an individual would achieve in the absence of competition. A similar density per individual biomass relation seems to hold for interspecific competitive situations.

White and Harper argued that differential mortality resulted from the changing abilities of individuals to obtain the limiting resource(s) as it was depleted by growing individuals. If light is an important resource, taller individuals in a dense stand will be able to obtain sufficient light energy while, at the same time, reducing the light energy available to smaller plants. As individual canopies overlap, the proportion of individuals being shaded increases and some begin to die. A hierarchy of resource exploitation ability, then, generates a differential growth rate pattern among individuals in the population and the smaller individuals die first.

White and Harper (1970) point out that the distance between nearest neighbors is

$$r = c/d^2 \qquad (11\text{-}10)$$

where r is the distance to nearest neighbor, d is the density of the plants, and c is a constant that depends on the distribution pattern of individuals [for example, $c = \tfrac{1}{2}$ if the distribution is random (Clark and Evans, 1954)]. If the area has a 100 percent canopy cover (that is, is filled with individuals so that they are competing), then the r value should be a linear estimate of the size of the individuals. In this case the volume of an individual can be approximated by r^3 and equation (11-8) can be written as

$$k_1 v = cd^{-3/2} \qquad (11\text{-}11)$$

where k_1 is a constant that depends on the shape of the plant. Since volume is an index of weight w, $v = k_2 w$, which can then be substituted into equation (11-11) to yield

$$k_3 w = cd^{-3/2} \qquad (11\text{-}12)$$

where k_3 is a constant relating k_1 and k_2. Rearranging this equation slightly produces

$$w = Kd^{-3/2} \tag{11-13}$$

which is equivalent to our previous equation (11-8) relating density and individual biomass. Here K is a constant that depends on how much space is available to each individual in competitive circumstances.

The hierarchy of resource exploitation can be determined at various stages in the life cycle of the plant, depending on which stages are important in determining the ability of the individual to exploit the limited resource. Black (1958) found that seed size in *Trifolium subterraneum* was important in competitive ability because it influenced the initial stages of growth, especially height growth. After 12 weeks in dense plantings of large and small seed types, plants derived from the larger seeds intercepted 97 percent of the light, leaving only 3 percent for individuals from the smaller seeds. Mortality rate was much higher in individuals from small seeds.

In initial seedling growth the redistribution of stored seed energy to various growing organs within the seedling may be important. Individuals that face strong competition for light will be favored if they channel more energy into elongation and structural support to overtop the competitor. On the other hand if the limited resource is a critical nutrient in the soil, individuals that produce large root systems may be favored. We might expect different energy channeling strategies among populations of a single species depending on the competitive conditions (McNaughton and Wolf, 1970).

Monk (1966) suggested that higher root-to-shoot ratios in herbaceous perennials than in annuals might be an important influence on the ability of perennials to displace the annuals competitively in a successional sequence. Harper and Ogden (1970) also showed that the relative amount of energy stored in roots in comparison to other plant organs could be influenced in *Senecio vulgaris* by planting density. At high density (Fig. 11-8) there was a relatively large amount of energy in roots and smaller amounts in shoots and reproductive parts. In addition, the timing of appearance of reproductive parts was correlated with changing energy distribution within the plant.

The initial growth rate of a seedling, regardless of the differential channeling of energy to organs, may be critical to later survival. Once a seed has reached its growth site, growth rate will then be determined partly by site characteristics (Harper et al., 1970) and partly by energy content of the seed. Selection should optimize the channeling of energy into individual seeds relative to early growth requirements as a mechanism for insuring the highest probability of seedling survival. Many plants reduce the total seed output per individual as plant density increases, but there is little change in the average seed size (Palmblad, 1968; Harper et al., 1970). A few species, such as sunflower (Clements et al., 1929; Fig. 11-9) and brome grass (Palmblad, 1968), reduce the average seed size at high plant densities.

For many organisms reduced size leads to reduced reproduction. For example, Harper and Ogden (1970) tested the reproductive output of *Senecio vulgaris*, a common weed, in a variety of crowding conditions. They reported that reproductive output remained at about 18 to 24 percent of the total energy stored in the

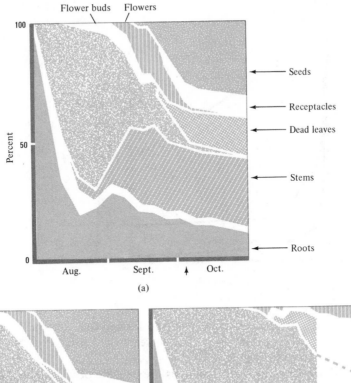

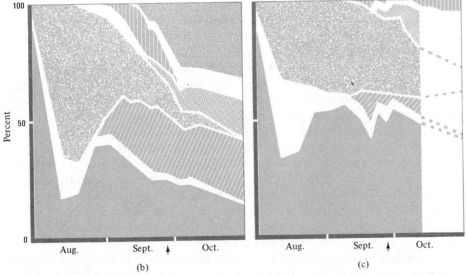

FIGURE 11-8 Percentage of energy channeled into various structures in *Senecio vulgaris* as a function of time since planting and density of potted plants. Energy is estimated here from dry weight biomass. (a) represents low density; (b) is moderate density; (c) is high density of individuals per pot. Arrows indicate when maximum total calories in the plants were recorded. (After Harper and Ogden, 1970.)

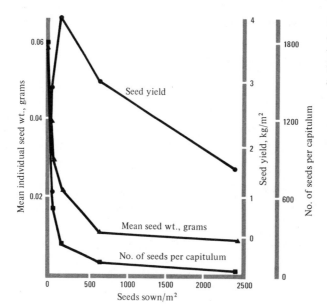

FIGURE 11-9 Variation in seed output in sunflowers as a function of the density of seeds planted. (After Clements, et al., 1929 and Harper, 1961.)

plant; that is, that smaller plants transformed the same percentage of energy into seeds as did larger plants. Absolute seed yield decreased, but proportional yield was approximately constant.

Many plants and some animals are capable of vegetative reproduction, so that the offspring are of the same genotype. An obvious advantage to the strategy is that offspring of a successful genotype can be produced without waiting for reproductive maturation of the parent plant or, especially in the temperate zones, without waiting until the next growing season for the seedlings to start growing. Vegetative reproduction diminishes the effect of gene flow on competitive ability and also reduces the time required for the competitor to add new individuals to the population. Nearly all the plants of *Rumex acetosella* that were added naturally to experimental field plots were derived from vegetative reproduction and not from seed set by the mature plants (Putwain et al., 1968). The *Rumex* species reproduced by both mechanisms, but the predominant input to the natural populations was from the vegetative rather than sexual (seed) reproduction.

The importance of this size plasticity is not universal in plants as shown in studies of *Papaver* in Britain (Harper and McNaughton, 1962). Individual poppies showed predominantly a mortality response to increasing density rather than size plasticity. There was some decrease in average size as density and time increased, but the predominant result was that there was an upper limit to the number of plants produced from increasing numbers of seeds per unit area (Fig. 11-10).

Similar types of differential growth characteristics and abilities to withstand reduced resource input without mortality are important in animals, especially sessile individuals such as adult marine invertebrates that remain fixed to the inshore rocky substrates and have larval and adult forms of limited mobility. Thus larval forms

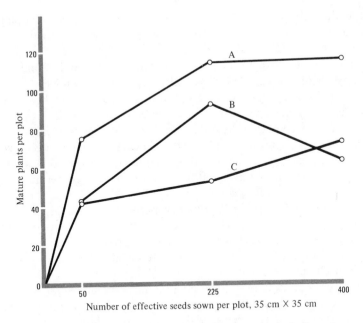

FIGURE 11-10 Influence of seed density on survival of mature plants of five species of *Papaver*. Survival is judged at final harvest. Each curve presents data from a different portion of the experimental garden to illustrate the influence of a heterogeneous environment on this type of experiment. (After Harper and McNaughton, 1962.)

of flies that are normally reared in an environment in which the resources occur as scattered clumps, such as carrion or dung piles, are unlikely to find a suitable new environment if they wander away from a site of intense competition. These forms generally should evolve responses similar to those of the stationary plants.

Extrinsic Mechanisms

Extrinsic mechanisms of competitive interaction involve actions by an individual that increase its probability of survival and reproduction by acting on another individual to reduce the impact of that competitor on the limited resource. These interactions in animals and plants might involve direct interference in obtaining the resource or a general reduction in the competitor's ability to use the resource. A general trend in plant succession is higher canopies in later stages. This is probably related to the importance of light as a limiting resource (Harper, 1967). Taller individuals maximize their own photosynthetic capabilities and shade shorter ones.

An individual that is fixed in space because the resource is clearly fixed in space can also evolve mechanisms to insure that the resource is not exploited at all by competitors. To do this requires eliminating the competitors from the resource zone of the individual. In animals territorial behavior may restrict competitors from entering and using a spatially delimited area (Brown, 1964; Watson and Moss, 1970). Plants can also maintain distinct zones in excess of their actual area of occupation by producing chemicals that inhibit the growth of other individuals in the surrounding area (Muller, 1966; Whittaker, 1970a; Whittaker and Feeny, 1971). McNaughton (1968a) and others have noted that chemicals may be detrimental to individuals of the same species (*autotoxic* effects) as a mechanism for limiting population size as well as to individuals of other species (*allelopathy*). Autotoxicity is most likely to be important in plants that invade relatively specialized sites in which they quickly reach nearly monoculture conditions. As species richness increases and the interspecific competitive ability of the species becomes more important because of multispecific interactions, presumably there will be selection against autotoxic detrimental metabolites. Many workers (for example, Blum and Rice, 1969; Whittaker and Feeny, 1971) have noted that the production of certain chemical compounds by species of plants characteristic of early successional stages may be important in determining the rate of successional change. We shall consider this in more detail in Chapter 13.

In summary, organisms may have intrinsic or extrinsic responses to increasing competitive pressure. Plants and animals, especially sessile forms, clearly show mortality and size plasticity, whereas other organisms show some sort of extrinsic response. The dichotomy is blurred by the fact that an intrinsic response, such as taller shoots, may lead to overtopping of surrounding individuals, an extrinsic interaction. The evolution of response types depends on the ability of the individuals to influence the competitive situation by site choice and on the characteristics of the limited resource. Light must be competed for by reaching the energy and keeping others from using or attempting to use it. Nutrients are probably best competed for by exclusion of individuals from the supply, rapid uptake, or ability to withstand low nutrient availability.

OUTCOMES OF INTRASPECIFIC COMPETITION

The only way competition may be diminished if competitors are members of the same species is to reduce population demand for the limiting resource. This may be accomplished by decrease in population density, increase in the range of resource exploitation, increase in individual efficiency, or substitution of another resource; then the population will be below a redefined carrying capacity, at least until the population increases to this capacity. Substituting other resources depends in large part on the availability of further resources, either absolutely or relative to the use of those resources by other populations. Expansion onto a resource requires that it be superabundant relative to current utilization or that the expanding species be the better competitor.

Islands sufficiently removed from the mainland or other island source areas probably approach an equilibrium number of species that is somewhat below what the island actually could support (MacArthur and Wilson, 1963, 1967). Because of the reduced number of potential interspecific competitors, the range of resources

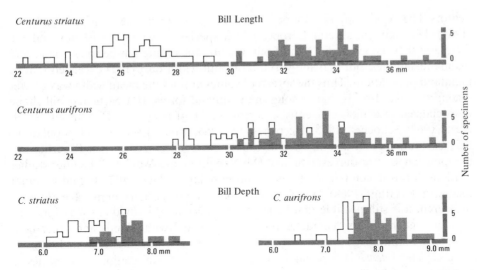

Centurus striatus   Bill Length

Centurus aurifrons

C. striatus   Bill Depth   C. aurifrons

FIGURE 11-11 Variation in bill length and bill depth in two species of *Centurus* woodpeckers. *Centurus striatus* is an island species; *C. aurifrons* is from a mainland population. Colored histograms are males; open histograms are females. (After Selander, 1966.)

potentially available to a population may be somewhat greater on islands than on the mainland. Selander (1966) has shown that on the island of Jamaica, where there is only one woodpecker species, the bills of the male and female woodpeckers are more differentiated than those of many mainland species that co-occur with other trunk-foraging species (Fig. 11-11). The bill differences reflect differences in the foraging techniques employed by the two sexes (Fig. 11-12).

A similar result was reported by Ligon (1968) for the red-cockaded woodpecker (*Dendrocopos borealis*) in the pine woods of the Florida peninsula. The usual explanation for this dimorphism is that it provides a means of expanding the food exploitation range of the species as its population expands in the absence of com-

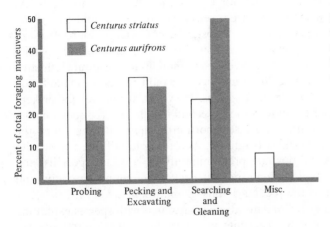

FIGURE 11-12 Comparative foraging maneuvers of the sexes of *Centurus striatus,* an island species of woodpecker. (Data from Selander, 1966).

petitors. The expansion need not be sex-related, as shown by the results of Van Valen (1965). He found in several European bird species occurring on islands and the mainland that the average width of bill was not much different between island and mainland samples, but that the variation of bill width on islands exceeded that of mainland populations. Thus the spread of values around the mean width was greater among the island birds than among the mainland forms. If a particular bill shape is specialized to exploit efficiently a narrow range of resources (Hutchinson, 1959; Kear, 1962; Schoener, 1965; Hespenehide, 1966; Root, 1967), island populations may be comprised of a series of individuals, each specialized for small portions of the total range of resources exploited (McNaughton and Wolf, 1970). Other studies of island forms in comparison to mainland populations have failed to find a similar relationship (Grant, 1968). However, since behavioral specializations have not been measured, it is still possible that behaviorally specialized individuals may form the basis for much of this "character release" associated with islands and other non-saturated community systems (Schoener, 1965; MacArthur and Wilson, 1967).

Character release is a broadening of the range of exploitation of a species in the absence of better competitors. This outcome is presumably predicated on the fact that competition is for such a limited resource that an efficient strategy for reducing competition is to switch to another. This will most likely occur when the present population is near or at carrying capacity; that is, when the intraspecific competitive pressure is strongest. An alternative, but not exclusive, possibility is that individuals may compete for the better subunits of a resource dispersed throughout an area or for a resource for which the efficiency of exploitation is important to the reproductive success of the individual even though the resource is not limited in an absolute sense. For some populations this competition could produce increased specialization. Previously, the example provided was of hummingbirds in different quality territories.

Gill (1971) studied two species of white-eyes (birds in the genus *Zosterops*) that were derived from mainland species, probably generalists (or opportunists) eating some fruit, some insects, and some nectar. After the arrival of colonists of the first species on Reunion Island in the Indian Ocean, the lack of competitors for nectar made it the most efficient energy source for exploitation by the individuals. As the population of the first invader increased, selection pressure would shift from generating increased specialization on the most efficient energy source toward the broader range of resources previously exploited by the mainland forms. If no competitors for the other resources arrived soon enough to constitute effective competition on the island, the first *Zosterops* species would reacquire exploitation range that was shown by the continental species. However, if another factor kept the population of the first species below K or if a second invader arrived soon enough and was more efficient on the fruit and insect portion of the food range, then the first species might retain the specialist condition. According to Gill the two species of *Zosterops* on Reunion Island now are predominantly (a) a nectar specialist and (b) a species with a broader exploitation range distributed among nectar, fruits, and insects in a manner similar to the feeding habits of the mainland forms.

For species that occur in environments already saturated with species exploiting the same resource gradient, it is unlikely that major evolutionary adjustments can be made to expand the exploitation range under the impact of intraspecific competition. In this case the populations can be maintained at or near the carrying capacity of the environment as modified by the presence of competing species or the population can fluctuate about this level. The degree to which fluctuations occur will depend in part on the selective background associated with long-term population trends (see Chapter 12). For species that are principally interference competitors, the level and severity of detrimental encounters will increase at or near the carrying capacity of the habitat, perhaps holding the population relatively stable. Carl (1971) noted that surplus individuals were forced to emigrate from high-density ground squirrel populations. Emigration exposed excess individuals to higher predation and reduced their numbers.

MECHANISMS OF COEXISTENCE

Two species that are complete competitors can coexist if, in a sufficiently short time, they can evolve ecological differences providing each with a subniche refuge in which the other species is absent or a sufficiently poor competitor. In other words they can coexist if they evolve until there is no longer complete competition between them but rather an overlapping *n*-dimensional space such that individuals continue to occur in the overlap section of the space by reinvasions from their exclusive subniches. An alternative possibility is that the niche of a particular species is completely included within the larger niche of another species. In this case survival of the species with the smaller niche depends on its ability to outcompete the larger niched species on critical overlap parameters.

It is likely that evolution driven by competition increases the differences between similar species in areas of geographic overlap. Brown and Wilson (1956) called this evolutionary response "character displacement." Although it is assumed that character displacement has interspecific competition as its driving force, relevant data to test the hypothesis are meager (Miller, 1967).

The most easily measured form of character displacement is morphological divergence of characteristics associated with exploitation of definable limiting resources. Hutchinson (1959) suggested that the morphological similarity of the feeding structures of sympatric congeners might be an index of the relative overlap of food portions of the niches. For several pairs of bird species, Klopfer and MacArthur (1961) found that the degree of similarity of bill length was greater among tropical than among temperate zone congeners. They suggested that the species had more similar niches in the tropics than in the temperate zone, possibly related to the more predictable food resources of the tropics and the increased ability of the species to specialize on these predictable resources.

Schoener (1965) examined many more bill length ratios of sympatric congeners and discovered that there was a relation between this ratio and estimated food availability. The ratio of the length of the longer bill to the shorter bill is generally larger than 1.14 among pairs of species that tend to feed on food in low abundance relative to population or individual body sizes. Low food abundances require that the individual exploit a wide range of available food more efficiently than a competitor species and will tend, evolutionarily, to separate characteristics of the feeding apparatus, thus permitting the species to specialize on different portions of the food range. Schoener argues that abundant food generally will permit more similar morphological feeding specializations. The more similar species become in morphological specialization, the more precise must be their food discrimination ability. At some degree of similarity, discrimination and distinctness can be achieved better by habitat or foraging behavior specializations than morphological distinctness.

In a study of five sympatric species of warblers (birds of the genus *Dendroica*) that had similar bill lengths (12.47 to 13.04 mm) MacArthur (1958) found that they also used the same gross habitats, but that each species tended to apportion its hunting time in somewhat different areas of the same species of tree (see Chapter 4) and even to forage in different ways (Fig. 11-13). It seems likely that these habitat and behavioral differences would cause birds to encounter somewhat different prey populations, thereby reducing competition between species.

Competing species also may coexist if relative interaction within and between species varies through time. The species in this case are sufficiently different along critical niche parameters for the competitive interaction to favor first one species and then the other because of changes in the environment through time. One obvious change is in relative frequency of the two species as the competitive interaction leads to exclusion of one of the species. If the value of the competition coefficient depends on the proportion of the total population contributed by each species, it is possible that high densities increase the influence of intraspecific competition and reduce the relative interspecific competitive ability. Alternatively at low relative densities the intraspecific interactions are reduced and a premium is placed on interspecific competitive ability (McNaughton and Wolf, 1970). Competitive ability can become frequency dependent and this may lead to an oscillating ratio of individuals in the two species. Elimination of one or the other species occurs only by chance as one species approaches the nadir of its relative abundance or encounters an exceptionally detrimental environment.

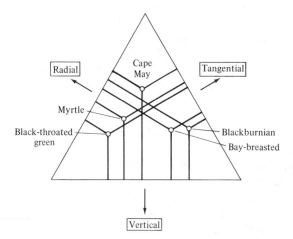

FIGURE 11-13 Foraging behavior of five species of warblers in spruce trees. The length of the line from the point to the perpendicular axis indicates the proportion of time each activity was used. Vertical signifies moving up and down in the tree; tangential signifies moving along the circumference with the trunk as the center of the circle; radial signifies moving along the axis from the center of the tree to the outside. (After MacArthur, 1958.)

Frequency-dependent competitive advantages result from shifting alpha values as relative abundances of competing species change. To maintain coexisting populations of each species, intraspecific competition at high densities must reduce the alpha of the most abundant species sufficiently for the other species to have the advantage in interspecific competition. This could occur if some other interaction, limited to conspecific individuals, became intense at high density. If autotoxins that were fairly specific in action on conspecifics became common at high densities and reduced the growth rate of the population, it could give the other species a temporary advantage. Behavioral interactions, such as intraspecific territoriality that may reduce population growth rate at high densities (Watson and Moss, 1970; Wynne-Edwards, 1962), may also give the other species a competitive advantage. The theoretical equilibrium density maintained reflects the different competitive abilities of the species and the detrimental effect of intraspecific interactions at high densities.

One way to view data that show frequency-dependent competitive interactions is to plot the relative output of offspring against relative abundances of the two species in the parental generation (Fig. 11-14). If all values fall along or are parallel to a line representing equality of output (45 degrees from the origin), there is no frequency-dependent advantage. If the points fall above the line, then one species is favored at all densities; if the points fall below the line, then the other species is favored. Points that fall directly on the line indicate ecological equivalence under conditions of the experiment such that neither species is favored. Frequency-dependent competitive advantage is indicated if the points fall below the line at high relative abundance ratios and above the line at low relative abundance ratios. The relative abundance ratio at the point where data cross the equality line is the theoretical equilibrium of relative abundances. Marshall and Jain (1969) tested two species of oats (*Avena fatua* and *A. barbata*) in single species cultures and also in mixed cultures. Their mixed cultures were tested at a variety of absolute densities and within each density class the relative abundance of the two species was varied. The results (Fig. 11-15) clearly showed that there was a frequency-dependent response in which reproduction of the less common species tended to be relatively higher than that of the more abundant species. However, the response varied with absolute densities, leading to the conclusion that frequency-dependent competitive advantage depends on the strength of the competitive interaction (that is, density) and the differential responses of the two species to intraspecific and interspecific competition.

Viewed in a slightly different way this same idea leads to another mechanism of coexistence. Assume that the competition coefficients remain constant but that each species is a better competitor in a slightly different set of environmental conditions. Under constant conditions, one species or the other would eventually win out in a clear case of competitive superiority. Identity of the winning species will be dependent on competition coefficients in the experimental environment. If competitive exclusion requires a finite, but relatively long, period of time to succeed, it is possible for environmental conditions to shift through time such that first one species then the other is favored. Temporal patterns of environmental change will determine whether either species can increase sufficiently to win the competitive interaction before conditions change to favor the other. This possibility is quite

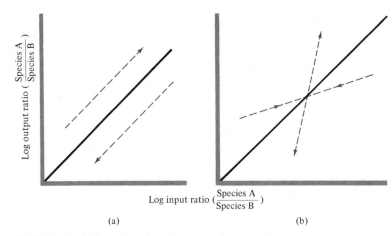

FIGURE 11-14 Plots showing frequency independence (a) and frequency dependence (b) of output of two species that are potential competitors. The solid line in both graphs has slope 1.0 indicating that the output ratio equals the input ratio and that there is no change in relative abundance. The dotted lines in (a) show that the output ratio is consistently higher or lower than the input ratio. One or the other species will eventually be eliminated. In (b) the output ratio (slope $\neq$ 1.0) varies as a function of the input ratio. If the slope is less than 1.0 (dashed line with arrows pointing toward the solid line) species A increases in relative abundance when rare and decreases in relative abundance when common. Theoretically, the relative abundance should tend to an equilibrium at the point where the dashed line crosses the solid line (output = input). If the slope in (b) is greater than 1.0 (dotted line with arrows pointing away from solid line), A increases in relative abundance when common and decreases when rare. Depending on the starting relative abundance ratio, A either displaces B (start above solid line) or goes extinct (start below solid line). (After Harper, 1967.)

similar to the previous example of fluctuating alpha values with changing density; in both cases the environment is changing. In the present case environmental characteristics are time-dependent. In the former example environmental characteristics are dependent on relative densities. In either case the major requirement for coexistence is that the two species be sufficiently different for one species to be favored at one end of the changing continuum and the other to be favored at the other end.

A classic, hypothetical example of the impact of temporal environmental changes comes from Hutchinson's (1961) analysis of the "paradox of the plankton." The number of plankton species in a small sample of water is much greater than might be expected if the habitat is homogeneous and the competitive interactions are intense. Hutchinson hypothesized that co-occurrence of planktonic species was dependent on relatively rapid seasonal shifts in environmental conditions favoring first one and then another plankton species. Richerson et al. (1970) hypothesized

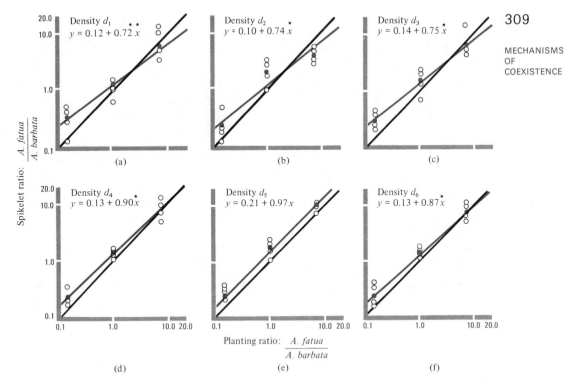

FIGURE 11-15 Input-output ratio diagrams for *Avena fatua* and *A. barbata* showing influence of density of plants on frequency dependent reproduction and growth. Output ratio is number of reproductive stalks for each species. Densities in these experiments doubled successively from 8 plants per pot (d_1) to 256 plants per pot (d_6). Asterisks indicate that the slope of the line is significantly different from 1.0. See Figure 11-14 for a theoretical explanation of these graphs. (After Marshall and Jain, 1969.)

on the basis of studies of Lake Tahoe, California, that a large amount of spatial heterogeneity existed in physical and chemical characteristics of the lake and that the distribution of species was in scattered areas throughout the water. They reasoned that even with continual mixing, there can be spatial areas within the habitat that favor different species. They stressed that the two hypotheses to explain plankton diversity are not mutually exclusive but could reinforce each other in generating diversity.

Let us consider the influence of seasonal changes on competitive advantage in intraspecific competition. Dobzhansky (1943) found that relative frequency of the gene inversions in two California populations of *D. pseudoobscura* changed in a predictable fashion associated with seasonal habitat changes (Fig. 11-16). However, it is difficult to know whether the changes reflected (a) differences in reproductive potential in relation to environmental conditions, in which there is no competitive interaction between carriers of the two morphs or (b) whether all or part of the frequency changes reflected different competitive abilities of the morphs

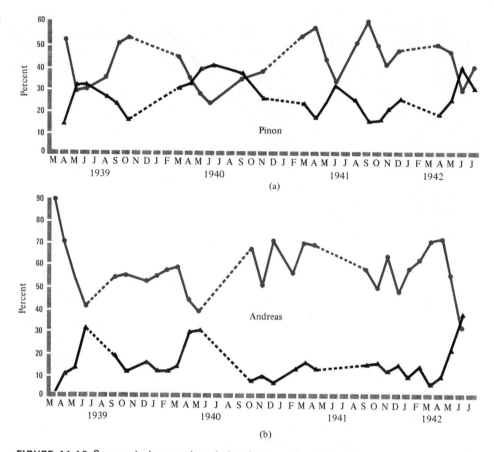

FIGURE 11-16 Seasonal changes in relative frequencies of two chromosome patterns in populations of *Drosophila pseudoobscura* at Pinon Flats and Andreas Canyon, California. (After Dozhansky 1943.)

in the changing environment. It is also possible that the results derived from differential migration into and out of these populations. Beardmore (1970) reported that the frequency of the fast allele at the esterase six locus of *Drosophila* increased in frequency at low temperatures in a fluctuating or constant temperature environment ranging from 20° to 30°C. He also found that frequency of this allele increased with population density.

A possible mechanism for increasing relative growth rate of a population subject to interspecific competition is an increase in the total number of offspring produced. If this can be accomplished at low relative densities, the rarer species could be at an apparent competitive advantage. One possible way to achieve greater reproductive output is to increase the production of females relative to males at low densities. If one male can mate with many females, less of the critical resource would be channeled into nonreproductive individuals. On the other hand, Harper (1967),

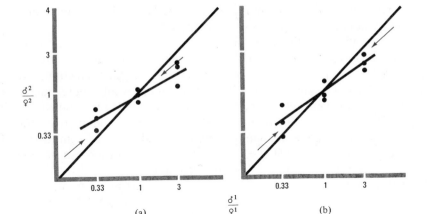

FIGURE 11-17 Male to female ratio of *Rumex acetosella* planted in the spring $(\delta_1/\female_1)$ plotted against the ratio the next autumn $(\delta_2/\female_2)$. Dry weight of above ground parts is shown in (a); (b) indicates the number of inflorescences. Values for three replicates are plotted individually. Arrows show direction of frequency changes relative to initial frequencies. (After Harper, 1967.)

reporting data collected by Putwain, noted that the output frequency of male to female plants in *Rumex acetosella* changed with varying input ratios to tend toward equality of sex ratio (Fig. 11-17).

COEXISTENCE IN POPULATIONS MAINTAINED BELOW K

Competitive displacement of coexisting species could occur if the fundamental niches of the species overlapped sufficiently and population size reached a point where the resource(s) involved in the overlap became limiting (reach K). In this case one key to coexistence is that the population size may be kept sufficiently low for competition to occur rarely, if ever, and for the species to coexist even though they are potential competitors. Paine (1966) studied intertidal invertebrates that coexist as sessile organisms along the coast of the state of Washington. In the presence of *Pisaster ochraceous,* a predatory starfish, the abundance of the best competitors was kept sufficiently low for other species of invertebrates to remain as part of the community. When he experimentally removed starfish from a section of the community, the number of prey species dropped from fifteen to eight in about one year. Paine thought that the eventual outcome would be nearly complete dominance of the rock faces by the one or two species of mollusks that could grow best under those conditions. He concluded that predators may sometimes serve to keep potentially competitive prey at sufficiently low population levels to prevent the complete elimination of the poorer competitors. In his experiments the continued existence of several species depended on the preferential predation by the starfish on the best competitors. If for some reason starfish preferred to prey on poorer competitors, the system would move even more rapidly than in the absence of predation toward a community dominated by the best competitor(s).

It seems probable, in this case, that growth characteristics associated with the best competitors were also important in determining the predator preference. If the predator showed no preference for prey species, the rate of competitive exclusion would be slowed. Presumably the better competitors would more than replace themselves in subsequent settling periods, eventually leading to competitive exclusion. Only with preferential predation, "switching," on common or good competitor species could the predator maintain complete coexistence of competitors. This preference must be reversed if the preferred prey becomes too rare to prevent it from eventually being eliminated from the community.

We will return to the problem of how species coexist in nature in Chapter 19.

LIMITING SIMILARITIES OF COEXISTING SPECIES

Throughout this chapter we have assumed that two species can coexist if they are "different enough." However, we have yet to discuss what we mean by "different enough" and whether, in fact, this is a definable entity. A similar question is what might happen in a saturated *n*-species system to which another species is added. Finally, we come to the critical question of what determines the numbers of coexisting species in a community. This last question will be considered from other standpoints in Chapters 13 and 19.

Although we can never precisely answer questions such as how similar is too similar to coexist or how different is too different, we nonetheless can suggest hypotheses about variables that may influence the ability of two species to coexist. If an initial assumption is made that we are looking at a single resource gradient and the ability of different phenotypes (individuals or species) to exploit resources along the gradient, it is possible to predict theoretical situations in which a third phenotype may invade or may be excluded.

Assume that the ability of two phenotypes to exploit a resource gradient is somewhat different. A graph can be drawn as in the left portion of Figure 11-18a that represents the ability of a range of phenotypes to exploit two portions of the resource gradient (MacArthur, 1965). The phenotype that is most successful in depleting the resource, presumably a relative specialist on that resource, will outcompete all other phenotypes on that resource. This would be similar with a second phenotype on a second resource. Each phenotype along the horizontal axis reduces each of the resources represented by the U-shaped curves to a specified level. In the graph, phenotype I can deplete resource one to a moderately low level and resource two only to a very high level. The ability of a phenotype to reduce a resource depends on its adaptations to exploit that resource. Thus if the two resources are relatively different in terms of phenotype differences, a specialist on resource one will have difficulty exploiting the second. These differences in resources are related to the range of exploitation efficiencies of different phenotypes and the position of the resource along the gradient. If a change of one unit along the phenotype axis brings a marked change in the ability of the new phenotype to exploit the resource,

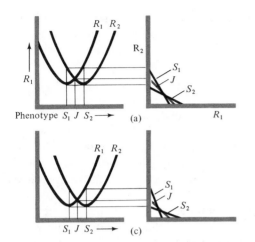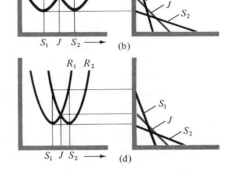

Phenotype S_1 J S_2 ⟶ (a)

S_1 J S_2 ⟶ (b)

S_1 J S_2 ⟶ (c)

S_1 J S_2 ⟶ (d)

FIGURE 11-18 The relationship between phenotype specialization, difference between two resource types, and the ability of phenotypes to co-exist. Left-hand curves plot minimum resource amount that will support a population along the ordinate and phenotypes along the abscissa. In (a) phenotype S_1 can survive when resource 1 is at a lower level than it can on resource 2; the reverse holds for phenotype S_2. Curves of the same shape with nadirs relatively more separated (b) clearly are more difficult for a single phenotype to exploit. A similar result holds as the curves steepen with nadirs in the same position relative to the two phenotypes. A third phenotype J, the so-called jack-of-all-trades, because it is less specialized than either S_1 or S_2 can outcompete either specialist in these cases only when it can reduce both resources to levels below which neither specialist can maintain its population. The right-hand graphs transpose the curves to resource axes as illustrated in (a). J outcompetes S_1 and S_2 when the line for its resource utilization somewhere lies inside the crossing of lines S_1 and S_2. (After MacArthur, 1965.)

then the shape of the U is very steep. Alternately, if the U is shallow, a one-unit change in phenotype will represent little change in exploitation efficiency. The relative resource exploitation ability of two phenotypes separated by some arbitrary number of units on the X-axis depends on how different the resources are and the exploitation plasticity of the phenotype.

From these graphs we can ask whether a specialized or more generalized exploiter of the two resources will be favored in competition (MacArthur, 1965). The U-shaped graphs are transformed into new graphs in which the axes are the level of the resource of each type and a line represents the ability of one phenotype to reduce each resource (Fig. 11-18). If we introduce a third, intermediate phenotype, it is possible to predict from these graphs when the intermediate and hence more generalized phenotype (a) will be able to outcompete the two specialists and (b) when the two specialists will outcompete the generalist.

When the resources are sufficiently similar and the two specialists are not too different, the intermediate form will be able to reduce resource levels below the one resulting from the combined activity of the two specialists (Fig. 11-18a). The gen-

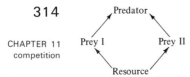

FIGURE 11-19 Theoretical system of predator-prey resource interactions that may or may not be stable depending on the number of limitations of population growth for the species. See text for further explanation. (After Levin, 1970.)

eralist will replace the specialists by being relatively good on two similar resources, whereas the specialists are too specialized for their particular resource. As the resources become more different relative to the phenotypic response, the specialists do better, each on its separate resource (Figs. 11-18b, d). The more the specialists can reduce the resources, the less likely that the intermediate will win the competitive interaction (compare Figs. 11-18a and c).

These graphical analyses suggest that the number of species coexisting in an equilibrium system is to some extent regulated by both exploitation ability of phenotypes and degree of resource difference along the gradient. As resource predictability at any point increases, the probability of specialization probably also increases. Similarly, the possibility of specialization on rare resources is decreased. In essence the argument is that more abundant and more predictable resources allow a phenotype to specialize and to maintain a viable population. As predictability and availability decrease, the size of the population that can be supported decreases. An important factor may be resource divisibility. Predators taking insects could specialize on size categories, on taxonomic units, or on habitat parameters (Root, 1967), whereas plants competing for light will be much less able to subdivide the parameters distinguishing photons (wavelength, intensity).

A theoretical analysis suggested to Levin (1970) that a community in equilibrium could exist provided the number of species was identical to the number of factors limiting the abilities of separate populations to grow. In a simple predator-prey system (Fig. 11-19), the two prey species can coexist if they are limited by separate combinations of predator and resource limitation. However, if the two species are limited only by the predator and are feeding on a superabundant resource, one species or the other will eventually be eliminated by predation. This suggests that communities with fewer species than those that potentially limit the resource combinations have not reached a species abundance equilibrium. According to this view the species diversity of a community depends on the number of combinations of independent limitations present (McNaughton and Wolf, 1970).

FIELD ESTIMATES OF COMPETITION COEFFICIENTS

Competitive interactions between two species have been defined in terms of the relative impact of one species on the ability of the second to increase. This was cast in terms of the changes that could be expected in the logistic description of population growth because of dependency on competitive interactions. Application of the theory to natural situations requires that alpha values be estimated from field data. The logistic equation normally is viewed as too simplistic an explanation of population growth in natural situations (see Chapter 9) and hence modifications of the logistic equation to reflect competition would be too simplistic. However, the idea that competition acts to reduce ecological success can be cast in terms of successful reproduction and the relative impact of several species on limited or variable quality resources. In this case some approximation of alpha becomes important for predicting the relationship between two species.

It is probably impossible, however, to estimate alpha values from field data without first knowing the K values for each species (virtually impossible to measure) and the equilibrium abundances of each species (a practical impossibility since it is hard to know when equilibrium is achieved for species that theoretically should not coexist if they are both regulated by the resource). In measuring relative abundances in the field we must assume that each individual is equivalent. The implicit assumption then is that for species limited by the same resource the alpha is such that the equilibrium (if there is one) is unstable and will eventually lead to the extinction of one or the other species. The conclusion is that present theory does not permit us to generate meaningful alpha values from observational field data. However, it should be possible to perform field experiments in which individuals of a competing species are added to the community. The changes in population sizes of the competitors through time should provide an estimate of the competition coefficients. This type of experiment is equivalent to those of Gause, Ayala, and Vandermeer mentioned earlier in which alpha values were calculated from two-species laboratory experiments.

GENERAL CONCLUSIONS

1. Competition is a process of population interaction when two or more individuals make simultaneous demands on a type of resource, or quality of resource, in insufficient supply to meet the needs of all individuals.
2. Competition influences the ability of an individual to survive and reproduce, and hence the ability of the population to maintain itself or to increase through time.
3. Competition may occur between individuals of the same species or individuals of different species. Both short-term and long-term outcomes of competition are influenced by the degree of genetic similarity and probabilities of gene exchange among competitors.
4. Except for ecologically identical or very similar populations, there are four possible outcomes of interspecific competition in a two-species system: (a) one species is the better competitor and always wins; (b) the other species is the better competitor and always wins; (c) the species coexist indefinitely in a stable equilibrium; (d) there is an unstable equilibrium, although the eventual outcome is that one species or the other always wins. In the last case both species compete better against the other species than against individuals of their own species. The eventual winner will depend on which species achieves numerical superiority first.

    Coexistence in a stable equilibrium depends on competitors being "different enough" in ecological requirements for each to always have some refuge within its niche where it is competitively superior.
5. The two major types of competitive interaction, interference and exploitation, are characterized by specific mechanisms of resource exploitation. Each depends on

the characteristics of the resource that is determining K for the competing populations.

6. Individual mechanisms affecting probability of reproduction in a competitive situation will depend in large part on the degree of mobility of the organism and the type of resource setting K. Nonmobile organisms must evolve mechanisms for reducing the probability of a competitor occurring nearby or for increasing the probability of survival in competitive situations. Many sessile organisms show size plasticity and ability to survive times with little input of some critical resource. Mobile organisms may have similar mechanisms, but have the added capability of searching for a less competitive situation.

7. Competitive interactions provide strong selection for differences allowing individuals to coexist. For intraspecific competition this usually involves increasing the range of resource exploitation or exploitation of new resources so that individuals specialize on different resources. The outcome will appear as a broadening of the resource range for the population. Interspecific competition will tend to increase the specialization of competing species so that each has a resource refuge where it is the most efficient exploiter.

8. Nonequilibrium coexistence is possible if (a) competition coefficients change through time or (b) the environment changes through time to favor first one species and then the other.

SOME HYPOTHESES FOR TESTING

A continuing controversy surrounds the role of competition in defining the similarities and differences of coexisting species in natural communities. It is important to know what the role of competition is in generating the abundances and biological characteristics of species in communities. Experimental tests can be designed that either increase the level of competition by increasing population sizes or decrease competition by selective removal of individuals. With appropriate controls these tests should permit some quantification of the role of competition in coexistence.

As competitive interactions are shown to be important in ecological interactions (see Connell, 1961), a further, and very important, question is to identify resource gradients along which species compete. Plants obviously compete for light, water, and nutrients in many habitats. But the mechanisms of differentiation along gradients of these resources are poorly understood. How do species subdivide the resource gradients?

The ability to subdivide resources will play an important role in defining the number of species coexisting in a community. Each of the gradients could provide a potential resource on which a particular species might specialize. The occurrence of the species in nature then would reflect the distribution of resource specialties throughout the habitat space coupled with the occurrence of competitors. In the absence of competitors for a resource on which a species is less specialized, habitat occupancy of the species can expand. In species-dense communities the relative abundance or geographic distribution of the species will reflect similar properties of the resource.

Finally, it is important to recognize the possibility that some general statements can be made about the importance of resource limitation in setting population sizes. Hairston et al. (1960) started a continuing controversy (Murdoch, 1966*b*; Erlich and Birch, 1967) by suggesting that populations in some trophic levels are probably regulated by the availability of resources and at other tropic levels by the impact of predators.

population regulation

12

For years ecologists have debated the idea of regulation of numbers in natural populations (see recent articles by Brown, 1969a; Murdoch, 1970; Wynne-Edwards, 1970). Continuing discussions of terms such as *density dependence* (approximately equal to the idea of K-responsive) and *density independence* (similar to K-unresponsive) occupied important places at meetings and in the literature (*Cold Spring Harbor Symposium Quantitative Biology*, No. 22, 1957). The importance of competition, weather, and behavior in determining population numbers has stimulated many publications (see Andrewartha and Birch, 1954; Lack, 1954, 1966; Wynne-Edwards, 1962). In addition numerous studies and review papers have continued to ask if and how populations are regulated in nature. The controversy shows little sign of abating, but there have been some recent changes of direction that may produce significant results.

A major difficulty with many discussions of the problem of population regulation has always been in defining the term *regulation,* or *regulated,* especially in a way that allows appropriate tests. Many ecologists now would agree that a regulated

population is one that regularly tends toward a density approximating the ability
of the environment to support individuals (Murdoch, 1970). The capacity of the

environment to support a population changes through time in relation to the availability of critical resources, changing age and genetic structure of the population, and changing external sources of mortality. A regulated population might change in size in a pattern reflecting "tracking" of changes in carrying capacity of the environment. Tracking precision depends on the time it takes the population to respond to environmental changes. Precise tracking requires a strong feedback relation between organism and environment. The strength and the rapidity of the feedback relationship will determine the speed with which a regulated population changes in response to environmental fluctuations.

Populations with the same degree of regulation, as measured by the strength of the feedback relation, could exhibit quite different rates at which they converge on a new carrying capacity. The shape of the trend curve also could be quite different, oscillating or smooth, depending on how the population grows in response to changes in environmental capacity. One regulated population might approach the new environmental capacity gradually, whereas another might approach the new level with a series of oscillations of decreasing amplitude. These populations could be equally regulated, but differ in how the individuals respond to the negative feedback. On the other hand an equal negative feedback might take longer to translate into a changed population size depending on the age structure of the population, the breeding seasonality, the proportion of females, and so on. And, as pointed out in the population growth chapter, the characteristics of the response can vary within a population depending on the kind of factor setting the limit. Although different responses to changing environmental capacity make the process of regulation difficult to study, they do not negate the phenomenon of population regulation. Regulation will depend on the negative feedback between population characteristics and environmental capacity, whereas the rate of regulation will depend on the response characteristics of the population.

This, however, merely defines what is meant by a regulated population and does not provide a mechanism for testing for the presence or the absence of regulation in any given population. It is not sufficient merely to show that a population tends to expand to a certain level and to fluctuate around that level when the concurrent fluctuations in the environment are not accurately known. The unknown environmental effects lead to the concept of regulation without evidence that the population size is being controlled by some feedback relationship between the environment and the ability of the population to expand numerically. This is not to say that a population at a more or less steady state of numbers is not regulated, but only that to show regulation we must also know the ability of the environment to support individuals.

In Chapter 9 we discussed the mathematical limitation of using correlations of numbers at consecutive censuses to identify regulated populations. Both Maelzer (1970) and St. Amant (1970) concluded that sampling errors and inherent mathematical problems do not permit a rigorous test of regulation from census data, and they recommended experimental manipulation of populations as perhaps the most effective test of regulation.

The most appropriate experimental analysis involves the manipulation of populations, with appropriate controls, both above and below the apparent equilibrium level (Murdoch, 1970). In a regulated population experimental densities higher and lower than the controls should return to the control level if the population is regulated. The rate of return will depend on the type and intensity of the regulating feedback and the growth and mortality characteristics of the population.

Eisenberg (1966) is one of the few workers to attempt an actual experimental test of regulation—in a population of pond snails (genus *Lymnaea*). He increased the snail density in some enclosures to population sizes of 5000 per enclosure and reduced density in other pens to about 500 per enclosure. Controls remained at about 1000 individuals per enclosure. There was little proportional difference in adult mortality in the three types of enclosures (Fig. 12-1), so the experimental differences in adult densities were maintained throughout the period of the experiment. Control of density thus was not a function of differential mortality of the adults in relation to adult density. However, there were major differences in production of young in the three situations (Fig. 12-2), so the populations of young snails that would comprise the next breeding generation were nearly equivalent in all three situations (Fig. 12-3). The impact of changed densities on reproduction was so immediate and so strong that density in experimental enclosures converged on controls by the end of the reproductive period for that year. Rapidity of the response to changed density surprised even the investigator, "Regulation . . . was found to be more complete and to occur more rapidly than had been thought possible." (Eisenberg, 1966:903).

The primary concern of most ecologists investigating population regulation has been with factors that might be responsible on the assumption that regulation was obvious and important in the ecology of many organisms. But disagreement is widespread about the importance of regulation in nature. There also is disagreement as to which factors might be important in regulating populations, and as to whether

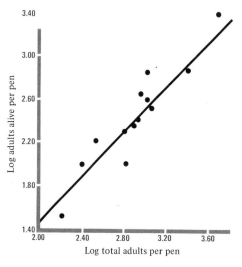

FIGURE 12-1 Relation between total number of adult snails per pen at the start of the experiment and the number of living adults per pen 5 months later. Each point represents a pair of pens. (After Eisenberg, 1966.)

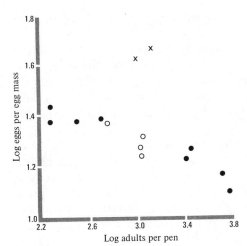

FIGURE 12-2 Relation between the number of adults per pen and the average number of eggs per egg mass laid in the pens. The low density pens have higher egg output per mass. X's indicate pens provided with supplemental food; circles are control pens; dots are pens in which densities were increased or decreased. (After Eisenberg, 1966.)

an apparent regulatory factor actually is affecting the population or is merely correlated with another, more ultimate (in the evolutionary sense) regulatory factor. There are numerous classifications of regulatory phenomena (for example, Andrewartha, 1957; Chitty, 1960; Nicholson, 1957) that attempt to differentiate factors by how they influence populations, by whether they are factors external to the populations or within a population itself, or by the kind of response produced.

To act as a regulatory mechanism any factor must influence reproduction, mortality, migration, or all three factors in such a way that the population tends to increase when below the environmental capacity and decrease when above. The major environmental cues available to trigger a response are population density or some effect of that density on the environment of an individual. The regulatory factor must influence, either directly or indirectly, the individuals in relation to density. Provided the effect is stringent enough, it need not be graded as the environment approaches its full capacity to support more organisms; however, the response of the population will depend on whether the effect is graded or a threshold (see

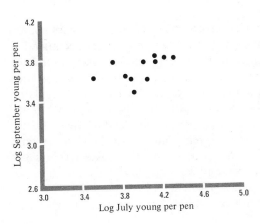

FIGURE 12-3 Relation between the number of young snails alive in the experimental pens in July and the number alive in September. Note the relative constancy of numbers alive in September. (After Eisenberg, 1966.)

Chapter 9). We are assuming that the environmental capacity is K and that the individuals in the population are responsive to K.

In the following discussion, factors that have been implicated or suggested as important in regulating populations will be presented. *External* factors are environmental agents acting to influence numbers. *Internal* factors are changes within the organisms as density changes. External factors include physical factors, principally climate, competition, disease, and predation. Internal factors include physiology, behavior, and genetics.

As previously noted there have been relatively few attempts to test the regulation of a population experimentally. Most of the work has been on populations that maintain a relatively steady population density or else show predictable and fairly regular changes in population size through time. Changing populations actually may provide the best nonexperimental clues as to the regulatory factors for populations if the cause-and-effect relationship can be ascertained between density and a factor that controls it. Much of what we say about some of the possible regulatory mechanisms will depend on studies that have been done on fluctuating populations, such as rodents and insects. The major problem with this approach is that some factors that occur with regularity may be associated with density of the population, but only because they influence the length of time the population has to reproduce and increase to a certain density or because they influence the total reproductive output of individuals regardless of population density. These factors presumably would generate similar responses independently of population density and hence cannot be termed regulatory factors. Strongly fluctuating populations can occur with regularity if favorable conditions for population growth occur at regular intervals and are followed by periods of harsh conditions that cause the population to decline (Fig. 12-4). The increase and decline will occur despite population density. The height of the population peak will be related to the length of time the conditions are good and bad. The key is to show that the fluctuation is caused by the negative feedback on population growth of some factor(s) in the environment that is important for determining population density.

COMPETITION

In Chapter 11 we clearly separated two types of competitive interaction—exploitation and interference— that reflected two fundamentally different ways of affecting other populations drawing from limited resource pools. In exploitation competition each individual, acting independently, attempted to accumulate as much of the resource as possible. In organisms in which the amount of accumulated resource regulated the probability that the individual would reach maturity and also the number of offspring subsequently produced, exploitation interactions could lead to widely fluctuating populations (such as the blowflies in Chapters 9 and 11) if the resource limit is regularly encountered. Unless the response to decreasing resource levels is felt almost instantaneously in decreasing survival and reproduction by the extant population, there is little chance that the population will stabilize at a predictable level relative to the rate of availability of the limited resource. Presumably, in a population

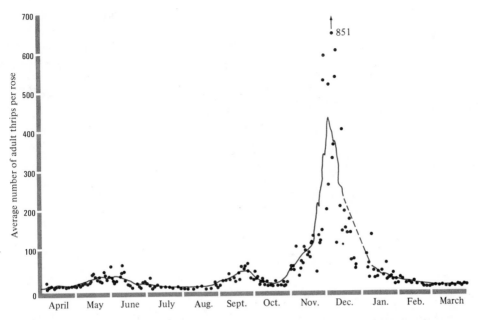

FIGURE 12-4 The population of *Thrips imaginis,* measured as number per rose, occurring on roses near Adelaide, Australia, in 1932–1933. The curve is a 15 point moving average to smooth small fluctuations in the daily records. The peak in abundance is associated with the season of rainy weather in this area. (After Davidson and Andrewartha, 1948.)

that regularly encounters a resource limitation (for example, in the laboratory), exploitation competition will generate an oscillation about the level at which a steady state population could be maintained, but it will rarely maintain numbers at that level for any length of time and then probably only by chance.

The ability of a population to return to a fairly definite control level after a density perturbation will presumably be based on assuring the availability of sufficient resources for the steady state number of individuals. At experimental levels below the control population size the population will gradually return to the control level at a rate dependent on the growth rate of the population. Populations shifted to levels higher than the control level will show decreased survival and reproduction until the steady level is regained.

The experiments with pond snails by Eisenberg (1966, 1970) showed that density perturbations about the control level generated changes in reproduction, possibly by affecting energy available for reproduction. He tested this hypothesis by adding extra food (boiled spinach) to some populations, and found that reproductive output and size of the adults increased as the amount of added food increased. The snails responded by changing growth rates and total size attained and also by changes in number of eggs per mass and total number of eggs produced (Fig. 12-5). In the 1966 report, however, the major change was in the production of young because the experiment started after adult size had been attained. Eisenberg concluded that the

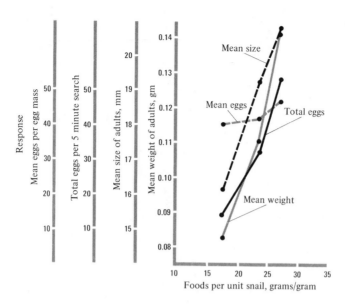

FIGURE 12-5 Relation between the amount of food provided per snail and growth and reproduction characteristics of snails in experimental pens. (After Eisenberg, 1970.)

type of population response depended on the timing of the resource limitation relative to the life cycle of the individuals. Even for populations that practice exploitation competition it is possible that density might be regulated fairly closely depending on the influence of the regulating resource and the timing of its impact. Nicholson (1957) reported that blowflies that initially were cycling strongly with limited larval food intake and unlimited adult food availability would reduce the amplitude of the population oscillations when the adult food intake was also limited (Fig. 12-6). If the resource limitation acts quickly on the survival and reproduction of the adults, it is possible to dampen the oscillations as in the blowfly experiment. The damping effect will decrease as the time lag between reaching the resource limit and the impact of that limitation on the growth of the population increases. In Nicholson's experiment when the larvae had sufficient food and only the adults were limited, there was a time lag in the decline of adult members of the population since the limitation was on their ability to reproduce and not on their ability to survive. The limitation in resource would not be evident until a generation later; then it would be reflected in a violent decline in adults since the resource was parceled among them so that none got enough to reproduce at the normal level. If some adults had obtained sufficient protein and others none, presumably the decline would have been less severe.

Organisms that interact via interference competition can reserve sufficient amounts of a limited resource to maintain the population at or below the potential carrying capacity. Populations that fall below this level will increase and populations above this level will decrease, but potentially with less violent oscillations than in populations that interact via exploitation competition. Some birds that have evolved territorial behavior as an overt behavioral result of interference competition have reproductive responses closely correlated with increasing density.

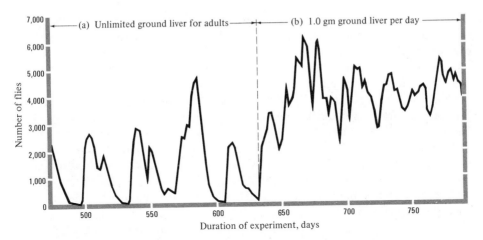

FIGURE 12-6 Changes in numbers of a laboratory blowfly population through time. During the first half of the experiment the adults had unlimited food available, but the larvae were food limited. During the second half, both adults and larvae were food limited. Note stabilization of the population fluctuations when both life stages were food limited. (After Nicholson, 1957.)

The clutch size of the great tit (*Parus major*) in Wytham Wood, near Oxford, England, declines nearly linearly with increasing density of adults (Fig. 12-7). However, whether territorial behavior is important in regulating population size is a matter of continuing debate (Watson and Moss, 1970; Brown, 1969*a*).

Brown notes that there are essentially three levels of response among territorial breeding individuals as density increases: (a) the population size is below what can be supported by that habitat and there is no exclusion of breeding adults; (b) there are nonterritorial, nonbreeding adults, but only males; (c) nonterritorial individuals include some females that are excluded from breeding. In the first two cases there is no absolute limitation on breeding imposed by the resource level since all potential

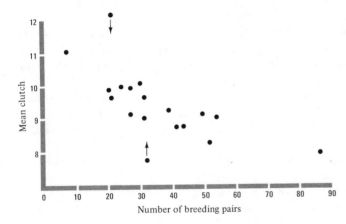

FIGURE 12-7 Relation between clutch size and the density of breeding pairs of Great Tits (*Parus major*) in Marley Wood, near Oxford, England. Each point represents a separate year from 1947 to 1964. The arrows indicate unusually early or late seasons. (After Lack, 1966.)

reproductives in the population are mated and resident on a territory. In the second case, however, the reduced output of young that may accompany increasing density can still reduce the total reproductive output. Finally, in the third case the exclusion of some females from breeding in the area potentially restricts the reproductive capacity of the population.

There are many cases of adult male birds that have been removed from a breeding territory being quickly replaced by other males (Hensley and Cope, 1951; Orians, 1961; Watson and Moss, 1970), which suggests that territorial behavior in fact can limit the number of territorial males. There are relatively few studies (for example, Carrick, 1963; Delius, 1965; Ribaut, 1964) that document the exclusion of females from breeding. As Brown (1969b) noted, however, excluded females may settle finally in a suboptimal habitat where some reproduction is possible. Their contribution in offspring may increase the population output to higher than would be expected if they had settled in a dense population in an optimal habitat, provided the negative relationship between number of pairs and total offspring per pair is sufficiently stringent (see Chapters 9 and 11).

PREDATORS

Hairston et al. (1960) suggested that carnivores are probably food limited, whereas herbivores are not. This hypothesis would mean that prey density is kept relatively low by predation so that prey do not reach a food-limited carrying capacity. Studies of natural populations showing that predators do regulate prey density are limited. The classic examples come from biological control work in which the predators are introduced to reduce the density of the pest species. A good example is the control of *Opuntia* cactus by the moth, *Cactoblastis,* in Australia (Chapter 10). So far effective use of predators in biological control has depended on finding predators that are specifically adapted to one or a few species of prey. In this situation the ability of the predator to control its prey depends on the relative ability of each population to reproduce, grow, and colonize new areas. If the predator is too efficient, it will quickly wipe out or drastically deplete the prey and become extinct itself unless it has evolved the ability to withstand low prey availability or switches to other prey species (for example, *Woodruffia* as a predator on *Paramecium;* see Salt, 1967 and Chapter 10). Hassel and Varley (1969) have suggested that predator species with sufficient internal controls of their own populations may be ineffective as control agents on prey. In this case it may require several predator species to reduce the prey to "acceptable" levels (see Chapter 10).

Pearson (1966) reported that predators, principally feral house cats, help determine the population size and the periodicity of cyclic fluctuations of the California meadow mouse (*Microtus californicus*) in the hills behind Berkeley and Oakland, California. The cats alone were unable to regulate numbers since the mice could easily outreproduce the cats. However, the mice were forced to stop breeding for about six months every year due to regular seasonal fluctuations in climate, and this allowed the cats time to crop the mouse population heavily. The ability of the cats to influence mouse density also depended on the numerical response of the

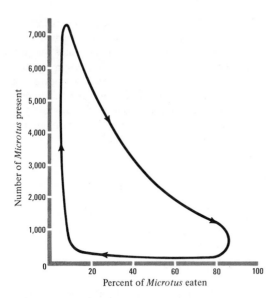

FIGURE 12-8 Theoretical curve of percent predation on a prey population by a predator showing a time delayed numerical response. After the population of prey increases there is a time lag before the predator immigrates into the area in sufficient numbers (or reproduces *in situ*) to increase percent predation. There is another lag, during which the predators switch to other prey species, as the primary prey population declines and reaches the lowest levels. (After Pearson, 1966.)

predators. As the mouse population increased, the cats moved into the area and preferentially preyed on the mice. As the mouse population decreased, the predators stayed around for some time but switched to other prey, such as wood rats (*Neotoma*) and gophers (*Thomomys*), until the combined populations of available prey reached such low levels that many of the predators emigrated. This continued presence of predators even at relatively low mouse densities reduced the mouse populations to very low levels. As the predators left, the mouse populations began to recover, presumably generating another cycle of prey abundance and increasing predator pressure (Fig. 12-8).

Murdoch (1969) suggested that to regulate populations of several prey species a predator must be able to change its diet so that the more abundant or better competitor of two prey species is preferentially eaten. Murdoch assumes that the desired regulatory impact can be achieved only when the proportion of prey taken is a sigmoid function of relative prey availability (Fig. 12-9), denoting what he defines as "switching." Although the phenomenon of switching is inherent in Holling's vertebrate (Type III) functional response and the specific search image concept (see Chapter 10), there have been few direct tests for switching.

Murdoch found that a predatory snail, *Thais,* exhibited a strong, apparently unmodifiable preference for one species of mussel and that this prey species comprised a high proportion of the diet even when another species was more abundant. He also found that the preference was maintained even after a training period of several weeks when *Thais* were given only another species of prey. Murdoch trained another predatory snail species, *Acanthina,* which showed a weak preference for mussels (*Mytilus edulis*), by offering only barnacles and then presenting barnacles as the more abundant of two alternative prey items, including mussels. The predator, *Acanthina,* then ate proportionately more barnacles than would be expected from the results

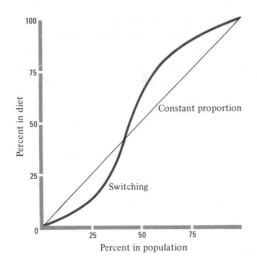

FIGURE 12-9 Theoretical illustration of switching in a predator offered two prey species. The straight line indicates that the predator takes prey in proportion to their occurrence, and the dashed line shows that uptake percentage exceeds representation of prey at high proportions and is proportionately less than representation at low proportions. (After Murdoch, 1969.)

obtained after no training and offering 50 percent barnacles. In this case with a continued reward of barnacles *Acanthina* exhibited a disproportionate preference for them and hence showed switching by Murdoch's definition. However, *Acanthina* trained on barnacles and presented with mussels as the more abundant prey quickly showed a decline from preferring barnacles to mussels. With no training *Acanthina* showed a marked preference for mussels when barnacles and mussels were equally available. Thus *Acanthina* could effectively regulate the populations of *Balanus* and *Mytilus* based on the criterion of switching—that is, provided other essential features of the interaction were present, such as sufficient densities of *Balanus* to provide "training" for *Acanthina* and patches of *Balanus* to provide the appropriate reward pattern.

These examples indicate that whether a predator can serve as a regulator of prey populations may depend on the predator's food preference. Predators that are specialized to a single prey species, such as *Woodruffia* or *Cactoblastis,* presumably continue to take individuals of the prey species until the energy intake is insufficient to support further predation. *Woodruffia* then encysts and waits for the population of *Paramecium* to increase. *Cactoblastis* individuals either emigrate to a new locality with sufficient *Opuntia* to support a population or they die. The dispersal potential of the adult moths makes this strategy available to *Cactoblastis,* but *Woodruffia* may be forced to encyst because of the difficulty for a protozoan to evolve appropriate emigration techniques.

In Huffaker's (1958) laboratory experiments with two species of mites, one predatory on the other, the dispersal ability of the prey in a complex environment produced cyclic fluctuations in predator and prey numbers. Presumably with the correct balance between predator and prey dispersal abilities and the appropriate reproductive potentials of each species, a more nearly steady state condition could be achieved. The cactus and moth populations in Australia reached essentially steady state densities if the populations were counted over a very large area. Local popula-

tions obviously appear and disappear, since the moth essentially decimates the cactus
population in a fashion similar to Gause's laboratory experiments with *Didinium*
and *Paramecium* (Chapter 10), but the sum of all local populations remained fairly
constant and at a low level.

PHYSICAL FACTORS

Physical factors can influence both mortality and reproduction of individuals in a
population. The critical question relative to regulation is whether physical factors
influence the population in such a manner that the effect will be to change the growth
rate in direct relation to the total population density or whether the effect is independent of density. Only if a physical factor predictably will drive populations toward
a steady state level will it serve as a regulatory factor for that population. In essence
this requires that the level of effect of the physical factor must be associated in some
way with the population density.

Many attempts have been made to correlate weather conditions with population
density and especially with population fluctuations. Sunspots apparently occur
cyclically, the phases of the moon near the onset of the breeding season may be
cyclic, temperature averages can cycle, and so on; to date, however, there has been
no good evidence that any of these are responsible for regular population fluctuations
(see Lack, 1954). An obvious test would be of growth characteristics in two adjacent
populations in a habitat as nearly uniform as possible, so that the climatic conditions
can be assumed to be the same. If the populations cycle out of phase, then weather
probably is not an important regulator of numbers; if they cycle in phase, however,
the weather cannot be eliminated even though it has not been identified as the causal
factor. Chitty (1960) reported that adjacent populations of voles (*Microtus*) at Lake
Varny, Wales, fluctuated independently of each other despite being under similar
climatic conditions. Similarly, DeLong (1967) showed that population sizes of house
mice (*Mus musculus*) in adjacent fields in central California were not correlated,
although some of the differences could be attributed to his experimental increases
of food supply for several of the populations.

Fluctuations of adjacent populations could be out of phase if the initial peak
had depended on locally favorable climatic conditions and adjacent peaks were
produced by movement of individuals from one local population to another as densities increased to levels not supportable by the local environment (Watt, 1968). Such
expanding population outbreaks (Miyashita, 1963) would generate asynchronous
populations but possibly not on a very local scale. Scale definition is important, but
the key is to be able to follow the peak population as it spreads. The peak at each
succeeding location may depend on local climatic conditions, but its presence
initially was dependent on good conditions at the geographic starting point of the
outbreak.

We saw in Chapter 9 that physical factors can affect the growth rate of the
population, but generally the effect is independent of density. Mullen (1968) reported
that an important influence on the reproduction of lemming populations at Point
Barrow, Alaska, was the length of the breeding season, which was primarily deter-

mined by weather severity. At high to fairly low densities the season was long and total reproduction was high when the weather was good. Mullen concluded that reproductive timing and success were primarily dependent on extrinsic factors independent of density. Obviously lemming density does not have an influence on the severity of the weather. However, the density might influence the degree to which the weather affects the lemmings.

Pitelka (1957) reported that in years of high lemming densities, the ground cover might be almost completely removed and at times the lemmings would even begin to turn over the earth to find food. If grass cover is an important refuge for lemmings in bad weather, loss of the protective ground cover at high densities could make the lemmings more susceptible to severe weather conditions (or predation). By reducing the nutritional level of individuals, high population densities may also lower their resistance to severe climatic conditions, although Mullen could not find good correlations between food quality and length of breeding season. In these examples the climatic conditions are secondary influences and other factors more directly influencing density are required to produce changes in population growth rates.

The influence of certain climatic conditions also depends in large part on how a species responds (see Chapter 6). Homeotherms, for instance, channel major expenditures of energy into maintaining a fairly constant internal environment in the face of fluctuating weather conditions, whereas heterotherms, which include plants, deal with environmental fluctuations in other ways. The effects of temperature on reproduction and growth of heterotherms are obvious from the discussion in Chapter 9. Many authors (for example, Burky, 1971), especially those working on invertebrates, have noted that lower temperatures reduce population growth rates through reduction of egg output and the rate of maturation of young individuals. Severe weather conditions can also increase mortality and cause a decline in population numbers.

Climatic conditions can affect the rate of population growth indirectly by affecting the availability of other critical resources. Mullen (1968) noted that in short growing seasons the lemmings stopped breeding earlier. Although this may have been partly a response to the lowered temperatures, a season of reduced net productivity of food plants also reduced the availability of high-quality forage. For many insectivorous birds that breed in the northern temperate region the timing of the onset of breeding is apparently related to the timing of the availability of energy—that is, insects—by which the female produces eggs or the adults feed the young (Lack, 1966). In years of poor weather the output of young of the common swift (*Apus apus*) is somewhat reduced relative to better years, probably by the reduction in the availability of flying insects for food for the adults and young. If the severity of the weather correlates with increasing population density in a regular fashion, it would appear that weather could be a causative agent for the changing population growth rates; in fact, however, the effect is mainly through the influence on food availability. The influence of cloudy weather on the availability of energy for green plants is obvious since the corresponding drop in temperature and reduction in light energy both may decrease photosynthesis. The maintenance costs of the plants are also somewhat reduced, but less energy is available for reproduction.

At the peak densities of lemmings at Point Barrow, Alaska, and even after the population began to decline, there was no apparent reduction in the reproductive output of females (Mullen, 1968). However, there may be a north–south gradient in the importance of reduced reproduction in generating population declines of small mammal, especially rodent, populations. Reproductive output of the lemming population at Point Barrow may be primarily tied to climatic conditions, for they influence the length of time during which breeding can occur as well as the length of the growing season and total net primary productivity supplying food for the lemmings. The north–south gradient of declining female reproductive output may reflect the severity of the climatic conditions. In severe climates lemmings may find themselves well below K, on the average, having been through a harsh period of reduced reproduction or increased mortality, or both. Presumably these populations would be selected for high reproductive output. In more equitable climates the influence of physical conditions on reproductive output of the population is not as strong; the population more frequently is at or near K, leading to selection which would produce intrinsic regulators that reduce reproduction in high-density populations. However, Keller and Krebs (1970) noted no decline in litter size in declining *Microtus* populations in Indiana. They did report lower total reproduction, especially during fall and winter, in declining populations.

PHYSIOLOGICAL MALFUNCTION

Some authors have postulated that the physiological functioning of an individual is influenced by the availability of a critical resource or by population density (Christian and Davis, 1964). An individual that is sufficiently upset physiologically may be less able to reproduce or may be subject to a greater probability of death, both of which influence population growth. If the deleterious effects associated with physiological malfunctioning are directly correlated with increasing population size, then they can serve as a density regulator, the effectiveness of which will depend on the degree of coupling with population changes and the rapidity with which the changed physiological state affects individual reproduction and mortality.

Christian and his co-workers have been active in studying this area of population regulation (Christian, 1971). They suggest that increased population size influences the behavioral interactions among individuals in such a way as to cause changes in the functioning of the adrenal cortex and pituitary gland. Resulting changes in hormonal balance influence an individual's ability to reproduce and to combat disease. As the principal cause of the adrenal-pituitary malfunction they suggest an increase in aggressive behavior at high densities, which in a feedback loop through the hormonal system influences the output of ACTH (adrenocorticotropic hormone) from the pituitary and hence the output of hormones from the adrenal, principally the cortex. As evidence they cite increased adrenal weights and changes in the structure of the adrenal cortex that are correlated with population density (Fig. 12-10). They also note that these internal changes tend to be correlated with the dominance status of the individual in the population hierarchy (Fig. 12-11). The most dominant individuals, those most capable of winning aggressive encounters,

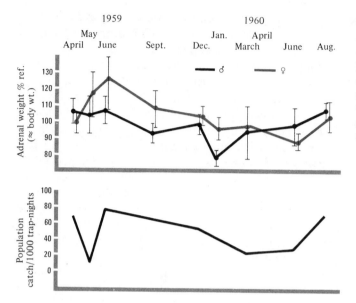

FIGURE 12-10 Relation between weight of adrenal glands of male and female deermice (*Peromyscus maniculatus*) in south-central Pennsylvania and population sizes of the mice. (After Christian, 1971.)

tend to show little or no change in adrenal function and morphology, suggesting that their dominance effectively keeps the frequency and severity of aggressive encounters low even at higher population densities. The major physiological effects of high density and increased aggression are found among the individuals in the middle of the dominance hierarchy. These animals show enlargement and presumably increased functioning of the adrenal cortex. Very subordinate animals, provided their position is relatively stable, are also little involved in aggressive encounters and show little change in the adrenal cortex with changing population size.

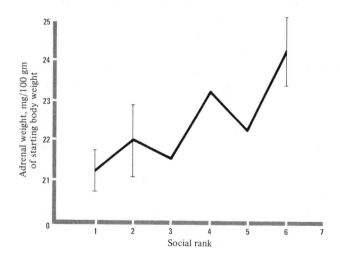

FIGURE 12-11 Relation between social rank and relative adrenal weight in laboratory populations of wild-caught house mice (*Mus musculus*). Rank I is dominant individuals and rank VI is the most subordinate individuals. (After Christian, 1961.)

Most of the studies so far are of laboratory or semiwild populations, often at much higher densities than are found in wild populations. One criticism of the physiological malfunctioning theory is that studies of natural populations do not show the relation between adrenal state and population size that Christian reports. Krebs (1964), for example, reported no significant change in adrenal weights throughout a lemming cycle at Baker Lake, Canada.

However, as Christian and Davis (1964) point out, the relation that they report for adrenal weights and population density is further dependent on age, sex, dominance status, and probably other factors. To show a significant correlation of adrenal weights and population density requires large samples and precise grouping of individuals on the basis of the other variables. It appears unlikely that adrenal activity is sufficiently tied to population density in all species of vertebrates with regulated populations to produce the necessary feedback that occurs with an increase in numbers. However, it is possible that adrenal activity might act as a regulatory mechanism in some species or populations.

Current work on rodents (Bronson, 1971) supports hypotheses that certain changes in aggression and adrenal characteristics may be mediated by *pheromones,* chemicals produced by one individual which influence other individuals. If production and release of these substances are positively related to population density, they could initiate appropriate influences on the population growth rate and thereby work as chemical regulators through physiological processes.

Recent evidence indicates that pheromones can influence the reproductive output of a female rodent (Whitten and Bronson, 1970; Wilson, 1970). Female mice placed with a strange male are more likely to suffer pseudopregnancies or embryo resorption, depending on whether they were pregnant at the time of introduction of the strange male, than if kept with familiar males. The response may be mediated by a pheromone since it can be reduced by blocking the olfactory receptors of the females. Nesting material impregnated with the urine of a strange male elicits the same response, strongly implicating pheromones produced by the strange male. Another regulatory pheromone may be produced by female mice. Significantly more females develop pseudopregnancies in the absence of a male if they are in groups of four or more than if they are in smaller groups or kept alone. The effect apparently can be countered to some extent by introducing a male to the group. The presence of the male, or simply urine produced by the male, induces and accelerates estrous; the effect is partly a function of the strength of the chemical signal contained in the urine (Fig. 12-12).

Pheromones and their associated effects should be most active as population density increases. Pseudopregnancies or embryo resorption would produce a negative relation between population density and total reproduction by the population, as required by a regulatory mechanism. The positive effects of familiar males on reproduction should oppose this negative feedback. However, the importance of these potential, chemically mediated regulators has yet to be documented in natural populations.

A second influence of a malfunctioning pituitary-adrenal pathway is an increased susceptibility to disease. However, if this relation is to be important in regulating

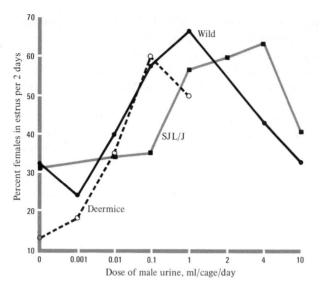

FIGURE 12-12 The relation between percent of females coming into estrus and the rate of application of urine containing male pheromone. Note change in scale on abscissa. (After Bronson, 1971.)

population density, disease organisms must be available at appropriate times in the population cycle and must be strongly correlated with the occurrence of population declines. Chitty (1960) reported that pneumonia occurred during a decline of one vole population, but adjacent populations also declined that were not affected by the disease. DeLong (1967) found that one mouse population declined coincidentally with the appearance of high numbers of diseased animals, but he could find no similar relation for several nearby populations that also declined. Although high-density populations may be more subject to infectious diseases, parasites, or predators that could cause population declines (MacArthur, 1965; Janzen, 1970), it would appear that disease in most populations is too irregular in occurrence and intensity to function as an adequate regulatory mechanism.

BEHAVIORAL INTERACTIONS

Interactions among individuals of one or more species may limit the number of individuals that survive and reproduce in an area. Behavioral interactions can (a) lead to reduced survival and reproduction by extant adults, (b) reduce recruitment of new juvenile individuals into the population, or (c) produce dispersal of members of the population. Each possible impact on population numbers will reduce the total number in a specific area, but the impact on surrounding areas would vary. So far the only consideration of behavioral population regulation has been at the local level where such behavior tends to influence the number of individuals in a relatively small and circumscribed area.

If behavior is to be established as a regulatory mechanism, behavioral changes must be shown as causally related to the variations in population numbers through time. So far most studies merely show that a class of behavior such as territoriality is correlated with the density of individuals, and that some individuals are excluded from the population. However, there is limited evidence that such behavior has an impact on the growth characteristics of the population. If behavior is to have a major impact on reproductive output, it must reduce either the number of offspring produced per female or control the number of breeding females in the area (assuming males can mate with more than one female).

In a confined rat colony certain areas of the habitat were preferred spatial positions (Calhoun, 1962). Rats preferred center areas over ends and preferred low to high burrows. As population density increased, more and more individuals seemed to change behavior, so that they visited feeding dishes coincidentally with other individuals. This gradually led to an aggregation of feeding individuals at the feeding trough in one of the center sections of the pen. It is likely that this represented a congregating point for subordinate individuals. The high frequency of encounters at a food source leads to aggressive relations that eventually are translated into "abnormal" behavior patterns. Among the results of this behavior are decreased survival of adults and reduced reproduction. It appears that high density of the confined rats led to reduced reproduction, a potential regulatory mechanism.

Several field studies have shown the relation between population density and individual reproductive output. As mentioned before, Lack and his co-workers (Lack, 1966) found that in breeding titmouse (in Wytham Wood, near Oxford, England) the number of eggs per individual decreased as a nearly linear function of density (Fig. 12-7). They attributed this partly to the increased time and energy spent in aggressive encounters in higher-density breeding populations. However, Watson and Miller (1971) reported that in the red grouse (*Lagopus lagopus*) territory size and aggression were positively correlated (Fig. 12-13), suggesting that more aggressive birds could hold larger territories. If larger territories also contain more food for the young, then more aggressive birds could leave more offspring than less aggressive birds forced onto poorer quality territories. From 1958 to 1961 grouse holding larger territories tended to survive better over the winter, but the trend was significant only for 1961 (Table 12-1).

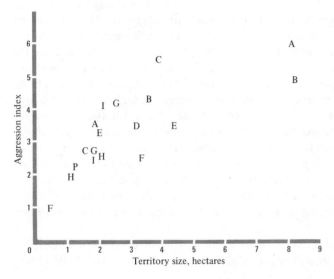

FIGURE 12-13 The relation between the aggressive index for a male and the territory size of that male for several individuals of Red Grouse (*Lagopus scoticus*) in Scotland. Higher values of the aggression index indicate more aggressive males. Letters refer to individual males; repeated letters are for several years. (After Watson and Miller, 1971.)

TABLE 12-1 Comparison of territory size between red grouse males that survive over winter (until May of next year) and those that disappear from the population (emigration or mortality)[a]

Year	Fate	No. of territories	Mean territory size (ha)	Significance of difference in territory size
1958	Survive	36	1.67	
	Disappear	12	1.25	Not significant
1959	Survive	31	3.87	
	Disappear	5	2.58	Not significant
1960	Survive	31	4.70	
	Disappear	2	2.96	–
1961	Survive	89	1.79	
	Disappear	11	0.91	$P < 0.001$

[a] Although the remaining birds always had larger territories, the difference was significant only in 1961.
Watson and Miller, 1971

Although most of the behavioral interactions that are supposed to regulate population size are aggressive, the environmental conditions that select for aggressiveness are not always clear. Although a behavioral response might influence the density of a population, there must be an underlying reason why natural selection preserves the behavior at certain densities and under certain stimulus situations.

Several investigators have found that the aggressive level of individual mice varies in laboratory (Krebs, 1970) and seminatural field situations (Healey, 1967) in relation to population density. Krebs found that *Microtus pennsylvanicus* from increasing and peak density populations showed greater aggressive responses in staged encounters in the laboratory than did individuals from declining populations. He was not able to discover whether these were short-term acclimation changes due to changing population densities or genetic changes resulting from selection for different aggressive levels at high and low population densities as Chitty (1960, 1967) suggests. However, Tamarin and Krebs (1969) did find that short-term genetic changes were occurring in his fluctuating *Microtus* populations, at least at one enzyme locus (Fig. 12-14). The genetic changes were not correlated with aggressive changes in the *Microtus,* but this does not eliminate the possibility that the behavioral changes might result from differential survival and reproduction of individuals carrying genes for varying degrees of aggressiveness. It is possible that highly aggressive individuals are also less successful reproducers; so at low population densities there would be strong selection against high levels of aggression, whereas at high densities the high aggressive levels preferentially generate survival rates that favor the lower reproducers. The genetic input to the next generation from the less aggressive, but higher reproductive individuals would be correspondingly reduced. The influence of this genetic change then would depend on the relation of generation time to the timing of population changes. For small mammals that can breed one or more times per season there is the obvious possibility that the genetic structure of the population could change rapidly if the reproductive differential of the two extreme genotypes is sufficient.

Studies of behavioral limitation seem to be looking at overt population phenomena rather than the underlying causal mechanism, unless as Chitty (1960)

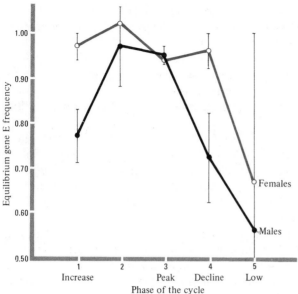

FIGURE 12-14 Equilibrium 337
values of the frequency of
allele E at the transferrin   BEHAVIORAL
locus in *Microtus ochrogaster*   INTERACTIONS
in a population in southern
Indiana. Vertical lines indicate
95 percent confidence intervals. (After Tamarin and Krebs,
1970.)

suggests, the occurrence of high aggression is mutually exclusive with high reproductive levels. Although this may be true, it is also possible that increased aggressive
levels are selected in relation to some other environmental factor that varies with
population density. It is likely that as the population increases in size, the relative
or absolute availability of a critical factor may decrease. Assurance of an adequate
supply of the resource to an individual would then be through aggressive interactions
that would influence the probability of an individual having some of the resource
available (see Chapter 11). In terms of the evolution of reproductive rates, this may
select at the same time for reduced reproductive output in terms of total energy
or the number of young; and of even greater importance, the competitive interactions
will have an influence on the number of young recruited to the next generation.

In this interpretation behavior overtly influences growth characteristics of the
population, but is itself a response to other environmental characteristics to which
the organisms respond. It seems unlikely that behavioral changes result from an
increased probability of encounters in high-density populations. High-density populations do show behavioral changes, but these normally can be related to changing
availability of one or more critical resources. Aggression is not selected for by the
probability of encounter with another individual but by the probability of some gain
being achieved by the aggression.

LOW-DENSITY "REGULATION"

The discussion of population regulation so far has concentrated on the ability of
populations to respond to approaches to carrying capacity by decreasing the rate
of addition of new individuals. For many populations, however, an equally important

consideration is to insure that fluctuations in population density do not lead to extinction. Have adaptations evolved that can reduce the probability of a population dropping below a minimum density threshold and going extinct? This threat to local extinction must be solved by either *in situ* responses or emigration if the organism is mobile. Thus stationary organisms must evolve mechanisms for reversing the downward trend of population numbers or mechanisms for surviving through periods of low density. Mobile organisms that emigrate must have well-developed abilities to locate suitable new habitat.

For a predatory species such as *Woodruffia* that has a specific prey preference, extinction could also occur not as a result of small predator population size but as a result of low prey availability. As we discussed in Chapter 10 *Woodruffia* has evolved encystment as a mechanism to survive periods of low prey availability (Salt, 1967). *Cactoblastis,* on the other hand, has good dispersal abilities and depends on adults locating a new food source when the local population of *Opuntia* cactus has been destroyed. Other predators might change the size of the territory that they defend as the availability of food per unit area decreases. Territory size in a small ground-foraging warbler, the Ovenbird (*Seiurus aurocapillus*), is inversely related to the availability of litter invertebrates that make up the bulk of its food during the nesting season (Stenger, 1958).

Stationary organisms can also evolve dispersal systems to counter the threat of extinction, but these usually involve dispersal of progeny such that the parental genes are perpetuated, but not the parental organism itself. Thus the local population might go extinct, but the genetic continuity of individuals within the population can be assured by adequate dispersal of offspring to new habitats.

The most common mechanism for enduring periods of harsh conditions that might lead to extinction of the population involves various methods of reducing the requirements on the environment. Diapause is a common metabolic and physiologic retreat from conditions that might otherwise lead to death of the individual and perhaps of the population. Many invertebrates can essentially shut down their metabolism in harsh conditions and emerge viable at some later time when the environment is more benign. Seeds of annuals are a means of storing energy to be used after a period of inhospitable conditions. Rhizomes serve as underground energy stores for many perennials.

GENERAL CONCLUSIONS

1. To be regulated a population must respond either intrinsically or via extrinsic factors to change population growth rates in relation to the ability of an environment to support a particular density. A regulated population numerically displaced from its K level will return toward that level. If the capacity of the environment to support a population changes through time, then the regulated level for the population also changes. Fluctuating populations may be tightly regulated, depending on how closely they follow changes in the ability of the environment to support individuals.

2. Competition, physical characteristics of the environment, predators, physiological

malfunctioning, and behavioral and genetic phenomena have all been suggested
as factors regulating populations. It is not always clear whether the coincident
changes in the supposed regulating factor and the population size represent
regulation or merely permission. To show regulation requires demonstration of
negative feedback coupling of population size to the environmental parameter.
Although the proponents of each regulatory mechanism are adamant about the
importance of "their" factor, there is little evidence at this time to verify their
faith.

3. Organisms also have evolved techniques of reducing the probability of extinction
of small populations. These techniques include diapause, aggressive interactions,
and dispersal. *Seed waits*

SOME HYPOTHESES FOR TESTING

In view of the paucity of data clearly determining the regulatory factors of natural
populations, research should be devoted to this question (see Maelzer, 1970; St.
Amant, 1970; and Murdoch, 1970). For a few populations there is evidence that
one or more factors can regulate population size; but for most populations that
appear to be regulated appropriate experiments have yet to be done. Without these
experiments the identification of regulatory factors must still depend on correlative
evidence, with the inherent difficulties of making cause-and-effect statements from
correlations.

Regulated populations occur among species that have an exploitation strategy
which permits the populations to approach or remain near K at regular and frequent
intervals. It is assumed that only factors that can function as K determinants are
also able to act as population regulators. Other factors influencing population growth
will presumably act more or less independently of population size, at least in a clearly
predictable fashion. Thus disease can reduce population sizes and the absence of
disease will allow populations to expand. Although there is some relation between
population size and the rate of spread of a disease, it is unlikely that the incidence
and infectiousness of a disease are regularly deterministic functions of population
density. Similarly, most weather factors probably fall into the permissive category
of environmental factors even though the impact of harsh conditions can be a
function of population density. The relation to population density probably is
mediated through other environmental characteristics that also are functions of
population density—for example, numbers of sites safe from predators, refuges to
escape the inclement weather, and the productivity of the energy source for the
population.

The most likely environmental conditions to regulate population density are those
that influence density by the amount of the resource present or by the rate of
movement of the resource into the resource pool of the population. We are thus
back to understanding K-determining resources and the impact of predators and
competition on these K factors—phenomena already discussed in Chapters 9, 10,
and 11. This circularity is not surprising since regulation is a population phenomenon
that is partly dependent on other species that co-occur with the population being
considered.

succession

13

Any landscape typically encompasses a variety of different ecosystems. In rural locations there often are areas which were once cultivated and then abandoned. Forests may contain sites burned or logged at various times in the past. Even in city centers there may be lots where razed buildings have not been replaced (Fig. 13-1). An examination of different-aged sites reveals that they often differ conspicuously in the plants and animals present. Thus over a period of years a given site would have different ecosystems at different ages. Some species typically appear on a site soon after disturbance, then are replaced by others as the site ages. Changes in the habitat accompany changes in species composition. In a vacant lot, for instance, soil temperatures would tend to be reduced as the soil was progressively shaded by developing vegetation. The process of biotic and abiotic change that ecosystems undergo as they age is called *succession*. It is a consequence of time-dependent changes in the processes examined in previous chapters: energy and materials flow, population growth, competition, predation, and evolution.

Succession begins on sites newly formed by geological processes or on disturbed

FIGURE 13-1 This vacant lot in an urban area is in the early stages of succession. (Peter G. Basich)

341

INTRODUCTION TO SUCCESSION

sites within previously existing ecosystems. Succession beginning on sites formed *de novo* as, for instance, river deltas are formed or glaciers recede, is called *primary succession;* succession arising from the destruction of previous ecosystems by, for instance, fires, floods, or the abandonment of cultivated land, is called *secondary succession.*

Secondary succession may be either autotrophic or heterotrophic. Whereas primary succession must be autotrophic since there is no readily available energy source except the sun, secondary succession may be based upon preformed organic energy sources as well. Heterotrophic succession is basically an import process, dependent on energy renewal from an external source; such ecosystems are likely to be more important in aquatic habitats, where the medium can act as the import agent, than in terrestrial ones.

Ecosystem changes with age typically involve a series of species replacements as certain populations appear on the site, grow, then decline and become locally extinct (Table 13-1). In this chapter we shall examine changes in community diversity, energy flow, mineral flow, organism properties, and spatial organization as succession proceeds. Particular emphasis will be placed on the causes of successional change and on the relationship of community diversity to ecosystem age.

The time scale of succession may vary considerably depending on the organisms involved. For example, there are decomposer successional sequences occurring constantly in the litter of forests which may require five to seven years to complete as plant tissues are successively colonized by different decomposer populations

TABLE 13-1 Bird communities in secondary succession in Georgia during the breeding-nesting season

Bird species	Age (years)	Forbs 1–2	Grass 2–3	Grass-shrub 15	Plant community 20	25	Pine forest 35	60	100	Oak-hickory forest 150–200
Grasshopper sparrow		10	30	25						
Meadowlark		5	10	15	2					
Field sparrow				35	48	25	8	3		
Yellowthroat				15	18					
Yellow-breasted chat				5	16					
Cardinal				5	4	9	10	14	20	23
Towhee				5	8	13	10	15	15	
Bachman's sparrow					8	6	4			
Prairie warbler					6	6				
White-eyed vireo					8		4	5		
Pine warbler						16	34	43	55	
Summer tanager						6	13	13	15	10
Carolina wren							4	5	20	10
Carolina chickadee							2	5	5	5
Blue-gray gnatcatcher							2	13		13
Brown-headed nuthatch								2	5	
Wood pewee								10	1	3
Hummingbird								9	10	10
Tufted titmouse								6	10	15
Yellow-throated vireo								3	5	7
Hooded warbler								3	30	11
Red-eyed vireo								3	10	43
Hairy woodpecker								1	3	5
Downy woodpecker								1	2	5
Crested flycatcher								1	10	6
Wood thrush								1	5	23
Yellow-billed cuckoo									1	9
Black and white warbler										8
Kentucky warbler										5
Acadian flycatcher										5
Totals (including rare species not listed above)	N	15	40	110	136	87	93	158	239	228

Data are number of breeding pairs or occupied territories per 40.5 ha (100 acres). Boxed values indicate the major species at a given successional stage as estimated by biomass, determined by multiplying the densities given here by average biomasses of adults of the species.
Johnston and E. P. Odum, 1956

(Frankland, 1966). There are also seasonal successional sequences in aquatic ecosystems that are fully as long—relative to the life cycles of the organisms involved—as terrestrial succession lasting several decades. The latter patterns, however, differ from most succession in that they are cyclic, occurring over and over again on the same site. Most succession, in contrast, is directional, tending to terminate in a stable species combination that is long lasting in comparison with previous species combinations. This persistent terminal ecosystem is called the *climax*.

A German, Anton Kerner (1863), was first to provide an extensive description of a successional sequence. He wrote about the pattern of community occupation and replacement on the Danube River plain and pointed out that, "One can distinguish easily the gravel left by the last high water mark from that of 10 years before, and this from that of 100 years ago. Similarly, one can follow step by step the whole series of successive generations of plant immigrants from the first sparse settlement on the most recent gravel, to the dense old elfinwood which has remained unchanged longer than human history records."

Henry Cowles, 36 years later (1899), described the successional changes occurring in the vegetation on sand dunes at the southern end of Lake Michigan (Fig. 13-2). The first plants to colonize unstable dunes are annuals, often species with succulent or creeping growth habits that tend to stabilize the sand against wind erosion. A year or two after the annuals colonize a dune it is invaded by grasses. The grasses replace the annuals and then in about 20 years they are invaded by cottonwood tree seedlings. The cottonwood groves serve as loci for subsequent invasion by pines. Eventually this pine bunch-grass community is replaced by the climax, a black oak forest. Climax species begin to appear 150 to 200 years after the dune is first formed and may remain on undisturbed sites for over 10^4 years (Olson, 1958).

We can draw an idealized diagram of population growth on a site which summarizes the general features of succession (Fig. 13-3). The initial populations increase, reach a maximum, and then decrease. Growth rates of early colonizing species are therefore in the form of a wave of alternating sign. Growth rates are initially positive but become negative later as the species go extinct locally. The periods of these cycles increase as succession proceeds until finally species appear that do not become extinct locally. These climax species grow to populations that are maintained on the site for extended periods. The early colonizing species must be subject to substantial

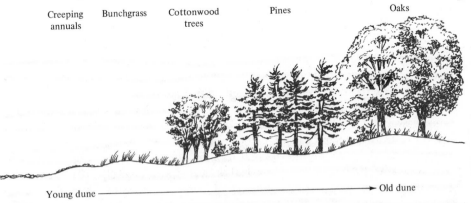

Creeping annuals   Bunchgrass   Cottonwood trees   Pines   Oaks

Young dune ⟶ Old dune

FIGURE 13-2 Diagram of the general pattern of succession on dunes at the southern end of Lake Michigan. Like most successional sequences, this one was traced by examining progressively older dunes at progressively greater distances from the lake. Such a retrospective reconstruction will be in error to the degree that unusual environmental variations, such as climatic change, divert the succession. (After Cowles, 1899)

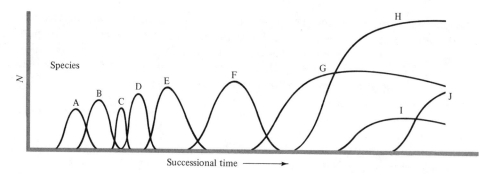

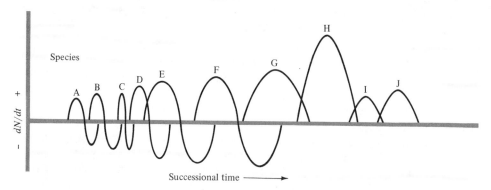

FIGURE 13-3 A general model of species replacements in succession with the species labeled A to J. In the upper graph, population size is diagrammed against time. In the lower graph, population growth rate is diagrammed against time showing that the early species go through a negative growth phase, that is, extinction, on the site while later populations reach some equilibrium size at which growth rate becomes zero.

selection for reproductive output and dispersal ability, whereas the climax species must be subject primarily to selection for competitive ability and capacity for persistence in a resource system saturated with organisms.

The *basic features* of succession are (a) it is directional—species appear and disappear with time; (b) the species replacement sequence is repeated on similar sites; (c) it involves reciprocal interactions between the organisms and the habitat with populations modifying the habitat to allow the invasion of subsequent occupants; and, to distinguish it from seasonal or decomposition cycles, (d) it terminates in a comparatively long-lasting species combination. Succession, then, is the time-dependent change in biological composition and habitat properties terminating in a persistent dynamic equilibrium between organisms and habitat.

A comprehensive framework of succession was first provided by F. E. Clements (1916), who pointed out that although it is a process in itself, it also is the outcome of several *subprocesses*. These are (a) immigration, (b) establishment, (c) site modification, and (d) competition, all of which lead to (e) ecosystem stabilization. Of the first four component processes, immigration rate is probably relatively constant through time, whereas establishment and site modification decline and interspecific competition increases with time.

Succession arises through the appearance of unoccupied habitats. Ecosystems initiating primary succession on sites newly formed by such geological processes as vulcanism, erosion, or sedimentation are initially devoid of previous organismal influence. Secondary succession, in contrast, originating within temporarily disrupted ecosystems, must be influenced by the site conditions remaining from the effects of previous occupants. Fire, for instance, one of the common agents initiating secondary succession, may enrich the soil substantially by releasing chemicals stored in biomass and litter. In Illinois the nutrient content of foliage developing on previously burned sites was generally conspicuously higher than on comparable control sites (Table 13-2). Such an enhancement of available nutrients following disruption may have a significant effect on the course and rate of succession.

Succession performs a function in ecosystems much like wound-healing and development in individual organisms. Replacement of a forest following a forest fire depends on succession. Small-scale succession frequently occurs even within relatively stable ecosystems as, for instance, on small forest areas blown down by high winds. The course taken depends considerably on initial site properties. Succession on small disturbed areas may be completed relatively rapidly. On the other hand some forest fires in mountainous areas have so completely destroyed the soil that there has been little revegetation after 50 years.

Our analogy with individual processes should not be carried too far. Succession arises out of the historical events governing site properties, organism properties, and dispersal probabilities. It is much less deterministic in course or consequences than individual physiological processes.

SOME GENERAL FEATURES OF SUCCESSION

Godwin (1923), when examining the flora of ponds of different ages, found that the number of species in a pond tended to increase with age (Fig. 13-4) but that the relationship was curvilinear. Species were added rapidly over the first few years

TABLE 13-2 Effect of burning on nutrient content of foliage growing on a successional field in Illinois

Treatment	% N	% P	% K
Control (unburned)	1.63	0.33	1.83
Burned	2.67	0.58	1.56
Control + ash	1.70	0.29	1.75
Clear cut	2.25	0.49	2.67
Clear cut + ash	2.44	0.51	3.07
Clear cut + litter	1.76	0.30	1.84

Old, 1969

and progressively less rapidly as the ecosystem aged. By converting the abscissa to a logarithmic scale, we find a good fit (student's $t = 8.98$; $P < 0.01$ that $b = 0$) to

$$S = 34.2 \log t - 38.4 \tag{13-1}$$

where S is the number of species in a pond and t is age of the pond in years. Many of the changes occurring during succession can be converted to a straight line by plotting time on a logarithmic scale. This suggests that modification of the ecosystem is much more rapid in early successional stages than in later stages, or that early species are more sensitive to small modifications.

Most studies of succession, like Godwin's, depend on examining similar sites of different ages to discern successional patterns. Since succession, as we have seen, may take a century or more to run its course, ecologists are limited to this approach for the later changes. Johnston and E. P. Odum (1956) utilized this approach to examine the succession of bird communities on abandoned agricultural fields in the southeastern United States (Table 13-1), recording individual densities (number of breeding pairs per hectare) and species densities (S per hectare). Both of these are straight lines when plotted against $\log t$ (Fig. 13-5) with

$$N = 2.515 \log t - 0.371 \tag{13-2}$$

and

$$S = 0.262 \log t - 0.062 \tag{13-3}$$

So both individual density and species density increased rapidly in early successional stages and then became asymptotic toward the climax. This suggests that the site became saturated with both individuals and species as succession proceeded. From the individual density data and biomass estimates (Norris and Johnston, 1958; Wolf, unpublished) we can estimate the biomass of the bird communities in different successional stages. This also is a straight line against $\log t$ with

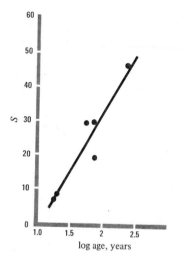

FIGURE 13-4 Increase in the number of plant species (S) present in a pond with time since the pond was formed. Species are added rapidly at first and then less rapidly as the pond ages (see Fig. 13-9b for an idealized linear plot of this relationship). (Data from Godwin, 1923)

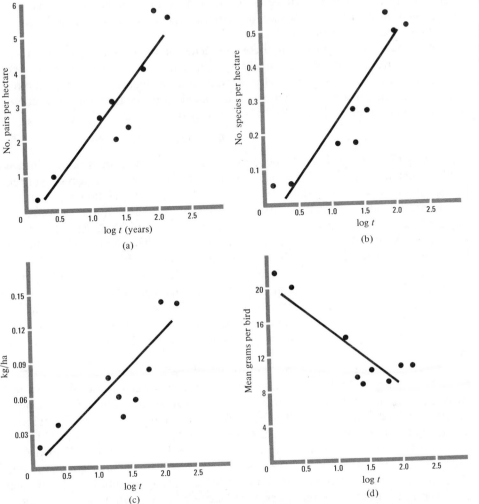

FIGURE 13-5 Patterns of change in successional bird communities in the southeastern United States; (a) number of breeding pairs per hectare, (b) number of species per hectare, (c) biomass per hectare, and (d) average size of an individual bird. Note that curves a to c are of the same general form as Figure 13-4—large increments at first and progressively smaller increments as the ecosystem ages. Average bird size, in contrast, declines rapidly at first and then levels off. (Data from Johnston and E. P. Odum, 1956)

$$B = 0.055 \log t \qquad (13\text{-}4)$$

where B is biomass in kilograms per hectare. Once again there was a relatively rapid initial saturation of the site with biomass and a subsequent deceleration of the rate of biomass addition. In the first ten years of succession about 2.515 breeding pairs, 0.262 species, and 55 gm of birds were added per hectare of the old field. According to the best fit lines, another 90 years is required to add similar increments, and another 900 years after that would be required for another similar addition of individuals, species, and biomass.

Among the major species on these old fields in biomass rank (Table 13-1) meadowlarks were replaced successively by field sparrows, towhees, cardinals, and wood thrushes as the ecosystem developed. Numerous warblers and other small birds were also added, so the birds that invaded tended to be smaller, on the average, than the species they replaced (Fig. 13-5). The best fit line is

$$\bar{B} = 21.15 - 6.18 \log t \tag{13-5}$$

where $\bar{B}$ is the average biomass per bird in grams. This was in striking contrast to the plants in these old fields which went from herbs to shrubs to forest over the period considered here (Oosting, 1942). The sequence of vegetation replacement was crabgrass and horseweed the first year, aster the second year, and broomsedge the third year. The broomsedge was followed by a shrubland that lasted for about 20 more years and then was replaced by a pine forest. This finally was replaced by an oak-hickory forest 150 to 200 years after abandonment. Whereas the vegetation proceeded from crabgrass to oak trees, the average size of a bird on the site was declining 6 gm per order of magnitude of successional age. We saw earlier, when examining energy flow, that respiration per unit biomass is an inverse function of body size. From this observation and equation (13-5) it may be a reasonable hypothesis that bird energetic efficiency declines as succession proceeds in these old fields.

Patterns of increasing individual and species densities with increasing age occur at all trophic levels. Mallik and Rice (1966) found that decomposer densities increased substantially in flood plain succession in Oklahoma. The soil of pioneer stands contained (per gram) 14,000 bacteria, 505 actinomycetes, and 12 fungi. An intermediate successional soil had 52,000 bacteria, 1600 actinomycetes, and 36 fungi. And each gram of climax soil averaged about 101,000 bacteria, 2900 actinomycetes, and 84 fungi. The number of fungal species present was 29 in the pioneer community, 47 in the transitional community, and 46 in the climax community. As in the producer and consumer communities, there was a relatively rapid increase in individual and species densities early in succession and a leveling off as time passed.

During the time that the individual and species numbers are changing, there will be considerable turnover in the individuals occurring in an area. Cooper (1923) mapped the individuals and recorded their species identities on a glacial outwash in Alaska. He reexamined the same area five years later and found that of the 673 individuals initially recorded, 417 had died during the intervening five years; however, 535 new individuals had become established for a net increment of 118 individuals. Although the number of individuals present had increased less than 20 percent, over 60 percent of the individuals had been replaced by other individuals. The individual colonization curve is considerably more dynamic than the mere change in density would indicate.

The rate of species addition may vary considerably according to the initial

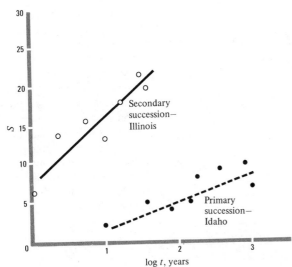

FIGURE 13-6 Species-time curves for two successional plant communities showing that species are added more rapidly in the mild climate-secondary successional sequence than in the harsh climate-primary succession. (Data from Bazzaz, 1968 and Chadwick and Dalke, 1965, respectively)

349

SOME
GENERAL
FEATURES OF
SUCCESSION

environmental properties of the site and, in general, the rate of addition will be considerably higher in secondary succession than in primary succession. Comparing primary succession on sand dunes in Idaho (Chadwick and Dalke, 1965) with secondary succession on abandoned fields in Illinois (Bazzaz, 1968), the lines are

$$S = 3.452 \log t - 1.275 \tag{13-6}$$

in Idaho and

$$S = 7.772 \log t + 8.825 \tag{13-7}$$

in Illinois (Fig. 13-6). The slopes of the two lines are significantly different (student's $t = 5.62$; $P < 0.001$) with species being added nearly twice as rapidly in the secondary successional system in Illinois as in the primary successional ecosystem in Idaho. The radically different climates of the two sites may also affect the rate of species addition in this comparison, since colonization rate would be affected by the general severity of a habitat site.

Williams et al. (1969) performed an experiment in which they examined the effect of succession on the evenness of distribution of abundance among the species, assessed as H' (see Chapter 3), with

$$H' = -\Sigma p_i \log p_i$$

where p_i is the relative abundance (as a fraction of total abundance) of the ith species in the sample. They bulldozed the vegetation off an Australian forest site and gridded the area into subplots. The relationship between H' for the total area and average H' based on the subplots was examined. We would expect that H' would be lower in any subplot than in the total area, and such a relationship was found (Fig. 13-7). H' for the total area, and the subplot averages, increased rapidly over the first five

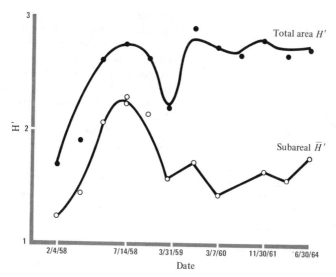

FIGURE 13-7 Change in H' of an area of Australian forest land subsequent to clearing and average H' of individual subplots. The fact that average H' of the subplots declines to a stationary level lower than total H' suggests that the area becomes "patchy," that is, that different species are found in different subplots. (After Williams et al., 1969)

months of the study, reaching a peak from which it subsequently declined. In the subplots the peak value was never reattained, but in the area as a whole the value of H' subsequently regained a level near the peak and maintained it for the four subsequent years of observation. As in the patterns seen earlier the change was strikingly nonlinear in time. H' increased rapidly during the first few months of the study and then became asymptotic.

The fall in subplot H' can be interpreted in two ways: (a) competition following saturation limited the number of species in any given subplot or (b) certain species dominated local areas and increased in abundance, while other species declined or, at best, merely maintained themselves. The fact that total H' regained a substantially higher level suggests that the vegetation was becoming "patchy"; that is, the species composition varied considerably in different subplots, so total H' increased again as the plots became different from one another. For the first five months of the study the slope of total H' on subplot average H' was 0.968, but this fell to 0.250 during the last four years. These slopes are significantly different from one another ($t = 3.074$; $P < 0.05$), indicating that during the initial colonization period, total sample H' increased almost as rapidly as within site H', but that after the initial rise and drop, most variation was added by the increasing "patchiness." Thus one of the consequences of succession was to increase the variability of the community from point to point; that is, there is an increase in the degree of pattern in the community as succession proceeded (Pielou, 1966b).

Another property of the ecosystem that may change during succession is productivity. As we saw earlier from the bird studies of Johnston and Odum, community biomass increases with time. Odum (1960) and Golley (1965) presented data on aboveground productivity of abandoned old fields in South Carolina during the first ten years of abandonment. Productivity was a negative function of log t (student's $t = 5.916$; $P < 0.01$) such that

$$P_n = 0.502 - 0.240 \log t \qquad\qquad (13\text{-}8)$$

where P_n is in kilocalories per square meter per year (Fig. 13-8). Thus during the first ten years of abandonment productivity declined substantially. Obviously this trend cannot continue indefinitely since at the end of 124 years ($\log t = 2.09$) $P_n = 0.502 - 0.240(2.09) = 0$. So either (a) this relationship must be truly asymptotic, or (b) the biomass must become stable by the time the community is 124 years old, or (c) equation (13-8) must not be accurate for later successional development.

One of the most important, but least understood, aspects of successional change is the relationship between energy flow and ecosystem age. The decline of primary productivity observed in these South Carolina old fields may be a consequence of transfer of initial free-nutrient pools in the soil into organismal compartments, with productivity subsequently limited by the rate of soil weathering and chemical cycling. On primary successional sites, as we shall see later in the chapter, nutrient supply often is low early in succession and enrichment is one of the principal effects of biological colonization. It is likely that net productivity on such sites increases with succession. Numerous qualifications must be attached to statements of patterns of energy flow accompanying succession until more information becomes available.

These problems are among the most important from man's standpoint. Because we depend on the useful energy yield (P_n) of ecosystems, it is important for us to understand techniques for optimizing this yield. Man currently depends primarily on early successional ecosystems for his existence; his agricultural fields are species poor and incapable of maintaining themselves over any reasonable time period in the absence of continuous human manipulation. Climax ecosystems, in contrast, are able to exist with little change over long time periods. Is climax productivity comparable to that of earlier successional stages? Certainly this is an important area of ecological research about which too little is known.

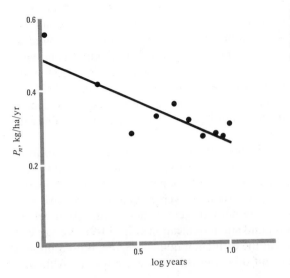

FIGURE 13-8 Decline of net primary productivity in old field succession. (Data from E. P. Odum, 1960 and Golley, 1965)

There is evidence to indicate that some successional changes may involve transitions of a more abrupt nature than the fits to log t would suggest. Most of the relationships we have examined have shown a good fit to log t, but we should be aware of the possibility that the points are not as continuously distributed as the regressions would suggest. In the old field sequence with which the birds studied by Johnston and Odum are associated, for instance, there are several changes in plant growth form as the fields progress from annuals, to perennial grass, to shrubs, to trees. There may well be relatively abrupt transitions involved in the conversion from grass to shrub or from shrub to forest communities. At present, however, there are insufficient data on these transitions to justify rejecting the regression lines. Important problems in successional study are (a) the degree to which time-dependent trends become truly asymptotic and (b) the potential occurrence of abrupt transitions between certain stages of succession.

COLONIZATION

The initial stage of succession is *colonization,* in which a habitat becomes occupied by organisms. As we have seen, the rate of species addition will vary considerably according to the nature of the initial site; similarly, the rate of biomass addition will also depend on environmental severity. In the Illinois old fields mentioned earlier (Bazzaz, 1968) vegetation covered 70 percent of the ground at the end of ten years. In the Idaho dunes (Chadwick and Dalke, 1965) the cover was only 15 percent at the end of ten years of colonization.

MacArthur and Wilson (1963) pointed out that colonization is itself an outcome of two separate processes occurring on a site—immigration and extinction. *Immigration* is the process by which individual organisms arrive, passively or actively, upon a site. It may occur passively through environmental carriers as, for instance, wind-blown seeds carried by air currents and aquatic invertebrates carried by water currents; it may occur actively as birds and mammals arrive on available sites through their own activity patterns. Since all of the newly arrived individuals are rarely of the same species, new species will be added to the community as colonization proceeds. *Extinction,* of course, is the process by which individual genotypes and species are eliminated from the site. As we saw from Cooper's (1923) study of succession on glacial outwash, there is considerable turnover in the individuals on a site. This turnover may be accompanied by extinction if subsequent colonists are better competitors, or if habitat properties become so modified as to eliminate favorable sites for certain species.

MacArthur and Wilson (1963, 1967) have discussed the nature of colonization on islands and we may, for current purposes, regard successional ecosystems as a type of island. They hypothesized that the species addition curves already considered result from the interaction between an externally driven immigration process that adds species to the ecosystem and an internally driven extinction process in which species are eliminated by competition and site modification (Fig. 13-9). We can call I the immigration rate ($+dS/dt$) and E the extinction rate ($-dS/dt$). Both represent some change in species number per unit time, but the signs are opposite. Although

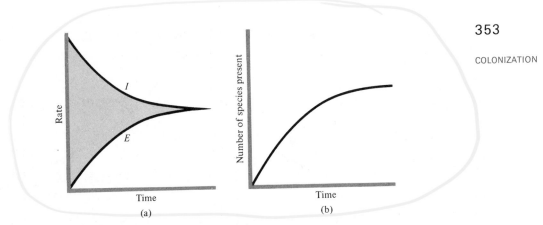

FIGURE 13-9 Diagrammatic colonization model with the species addition curve (b) a consequence of integrating immigration and extinction through time (a). (After MacArthur and Wilson, 1963, 1967)

we are considering the species addition curve, it should be borne in mind that the functional curve is the individual addition curve which, to a certain extent, "drags" the species addition curve along with it. The overall rate of species addition to a community is $I - E$; that is, the colonization rate is the immigration rate minus the extinction rate. Any change in species number with time results from inequality of the two processes. Hence

$$S_t = \int_0^t (I - E)dt$$

where S_t is the number of species at time t and the colonization equation is integrated over time dt. The fact that both S and N are functions of log t suggests that I is large at an early time in colonization and decreases with time, whereas the converse is true for E. This may depend, however, on how I is defined. If I is defined merely as the rate at which organisms enter a site, this is probably a constant with some random variation depending on chance variations in the environment. If, however, I is defined as the rate of addition of *new* species to the site, then it takes the form given in Figure 13-9, declining with time. This is essentially a probabilistic model of colonization in which colonization is the difference between the probability of establishing a new species and the probability of extinction of an existing species. Constant diversity with time arises out of the convergence of E and I.

A direct test of the equilibrium colonization model was provided by Cairns et al. (1969), who examined succession of protozoan species on polyurethane foam sponges (3 cm × 3 cm × 15 cm) floated in a lake in Michigan. A plot of the number of species present against the number of days sponges were in the lake (Fig. 13-10) fitted ($F = 422$; $P \ll 0.005$) an exponential curve of the form predicted. In addition, the colonization rate, expressed as the number of new species recorded at a sampling date divided by the time from the preceding date, declined with time ($F = 5.27$;

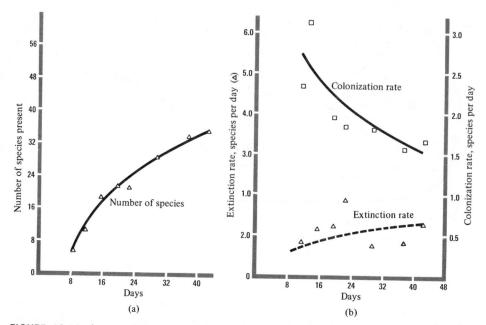

FIGURE 13-10 A test of the colonization model in Figure 13-4 through examining the succession of protozoan species on spongy blocks suspended in a lake: (a) the species addition curve and (b) colonization and extinction rate curves. Although there was no significant increase in extinction rate over the experimental period, suggesting that site modification and competition were at relatively low levels, the rate at which new species "found" the blocks did decline. (After Cairns et al., 1969)

$P < 0.1$) in accordance with the model. However, there was no evidence of an increase in extinction rate during colonization ($F = 1.68$; $P = 0.25$). This suggests that these protozoan ecosystems were still far from saturating their resources sufficiently to generate competitive extinction by the time the experiment was terminated.

The form of the relationship between S and $\log t$ suggests, in fact, that E may never converge completely on I, but that it comes so close to convergence that extremely long times—the type suggesting evolutionary time periods—are required for each additional species to penetrate the ecosystem. It is possible that "colonization" on the right-hand end of the $S/\log t$ curve represents species formation as well as immigration from remote populations.

Maguire (1963) examined the colonization of jars containing aqueous nutrient solutions placed at different distances from a freshwater pond in Texas. The total number of species recorded was a function of $\log t$, as we have seen in previous successional sequences, but the rate of species addition tended to decline with distance from the pond. We found earlier, in examining barn owl migration from fledging sites, that recovery of banded birds was a negative function of the logarithm of distance from the banding site. Similarly, in the miniature ponds set up by Maguire, the number of species tended to decline with distance from the pond.

Microorganisms may be dispersed in the air as spores and, in addition, may be carried by birds, insects, and other animals. We might expect that dispersal would show a logarithmic decay from the source pond, just as barn owl dispersal declined away from nests (Chapter 4). In Maguire's experiment, after 25 days of exposure, the number of species of microorganisms in a jar was a weak ($t = 2.09$; $P < 0.1$) negative function of distance from the pond with

$$S = 26.69 - 8.69 \log d \tag{13-9}$$

where d is distance of a jar from the pond in meters (Fig. 13-11). We would generally expect the initial immigrants into a successional area to originate in adjacent ecosystems. The probability of an organism reaching a given site decreases with log of distance from the nearest population. The slope of the graph depends on dispersal mechanisms. However, whether such immigrants would become established is dependent on site properties. The pattern of repeated species replacements in successional sequences suggests repeated changes in site suitability for colonization as succession progresses.

In certain ecosystems there may be a reservoir of potential colonists that is suppressed by the prevailing community. Kellman (1970) examined the seed content of soils under a coniferous forest in British Columbia. In the soil of a 100-year-old Douglas fir and hemlock forest, 69 percent of the seeds germinating were red alder. In the 1.1 ha stand only two red alder trees were present. No seeds of fir or hemlock germinated, and 28 percent of the seeds germinated were from species completely absent from the forest. With 1016 viable seeds per square meter, almost all were derived from early successional species rather than climax species. Kellman suggested that much of the seed may have been deposited in the litter during succession and more may have been blown into the stand from adjacent successional communities.

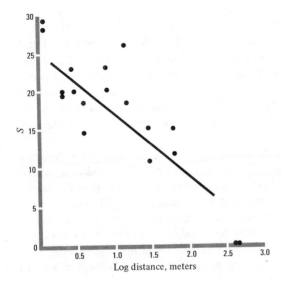

FIGURE 13-11 Relationship between the number of microbial species in a jar of water after 25 days of succession and distance of the jar from a pond. (Data from Maguire, 1963)

The soil of this forest contained a large reservoir of potential pioneer occupants should disruption occur. It is possible that climax species may be suppressing a pool of colonizing species already on the site in a dormant state with considerable longevity and may not be dependent on immigration for their occurrence. In this case the instantaneous immigration rate intercept at time zero is equal to the ecosystem's stored supply of colonizers.

In addition to a potential for cryptic colonization in dormant forms, colonizing species may have particularly efficient mechanisms of long-range dispersal. In plants this may take the form of seed modifications that facilitate transport by animals, such as spines that become snagged in fur or feathers (Salisbury, 1970). Colonizing species are also likely to have a large reproductive potential, r_m (Lewontin, 1965). Jones (1968) examined the reproductive capacity of plant species that were important in different stages of succession in African grasslands. The average reproductive capacity in propagules per plant was 12,400 for pioneer colonizers, 3560 for the initial grass species, 48 for the secondary grass species, and only 2 for the climax species. The potential reproductive output of component species declined about one order of magnitude with each successional stage. Old (1969) found that burning old fields resulted in a subsequent radical shift in the partitioning of biomass between vegetative and reproductive structures of plants. In undisturbed fields only about 7.2 percent of the aboveground biomass was present in flowering structures. In burned fields, in contrast, 52.6 percent of the biomass was incorporated into reproductive structures.

Proctor (1959) has shown that migratory water birds may act as efficient agents of long-range dispersal of algae. DeVlaming and Proctor (1968) performed a series of experiments in which they fed seeds of aquatic colonizing plants to ducks and killdeer. Many seeds passed through the digestive tract still viable after being in the bird for periods up to 120 hours. Given flying speeds of 40 to 50 miles per hour, 9 of the 24 seed species tested were retained long enough to be transported over 1000 miles by birds. Two species were retained long enough to be carried over 2500 miles. Vlaming and Proctor point out that although "Wind may serve as the major dispersal mechanism for a number of species . . . this method is highly random." Dissemination by birds, in contrast, may be much less random according to the tendency of the bird to seek out a particular type of habitat. Vlaming and Proctor observe that "Water birds tend to rise from one body of water and fly directly to some other aquatic habitat." Colonizing species with adaptations for dispersal by such an agent may be much more efficient colonizers than species depending on random mechanisms.

In addition to dormancy and efficient means of long-range dispersal, colonizing species may be highly polymorphic in response to environmental conditions terminating dormancy. Cavers and Harper (1966) examined the germination properties of seeds from two colonizing species and found that seeds had highly polymorphic germination behavior. In particular, when the seeds were exposed to favorable conditions, germination tended to be intermittent. Salisbury (1970) also found that seeds of colonizing species tended to go through periods of germination spread over a relatively long time. For species with a low probability that seedlings will be in an appropriate environment for sustained growth, intermittent germination may allow the maintenance of a reservoir of individuals in the ecosystem. Such a species makes a "bet" with the habitat that some of these individuals will survive to reproduce. Cavers and Harper found that seed from different individuals in the same population, from different populations, and even from different positions on the same individual varied in germination properties.

The unique difference between colonizing and climax species is that the former are always doomed to extinction on any given site and must depend on colonization of another appropriate site to survive. The grasshopper sparrow in fields in the southeastern United States has a duration of less than 20 years on an area after it has gone out of cultivation. Such a species will be under strong birth selection, with evolution driving toward large brood size, early age of first reproduction, and multiple litters per growing season, as well as effective dispersal.

Two major ecological effects that man has had on the earth's ecosystems are an expansion of the area available to colonizing species and an increase in the longevity of such habitats. Many of the "weed" and "pest" species that plague man are, like many of his crop species, colonizing species whose habitats are expanded and perpetuated by man's activities (Baker and Stebbins, 1965). Seed supplies of many agricultural crops are infested with seeds of associated weeds; probably the most successful agent of weed dispersal is man himself.

The world's most widely cultivated crop is wheat (Baker, 1965), which might be regarded as a garbage grain since it was initially brought into cultivation in Asia Minor where it colonized refuse heaps. Man has depended on his own energy, the energy of beasts of burden, and the energy of fossil fuels to convert much of the earth into a state of arrested colonization so that he can exploit the high productivity and r_m of colonizing species. One wonders how long such a strategy can be sustained. Understanding the mechanisms of succession eventually will allow man to manage ecosystems more efficiently for his own benefit.

SITE MODIFICATION

Occupation of a site by organisms modifies the properties of the site substantially. This is believed to be one of the principal driving forces in successional change. The initial colonists in terrestrial successional sequences are often soil microorganisms, including mosses (Cooper, 1931) on glacially derived soil and lichens (Jackson and Keller, 1970) on volcanic rocks. These microorganisms have substantial effects on the properties of the rock particles they colonize. Silverman and Munoz (1970) examined the effect of a fungus, *Penicillium simplicissimum,* on the properties of finely ground rock particles in a growth medium. The fungus was isolated from the surface of weathering basalt and subsequently grown in a standard growth medium containing 500 mg of powdered sterilized rock particles. Controls consisted of sterile medium, sterile medium with rock, and innoculated medium without rock. Cultures were incubated for seven days in 50 ml of medium at 30 degrees and, following centrifugation and filtering to remove rock and fungal particles, they examined the silicon, aluminum, iron, and magnesium found in the medium. The amount of these elements was always substantially larger in the presence of active fungus than in the controls. The amount of silicon dissolved out of the rock ranged from 0.3 to 31 percent of the total initially present. Different proportions of various elemental constituents were dissolved in different minerals. The solubility of silicon was promoted by magnesium content of a mineral. The range of aluminum freed from the rock was from 0.7 to 11.8 percent. Both iron and magnesium were released in large quantities in the presence of the fungus. From 25 to 50 percent of the iron and 25 to 56 percent of the magnesium originally present in the rock appeared subsequently in the medium when the fungus was present.

The fungus dissolved rock components by releasing organic acids which dissolved part of a rock material. The pH of the medium used by Silverman and Munoz declined from 6.8 to less than 3.5 during the seven-day growth period. Citric acid was the major acid excreted into the medium by the fungus. A medium in which the energy sources (glucose and yeast extract) were completely consumed by the growing fungus continued to dissolve rock components after removal of the fungus. Comparable solubilities were achieved in fresh, uninoculated medium to which citric acid was added.

Similarly, in a classic study of the habitat changes accompanying succession (Crocker and Major, 1955), the pH of glacially formed soil changed rapidly during an intermediate successional stage (Fig. 13-12). Even 30 years after glaciation some areas remained bare of vegetative cover and pH declined much less rapidly in such bare patches than it did in areas covered with vegetation. Most of the reduction in soil pH occurred while an area was colonized by alders. During 35 to 50 years of occupation by alder, the soil reaction declined from a pH of near 8 to near 5. In a soil transect across an alder thicket from the oldest center portion to adjacent bare soil, Crocker and Major found pH was 7.9 for bare soil, 7.2 for 9-year-old alder, and 6.5 for 18-year-old alder.

Alder, in fact, constitutes what might be called a *reactive species* in this sequence of primary succession, since the major changes in soil properties are associated with alder occupation. The principal successional stages are mosses and herbs followed by willows, then alder (which appears after about 10 to 15 years), and finally, after 80 to 100 years, spruce and hemlock (Cooper, 1931). The greatest increases in litter, organic carbon in the soil, and total nitrogen occurred while alder thickets occupied the site (Fig. 13-12). The rate of nitrogen accumulation averaged 62 kg/ha/yr under alder. This rapid enrichment of the ecosystem nitrogen pools results from the association of symbiotic nitrogen-fixing microorganisms with the roots of alder. The subsequent depletion of nitrogen by the climax spruce-hemlock forest suggests that alder builds up a reservoir of nitrogen which is subsequently drawn upon by the climax forest. It is quite likely that nitrogen depleted from the soil in the climax does not disappear from the ecosystem but is merely transferred into the organic compartment. The rate of organic carbon accumulation in the soil averaged about 15 gm/m^2/yr during the successional sequence, but this accumulation terminated in the climax forest and the compartment size became stabilized at about 5 kg/m^2.

As we saw earlier when examining materials flow, one of the consequences of agricultural exploitation of land is a depletion of stored nutrients, particularly

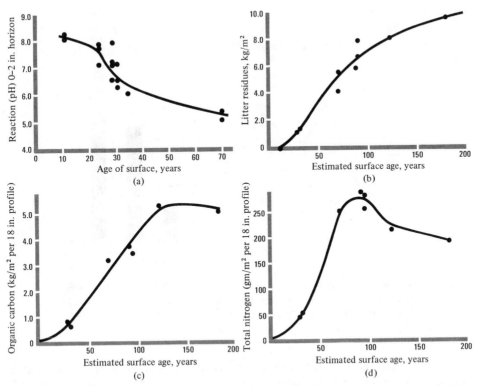

FIGURE 13-12 Changes in soil properties during succession on glacial debris: (a) decline in soil pH, (b) increase in the litter compartment, (c) increase in soil carbon, and (d) increase, and a subsequent decrease, in the soil nitrogen compartment. Note the different time scale in (a). The period of most rapid change in soil properties is while alders are the principal plants present. (After Crocker and Major, 1955)

nitrogen. As in primary succession part of the site modification occurring in secondary succession on abandoned agricultural fields consists of increasing the size of nutrient compartments associated with the soil. Odum (1960) found that there was a threefold increase in the litter compartment during the first three years on abandoned fields in South Carolina (Fig. 13-13). Other soil patterns were much less definite, however, and no clear trends in soil phosphate, potash, or organic matter were evident during the same time period. Legumes, which also have symbiotic nitrogen-fixing microorganisms associated with their roots, may be particularly important early in succession (Campbell, 1927). Total soil nitrogen increased with time from cultivation, whereas, conversely, the percentage of legumes in the community decreased over the same period.

Mellinger (1972) examined succession on abandoned hayfields in central New York and found that soil concentrations of phosphorus, potassium, magnesium, and calcium were related to successional age (Table 13-3). Soil phosphorus declined

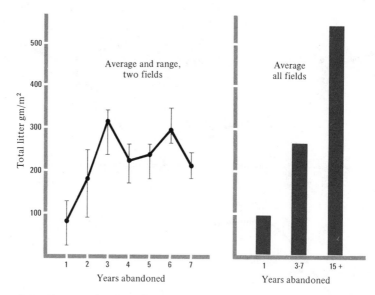

FIGURE 13-13 Increase in size of the litter compartment on abandoned fields. In the left-hand graph, yearly averages and ranges are shown; in the right-hand graph, the sequence is extended to 15 years after abandonment. (After Odum, 1960)

precipitously in the early stages of secondary succession, whereas there was a general increase in potassium, magnesium, and calcium with time. Nitrate nitrogen proved to be quite variable within any stand and although there was a regular decline in this nutrient with successional age, the data were sufficiently variable that there were no significant differences among the fields. These data suggest that the pattern of soil modification in secondary succession may be much less directional than in primary succession. In the latter process soil particles are generally dissolved and

TABLE 13-3 Soil properties in a series of successional fields in central New York. Nutrient concentrations are ppm averaged over the growing season

Soil property and F for differences	Field age from last cultivation (year)		
	5	16	36
pH ($F = 9.68$; $P < 0.01$)	6.35	6.14	6.75
Phosphate ($F = 23.7$; $P < 0.01$)	3.92	0.65	1.10
Potash ($F = 21.4$; $P < 0.01$)	36.37	46.00	66.00
Magnesium ($F = 13.54$; $P < 0.01$)	152.7	139.4	231.6
Calcium ($F = 7.10$; $P < 0.01$)	1654.0	1445.0	2240.0
Nitrate ($F = 2.07$; $P = $ n.s.)	11.02	7.52	5.80

Mellinger, 1972

there is nutrient enrichment of the site. In secondary succession, in contrast, certain nutrients may be depleted while others may be enriched. Mellinger's data suggest that phosphorus may be such an important limiting factor on these sites that much of it becomes immobilized in the organic compartment.

Another pattern of site modification resulting from the presence of organisms is an increase in spatial variability of the habitat. Lawrence et al. (1967) examined the soil properties associated with *Dryas drummondii,* the most important mat-forming plant in early succession at a site on Glacier Bay, Alaska. They found that *Dryas* mats established conspicuous gradients of soil properties. Soil organic matter, in grams dry weight per square centimeter, varied from 0.223 in the center of mats, to 0.154 at a midradial position, to 0.074 at the mat periphery for a highly significant ($F = 27.93$; $P < 0.005$) patterning of soil organic matter. One of the most important effects of organisms on a site may be the generation of habitat heterogeneity, creating spatial gradients in a variety of environmental factors.

In addition to creating habitat heterogeneity, vegetative cover generally has an ameliorative effect on atmospheric evaporation, an effect that may be important to the character of both plant and animal habitats on a site. Cain and Friesner (1929) found that loss by evaporation from porous ceramic bulbs was 14.6 cm³/day in an old field and declined through successional stages to only 4.7 cm³/day in climax forests. Water loss in the early successional stage was over 2.5 times the water loss in the climax community. This lower loss from evaporation may have resulted from the greater water loss, via transpiration, of the vegetation itself. Since the greater transpiration rate of the forest tends to create high internal humidities, the water loss of organisms that live within the forest may be diminished. In addition, air temperatures and air movement were generally lower in the older successional stands. Fluctuation in evaporation loss from day to day was also much greater in the old field than in the forest, suggesting a general stabilization of environmental properties within the community as succession proceeds.

An important factor affecting animal habitats in succession is modification of environmental cues by the developing vegetation cover. Activity patterns of mosquitoes in different types of vegetation differed conspicuously (Haddow, 1961). Using animals as bait and recording biting activity, it was found that there were conspicuous postsunset and presunrise peaks of activity under the forest canopy (Fig. 13-14). In a banana plantation, in contrast, activity was much less periodic, with a slow rise in the afternoon hours to a relatively constant level of activity through most of the night and then a gradual decline as dawn approached. Since light is an important component in the regulation of these activity patterns (Lumsden, 1966), different light climates in the forest and plantation probably regulate the biting behavior of these mosquitoes.

We have seen from an earlier consideration of microbial populations that their densities increase with successional time. The accumulation of litter and organic matter in succession "drags" decomposer succession in the wake of plant succession. Similarly, modification of environmental cues to animals by the developing vegetation may constitute a driving force in the succession of animal communities.

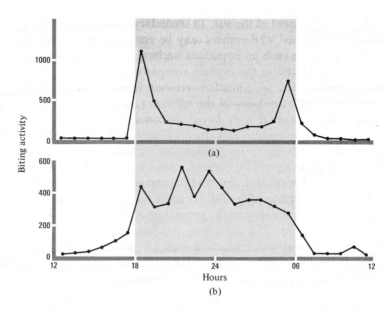

FIGURE 13-14 Diurnal biting pattern (mean number of bites per hour on a "bait" animal) of a mosquito (*Mansonia fuscopennata*) in (a) a tropical forest and (b) a banana plantation. Shaded region is night. (After Haddow, 1961 and Lumsden, 1966)

MECHANISMS OF SPECIES REPLACEMENT

Our consideration of site modification during succession provides some insight into the mechanisms of species replacement. Given the ability of successional species to modify site properties, there may be a tendency for these species to modify their own environment out of existence. On the other hand, they merely may create conditions under which later successional species are better competitors. The discussion of site modification emphasized plant communities because they probably constitute the principal organismal driving force in succession. This point should be qualified, however, by recognition that consumers may have considerable impact on composition of the plant community. Studies of the effect of grazing on grassland composition (Nicholson et al., 1970) indicate that herbivores tend to increase the proportion of rare species and decrease the abundance of dominant species in grassland. Harper (1969) has pointed out that animals create habitat heterogeneity with trails and localized areas of disturbance that will affect plants. This suggests that species replacements in succession may be at least partially related to reciprocal habitat modification involving all trophic levels on the site.

A common trend in plant succession is an increase in canopy height as taller and taller species progressively dominate the site. The shaded understory therefore is confronted with the problems of maintaining itself and growing under low light. This is particularly important for tree seedlings which may be shaded by shrubbery as well as the tree overstory. In addition, shaded seedlings are more subject to attack by fungi (Vaartaja, 1962), and ability to resist this attack is an important component of species replacement patterns in plant succession.

Grime and Jeffrey (1965) found that ability of seedlings to survive under shaded conditions was related to seed weight (Fig. 13-15). Species with large seeds had a much lower death rate in the shade than species with small seeds. Death was often associated with increased fungal infection. The ability to grow under shaded conditions and to resist fungal attack is a decided advantage when growing under a vegetative canopy. However, ability to grow under shade is associated with low growth potential under sunny conditions (Grime, 1966). The relative growth rate (mg/gm/hr) in the sun was 1.22 in shade-tolerant tree species, 5.25 in shade-intolerant tree species, and 12.07 in early successional herbaceous species ($F = 35.4$; $P < 0.005$ with $df = 1, 17$). Low respiration rate was associated with low growth potential in the sun. Average respiration rate (mg/gm/hr) was 1.87 in shade-tolerant trees, 4.42 in shade-intolerant trees, and 5.32 in early successional herbaceous species ($F = 18.02$ for $P < 0.005$). This suggests that successional replacement may be in part a compromise between ability to grow rapidly under sunny conditions and ability to survive fungal attack under shady conditions.

Early successional species produce many more, but much smaller, seeds than climax species. They also have seedlings with a high potential for growth under sunny conditions. The high dispersability of small seeds gives them a high probability of "finding" a newly disrupted habitat. The high growth potential allows them to colonize such a habitat rapidly. Low survivorship in the shade, however, resulting from the seed property that assures high dispersal capability, renders these species poor competitors in an established community. Such species have been organized

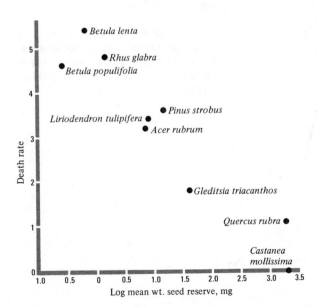

FIGURE 13-15 Decline in seedling death rate in the shade (mean number of fatalities after 12 weeks) with increase in the mean seed weight of nine tree species. (After Grime and Jeffrey, 1965)

by evolution for continual colonization of newly available areas. Later successional species, in contrast, are organized for invading existing communities.

Whereas Grime's studies provide considerable insight into the mechanisms involved in long-term replacement patterns, Keever (1950) has examined the first three years of succession on abandoned fields in North Carolina. These fields commonly are abandoned in the fall after the last crop has been harvested and are occupied the next year primarily by horseweed. The second year horseweed is replaced by an aster, which is replaced in the third year by broomsedge, a perennial grass. Surviorship, calculated as the density of the species in September divided by the density during March or June (depending on first appearance in the community), was 1.1 for horseweed in the first year and 0.26 in the second year; it was absent the third year. For aster comparable figures were 35.38, 2.98, and 0.16; for broomsedge they were 2.09, 1.75, and 61.7. The ability of horseweed to occupy the site the first year depends on its winter annual growth habit. The plants germinate the first fall, live as a rosette over the first winter, and then bloom during the following summer. Aster seeds mature in November and do not germinate until the spring after abandonment. As the survivorship data indicate, the aster population increases dramatically during this first year, whereas the horseweed population barely maintains itself. Aster replacement of horseweed during the following year suggests greater competitive ability; but direct experimental evidence of this is lacking. It would be interesting to grow the two species in the same pots and examine survivorship. Broomsedge seeds require exposure to cold winter temperatures to break dormancy, and the slow growth rate of seedlings (3.62 gm in five months) does not allow them to compete effectively during the second year with the rapidly growing aster (10.75 gm at five months). The perennial bunch growth of the grass, however, allows it to spread out from a center of establishment and the vigor of asters is inversely related to their distance from broomsedge bunches (Fig. 13-16).

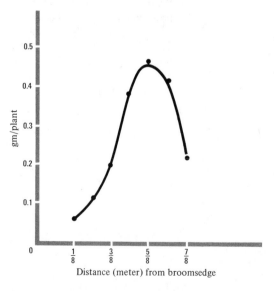

FIGURE 13-16 Relationship between average size of aster plants and distance of those plants from clumps of broomsedge. (Data from Keever, 1950)

gm/plant

Distance (meter) from broomsedge

The ability of broomsedge clumps to suppress asters allows these clumps to expand slowly so that by the third or fourth year broomsedge dominates the site; it then remains on the site until replaced in 10 to 15 years by shrubs. Species replacement on these old fields arises out of the interactions between colonizing ability (high in horseweed because of its fall seed germination pattern) and competitive ability (high in broomsedge because of its perennial bunch-type growth habit). Prior to agricultural development of the southeastern United States, horseweed was probably a rare species occupying small open areas created by fire or windthrows in forests. Agriculture, of course, opened up thousands of acres for colonization by this species.

In earlier consideration of the pattern of plant succession on Lake Michigan dunes, we mentioned Cowles' (1899) observation that cottonwood groves serve as loci for subsequent invasion by pines. Similarly, in the Glacier Bay successional sequence, mats of *Dryas* act as loci for invasion by cottonwood. We have seen that there are sharp gradients of organic matter associated with these mats; Lawrence et al. (1967) compared the vigor of cottonweed seedlings growing outside of the mats with those growing in association with *Dryas*. The tree seedlings were more vigorous when growing in the presence of *Dryas*. The cottonwood saplings associated with *Dryas* had an average annual height increment that was 20 percent greater and the biomass had an increment 47 percent greater than saplings not growing with *Dryas*. This suggests that the accumulation of nutrients in the area around individuals of early successional species establishes appropriate conditions for invasion by species that are less competitive on the site prior to this enrichment.

Evidence of this sort would seem to be decidedly teleological, since it suggests that the early successional species are organized to prepare the site for invasion by their replacements on the site. It is hard to imagine, however, how such properties could originate through natural selection since selection should drive any species toward perpetuating its own occupation of a given site. In terms of survival properties, however, we have seen that poor survival in an established community results from just those seed properties that allow rapid colonization of unoccupied sites. Why a perennial species, such as *Dryas,* should "prepare" the site for its own replacement is a more difficult question. The answer probably resides in too much of a good thing. The accumulation of organic matter associated with *Dryas* mats is an inevitable outcome of imbalance between litter input and decomposer activity. In the case of *Dryas* this imbalance eventually reduces vigor in the middle of the mat to the point where cottonwood seedlings can invade.

On badly eroded, infertile, abandoned farmland in central Oklahoma, succession consists of a short-lived weed stage that is replaced in two or three years by an annual grass community; this occupies the site for 10 to 15 years before being replaced by a perennial bunchgrass community; the bunchgrass occupies the site for up to 50 years before the climax prairie becomes reestablished (Booth, 1941). The annual grass stage is dominated by three-awn grass (*Aristida oligantha*), and the perennial bunchgrass community is dominated by little bluestem (*Andropogon scoparius*). Early annual stages are impoverished in nitrogen supply (Finnell, 1933) and bluestem has higher nitrogen and phosphorus requirements for growth (Fig. 13-17) than three-awn

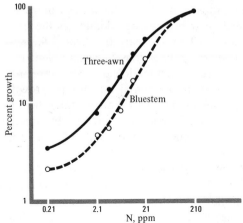

grass (Rice et al., 1960). Thus a gradual enrichment of soil nitrogen might allow bluestem to eventually displace three-awn grass. In addition, water-soluble inhibitors produced by the colonizing species are important in driving the succession (Rice et al., 1960). One of the important species in the initial community is the annual sunflower (*Helianthus annuus*) which produces compounds inhibitory to its own growth (Curtis and Cottam, 1950).

Wilson and Rice (1968) examined the effect of decaying sunflower leaves, an aqueous leaf leachate, and root exudate on the growth of several species from the initial colonizing communities and the three-awn grass that replaces them. Sunflower produced compounds that seriously inhibited the growth of the early successional species, whereas three-awn grass was insensitive (Table 13-4). Although the three-awn grass had a decidedly lower growth rate in control samples, it was insensitive to inhibition by the decaying sunflower leaves. In addition, germination of three-awn seeds was not affected by the chemical inhibitors, although germination of early successional species generally was inhibited. The short duration of the initial colonizing stages on these fields arises out of chemical inhibition involving sunflower as well as other species. Because three-awn grass is able to tolerate both low nitrogen levels and inhibitors deposited in the soil by the previous occupants, it becomes the dominant plant (Olmstead and Rice, 1970).

That certain plants would be selected for compounds that are self-inhibitory appears paradoxical, however, and we may well ask how these early colonists came to contain high concentrations of compounds that inhibit their own growth. These inhibitory compounds, however, also are potent deterrents to attack by many parasites and predators (Whittaker and Feeny, 1971). Natural selection in these colonizing species may have been for resistance to herbivore attack, with the compounds only secondarily being also toxic to plants. Of course, the presence of inhibitory compounds in the sunflower would give it somewhat of a competitive advantage against other colonizing species if, as the experiments with decaying sunflower leaves indicate, the inhibitor acts more effectively against potential competitors than against

TABLE 13-4 Effects of decaying sunflower leaves on growth and germination of seedlings

Test species	Expt. No.	Mean dry weight of seedlings (mg)		F	Germination[b]
		Control	Test		
Helianthus	1	44	22[a]	45.4	52
annuus	2	36	21[a]	10.7	40
Erigeron	1	54	19[a]	20.0	87
canadensis	2	32	16[a]	9.4	71
Rudbeckia	1	17	3[a]	36.0	95
hirta	2	12	2[a]	19.3	81
Digitaria	1	126	16[a]	67.8	106
sanguinalis	2	97	11[a]	84.4	97
Amaranthus	1	78	12[a]	23.7	56
retroflexus	2	91	16[a]	41.6	32
Haplopappus	1	13	8[a]	7.0	71
ciliatus	2	26	10[a]	12.6	64
Bromus	1	47	17[a]	114.0	97
japonicus	2	39	15[a]	71.8	94
Aristida	1	15	21	3.7	97
oligantha	2	19	23	2.2	102

[a] Dry weight significantly different from control.
[b] Expressed as percent of the control.
F is from analysis of variance
Wilson and Rice, 1968

the species itself. Under these conditions natural selection may favor production of the compound up to the point at which self-inhibition becomes more detrimental than beneficial. Once again we have an example of the subtle compromises that make the study of ecological mechanisms so challenging. The environmental factors that are potentially limiting to a successional species, including the habitat nutrient level, herbivore attack, competition, and ability to grow in shade, constitute a complex environmental association determining species replacements. Unraveling the operative factors is both challenging and frustrating, as Rice's studies indicate, since completely satisfactory answers are extremely elusive.

Structure of the vegetation, including the number of suitable singing perches, canopy height, and canopy structure, appears to be particularly important in regulating bird community structure. As these change in succession, the bird community also changes (Lack, 1933). Such structural characteristics may be more important than species composition of the plant communities in determining bird distribution (Pitelka, 1941). Terborgh and Weske (1969) measured the number of species in bird communities occurring in different successional stages in Peru. They also constructed foliage density profiles for the ecosystems and these varied considerably in different age stands (Fig. 13-18). In the mature forest there were foliage concentrations at the ground and at a height of 18 meters, but the foliage was generally evenly distributed from a height of 6 meters to the forest canopy surface at about 50 meters. In second-growth forest, in contrast, the foliage was very dense near the ground and fell off sharply to the canopy surface at 18 meters. From these foliage density

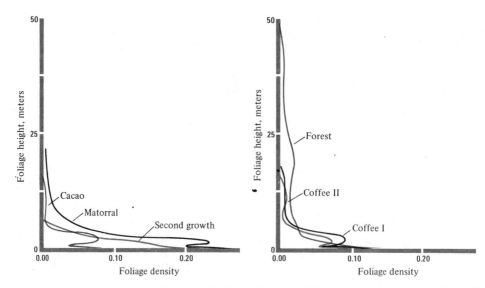

FIGURE 13-18 Vertical distributions of foliage in several Peruvian plant communities of different successional age from cultivated fields (cocoa and coffee) to mature forest. Foliage density is cover (m² of leaf silhouette per m² of ground surface). (After Terborgh and Weske, 1969)

profiles we can calculate a habitat diversity index based on the formula for the variance of values in a normal distribution,

$$Dh = \sqrt{\frac{\Sigma(f_p \cdot p^2) - (\Sigma f_p \cdot p)^2/\Sigma f}{\Sigma f}}$$

where Dh is habitat diversity and f_p is the density of foliage at a given distance p from the ground surface. This formula was applied to the foliage density profiles by using values at heights of 0.3, 1.5, 3, and 6 meters, and then 6-meter intervals to the canopy surface. Bird habitat diversity, as influenced by the foliage structure, can be expressed as an absolute variability from the above equation, or as a variability based on the mean foliage height:

$$Vf = Dh/\bar{p}$$

where $\bar{p}$ is the average distance of the foliage from the ground surface. From the Peruvian data, then, we can ask whether the number of species in a bird community is related to the total foliage height, the total foliage density, the absolute variability of the foliage density profile, or the relative variability of the foliage density profile. The most important factor affecting bird species number was absolute foliage height diversity (Fig. 13-19). This vegetation property explained almost 95 percent of the variation in bird species diversity among these successional stands. Similarly, total canopy height was related closely to the number of bird species. In contrast, bird diversity was related neither to relative variation in foliage height density nor total

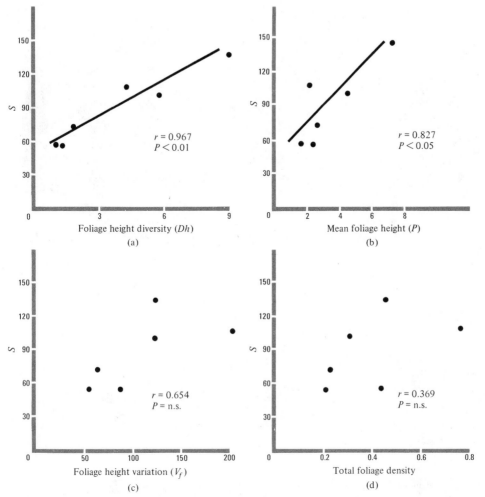

FIGURE 13-19 Relationship of several structural properties of the plant community to number of bird species present; (a) foliage height diversity, (b) mean foliage height, (c) foliage height diversity divided by mean foliage height, and (d) total foliage density. The principal vegetation property related to bird community diversity is the layering of canopy levels over a great vertical distance. (Data from Terborgh and Weske, 1969)

foliage density. The important factor related to bird diversity in succession was the spreading of foliage over a great vertical distance rather than merely the amount of foliage.

The plant community may be regarded as a structural matrix within which the bird community exists. Ecologists (Salt, 1957) have been able to place the foraging positions of bird species within this matrix (Fig. 13-20). The creation of three-dimensional "patchiness," or structural heterogeneity, by successional vegetation is

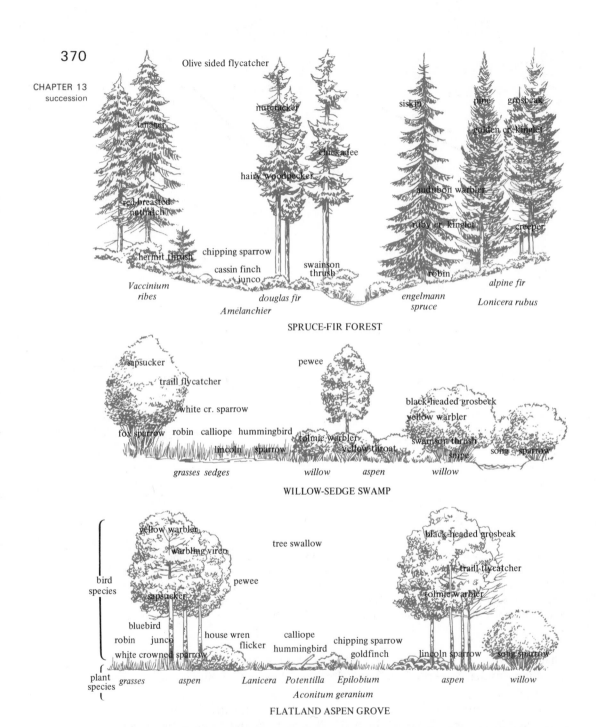

FIGURE 13-20 Vegetation profile diagrams for mature forest (spruce-fir) and two successional communities in Wyoming with the foraging positions of bird species indicated. (After Salt, 1957)

clearly an important mechanism in the $S/\log t$ curves of bird communities. If bird

S is a linear function of variability of the foliage density profile, as the data above
indicated, we may assume that this variability also is a function of $\log t$. Because
structural cues in the vegetation are important in defining animal habitats, the
turnover of animal species in succession must arise in part out of turnover in
structural properties of the vegetation.

Harris (1952) constructed artificial woodland and field habitats in the laboratory
to test for habitat selection by woodland and field ecotypes of the deermouse
(*Peromyscus maniculatus*). He found that the woodland race spent more time in the
artificial woodland habitat, whereas the converse was true for the field race. Wecker
(1963) established a pen on the boundary between a woodland and an abandoned
field with half the pen in an oak-hickory forest and half in a successional bluegrass-
goldenrod field. The pen was partitioned into compartments connected by runways,
some of which were equipped with electric sensing treadles that recorded passage
of a mouse over the runway. The activity patterns of the field ecotype of the deermouse was examined utilizing (a) 12 individuals caught in the field, (b) 13 laboratory-reared individuals that were 12 to 20 generations removed from parental stock
trapped by Harris, (c) 12 first-generation laboratory-reared offspring of the field-
captured individuals, (d) 13 field-reared offspring of the laboratory population
derived from Harris' study, (e) 7 woods-reared offspring of field-caught animals, and
(f) 9 offspring of the laboratory population reared in the woods. Animals originating
in the field showed a decided preference for the field habitat, whether reared in the
woods or laboratory (Table 13-5). However, the population that had been maintained
in the laboratory for several generations had lost its habitat preference and chose
woods and field with equal frequency. One conspicuous difference between the
laboratory and field populations was the decidedly variable habitat use of laboratory
animals. They were quite variable in their activity pattern, except when they had
been reared in the field. In this case they showed a decided preference for the field
habitat and were as uniform as the field animals in their behavior pattern.

Although the field race of deermouse showed a decided preference for the field
habitat, it is clear that the behavioral response arose partially out of *habitat im-
printing* (Thorpe, 1945) on a genetically determined preference. Rearing in the

TABLE 13-5 Relationship between habitat origin of field
deermice and their habitat preference as indicated by activity
patterns. V is coefficient of variation of the time measurements

| Habitat-origin | Percent Activity | | P | V |
	Field	Woods		
Field	84.3	15.7	0.01	19.9
Laboratory	47.8	52.2	n.s.	46.4
Field, laboratory reared	71.5	28.5	<0.01	40.0
Laboratory, field reared	84.6	15.4	<0.01	12.6
Field, woods reared	76.7	23.3	<0.05	28.8
Laboratory, woods reared	58.5	41.5	n.s.	66.8

Wecker, 1963

appropriate habitat was required to ensure the behavioral stereotype resulting in decided habitat preference. The fact that the first laboratory-reared generation continued to show this preference, as Harris also found, indicates that some portion of the "imprint" itself is inherited. Given the rapid reversal of erratic habitat selection by laboratory animals raised in the field, it would be extremely interesting to see a plot of percentage time an individual spent in the field against number of generations in the laboratory. These studies indicate that the response of successional animals to habitat change results from genetically determined cue responses to habitat that are, in fact, habitat reinforced. It is clear that the progressive replacement of an old field by invading trees would eventually result in displacement of the field deermouse from a site. Completing the complexity of the relationship is the fact that field deermice might accelerate the successional process through preferential feeding on field species (Williams, 1959). Although vegetation change is a mechanism driving herbivore succession, herbivores may cause vegetation change through feeding habits.

SPECIES STRUCTURE IN SUCCESSION

We have considered evidence of coordinated changes in habitat properties, ecosystem biomass, productivity, spatial organization, species composition, and species adaptive properties as succession proceeds. We also have seen that there is a consistent increase in the number of species during succession and that the rate of species addition may vary depending on initial habitat properties: more rapidly in secondary succession and in mild habitats than in primary succession and harsh habitats. Accompanying changes in species number are changes in relative abundances of species. Some species enter as minor species, become a predominant species for a time, and then disappear. Over the last few years there has been considerable interest in these patterns, much of which can be traced to a paper in 1955 by Robert MacArthur. He introduced ecologists to the Shannon formula for uncertainty from information theory and proposed that ecosystem stability might be measured by the uncertainty function. We have examined this formula

$$H' = -\Sigma p_i \ln p_i$$

several times previously, including Williams et al.'s (1969) application of it to successional communities in Australia.

One of the most venerable of ecological concepts is the idea of *diversity,* or of the degree of dissimilarity among individuals in a community. One measure of diversity, which we have been using consistently, is the number of species in the community, usually called *richness*. It also has been suggested for some time that diversity may be a function of the relative distribution of individuals among the species, that is, of relative abundance. H' encompasses both concepts, since it increases as richness increases and as species abundances become more similar.

To ask how H' changes with succession, we must specify the individuals and the classes, and assign the individuals to the classes so that the p_i may be calculated. The most obvious classes occurring in ecosystems are species, with the individual

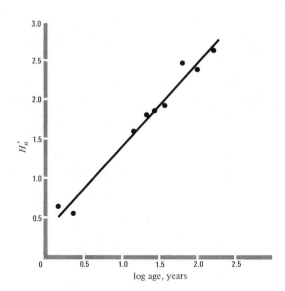

log age, years

FIGURE 13-21 Increase in H' of bird communities (based on density) with successional age of the ecosystem. (Data from Johnston and Odum, 1956)

SPECIES STRUCTURE IN SUCCESSION

organisms assigned to species. This has been the standard approach to evaluating H' in communities. Returning to Johnston and E. P. Odum's (1956) studies of bird succession on abandoned agricultural fields, we can ask whether H' changes with succession. Since we have already seen that species richness increases with succession, we would expect H' to increase also, which, in fact, it does (Fig. 13-21) with

$$H'_n = 0.337 + 1.081 \log t \tag{13-10}$$

where H'_n is based on density ($t = 8.521$; $P < 0.001$). Thus H' increased about one unit per order of magnitude increase in successional age. We have already seen, however, that S, species richness, increases as a function of $\log t$; we may ask whether H' is increasing merely as a function of S or whether there is a rearrangement in the relative abundance of species together with the increasing S. Since H' is maximal when the species are equally abundant, its maximum value for a given S is $\ln S$. H'_n can be plotted against $\ln S$ to see whether the distribution of relative abundances relative to the maximum evenness is changing with time. In fact (Fig. 13-22)

$$H'_n = 0.143 + 0.741 \ln S \tag{13-11}$$

Since the slope is less than one, evenness relative to the maximum possible is decreasing ($t = 2.463$ for $b = 1$; $P < 0.05$). This is an interesting result, because it indicates that although total diversity is increasing with succession, relative evenness is decreasing. In other words the individuals are not distributed as evenly among the species being added as they were among the initial colonists. As succession proceeds, the increase in "information" rises from adding new species rather than from redistributing the individuals among the species to make the distribution more even.

Another approach to the question of diversity, without explicitly considering relative abundances, is to examine the relationship between numbers of individuals

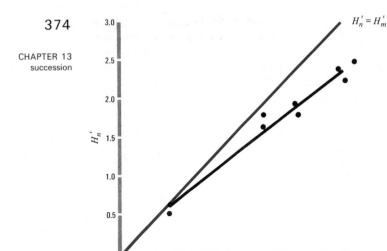

$H'_n = H'_m$

$H_m = \ln S$

FIGURE 13-22 Relationship between density-based H' of successional bird communities and the maximum possible H' (that is, if all species had equal numbers of individuals). Degree of departure from the equality line (a 45° angle) increases at higher values of H'. (Data from Johnston and Odum, 1956)

and numbers of species. Since diversity resides in the distinctness of elements constituting a system and since the elements of ecosystems are individuals, the distinctness of these individuals may be examined by the relationship between the species richness and the individual richness of the system. The ratio, $S/\log N$ (Fisher et al., 1943), increases with time ($t = 3.26$; $P < 0.02$) in the bird communities, according to (Fig. 13-23)

$$D_n = 3.678 \log t - 0.009 \qquad (13\text{-}12)$$

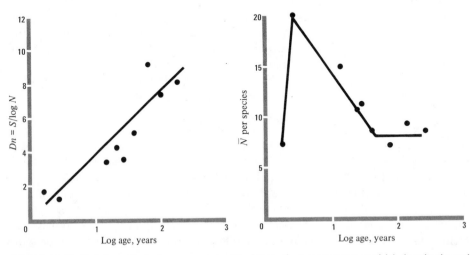

FIGURE 13-23 Relationship between successional age of an ecosystem and (a) density-based bird species diversity and (b) average number of individuals per species. (Data from Johnston and Odum, 1956)

where D_n is diversity estimated as $S/\log N$. This may be regarded as an inverse index of the average energy per pathway in the ecosystem, assuming that the species represent energy pathways and the individuals constitute the energy contents of those pathways. The fact that D_n increases with successional age suggests that the average energy content of food chains decreases as the system ages. This is, of course, a restatement of the relative abundance relationship discussed above, and the question may be asked more directly by examining the average number of individuals per species (Fig. 13-23). The pattern of $\bar{N}$ per species is complex against $\log t$, increasing drastically from the first year to the third year of the successional sequence, then going through a straight-line decline within the next five successional communities ($t = 10.36$ for $P < 0.001$) before becoming asymptotic in the last communities. Hence average population size increases rapidly in the first years of this succession, then declines until the later stages of succession, at which point it may stabilize again.

Another aspect of the development of species structure in succession is the change in abundance of the most abundant species in the community. There are two components of abundance in a community: absolute abundance and relative abundance. *Dominance,* or the tendency of certain species to be so predominant in abundance that they control the occurrence of other species, is often inferred from relative abundance, with the most abundant species designated as dominants. More appropriately, however, dominance is the appropriation of potential niche space of certain subordinate species by other dominant species (McNaughton and Wolf, 1970). We already know that total abundance, as assessed by biomass or density, increases with successional time. Our current purpose is to ask how the abundance of the most abundant species also changes. If the abundance of the dominant species increases more rapidly than total abundance, we may assume that dominance increases. If, however, total abundance increases more rapidly with succession than abundance of the major species, we may assume that dominance decreases. We can regard the populations as elements in an increasing energy pool (Fig. 13-24) and

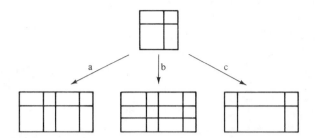

FIGURE 13-24 Alternate hypotheses relating the effect of increasing diversity to the relative abundance of the major species in a community. In alternative a, species are added but niche sizes are identical; in alternative b, increasing diversity results in a decline in niche size; in alternative c, niche size increases as species are added. (After McNaughton and Wolf, 1970)

examine the way individual elements change relative to the total pool content. It is possible for abundance of the major species to (a) remain the same as total abundance increases, (b) increase less rapidly than total abundance, or (c) to increase more rapidly than total abundance.

The plot (Fig. 13-25) of absolute abundance (as density of the major bird species) against log t gives no significant slope ($t = 1.85$; $P < 0.1$), indicating that the number of individuals in the major species neither increases nor decreases with successional time. Since the number of individuals in a population is a measure of carrying capacity, this indicates that this measure of carrying capacity of the major niche does not change as succession proceeds. We have already seen that the total number of individuals is increasing; therefore the proportion of the total abundance contributed by the dominant species must decrease with time. In fact ($t = 2.717$; $P < 0.05$)

$$p_1 = 0.714 - 0.271 \log t \qquad (13\text{-}13)$$

where p_1 is the relative abundance of the most abundant species. Therefore, for Johnston and Odum's bird communities, the proper model is alternative a in which the actual size of the major niche, as defined by individual density, remains constant while additional niches are added onto the community during succession.

The carrying capacity of the major niche is fairly constant with succession, although the nature of this niche must change, as changing species composition indicates. Furthermore, niches are added to the ecosystem by accretion, rather than by dividing up some finite energy supply. The sizes of the added niches decline with

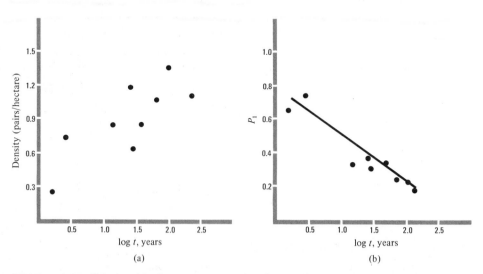

FIGURE 13-25 Relationship between successional age of an ecosystem and (a) absolute density of the most abundant bird species and (b) relative density of the most abundant bird species. This suggests that niches are added during diversification, but that the size of the major niche is constant, that is, that alternative a of Figure 13-24 occurs during succession. (Data from Johnston and Odum, 1956)

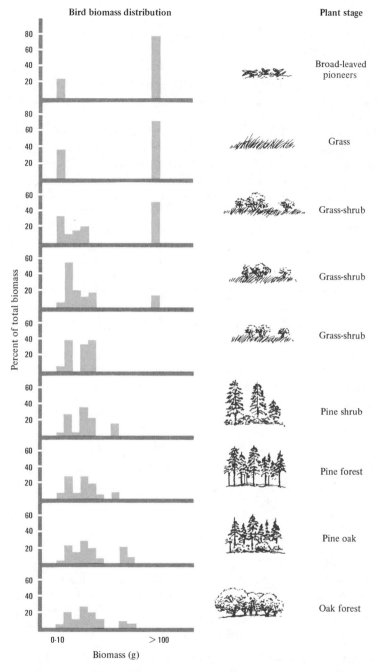

Bird biomass distribution

Plant stage

Broad-leaved pioneers

Grass

Grass-shrub

Grass-shrub

Grass-shrub

Pine shrub

Pine forest

Pine oak

Oak forest

Percent of total biomass

0-10   > 100

Biomass (g)

FIGURE 13-26 Distribution of bird species among biomass classes (10 gm intervals) in a successional sequence in the southeastern United States. Early communities are composed of birds of radically different size classes while as succession proceeds, the large birds disappear and more intermediate-sized species are added. (Data from Johnston and Odum, 1956, converted to biomass with data of Norris and Johnston, 1958 and Wolf, unpublished)

time. Species structure changes through the addition of species with new niches and through replacement of existing species by other species that occupy similar-sized niches.

This consideration of diversity and changes in relative abundance as an ecosystem ages has been couched exclusively in taxonomic terms, since the usual approach to evaluating community structure is through species composition. Because these concepts are based on partitioning individuals among classes, however, there are as many approaches to assessing structure as there are meaningful classes that can be devised to describe the structure. One approach which suggests ecological meaning is to examine the partitioning of biomass among biomass classes. In the South Carolina bird communities, adults can be assigned to biomass classes. An examination of these classes indicates that the distribution of relative biomass changes as succession proceeds (Fig. 13-26). In the initial stages of succession, biomass is partitioned among two distinct size classes, with birds being either very large or very small. As succession proceeds, the large size class becomes increasingly less important, and smaller size classes are added so that, as we saw earlier, the average size per bird decreases. Since our analyses of energy indicated that metabolic efficiency per unit biomass is related to size (decreasing as size decreases), the biomass classes may be more meaningful in terms of energy flow through the community than the species classes used above.

By examining the biomass patterns, we can ask a series of questions such as those asked earlier for taxonomic classes (Fig. 13-27). One invariant trend in succession is a straight-line fit of S to $\log t$, so we may ask first if the number of size classes increases with time for the biomass classification system. It does ($t = 5.04$; $P < 0.01$), with

$$C = 1.07 + 3.19 \log t \qquad (13\text{-}14)$$

where C is the number of biomass classes, in 10-gm increments, occurring in the community. This, of course, is in agreement with the pattern of species richness occurring during succession. Second, does the biomass-based H' increase with time? Yes ($t = 7.89$; $P < 0.001$), with

$$H'_b = 0.406 + 0.684 \log t \qquad (13\text{-}15)$$

where H'_b is based on biomass classes. However, for the third question, H'_b does not increase less slowly than the maximum possible in succession ($t = 1.683$ for $P > 0.1$ testing for $b = 1$). In this case the relative abundance does not decrease, with

$$H'_b = 0.873 \ln C - 0.047 \qquad (13\text{-}16)$$

where H'_b are as previously defined and $\ln C$ is the maximum possible H'. Hence with the classification based on biomass classes rather than species, relative abundance compared with maximum possible is constant. Fourth, diversity increases with time ($t = 3.756$; $P < 0.01$), with

$$Db = 1.2 + 1.253 \log t \qquad (13\text{-}17)$$

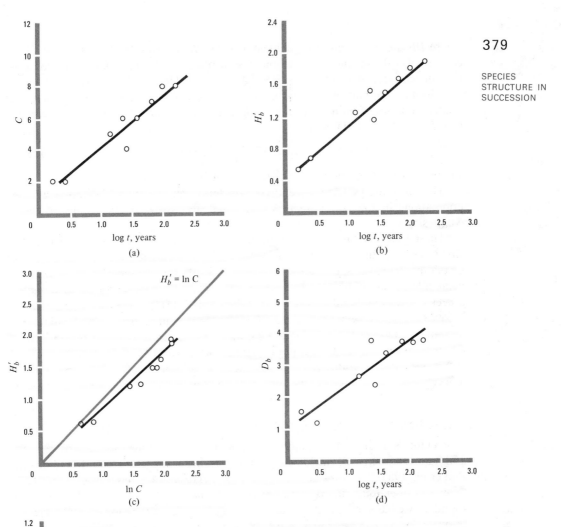

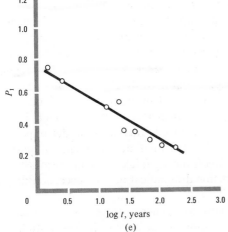

FIGURE 13-27 Biomass diversity in relation to succession of bird communities: (a) number of biomass classes increasing with age, (b) increase in biomass-based H' with age, (c) relationship between actual H' and the maximum possible ($\ln C$), (d) increase in biomass diversity (number of biomass classes divided by the common log of total biomass) with age, and (e) decline in relative abundance of most abundant biomass class in relation to time. These patterns indicate that diversity increases with time, however measured, and that the relative importance of the major ecological class also declines with age. (Data from sources of Fig. 13-26)

where Db is biomass diversity defined as $C/\log B$ (B being the total biomass of the community). Fifth, the absolute biomass contributed by the major biomass class remains constant with time ($t = 0.923$ for $b = 0$). That is, there was no change in the absolute amount of biomass in the dominant biomass class. Finally, then, the relative amount of biomass in the dominant biomass class declines with succession ($t = 5.846$; $P < 0.001$), with

$$p_1 = 0.797 - 0.265 \log t \qquad (13\text{-}18)$$

where p_1 is relative abundance of the largest biomass class.

Therefore the patterns of relative abundance and diversity are similar whether populations in the community are classified according to taxonomic or energy composition.

These successional patterns indicate that the nature of ecosystem organization changes substantially with age. The total energy content of the ecosystem increases by adding new niches, each of which tends to be slightly smaller than the preceding niche added; so although total abundance increases with succession, there may be a decrease in how evenly this energy is distributed. Diversity increases, abundance increases, dominance decreases, size of the average population decreases, and relative abundance may decrease or remain constant. We return to these patterns when we consider the patterns of community organization. The overall suggestion of these patterns is toward an increase in the variety of types of niches in an ecosystem and a decreasing importance of any one species in defining the nature of the ecosystem.

PROCESS PATTERNS IN SUCCESSION

E. P. Odum's studies on old field succession indicated that net productivity declined and biomass increased with age. It also has been suggested (H. T. Odum and Pinkerton, 1955) that respiration must increase relative to P_n, and that the importance of decomposer food chains increases (E. P. Odum, 1969). Mallik and Rice (1966) found, as we have seen, that the density of decomposer populations in soil increased in an Oklahoma successional sequence. They also found that the respiratory rate per unit of soil mass increased with age. In soil samples taken in March from pioneer, transitional, and late successional stands, oxygen uptake in microliters per gram of soil per 15 min at 29°C was 9 for the early stage, 22 for the intermediate stage, and 75 for the mature stage. These data suggest that energy and chemical flow rates through the soil compartment increase roughly eightfold during succession. Since the data on materials flow seen earlier indicated that nutrient turnover rate is a positive function of respiratory rate, substantial increases in flow through both materials and energy compartments can be deduced. Increases in decomposer populations with time are instrumental in the eventual development of quasi-steady-state soil compartments of the type found for organic carbon in the Alaskan glacial succession. One of the principal functional modifications in succession may be the closing of materials cycles.

Jones (unpublished) has examined the patterns of energy processes over succession in small aquatic ecosystems called *microcosms* (Beyers, 1963). In Jones' studies, flasks containing 200 ml of inorganic nutrient solution were innoculated with 1 gm of soil, and the patterns of photosynthesis, respiration, and biomass accumulation were examined (Fig. 13-28). Productivity was a parabola on successional time, with a time lag as the initial populations became established from germinating spores. This lag was followed by a rapidly increasing assimilation rate that peaked at an age of 20 to 25 days and then declined. Respiratory production of carbon dioxide during the dark showed a similar pattern which, however, lagged slightly behind the productivity pattern. The accumulation of biomass, which is obtained by integrating the area between the photosynthetic and respiratory curves, showed a sigmoid pattern with time, increasing only slowly until the ecosystem was about 10 days old, and then growing rapidly over the period from 10 to 100 days. For an ecosystem that does not respire its biomass, and so degenerate, R must be some function of P_n: if the relative importance of consumer populations increases with age, however, the ratio, R/P_n, will approach one toward the climax. Jones found that this ratio was in the form of an inverted parabola, declining from the earliest measurable values to reach a minimum at around 20 days and then increasing rapidly over the succeeding time periods.

With P_n declining and B increasing, the energy flow per unit biomass, measured by B/P_n and having the dimensions of time, should increase with succession. Such a pattern did occur with the ratio in the form of an accelerating exponential, so that B/P_n must approach infinity in very old ecosystems. Biomass in these microcosms grows according to a sigmoid form, and it is probably significant that these ecosystems cannot be completely related to natural succession because both the biotic and nutrient components are established at zero time by the original additions. Since the immigration rate drops instantaneously to zero, the opportunity for invasion by outside species not capable of growing in the initial conditions but capable of invading at a later stage is reduced to zero. This may be important in generating the form of the P_n and R curves and hence the biomass curve. Also important is the fact that total mineral content is fixed. In a natural successional system one of the site modifications we have observed is the dissolving of mineral nutrients out of the substrate so there is a constant flux, however small, of new nutrients into the ecosystem. The biomass curves in natural successional systems appear to be a straight line on log t. This may result from a decrease in the rate of nutrient flow into the system as it ages, but with the organisms always acting as nutrient traps by adding progressively smaller biomass increments over any given time period as successional age increases.

STABILITY OF SUCCESSIONAL ECOSYSTEMS

One of the most widely discussed ecological concepts is the idea of stability (Woodwell and Smith, 1969) and particularly a supposed increase in stability as succession proceeds. Clements (1936) argued that stabilization is the universal tendency of every ecosystem under controlling environmental conditions and that "Climaxes are characterized by a high degree of stability." Although "Change is constantly and universally at work . . . this is within the fabric of the climax and

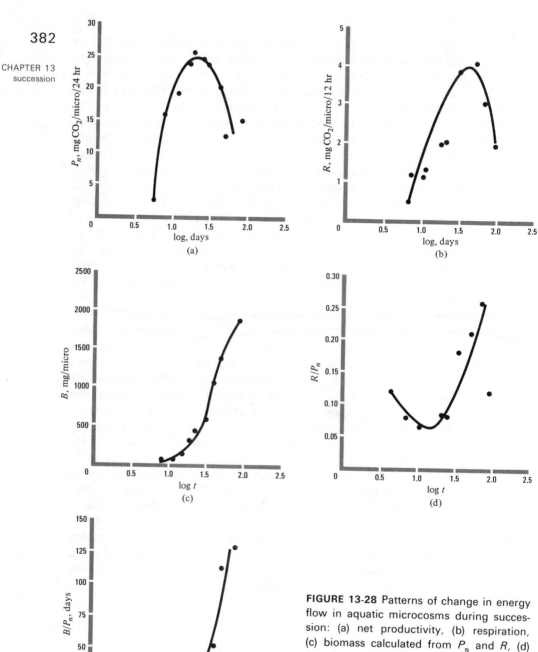

FIGURE 13-28 Patterns of change in energy flow in aquatic microcosms during succession: (a) net productivity, (b) respiration, (c) biomass calculated from P_n and R, (d) the ratio of respiration to productivity, and (e) turnover time, the ratio of biomass to P_n. Note that none of these patterns is a simple straight line against time, but are related in more complex fashion to successional age of the ecosystem. (Data from J. A. Jones, unpublished)

not destructive of it. Even in a country as intensively developed as the Middle West, prairies exhibit almost complete stability of dominants and subdominants in spite of being surrounded by cultivation." For Clements stability is defined in terms of persistence, a capacity for maintaining occupancy of a site for long time periods. Ability to occupy a site may depend on two functional attributes: (a) the magnitude of change occurring in site properties and (b) the ability of populations to cope effectively with changes that do occur. For instance, greater persistence may arise out of a decreasing rate of site modification as succession proceeds so that finally change becomes almost imperceptible. In a forest community, for example, significant modifications in site properties may require hundreds of years. On the other hand persistence may reside in the ability of populations to resist random variations in the environment.

It is possible that biotic stabilization of certain physical properties of a site may allow an increase in the proportion of the genome that may be directed by natural selection toward compensating for other site properties which are not biotically stabilized. We have seen, for instance, that soil properties change rapidly in primary succession during periods when the site is occupied by certain reactive species, such as alder on the glacial deposits. Finally, a relatively stable combination of soil properties is produced. We also have seen that environmental fluctuations (such as humidity) tend to be suppressed internally by vegetation so that the ecosystem's internal conditions are much less variable than external conditions. This tendency for organisms to extinguish environmental variation may be an important mechanism of stability in succession.

In addition to the effect that the biotic component may have in stabilizing potential variation of the abiotic component, however, how the biotic component responds to a habitat variation forced upon the ecosystem is an important component of stability. For instance, if there is an exceedingly cold and wet growing season one year, is it the early or the late successional stand that is most affected? That is, stability can be defined in terms of the ecosystem's response to an externally originated perturbation. Response to the perturbation will be expressed as a deflection of ecosystem properties from their initial state. Presumably the persistence of climax species must arise in part out of the ability of the ecosystem to "damp," or reduce the effect of, such perturbations. The ecosystem's *stability* can be characterized by three components: (a) the rapidity of the response to the external input, (b) the magnitude of the response, and (c) the time required for the response to decay back to the original, or some new, ground state.

Hurd and Mellinger (Hurd et al., 1971) examined the effect of a fertilizer perturbation on net productivities of plants and arthropod consumers on abandoned agricultural fields in central New York. The fields had been planted in hay 7 and 16 years before an experimental addition of fertilizer containing nitrogen, phosphorus, and potassium (each at the rate of 560 kg/ha) on May 1. The plants responded immediately. Productivity was higher in fertilized than in control plots and the producer's response to the perturbation was propagated upward through the arthropod food chains. Producer biomass increased linearly through the growing season until reaching peaks in late August and early September. Producer net productivity, exclusive of herbivore consumption, was calculated as the slopes of the biomass lines with time. There were two arthropod biomass peaks, one early in the season and one late in the summer, and the positive biomass increments were separately analyzed.

According to ecological theory (Margalef, 1963; E. P. Odum, 1969) the older successional ecosystem should show less response to the externally originated stimulus. This theory was supported at the primary trophic level, with the proportional stimulation of total aboveground and belowground net productivity being 22 percent on the old field and 66 percent on the young field (Table 13-6). Aboveground portions of the primary producers responded more dramatically than belowground organs, with the proportional stimulation of productivity being 70 percent and 96 percent aboveground and only 13 percent and 56 percent belowground for the old and young fields, respectively.

Arthropod herbivores, in contrast to the plants and to general ecological theory, were much less stable in the older ecosystem (Table 13-7). Although productivity was stimulated 15 to 30 percent in the 6-year-old field, it was stimulated about 200 percent in the 17-year-old field. And although the magnitude of the productivity fluctuation tended to be "damped" seasonally in the young field, it was fully as great in the late season peak in the older field. The pattern was similar in the arthropod carnivores, with the 6-year-old field showing little response after fertilization, whereas the populations increased 35 to 100 percent in the older field. There was a pronounced tendency for the perturbation to be seasonally damped in the old field carnivores, however.

We already have examined the relationship of H' to successional patterns and considered MacArthur's (1955) argument that H' is a stability measure. The per-

Table 13-6 Effect of a fertilizer perturbation upon productivity of shoots and roots of plants in successional fields of different ages

	Productivity (gm/m²/day)					
	Roots		Shoots		Total	
	6-Year field	17-Year field	6-Year field	17-Year field	6-Year field	17-Year field
Control	14.4	15.1	4.46	2.68	18.86	17.78
Treatment	22.5	17.1	8.76	4.56	31.26	21.66
Treatment ÷ Control	1.56	1.13	1.96	1.70	1.66	1.22
F_a	8.86[a]		1550[a']			
F_t	1.93		1662[a]			
F_{at}	3.23[b]		255[a]			

[a] $P < 0.01$
[b] $P < 0.1$
F_a is the effect of age upon net productivity.
F_t is the effect of fertilization upon net productivity.
F_{at} is the interaction between age and fertilization.
Mellinger, 1972; Hurd et al., 1971

TABLE 13-7 An analysis of the relationship between successional age and stability of terrestrial ecosystems upon perturbation by the addition of inorganic nutrients. Two abandoned hayfields were fertilized in May 1970 and diversity and P_n were examined during the subsequent growing season for herbivorous and carnivorous arthropod consumers. Diversity is the number of species present in standard samples, and P_n is given in mg/m^2/day. F_a is the age effect, F_t is the treatment effect, and F_{at} is the interaction effect between age and treatment. Where F_{at} is significant, fields of different ages responded differently to nutrient enrichment

	Herbivores				Carnivores			
	Early		Late		Early		Late	
	6-Year field	17-Year field	6-Year field	17-Year field	6-Year field	17-Year field	6-Year field	17-Year field
			Net Productivity					
Unfertilized	6.06	3.56	4.17	3.77	4.75	2.43	1.61	1.01
Fertilized	7.92	10.71	4.77	11.85	4.45	5.05	1.17	1.36
F_a	0.2		100.9^a		3.6		0.6	
F_t	93.6^a		172.0^a		6.8^b		0.0	
F_{at}	31.8^a		127.8^a		10.6^a		2.4	
			Diversity					
Unfertilized	3.40	3.65	3.35	5.00	1.85	1.75	1.50	2.50
Fertilized	4.20	5.50	5.45	4.45	2.35	3.05	1.30	2.05
F_a	17.7^a		1.4		3.4		17.3^a	
F_t	51.8^a		10.2^a		31.0^a		0.4	
F_{at}	8.1^a		26.9^a		6.1^b		0.4	

[a] $P < 0.01$
[b] $P < 0.05$
Hurd et al., 1971

turbation experiment provides an independent test of the latter proposition since it provides a clear stability measure—the degree to which a community responds to an external perturbation. As a test of MacArthur's proposition we can plot control H' against the treatment P_n divided by the control P_n. If the proposition is correct, H' should be a negative function of the proportional response to the fertilizer stimulus. In fact, there was no significant relationship between the two values (Fig. 13-29) and the slope was not different from zero ($t = 0.240$); that is, there is no relationship in these ecosystems between "information content" of a trophic level and the ability of that trophic level to "dampen" an external perturbation.

Jones executed perturbation experiments on his experimental microcosms in which the nutrient supply to the ecosystem was enriched with nitrogen, phosphorus, calcium, and potassium at various ages; photosynthesis was measured 10 to 15 days later to see what effect a perturbation had on the microcosms' energy gateway. Nutrients added at the time that the succession was initiated (at time zero) enhanced subsequent photosynthesis, but this enhancement was rapidly translated into an inhibition in subsequent successional stages (Fig. 13-30). The shape of this inhibition curve is roughly the inverse of the productivity curve presented earlier (Fig. 13-26), which suggests that the sensitivity of the ecosystem to a nutrient perturbation is a positive function of net productivity. If nutrients are added to the ecosystem when

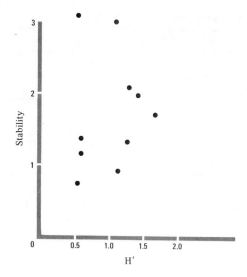

FIGURE 13-29 There was no relationship between stability of a trophic level and density-based H' of that trophic level in successional old fields in New York. Stability was defined as net productivity of nutrient-enriched plots divided by control productivity. (Data from Hurd et al., 1971, Mellinger, 1972, and Hurd, 1972)

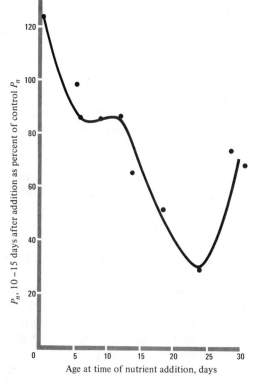

FIGURE 13-30 Relationship between age of successional microcosms at the time of nutrient enrichment and the effect of this enrichment on net productivity 10 to 15 days later. Note that this curve is essentially the inverse of the productivity curve of Figure 13-28a. (Data from J. A. Jones, unpublished)

net productivity is low, subsequent net productivity is not affected; but if the net productivity is high when the nutrient perturbation occurs, it is severely inhibited at a subsequent period. This experiment also indicated that stability of an ecosystem is not a straightforward function of successional age, but may be related to succession in a complex fashion.

CLIMAX AND STEADY STATE

One of the most controversial areas of ecological study and research through the mid-twentieth century was centered on the nature of the climax community. This was summarized by Whittaker (1953), who suggested that "Common to succession and the climax is the interplay of populations, which in succession is expressed in directional change and in the climax as fluctuations around an average." He concluded that "The climax is a steady state of community productivity, structure, and population," and "The balance among populations shifts with change in environment."

We have examined the patterns of succession over the early stages up to the climax, but have not considered changes occurring after ecosystem properties approach an asymptote. There are data from several sources that can be related to the occurrence of long-term steady states in ecosystems. One source of such data is deposits of decomposition-resistant organisms in lake sediments. Goulden (1969) examined the change in H' of detritus-feeding crustaceans which occur in the weed beds of lakes (Fig. 13-31). Goulden found that H' of these crustacean communities was linear against time for a 1000-year successional sequence in an Italian lake. There was no tendency for H' to become constant as the lake aged.

Another source of long-time-period data on community composition is pollen grains from surrounding vegetation that are deposited in lake sediments. Although certain types of pollen may be over- or underrepresented compared to the native vegetation (Whitehead and Tan, 1969), such deposits are a reasonable comparative measure of the changes in population abundance with time. Kendall (1969) examined the changes in pollen abundance occurring in sediments in Lake Victoria at the convergence of Uganda, Kenya, and Tanzania. In the immediate vicinity of the lake the adjoining ecosystems are swamps, but the prevalent current ecosystem types beyond the swamps are savanna, forest, and grassland. A sediment core 18 meters

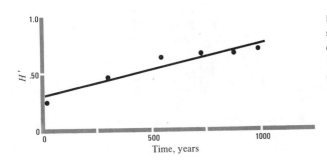

FIGURE 13-31 Increase in density-based H' of cladoceran communities in Lago di Monterosi, Italy, with age. (After Goulden, 1969)

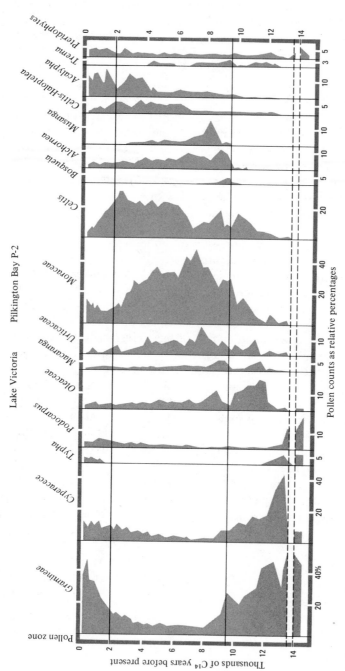

FIGURE 13-32 Relative abundance of plant species in the vicinity of Lake Victoria, Uganda, as estimated from pollen sediments over a 14,000 year deposition period. There was a more or less continuous change in the vegetation over this period. (After Kendall, 1969)

deep was taken from Pilkington Bay, at the northern end of the lake where half
of the Nile River originates. Carbon dating indicated that the core covered a time
span of 14,000 years. The relative abundances of different populations changed
continuously over the 14,000 years (Fig. 13-32). Grasses and sedges were abundant
in the earliest records and then declined in abundance to reach minima about 8000
years ago, from which time they began slow increases that continue to the present.
Coordinated changes occurred in pollen from trees of the genus *Celtis* and in the
family Moraceae, which increased as the grasses and sedges decreased, and similarly
began to disappear in the more recent collections. Various other pollen types showed
abrupt appearances and disappearances in the pollen record, indicating that the
composition of the vegetation was changing constantly.

Kendall interprets the pollen record to indicate that an initial grassland ecosystem
began to be invaded by forest species 12,000 years ago. Approximately 10,000 years
ago the forests declined somewhat, but then became rapidly reestablished to reach
a peak development shortly thereafter. Between 7000 and 6000 years ago the initial
evergreen forest was replaced by a semideciduous forest which remained abundant
until it began to be replaced by grasslands about 3000 years ago. Most of these
changes are believed to reflect changes in rainfall, with the ecosystem converted from
grassland to forest and back to grassland as the climate went through a moisture
oscillation. At any rate, these ecosystems constitute at most a quasi-steady state,
apparently stable on short-term observation but changing constantly in response to
climatic fluctuation.

At another level of resolution we can examine productivity patterns of woody
producers through the width of annual growth rings in areas with a seasonal climate.
Examination indicates that environmental variation may be translated into changes
in rate of ring growth. Estes (1970) found that oak trees in Illinois tended to deposit
rings at a rate of about 1.25 mm/yr, but that lumbering of the area around 1900
resulted in an abrupt tripling of the growth rate which returned to its previous value
by about 1930 (Fig. 13-33). This indicates that growth rings are sensitive indicators
of limitations on tree growth.

Data on the variability of growth rings of pines in undisturbed stands of California's
White Mountains (Fritts, 1966) provide direct evidence of the degree to which
productivity approaches a steady state in a "climax". Differences in absolute growth
rates associated with different sites, tree ages, and other factors are eliminated by
a transformation in which absolute ring width is divided by the slope of the regression
of ring width on time. This produces a "ring-width index" with a mean of one, and
a variance that will depend on the degree of steady state in ring production. For
instance, if ring width, as a reflection of primary net productivity, was the same
every year, the variance would be zero, and the steady state would be invariant.
Fritts found that the degree to which the trees approached steady state depended
on the site they occupied (Fig. 13-34). The index variance of the 104-year period
was 0.199 for trees in a moist forest, 0.241 for trees on a more arid southern exposure,
and 0.376 for trees on a dry, windswept ridge. The average variance of tree communities studied by Estes in the central midwest was 0.071, somewhat lower than
the California pines, but still not zero.

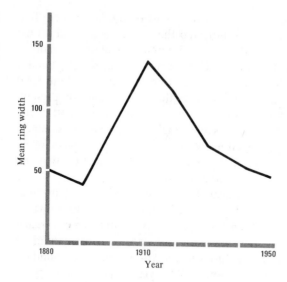

FIGURE 13-33 Growth release about 1888 of an Illinois oak sample resulting from selective logging. The vertical axis is 20 year mean ring widths in thousandths of an inch. Growth increased abruptly after adjacent trees were logged and then declined gradually, presumably as a result of increased competition as the forest regrew. (After Estes, 1970)

The idea of a climax appears difficult to sustain. The most that can be said for any ecosystem is that it is a quasi-steady state, in which the species composition will be changing constantly—imperceptibly in human terms—and productivity will be fluctuating around some mean, nonzero value. The overall tendency of succession, apparently invariable, is toward an increase in diversity; that is, an increase in the variety of components into which ecosystem energy is partitioned.

Certainly the climax is stable by comparison with the directionally modified early successional ecosystems. A particularly important distinction between early successional and climax ecosystems is that changes in the former are a consequence of directional modification of energy and material flow patterns by the organisms themselves. Directional changes of climax ecosystems, such as the shifts in African forests and grasslands documented by Kendall, are primarily a result of changes imposed by extrinsic variables. Referring back to the ecosystem model of Figure 7-19, succession is driven by biotic modification of intrinsic variables, while changes in the climax arise primarily out of changes in extrinsic variables.

Succession is among the most complex and difficult to unravel of ecological phenomena. It involves reciprocal reactions between population growth and site properties. Directional modification of energy and materials flow by growing populations represent feedback loops affecting the colonization process. We can summarize our consideration of succession by a model relating the different facets of the process (Fig. 13-35). The process is initiated on open habitats, originating by diverse causes. As the open habitat is colonized, the site is modified and competition occurs. Both of these effects, in turn, modify the colonization process, for they affect the ability of certain species to survive and of other species to invade. Competition, of course, has already been considered in detail in a previous chapter, so this chapter has concentrated on site modification as a principal mechanism in succession. It must be apparent, however, that site modification will affect the competitive ability of

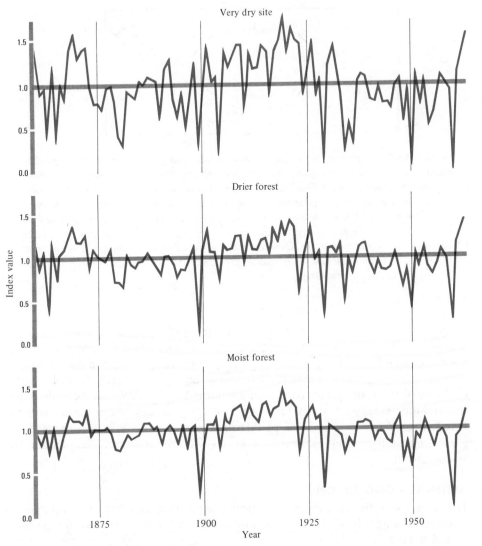

FIGURE 13-34 Tree ring index values for pines from three contrasting habitats. The average growth rate of each tree is set equal to one, and variations about this line reflect annual variations in wood production. The degree of variation decreases from the dry site to the moist site. (After Fritts, 1966)

species already present and of others immigrating to the area. For producers, principal components of site modification are changes in materials flow and spatial heterogeneity, as manifested in changing soil properties, and a denser and taller canopy. For consumers, changes in plant diversity, spatial heterogeneity, and food availability are among the major site modifications affecting colonization and persistence.

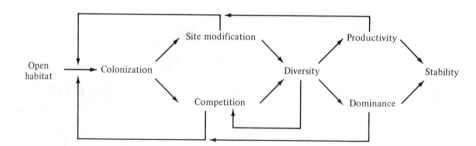

FIGURE 13-35 Summary of ecological succession. The process begins with an open habitat (that is, one with unexploited resources) and "terminates" in an ecosystem that is persistant through time, as regulated by changes in extrinsic variables. From our consideration of the process, you should be able to construct hypotheses in graphical form with successional age on the X-axis and each of the phenomena from colonization to diversity on the Y-axes, with lines relating each phenomenon to successional time. Each of these lines represent a hypothesis for testing. The outcome of some tests (for instance, colonization and diversity) would be more certain than others (for instance, stability).

Among the principal manifestations of site modification and competition are changes in diversity and relative abundance. Productivity also changes, but the pattern of change in this process is uncertain. Perhaps the most important consequence of succession is a change in ecosystem stability. When this is defined in terms of the persistence of species combinations through time, there can be little doubt that stability increases as ecosystems age. Less obvious, however, is the relationship between ecosystem resistance to perturbation and age. Because succession is the natural integration of all ecosystem processes, we are far from a complete understanding of it.

GENERAL CONCLUSIONS

1. Succession is the change in biological composition and habitat properties as ecosystems age. It results in a dynamic equilibrium (climax) between organisms and habitat.
   (a) Primary succession begins on sites devoid of influence from previous biological occupants. Bare rock, new deltas
   (b) Secondary succession begins on sites where previous occupants have influenced present site properties. fires, floods, abandonment
2. Diversity increases as succession proceeds. modifying species lead to their own extinction
3. Although itself a process, succession can be further subdivided into several subprocesses:
   (a) Colonization is a consequence of the balance between immigration of new species into a site and extinction of species presently on a site.
   (b) Site modification is a consequence of the effects of organisms upon properties of the habitat. A major effect of site modification is to increase the spatial

colonizing species are specifically evolved for rapid reproduction & dispersal

variability of the habitat and increase the proportion of free nutrients which are in organism rather than habitat compartments.

(c) Changes in competitive abilities as the habitat changes result in species replacements during succession. Changes in the plant community are important in regulating decomposer and consumer communities on the site and presumably vice versa.

4. As the number of species on the site increases, the relative importance of any one species declines.

5. Changes in the magnitudes of energy flow and chemical flow during succession are complex and poorly understood at present.

6. Stability of the ecosystem, in terms of the persistence of certain species combinations through time, increases as succession proceeds. However, the successional relationship of stability in terms of ability of the ecosystem to resist environmental perturbations is presently poorly understood.

7. Climax is a quasi-steady-state subject to change primarily as a result of changes in variables independent of the ecosystem, such as weather. Successional ecosystems, in contrast, are subject to change as a result of changes in ecosystem-dependent variables modified by activities of the organisms present.

SOME HYPOTHESES FOR TESTING

The opportunities for research on succession are so vast that it is difficult to focus on only a few questions. At least two areas, however, suggest that they might be particularly profitable to examine: (a) the importance of spatial heterogeneity in producing the increasing diversity with age and (b) the relationship of successional age to stability.

Several lines of evidence, including the decline of subplot H' in the Australian forests experiment, the tendency for cottonwoods to invade the center of *Dryas* mats in Alaska, and the correlation of diversity of Peruvian bird communities with variability of canopy structure, all suggest that increasing spatial heterogeneity, produced by the effects of organisms on the habitat, allows additional invasions. It would be interesting to manipulate site diversity artificially, by patchy fertilizer applications or creating topographic relief, and to examine the effect this had upon diversity. If high nitrogen in the center of *Dryas* mats allows cottonwood to invade, we might be able to create such invasion loci artificially by enriching local soil areas with fertilizer nitrogen. Similarly, we might be able to affect bird diversity by manipulating the canopy structure of successional stands; for instance, cutting sections out of a developing forest canopy might eliminate certain bird habitats and drive bird communities back to an earlier successional stage.

We have examined a few tests of the relationship between successional age and ability of ecosystems to resist environmental perturbations. The data from these experiments indicate no clear patterns related to age. Perhaps different kinds of perturbations would have different effects upon ecosystem structure. It would be feasible to provide different types of noncatastrophic perturbations and examine their effect upon subsequent ecosystem development. For instance, a series of aquatic microcosms at different stages of succession might be subjected to grazing by the introduction of a herbivore not naturally invading as they aged. Would later and more diverse microcosms be more resistant to destruction by grazing than early and species-poor ecosystems?

Much more information also is needed on the relationship of chemical and energy flow to succession. The tendency of biomass to reach some sort of steady state toward the climax suggests that total energy going into the system is almost totally consumed by the system as it reaches climax. That is, it suggests that total ecosystem respiration approaches gross productivity of the producers, as we saw for the microcosms. There are few data on natural successional ecosystems, however, that provide insight into such mechanisms in nature. Man's agricultural ecosystems, of course, are decidedly nonpersistent in time in the absence of continued energy input from man, draft animals, and fossil fuels. A knowledge of the mechanisms producing climax ecosystems in nature should provide important insight into ways man could increase the stability of his own ecosystems while diminishing the outside energy required to maintain them.

lake succession

14

A special type of succession which has become increasingly significant to human society is the process of lake succession called *eutrophication.* Eutrophication is the process of nutrient accumulation and ecosystem change that occurs in impounded waters, such as lakes, reservoirs, and ponds. The term was introduced to ecology in a description of bog succession and was later extended to lakes by a pioneer ecologist, Einar Naumann (Hutchinson, 1969). Naumann proposed a system of classifying lakes according to their nutrient content: oligotrophic, mesotrophic, and eutrophic. The suffix *-trophy* is from a Greek root meaning "food" and the prefixes mean "few," "intermediate," and "well" or "beneficial," respectively. Thus the sequence from oligotrophic to eutrophic is from poorly nourished to well nourished, that is, from low-nutrient to high-nutrient conditions.

In the overall successional pattern of lakes, progressive enrichment of the ecosystem with nutrients results in increasing productivity as a lake is converted from oligotrophic to mesotrophic to eutrophic. Gradually, sediment deposition diminishes the depth of the lake basin and the lake becomes a pond, with such submerged aquatic plants as *Potamogeton* and *Elodea* in the open water, floating plants such as *Nuphar* and *Polygonum* in shallower water, and marsh species like *Typha* and *Phragmites* along the shore. The pond eventually becomes a marsh, and the filled marsh will be colonized by adjacent land plants to undergo the terrestrial succession characteristic of the area. In any large lake, of course, almost all stages may be observed from oligotrophic deep water to eutrophic bays and marshy shallows. In this chapter we shall concentrate on the initial stages of lake succession leading to a eutrophic lake.

A lake is a functional and clearly visible ecological compartment of the type we examined earlier in materials circulation models. The tendency for a lake to become nutrient enriched will depend on the balance between nutrient input and nutrient output. Lakes are subject to nutrient input from rainwater directly on their surface and from run-off in their drainage basins with streams carrying dissolved ions leached from the soil. These nutrients will be subject to dilution in the lake itself, and some will move out of the lake if it has an outlet. We can write, then, a general model of enrichment as

$$E = \frac{I - O}{V}$$

(14-1)

where I is the nutrient input rate, O is the rate of nutrient output, V is the lake volume, and E is the enrichment rate. The rate of enrichment, or eutrophication, therefore, will depend on the balance between nutrient input and output and the total volume of water in the lake. An additional term, describing sedimentation of nutrients, may be important in many lakes. In addition to the input from the lake's drainage basin, there may be direct input through the activities of the organisms occurring in the lake. Many blue-green algae, for instance, are capable of converting atmospheric nitrogen into organic nitrogen, and this may serve as a significant enrichment mechanism for nitrogen if these algae are abundant (Howard et al., 1970).

Eutrophication may be an outcome of the general balance of physical factors influencing a lake, in which case we would refer to it as *natural* eutrophication. The process, however, may be accelerated by the activities of man, in which case it is called *cultural* eutrophication. It is on cultural eutrophication that much of our interest has been focused recently because of the rapid increase in the nutrient input to most of the world's lakes through the combination of population growth, urbanization, and industrialization. In this sense the concept has been extended to apply also to rivers and coastal marine waters subject to enrichment from man's activities.

Enrichment results in an increase in primary productivity, and a principal method for classifying lakes (Rodhe, 1969) is according to their productivity (Fig. 14-1). By far the most productive waters are those subject to cultural eutrophication, whereas oligotrophic waters, in contrast, are relatively unproductive. Lakes of meager productivity are likely to have a high content of dissolved oxygen, great depths, and populations of cold-water salmonoid fishes. Eutrophic lakes, in contrast, are likely to have low dissolved oxygen supplies in summer, shallow basins, and populations of warm-water fish such as carp. A lake such as Lake Superior in the United States with its coniferous forest drainage basin and, as a result, little nutrient input from

LAKE TYPE

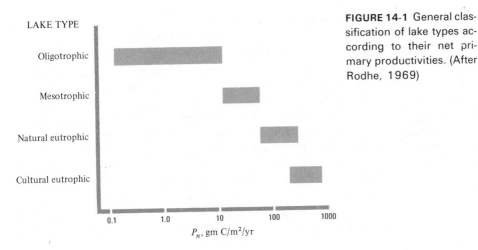

Oligotrophic

Mesotrophic

Natural eutrophic

Cultural eutrophic

0.1   1.0   10   100   1000

P_n, gm C/m^2/yr

FIGURE 14-1 General classification of lake types according to their net primary productivities. (After Rodhe, 1969)

drainage and with its great depths and cold water is probably permanently arrested in an oligotrophic state, maintaining a steady state for millenia in the absence of cultural enrichment. Other lakes, such as Lake Erie, which are comparatively shallow and subject to enrichment by run-off from prairie and deciduous forest lands, may be in a state of slow eutrophication, once again requiring millenia, but nevertheless more clearly directional than the succession in an oligotrophic lake. The effect of cultural enrichment is to accelerate the rate of succession rather than to change its direction. At the other extreme are such artificial reservoirs as Lake Mead formed by the Boulder Dam on the Colorado River. This man-made lake is estimated to have a lifetime of only 150 years (Sawyer, 1966) because of the deposition of the river's huge silt load in its still, quiet waters (Fig. 14-2).

FIGURE 14-2 Silt-covered basins in upper Lake Mead. Such silt deposition will shorten the lifetime of this reservoir considerably. (Bureau of Reclamation. Photo by E. E. Hertzog)

One of the centers of eutrophication study has been the University of Wisconsin where considerable insight into cultural eutrophication has come from studies of the lake system in Madison. Hasler (1947) was among the first to call attention to the importance of domestic sewage in cultural eutrophication by examining several European lake systems subject to accelerated succession by enrichment. Chief among these was the Zurichsee in Switzerland in which huge algal blooms were reported beginning in 1896. Two years later, the blue-green alga, *Oscillatoria rubescens,* never before seen in the lake, appeared in massive amounts. These eruptions were repeated in subsequent years and in 1902, 1.75 gm of wet biomass occurred per liter of water in the lake's outflow. The lake fishery had produced primarily trout and whitefish up until this time, but by 1915 they were replaced by cyprinoid (carplike) fishes. These changes in the Zurichsee occurred only in the lower lake basin which was subject to sewage input from a growing urban area. The shallower upper basin, which received no major urban drainage, maintained its oligotrophic character during the same period.

One of the principal symptoms of eutrophication is the large algal biomasses that develop in enriched lakes. Jonasson (1969) has presented data comparing the productivity of a mildly eutrophic lake (Lake Esrom) and one subject to cultural eutrophication (Lake Pedersborg) in Denmark. Annual productivity of Lake Pedersborg is almost three times the productivity of Lake Esrom (Fig. 14-3) and it reaches a striking peak of production in May—the so-called spring bloom characteristic of many waters but particularly exacerbated by cultural enrichment. The comparatively constant production rate of Lake Esrom throughout the warm months contrasts strongly with the decidedly time-dependent productivity in the highly eutrophic lake. In addition, the vertical distribution of algal productivity through the lake depth varies conspicuously. There is a pronounced surface concentration of algae in a highly eutrophic lake and a relatively even distribution in a less enriched lake. This combination of high production and surface concentration is responsible for many of the offensive qualities of eutrophic lakes since, as Jonasson says, "In the very rich lake (primary production) is reduced to a thick soup of algae at the surface." Such concentrations tend to destroy a lake for recreational uses and clog intakes of water systems.

Water can become opaque as a result of dense algal concentrations. One of the standard measures of eutrophication is the Secchi disk, a black and white disk 20 cm in diameter that is lowered into the water until it disappears and then raised until it reappears; both these depths are noted and their average is taken as a measure of transparency. Edmondson (1971) found that Secchi-disk transparency was a good index of water conditions leading to public complaints. In a lake he studied in Seattle, Washington, Secchi-disk transparency of the water during the summer decreased from 4.0 meters in 1950 to 1.0 meter in 1962 during a period in which sewage input to the lake increased steadily.

Findenegg (1966) pointed out that one of the trends occurring during eutrophication is a conversion of the primary producers from small unicellular algae to large, often filamentous forms. As a consequence, productivity per unit biomass tends to

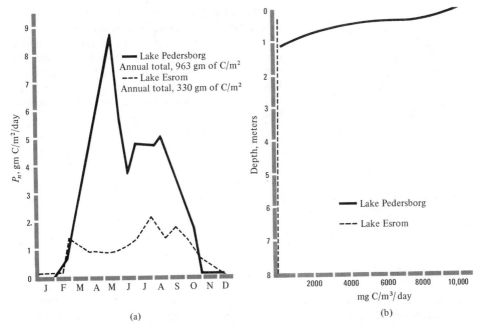

FIGURE 14-3 Net primary productivity patterns in a naturally eutrophic lake (Esrom) and a culturally eutrophic lake (Pedersborg): (a) annual cycles of P_n, (b) vertical distributions of P_n during mid-summer. The culturally eutrophic lake has a high productivity, which is concentrated in a thick algal "mush" near the lake surface. (After Jonasson, 1969)

decrease. In the oligotrophic Walensee in Switzerland, the assimilation ratio (milligrams fixed per milligrams present) was 0.07 in the spring bloom and 0.18 in the autumn bloom. Comparable values for the mesotrophic Lake Klopeinersee, Switzerland, were 0.03 and 0.07 and for the eutrophic Lake Worthsee, 0.025 and 0.024. These data suggest that in oliogotrophic lakes in which the primary producers are predominately small plankton the turnover of biomass may be relatively rapid, whereas in eutrophic lakes, with larger and filamentous primary producers, a huge biomass is accumulated and biomass turnover may be small once peak biomass is reached.

ENRICHMENT AND LIMITING FACTORS

A lake may be regarded as analogous to a favorite device for culturing microorganisms, the *chemostat* (Fig. 14-4). This is a culture device in which nutrients flow at a constant rate into a vessel containing growing organisms, most often a single population in microbial studies but easily convertible into a complex community (Margalef, 1967). All of the nutrients are in excess except one, and the population grows to the density sustained by the supply of that limiting nutrient. The population

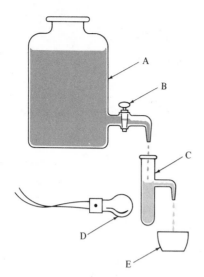

FIGURE 14-4 Diagram of an experimental lake, a chemostat. A nutrient solution flows from the reservoir (A) at a rate regulated by a valve (B) into a microcosm (C) illuminated by a bulb (D), with the excess volume emptying into a waste receptacle (E). Biomass and P_n sustained in C (the lake) will depend on the concentration and flow rate of nutrients from A.

density, then, will be the product of some proportionality constant that describes the conversion of resource to organism, and the nutrient flow equation (14-1). Brunskill and Schindler (1971) wrote an equation for water renewal time of a lake which considers the lake surface area Al, the land surface area in the lake's drainage basin Ad, annual precipitation per unit area P, the rate of water loss by evapotranspiration from both the drainage basin Ed and the lake El, and, finally, the lake volume V. The average amount of water leaving the lake annually, providing the lake level is constant, is

$$F = Al\,(P - El) + Ad\,(P - Ed) \tag{14-2}$$

and the water renewal time (Rw) is the volume divided by the flow (V/F). Brunskill and Schindler are part of a large Canadian research team studying artificial eutrophication of a system of lakes in Ontario between Lake of the Woods and Lake Superior. The study is centered at the Fisheries Research Board of Canada's Winnipeg research center. In the Experimental Lakes Area there are over 1000 lakes (Cleugh and Hauser, 1971), almost all of which are oligotrophic. The research team is involved in artificial eutrophication of the lakes on a controlled basis to provide careful experimental data on eutrophication. Such data are in short supply at present. Brunskill and Schindler presented data on water renewal time, Rw, of 15 lakes that differed conspicuously in the important parameters of equation (14-2) and reported a range of Rw from 2.6 to 38 years. These values, of course, are theoretical rates of water turnover based on volume and flow rate and some regions of a lake may be renewed several times during a given period while others may be relatively stagnant. But Rw does give an approximate estimate of the flow to volume ratio and hence the nutrient supply rate to a lake. In a chemostat, population productivity will be a function of flow rate since this determines the rate at which the limiting nutrient is available to the population. If a lake approximated a chemostat, then we would expect productivity to be an inverse function of Rw, since the longer it takes for the water to be renewed, the slower is the rate of nutrient addition per unit of lake volume. Schindler and Holmgren (1971), another team in the experimental lakes group, examined net primary productivity of several of the lakes and there were eight lakes in common between their survey and Brunskill and Schindler's estimates of Rw. For these eight lakes there was a significant ($t = 3.75$; $P < 0.01$) inverse relationship (Fig. 14-5) between maximum P_n and Rw such that

$$\log P_n = 2.92 - 1.81 \log Rw \qquad (14\text{-}3)$$

where P_n has the dimensions of milligrams carbon fixed per 100 m^3/day and Rw is in years. So for these oligotrophic lakes in Canada the more rapidly water flows through the lake and, we presume, the more rapidly nutrients become available as a result, the higher the net primary productivity of the ecosystem's phytoplankton.

Algal growth during the summer depletes nutrients in the water, particularly nitrate and phosphate (Fig. 14-6), and one of the critical questions about eutrophication has been what nutrients are most important in limiting the magnitude of algal peak biomass. One approach to determining these limiting factors has been artificial enrichment of lake waters to measure stimulation of primary production. Hutchinson (1941) was one of the first to do this in experiments on Linsley Pond in southern Connecticut, one of the most intensively studied bodies of water in the world. For one week periods, from late April until mid-July, 3.5-liter samples of lake water were resuspended in the lake after enrichment with the potassium salts of phosphorus or nitrogen, or both, at the rate of 1 mg phosphorus or nitrogen per liter (or both). Control bottles received distilled water additions of similar volume. After a week, they were then analyzed for chlorophyll concentration. Over the course of the period, chlorophyll content of the phosphorus-enriched bottles, as a percentage of controls, was 116, 207, 143, and 146. For the nitrogen-enriched bottles, values were 142, 143, 279, and 296. So stimulations of 16 to almost 200 percent were obtained by enrichment, with nitrogen being a more important limiting factor, particularly as the season progressed. When both phosphorus and nitrogen were added, chlorophyll percent-

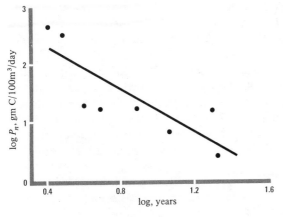

FIGURE 14-5 Decline in net primary productivity of lakes in the Canadian Experimental Lakes area with an increase in water renewal time. (Data from Brunskill and Schindler, 1971 and Schindler and Holmgren, 1971)

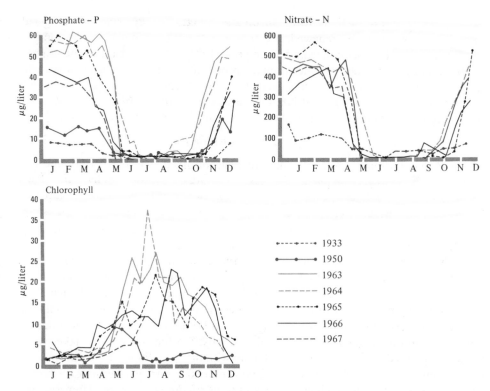

FIGURE 14-6 Seasonal changes in nutrients and chlorophyll concentrations of surface waters of Lake Washington during periods of increasing (1933–1963) and decreasing (1963–1967) sewage dumping in the lake. Increases in nutrients during the spring and fall were accompanied by increases in summer chlorophyll levels, a measure of algal biomass. (After Edmondson, 1969)

ages were 754, 927, 1343, and 2062. Thus bottles to which both nutrients were added showed enhanced algal growth substantially above bottles to which only one of the nutrients had been added. These data suggest that there is simultaneous limitation of phytoplankton development in Linsley Pond by both phosphorus and nitrogen. Hutchinson points out that although the phosphorus circulation system is relatively closed, nitrogen is open to the atmospheric reservoir through the activities of nitrogen-fixing blue-green algae and bacteria. One of the important algae in the spring bloom in this pond is a species of *Anabaena*, a genus in which nitrogen fixation is known. Hutchinson found that soluble organic nitrogen in the water increased from negligible to 1.1 mg/liter during the course of an *Anabaena* bloom's development and subsequent decay. It is interesting, however, that the blue-green alga most characteristic of cultural eutrophication, *Oscillatoria rubescens,* lacks nitrogen-fixing capacity (Fogg, 1965).

Sakamoto (1971) performed a series of experiments in which he added various nutrients to water from some of the Canadian Experimental Lakes and observed the effects on short-term (4 to 6 hours) uptake of radioactive carbon. In these oligotrophic lakes inorganic carbon concentrations were comparatively low (0.5 to 0.7 mg carbon/liter) and photosynthesis was a positive function of carbon dioxide concentration (Fig. 14-7). Enrichment of these waters by the addition of sodium bicarbonate resulted in a stimulation of photosynthesis that was an inverse function of initial carbon dioxide concentration. These data suggest that photosynthesis may be limited by carbon dioxide at concentrations lower than 5 mg carbon/liter. In the water from six lakes collected in mid-August, the addition of phosphorus stimulated photosynthesis an average of 17 percent, nitrogen caused a stimulation of only 7 percent, nitrogen and phosphorus together stimulated 17 percent, and the addition of iron with a chelating agent to keep it in solution stimulated photosynthesis an average of 43 percent. In these short-term experiments the availability of iron was an important factor limiting primary production of the phytoplankton. Phosphorus was somewhat less important and nitrogen was not a significant limiting factor.

Long-term studies of eutrophication of the lakes around Holstein, Germany, which are subjected to domestic sewage and agricultural run-off, indicated that phosphorus and sulfate were important agents of enrichment (Ohle, 1953, 1954). These lakes are naturally rather high in nitrogen, and the two enrichment agents were particularly important in stimulating growth of decomposers in the lake. It resulted in a greater rate of nutrient circulation, enhancing algal growth only secondarily. In fact, there is a concerted interaction between decomposers and producers in eutrophication. Thomas (1969) has shown that many bacteria are capable of producing vitamins required for growth of the green alga, *Cladophora glomerata,* that is often associated with enrichment. Enhancement of bacterial growth, then, might secondarily enhance growth of this alga. Addition of phosphorus to water from the lower Zurichsee in Switzerland greatly enhanced bacterial growth. Twenty days after the addition of sodium phosphate to surface water, 32,300 colonies/ml were present compared to 3653 in control samples. Similarly, in water samples from a depth of 20 meters, there were 1733 colonies/ml in nonenriched controls and 5225

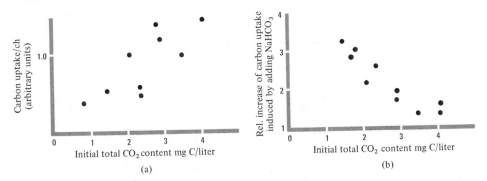

FIGURE 14-7 Data suggesting that photosynthesis of several of the Canadian Experimental Lakes is limited by available carbon dioxide: (a) increase in natural photosynthesis with increase in lake carbon dioxide content, and (b) decrease in the stimulatory effect of added carbon dioxide with increase in lake carbon dioxide content. (After Sakamoto, 1971)

colonies/ml when phosphate had been added. Thomas points out that phosphate enrichment, in addition to a potential for direct stimulation of algal growth, may also increase the availability of various bacterially produced growth factors required by algae. The cycle of interdependence between algae and bacteria is closed by glycolic acid, which is one of the major products excreted into the medium by algae (Fogg and Watt, 1966) and may serve as a respiratory substrate for bacteria. Phosphorus, nitrogen, sulfur, and perhaps other inorganic nutrients may cause eutrophication by catalyzing the exchange of organic compounds between decomposers and producers (Fig. 14-8).

Beeton (1965) collected information on the inorganic ion content of the Great Lakes during the nineteenth and twentieth centuries and found that substantial enrichment was associated with urbanization and population growth in the lake basins beginning in the early twentieth century. Sulfate, for instance, remained constant in Lake Superior from 1910 to 1960, whereas its concentration in Lake Ontario, after remaining constant during the late 1800s, increased dramatically from 1910 onward (Fig. 14-9). This sulfate enrichment has occurred during a period of substantially greater population growth, industrialization, and urbanization in the Lake Ontario basin compared with the relatively stable and predominantly rural population in the Lake Superior drainage basin. In other words the input term of the enrichment equation has been constantly increasing for Lake Ontario while not increasing for Lake Superior during the twentieth century. The Great Lakes have a slow rate of flow, relative to the volume, from Lake Superior in the west to the St. Lawrence River at the eastern end of Lake Ontario; Lake Michigan is a cul-de-sac between Lakes Superior and Huron. There is a general tendency for the ion content of the water to increase along the flow pattern and this tendency has greatly increased from 1910 to 1960, for the enrichment of the eastern lakes has been much greater over this period than has that of the western lakes. The sulfate content of Lake Ontario water was about 7.5 times higher than that of Lake Superior in 1910; by 1960 it was almost 15 times greater.

Lake Erie has shown an increase in inorganic ions over the twentieth century

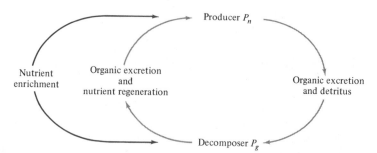

FIGURE 14-8 Generalized hypothesis of the nutritional interdependence of producer and decomposer communities. Nutrient enrichment, by stimulating both communities, is amplified around a positive feedback loop in which producer and decomposer activity promote one another.

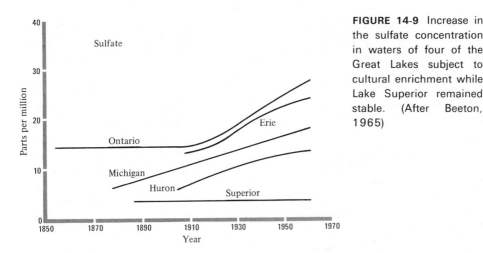

FIGURE 14-9 Increase in
the sulfate concentration
in waters of four of the
Great Lakes subject to
cultural enrichment while
Lake Superior remained
stable. (After Beeton,
1965)

405

ENRICHMENT
AND LIMITING
FACTORS

similar to that of Lake Ontario, but baseline values prior to 1900 are not available for comparison. The basin of Lake Erie showed the greatest population growth of any of the Great Lakes during the first 60 years of 1900, increasing from 3 to 10.1 million persons (Beeton, 1969). Urbanization is heavy in its basin also, with the cities of Detroit, Toledo, and Cleveland on its shores. The water filtration plant at Cleveland has been collecting data on algal density in intake water since 1919 and Davis (1964) used these data to document the incredible increase in primary producers over a 44-year period (Fig. 14-10). Algal density increased about fivefold during this period. Since phosphorus is one of the nutrients contributing to enrichment, Curl (1959) examined the sources of this element in western Lake Erie. He found that 71.5 percent of the phosphorus entering this end of the lake entered via the Detroit River and 22 percent entered via the Maumee River which drains the area

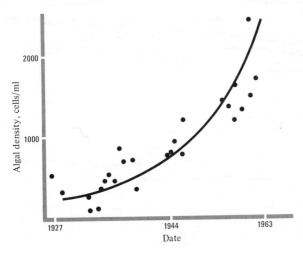

FIGURE 14-10 Increase in the density of algal cells in Lake Ontario waters at the intake of the Cleveland, Ohio, filtering plant during a period of rapid population and industrial growth in the lake basin. (Data from Davis, 1964)

of Toledo. The former source carried 405 metric tons/yr into the lake and the latter source carried 125 metric tons/yr. The concentration of phosphorus was lower, however, in the Detroit River than in any of the other sources, averaging only 2.6 mg phosphate/liter, compared with 43 in the Maumee, 22.8 in the Portage, 15 in the Raisin and Sandusky, and 11 in the Huron River. The huge volume of the Detroit River makes it an important bulk contributor, but the concentration of its contribution is relatively low. Because the volumes discharged by the Huron, Raisin, Portage, and Sandusky Rivers are relatively low, each contributes less than 3 percent of the total phosphate load on the western lake basin. With relatively high concentrations of dissolved phosphate, however, they may contribute to locally heavy algal blooms. Weibel (1969) examined the breakdown of the phosphate loads on Lake Erie between the upper lakes, rural drainage, and urban drainage. He found that the last class contributed 74 percent of the total phosphate load. Input from Lake Huron was less than 9000 kg/day, a similar amount came from rural runoff, and over 41,000 kg/day came from urban runoff. The latter could be further broken down into 32,000 kg/day from detergents, 14,000 kg/day from human wastes, 2700 kg/day from urban land runoff, and 2700 kg/day from direct industrial discharges. Insofar as phosphate contributed to the accelerated eutrophication of Lake Erie, by far the major source was household detergents.

SECONDARY ENVIRONMENTAL EFFECTS OF EUTROPHICATION

Eutrophication, as we have seen, is an enhancement of primary productivity and biomass at the base of lake food chains through an enrichment of the water content of growth factors that limit algae. The data on phosphate effects on bacterial activity in water of the lower Zurichsee in Switzerland suggest that enhancement of bacterial activity may be one of the initial effects of enrichment and that this secondarily enhances algal growth through the nutrient exchange loop connecting producers and decomposers. We have also seen, from the Zurichsee, that eutrophication is often accompanied by conversion of a cold-water fish community to a warm-water fish community. This change in fish communities arises out of changes in oxygen concentration of the water rather than out of changes in water temperature. One of the consequences of the high productivity of enriched waters is that the importance of detritus food chains becomes considerably increased. In the oligotrophic lakes tabulated by McCoy and Sarles (1969), plate counts of bacteria averaged 93 colonies/ml for water samples and 11,400 for bottom mud samples. Comparable data on eutrophic lakes were 438 and 170,100. In addition to the great increase in overall decomposer densities, the ratio of detrital to suspended bacteria was 120 in oligotrophic lakes and 388 in eutrophic lakes, indicating a considerable increase in relative as well as absolute importance of sediment decomposition. As a consequence of this increase in detritus, oxygen is severely depleted at the lower depths of eutrophic lakes (Fig. 14-11). As Birge and Juday (1911) emphasized in works laying the foundation of modern freshwater ecology (limnology), "As animals and plants sink after death, most of the decomposition goes on in the deeper water and especially close to and at the bottom. . . . After the oxygen is used up, anaerobic decomposition

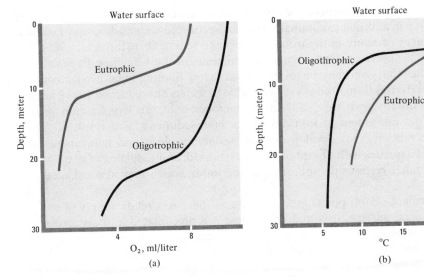

FIGURE 14-11 Comparison of the depth profiles for (a) oxygen concentration and (b) temperature in an oligotrophic and a eutrophic lake. Although the temperature profiles are similar, oxygen concentrations, particularly in deep water, are much higher in the oligotrophic lake. (Data from Jonasson, 1969 for eutrophic lake and Schindler, 1971 for oligotrophic lake.)

continues . . . [and] the lower water of a lake forms a zone of decomposition, whose processes are most vigorous at the bottom and decrease in intensity upward." When combined with the effects of temperature on water density, this depletion of oxygen in the lower waters of highly productive lakes is responsible for the conversion of fish communities upon eutrophication; that is, because water density decreases with increasing temperature, the coldest water in a lake tends to be near the lake bottom. When combined with warming of the upper lake surfaces by the sun, lakes develop a distinct thermal stratification during late summer (Fig. 14-11). Fish are distributed along this temperature gradient according to their preferred habitat temperature. Although the gradient itself is not affected by eutrophication, the distribution of oxygen in relation to temperature is radically affected as the colder water becomes extremely poor in oxygen. As a consequence habitat temperatures appropriate for salmonoid fish occur in water lacking sufficient oxygen to support fish, and they gradually disappear from the lake as their habitat is destroyed by detrital decomposition. The relatively minor oxygen flux required to support fish populations is diverted into decomposer populations by a massive detrital enrichment, and the fish that require cold water cannot survive. Fish capable of living in warm water become predominant in the lake since the surface layers of the lake maintain relatively high oxygen concentrations by diffusion from the atmosphere.

The three thermal regions that appear in lakes during late summer are (a) the *epilimnion,* which is the upper, warm region; (b) the *thermocline,* the middle region in which temperatures decline abruptly with depth; and (c) the *hypolimnion,* the lower region in which temperatures are uniformly cool. Depletion of oxygen in the hypolimnion, in addition to eliminating the niches of cold-water fishes, may facilitate the movement of many nutrient ions out of lake sediments (Mortimer, 1969). We saw earlier in Sakamoto's enrichment experiments in the Canadian Experimental Lakes studies that iron may be an important factor limiting algal productivity. In its oxidized (ferric) form, iron is very insoluble in water, but reduction to the ferrous form substantially increases solubility. Anaerobiosis of the hypolimnion greatly facilitates the movement of iron out of the lake sediments as a result of strong reducing conditions that develop in these sediments. Iron and manganese form insoluble precipitates with phosphorus when oxidized, and reduction of these during the late summer releases phosphorus into the water, which may also enhance productivity.

As Mortimer (1969) points out, "As long as there is a ready supply of oxygen to the sediment surface, iron and manganese are precipitated and accumulate at the surface as insoluble oxidized complexes. These complexes effectively scavenge the environment (water and lower sediment layers) for phosphate and other materials." The development of reducing conditions during late summer therefore generates a storage compartment of plant nutrients that can catalyze the fall algal bloom as the waters are recirculated in the fall. This secondary effect once again demonstrates the amplification networks that are involved in eutrophication as biological activity modifies the physical environment in such a way that the initial biological effects are subsequently amplified to larger and larger levels by nutrient precipitation and release. Inorganic nutrient enrichment enhances algal and decomposer activity, excess algal productivity amplifies decomposer activity, decomposer activity amplifies nutrient circulation, nutrient circulation amplifies algal productivity, . . . , and so on around the loop, until the rates are limited by some other factor such as light available to the algal populations or the supply of appropriate oxidizing agents to the anaerobic decomposer populations.

CULTURAL EUTROPHICATION: A CASE HISTORY AND SOME COST ESTIMATES

One of the most extended and thorough case histories of cultural eutrophication is available for Lake Washington in the city of Seattle. Stockner and Benson (1967) examined the morphological types of diatoms in sediments of the lake. There are two principal groups of diatoms, the *centrics*, which are radially symmetrical and are shaped like circular tablets, and the *araphidines*, which are bilaterally symmetrical and are shaped like flattened cigars. As eutrophication proceeds, araphidines become increasingly important in the diatom flora of lakes (Pennington, 1943). Stockner and Benson found that centric diatoms were the principal forms in Lake Washington sediments until the mid-nineteenth century, when they began to be replaced at an accelerating rate by araphidines (Fig. 14-12). The city of Seattle became established at about this time and by the early twentieth century sewage outflows were being dumped into the lake. Complaints about the deteriorating conditions of the lake resulted in the construction of a small treatment plant which replaced the raw sewage

409

CULTURAL
EUTROPHICA-
TION: A CASE
HISTORY AND
SOME COST
ESTIMATES

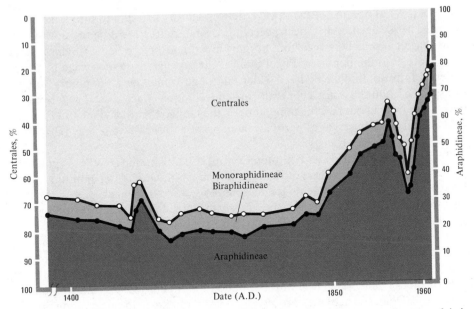

FIGURE 14-12 Changes in the balance among diatom groups in the sediments of Lake Washington. Note the drop in Araphidineae and increase in Centrales during a period of sewage diversion and subsequent reversal upon resumed dumping. (After Stockner and Benson, 1967)

with sewage receiving so-called primary treatment (solid material is settled out by gravity). Since this did not accomplish much, the city diverted its sewage into Puget Sound in 1926 and the lake then received sewage only during rainstorms.

In 1941 a secondary treatment plant (involving bacterial decomposition of sewage) began using Lake Washington as an effluent recipient. By 1959 there were eleven secondary treatment plants dumping into the lake (Edmondson, 1971). The content of araphidines, which declined following the initial diversion of sewage, shot abruptly upward as new treatment plants began to empty into the lake. In 1955 a large bloom of *Oscillatoria rubescens,* the alga famous to limnologists for signaling the eutrophication of the Zurichsee in Switzerland, appeared in Lake Washington. Sewage plant effluent at this time was flowing into the lake at a rate of 20 million gallons per day. Since the lake is an important recreational asset to the citizens of Seattle, a bond issue was approved in 1958 to divert sewage from Lake Washington. Diversion began in 1962 and was completed by 1968. Phosphate-phosphorus and nitrate-nitrogen concentrations during the early spring declined from 50 and 500 μg/liter, respectively, in 1962 to 10 and 300 μg/liter in 1969. Algal biomass, as indexed by chlorophyll concentration during the summer bloom, declined from about 25 μg/liter in 1964 to 5 μg/liter in 1969. Since the decline in algal biomass has been more closely correlated with reduction of phosphate than nitrate, Edmondson argues that the former is a more important enrichment factor than the latter in Lake Washington.

But Lake Washington is, of course, only part of the story. Sewage does not just disappear. What became of the sewage that previously had flowed into Lake Washington? The sewage that had been receiving secondary treatment before flowing into Lake Washington is now receiving primary treatment before being discharged, far from shore, into the depths of Puget Sound. Clearly, diversion of sewage in a more enriched form to another body of water, although it may be a temporary solution to a specific situation, is not a generally useful technique. Since coastal waters receive the sewage from about half the population of the United States (Ryther and Dunstan, 1971), additional diversion of poorly treated sewage to such waters may be a particularly inappropriate strategy.

In 1968 the Federal Water Pollution Control Administration (FWPCA) published a massive four-volume survey estimating the magnitude of water pollution in the United States, the requirements for reasonable sewage treatment, and the costs of such treatment. Although the data previously examined on the sources of phosphate enrichment in Lake Erie indicate that industrial addition was insignificant, the FWPCA data indicate that in terms of the amount of water used and the quantity of solid material added to the water, industry is over twice as important as domestic activity. In 1964 industry processed 49.5 trillion liters of water and added 165 mg to each liter of that water to dump 8.2 billion kg of solid wastes into the national waterways. Domestic activities processed 20.1 trillion liters, adding 198 mg/liter for a total load of 4 billion kg of solid wastes. Malhotra et al. (1964) were able to remove 99 percent of the nitrogen and phosphorus in sewage using a lime treatment that cost $10.80/million liters at 1962 prices. Assuming an average annual inflation rate of 2.2 percent, the 1972 cost would be $13.43/$10^6$ liters. Since most of the nitrogen and phosphorus originate in domestic sewage, assuming that all domestic sewage and 25 percent of the industrial sewage would have to be treated, the annual treatment cost in 1972 would be $2.42 billion. This, of course, does not consider the cost of capital investment required to build the necessary treatment plants; but once such plants were built, the operating costs for decreasing enrichment would be insignificant.

The FWPCA assessed the existing sewage plant capabilities and needs according to drainage regions in the 1968 survey. Nationally the value of the existing physical plant was $2.95 billion, and the additional investment required was $1.05 billion. This compares to 1969 Federal budgetary expenditures of 81.24 billion dollars for national defense, 3.79 billion dollars for foreign affairs, 4.25 billion dollars for space exploration, or 6.22 billion dollars for agricultural policy. The proportion of the needed physical facilities that was currently available ranged from a high of 80 percent in the Great Lakes drainage basin to a low of 59 percent in the western Gulf of Mexico basin (Table 14-1). Two other regions that were significantly above the national average were the southeastern United States and the Ohio Valley, whereas the Tennessee River basin and the lower Mississippi basin were significantly below national averages in level of plant development.

Although some areas of the country have a more severe requirement than others, it is clear that a substantial reduction in the level of cultural eutrophication is not beyond the financial means of the United States.

This examination of the mechanisms of eutrophication suggests that the causes

TABLE 14-1 The capital value of the existing sewage plant as a percentage of the required treatment facilities in 13 drainage regions of the United States

Drainage region	Existing plant capability (percent of required)
North Atlantic	70.7
Southeast	75.3
Great Lakes	80.6
Ohio	80.0
Tennessee River	59.5
Upper Mississippi River	73.1
Lower Mississippi River	62.9
Missouri River	72.8
Arkansas-Red River	67.1
Western Gulf of Mexico	58.9
Colorado River	65.6
Pacific Northwest	72.3
California and Hawaii	73.7
95% confidence range of mean	66–74.4

Data from FWPCA, 1968, in Edmondson, 1971

are complex and tend to amplify one another. The solutions to cultural eutrophication certainly require more information about the nature of these amplifications and the critical factors regulating them; but in addition to more knowledge of causes, a greater will to implement known solutions is required if our water resources are not to deteriorate further. Eutrophication, of course, is only part of the water pollution problem. We have seen evidence of contamination of aquatic ecosystems with such biocides as heavy metals. Eliminating such contaminants as these is fully as important as arresting eutrophication. Although eutrophication has its primary effect on the value of water in economic and recreational uses, contamination with toxic substances may actually destroy the capacity of contaminated water to support life.

GENERAL CONCLUSIONS

1. Eutrophication is the process of nutrient accumulation and succession in aquatic ecosystems.
    (a) Natural eutrophication arises out of the balance between nutrient input and output in the absence of man's activities.
    (b) Cultural eutrophication arises from human activities increasing the input term.
2. Primary productivity increases as eutrophication proceeds. Algal biomass is often concentrated close to the water surface in highly eutrophic ecosystems.
3. The nutrients contributing most to eutrophication will depend, of course, upon the nutrients limiting algal growth. Eutrophication will be accelerated by enrichment of the limiting nutrients.
4. Frequent secondary effects of eutrophication include increases in decomposer populations, depletion of dissolved oxygen, and conversion from cold-water to warm-water fish populations.

SOME HYPOTHESES FOR TESTING

Among the most critical areas for study in eutrophication is controlled identification of the responsible nutrients in culturally enriched waters. We have seen experiments indicating that a variety of algal nutrients can enhance primary productivity. Although phosphorus is by far the most commonly cited cause in the popular press, we have seen evidence that carbon dioxide, nitrogen, iron, and sulfate also may contribute to accelerated eutrophication. In the absence of good experiments on local water bodies, ordinances aimed at eliminating the use of phosphate detergents are certainly premature. They may actually be detrimental if innocuous detergents are replaced with cleaning agents containing limiting factors presently important in local waters.

From the standpoint of basic research much more information is needed on the relationship between enrichment rate and eutrophication rate. Sudden changes in algal populations over a one- or two-year period suggest the existence of nutrient thresholds in aquatic producer communities. Exceeding such thresholds may trigger conversion from one type of major producer to another. Controlled experiments on the relationship among import, export, and lake volume would provide insight into the significance of such thresholds.

PART 4

BIOME PROPERTIES AND COMMUNITY ORGANIZATION

Preceding parts have concentrated on the general functional and organizational properties of ecosystems. This part considers the properties of specific types of ecosystems with similar structural organization and functional properties. Its objective is to convey a general impression of the diversity of ecosystems on the earth. The concluding chapter in this part returns, once again, to the general properties of the biotic portion of ecosystems. In particular, it concentrates on the relationships among populations in a given community. Although the community chapter might immediately follow the succession chapters, understanding it will be facilitated by a previous discussion of the properties of communities in different ecosystems.

A group of ecosystems in which primary producers have similar growth forms and consumers have similar feeding habits is called a *biome.* Such a group is an abstraction based on the similarities of the ecosystems included. As an abstraction it involves many compromises with detail in the interest of conveying a general impression of the ecosystem-type. In addition, the biome concept generally has a regional connotation associated with continuity in space.

In Chapter 4 we saw that producers with similar life forms occur in similar climates although their taxonomic identification may differ. There are prairies in Eurasia and North America that differ in species composition but are functionally and structurally similar. In the chapters which follow, what we refer to as biomes, for brevity, are best considered biome-types, differing in detail from region to region.

The earth's biomes can be roughly classified according to type. Few ecologists would agree on the number or status given to different ecosystem-types, and our purpose here is to characterize broad types of ecosystem organization rather than construct a rigidly objective classification system. In addition, classification always has the inherent problem of ignoring transitions between categories. There are a number of such transitions between biomes, such as estuaries between freshwater and saltwater ecosystems, and marshes between aquatic and terrestrial ecosystems. It should be borne in mind that the categories were not designed to be completely objective, but rather to help the reader understand a complex subject. In the table on the following page, biome-types are italicized.

BIOME CLASSIFICATION

Aquatic
  Freshwater
    Lotic (rivers and streams)
      Rapids
      Pools
    Lentic (lakes and ponds)
      Littoral (shoreline)
      Limnetic (upper open water)
      Profundal (lower open water)
  Marine
    Littoral (shoreline)
      Rocky
      Sandy
    Neritic (continental shelf)
    Upwellings
    Coral reefs
    Pelagic (open sea)
      Epipelagic (upper)
      Mesopelagic (middle)
      Bathypelagic (middle)
      Abyssal (lower)
Terrestrial
  Desert
    Hot
    Cold
  Tundra
    Arctic
    Alpine
  Prairie
    Moist
    Dry
  Savanna
  Temperate coniferous forest
  Temperate deciduous forest
  Tropical forest
    Rainforest
    Seasonal forest

freshwater biomes

15

WATER AS A MEDIUM FOR LIFE

Every habitat presents organisms with certain adaptive challenges that they must meet if they are to survive. Aquatic habitats, in which life generally is assumed to have begun, present a series of particularly distinctive challenges as well as advantages that arise out of the unique properties of the medium's principal constituent, water. Chemically, water is a rather unusual molecule. It is, in fact, one of the few inorganic substances that is a liquid at the temperatures occurring on the earth's surface.

Various properties of water (Mack et al., 1956) contributing to its ecological effects as a habitat medium are the following:

1. The *specific heat* of water is higher than that of any other common substance. Specific heat is the caloric input required to raise the temperature of 1 gm of a substance 1°C, and the specific heat of water is much higher than most other substances. For instance, at room temperature it is 1, whereas the specific heats

415

of oxygen gas and sand are 0.0156 and 0.188, respectively. This means that water has the ability to absorb huge quantities of heat with relatively minor changes in temperature. As a result aquatic habitats tend to be buffered against the changes in temperature that come from seasonal or diurnal radiation patterns. Large bodies of water warm up and cool off slowly on either a seasonal or a diurnal basis.

2. The *heat of vaporization* of water is extremely high. Heat of vaporization is the caloric input required to volatilize 1 gm of a substance. For water this quantity is 539 cal/gm, compared to 204 for ethanol and 67 for kerosene. The ability of large bodies of water to influence local climate arises out of this high heat of vaporization; water can act as a "sink" for heat in winds blowing over its surface, thus generating an air-conditioning effect.

3. The *density* of water is maximal at 4 degrees, and ice is less dense than water. As a consequence of the complex relation of water density to temperature, seasonal temperature changes at temperate latitudes generate a mixing effect in water bodies. The lower density of ice causes it to act as an insulating coat on the water surface during winter.

4. Water has a relatively high *heat of fusion*, 80 cal/gm. This compares with 28 for potassium nitrate and 29 for magnesium sulfate, for instance. This is the heat that water can absorb in the change from ice to water without any increase in temperature. The ability of water to absorb a fairly large quantity of heat when changing state, once again, moderates temperature changes.

5. Water has a relatively high *dielectric constant,* 80, which means that the attractive force between oppositely charged particles, such as ions, is greatly diminished when they are dissolved in water. In other words, water is a good solvent, particularly for ions.

6. Water has a high *viscosity,* or resistance to flow. As a result it is a comparatively resistant medium within which to move but, conversely, it is relatively buoyant, tending to support bodies suspended in it against the pull of gravity. The viscosity of water is about 100 times that of air. Therefore it is both more difficult to move through and easier to remain suspended within than the atmosphere.

7. The *solubilities* of metabolically significant *gases* vary considerably in water. Liquids, of course, will dissolve only limited quantities of gases, and the quantities that water can dissolve vary radically from their concentrations in the atmosphere. At standard conditions 100 ml of water can dissolve 0.34 gm of carbon dioxide, only 0.007 gm of oxygen, and 0.003 gm of nitrogen. This compares with atmospheric concentrations of these compounds by weight of 0.045 percent, 23 percent, and 75 percent, respectively. In other words the solubilities of these three gases in water are in reverse order of their concentrations in the atmosphere. The relatively low solubility of oxygen, as we have already seen when considering lake succession, often results in extremely low oxygen availability in aquatic habitats.

Finally, something must be said about the comparative concentrations of substances dissolved in freshwater and seawater. Seas act as receptacles for substances eroded from the land as water continually cycles between precipitation and evaporation, washing minerals out of the land and leaving them in the sea. Even precipitation, of course, contains substances absorbed from the atmosphere, and freshwaters may vary considerably in the concentration of substances in solution, depending on the substrates they have passed over and through. Most freshwater contains 0.1 to 0.5 gm of dissolved solids per liter; the ocean contains over 3.5 percent mineral matter with over 75 percent of this being sodium chloride; Utah's Great Salt Lake contains 23 percent solids, and the salinity of the Dead Sea is even higher (Mack et al., 1956). The latter two habitats represent unusual conditions arising out of the evaporation of huge quantities of water in a desert climate.

Of more interest to us immediately is the contrast between freshwater and ocean water. As a consequence of different external concentrations of dissolved salts, freshwater and marine organisms confront different osmoregulatory problems. Marine invertebrates, in general, are iso-osmotic with seawater. Other organisms with dilute body fluids as a consequence of freshwater or terrestrial evolution have a tendency to lose water to the surrounding medium. The two principal adaptive mechanisms to cope with the continuous osmotic outflow involve drinking large quantities of water with salt excreted, as do many bony marine fish, or producing urine strongly hypertonic to the blood, as do many marine mammals (Vernberg and Vernberg, 1970).

Animals living in a dilute solution such as freshwater, in contrast, are confronted with a constant influx of water from the medium. Nonosmoregulating organisms exposed to such waters will swell up and gain weight from additional water. Freshwater-adapted organisms, however, have been selected for reduced surface permeability to water and production of a very dilute urine combined with active uptake of salts.

A comparison of freshwater and saltwater teleosts (bony fish) indicates that the former utilize the gills as a salt-uptake organ and secrete a low-salt urine, whereas the latter excrete salts from the gills and produce urine with a comparatively high salt content (Fig. 15-1). As a result of these different adaptive modes, bony fish are

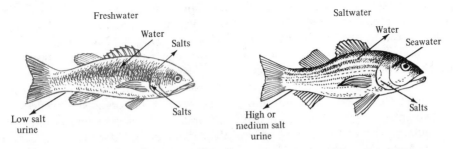

FIGURE 15-1 Freshwater and marine fish occupy habitat solutions that are hypotonic and hypertonic, respectively, relative to body fluids. The former have evolved salt-retaining mechanisms involving salt uptake at the gills and production of a low salinity urine. The latter have evolved water-retaining mechanisms involving drinking seawater and excreting salts in urine and at the gills.

able to maintain an adequate internal salt concentration in strikingly different osmotic media.

As we saw earlier when considering the hydrologic cycle, most of the earth's free water (97 percent) is in the ocean. With 2 percent of the free water in ice, the remaining 1 percent encompasses the two freshwater biomes. The total volume of flowing freshwater has been estimated as 3.83×10^4 km^3/yr and the total volume of freshwater lakes is about 12.5×10^4 km^3 (Foster and Harth, 1970). Of the lake water, over 75 percent occurs in the large lakes of Africa, North America, and Asia. In fact, over 20 percent of the earth's lake water occurs in a single body of water, Russia's Lake Baikal. The distinctiveness and fascination of aquatic biomes are emphasized by the fact that studying them has spawned sciences with separate names: *limnology* for freshwater study and *oceanography* for saltwater study. No comparable titles exist for study of terrestrial ecosystems.

THE LOTIC BIOME: GENERAL FEATURES

The *lotic,* or running water, biome is affected by the stream's fall or vertical change per unit of length. A generalized stream has its origin in some elevated water source, such as a spring or mountain lake, and flows downward at a rate affected by the altitudinal relief and the ratio between flow volume and stream bed volume. Flow rate gradually declines at progressively lower altitudes, and the volume of water increases until finally the enlarged stream becomes sluggish. Over this transition from a rapid, surging stream to a slow, sluggish river, water temperature will tend to increase, available oxygen will decrease, and the stream bottom will change from rocky to muddy.

This, of course, is an idealized presentation of the nature of the lotic biome, but the character of ecosystems in the biome is fundamentally regulated by flow rate. The critical velocity occurs around 50 cm/sec (Nielsen, 1950); above this velocity the stream bottom will consist of particles larger than 5 mm in diameter and as a result will have a firm, perhaps rocky, stream bed. Below this velocity, particles smaller than 5 mm are deposited so that the stream bed becomes soft and sandy or silty. In fact, all except extreme mountain and extreme flood plain streams are mixtures of rapids and pools, characterized by flow rates above and below 50 cm/sec as influenced by changes in the stream bed volume (Fig. 15-2).

At a given flow volume the deep wide expanse of a pool results in a slow water flow, with consequent silt deposition, whereas the shallow rapids are characterized by the rocky, slippery bottom well known to trout fishermen. Except for fish, which may move freely between rapids and pool ecosystems, the organisms occurring in the two areas are often completely distinct. Depth, however, is extremely important in regulating fish distribution in spite of their mobility (Sheldon, 1968). Species of darters studied by Forbes (1907) showed pronounced differences in the type of stream bottom they preferred. Also fish seem to be very restricted in the area over which they range (Gerking, 1959), so a stream contains a series of relatively distinct fish populations along its length. Young fish are usually found in shallow, comparatively unturbulent water, and many pool species spawn in rapids.

In the rapids, primary producers typically consist of diatoms and gelatinous green

FIGURE 15-2 Two examples of the lotic biome: A mountain stream with well-defined rapids and pools and the Mississippi River near Prescott, Wisconsin. Note the contrast between the silt-laden Mississippi and the St. Croix River at their junction. (*Top:* Bureau of Sport Fisheries and Wildlife, Dept. of Interior. Photo by Ben Schley. *Bottom:* Wide World Photos)

and blue-green algae which form a slimy community, called *aufwuchs,* on the rock surfaces. In a pool, higher plants rooted in the soft ooze at the bottom may be present in addition to aufwuchs. First-level consumers in rapids are largely insect larvae capable of maintaining their position in the swift current because of their flattened streamlined bodies and organs such as hooks (blackfly and caddisfly larave) and suckers (water penny beetle) which attach them to rock surfaces (Nielsen, 1950). Other adaptations to the rapids habitat include the almost universal occurrence in organisms of *positive rheotaxis*—the behavioral tendency to orient and swim upstream—and *positive thigmotaxis*—the behavioral tendency to "grasp" any surface that the organism contacts. First-level consumers in a pool, in contrast, are likely to be burrowing worms and insect larvae. There may be considerable difference in the oxygen content of the surging, highly oxygenated rapids and the still, less oxygen-saturated water of the pools. We shall see later in the chapter that organisms of the rapids may be more sensitive to reductions of dissolved oxygen than are pool organisms. The rainbow trout, a species native to rapidly flowing streams, is among the most sensitive fishes that have been tested for ability to withstand low oxygen concentrations. This species, in fact, has become somewhat of a standard for bioassays of oxygen content of waters (Herbert, 1965).

A comparison of the rapids, or torrential, biota with the biota of sluggish, lowland streams accentuates the differences between rapids and pools. The bottoms of such streams, because there is little vertical transition and hence a slow flow rate, are covered with a layer of silt. Rooted aquatic plants may be present in quieter stretches and here the aufwuchs of rock surfaces will be replaced in the slowest pools by planktonic algae. Much of the food energy in such a stream will be detritus from upstream and adjacent land. Among the first-level consumers, there will be more large organisms such as crayfish and mollusks, and cold-water fish will be replaced by such warm-water inhabitants as catfish and carp. Much of the bottom fauna will possess adaptations for burrowing, rather than the attachment adaptations of the torrential fauna. The fauna of sluggish streams must be capable of tolerating turbid and silt-laden waters, lower oxygen concentrations, and warm, variable temperatures (Russell-Hunter, 1970).

THE LOTIC BIOME: TWO EXAMPLES

MacKay and Kalff (1969) examined the association of stream insects with bottom types in a brook 1.7 km long in Quebec, Canada. The fauna was overwhelmingly insects in this mountain stream. Standing crop varied considerably according to bottom structure as determined by current velocity. Mean monthly standing crop in grams per square meter was 0.68 on gravel, 0.82 on detritus, 1.3 on sand, 1.4 on leaves, and 1.43 on stones. Densities were 920/m² on sand, 1300 on gravel, 2130 on stones, 3480 on leaves, and 5680 on detritus. The contrast between standing crop biomass and standing crop density suggests that detritus is occupied by many tiny insects, whereas sand is occupied by fewer but larger insects. Not only did the relative abundances of different insect larvae differ on a given substrate, but a given insect group often showed a decided preference for certain types of stream bottom (Fig. 15-3). Flies and mosquitoes were the predominant fauna on sand and gravel, but

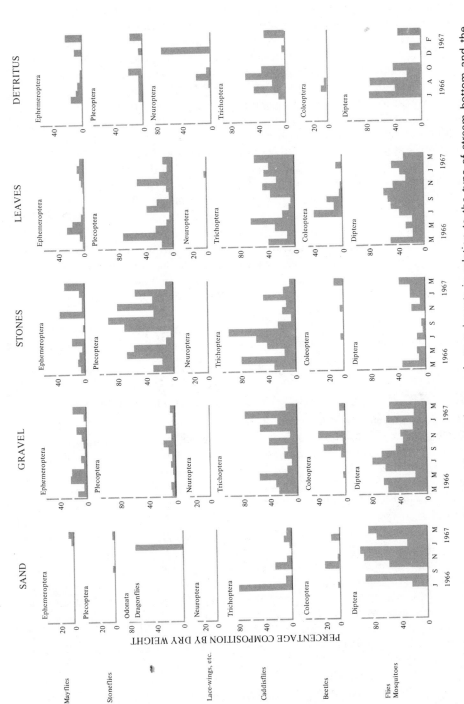

FIGURE 15-3 Changes in the relative abundances of stream bottom insects in relation to the type of stream bottom and the season. By scanning down the columns, it is clear that the composition of each community is distinct on each substrate type: flies and mosquitoes predominate on sand and gravel bottoms, caddisflies and stoneflies predominate on stone and leaf bottoms, and so on. (After Mackay and Kalff, 1969)

were largely absent from stony bottoms where caddisflies and stoneflies became more important. MacKay and Kalff point out that the sandy bottom is comparatively poor in microhabitat variability and supports the fewest species as well as the lowest individual densities. Variations in taxonomic composition of the faunal associations on different bottom types demonstrate the way current velocity, by sorting bottom particles according to size, affects the animal component of lotic ecosystems.

To contrast the mountain brook studied by MacKay and Kalff to a lowland stream, we can consider the bottom fauna in the central Mississippi River as examined by Carlson (1968). He sampled ecosystems in the navigation channel on the Illinois side of the river above Keokuk, Iowa. The bottom of the Mississippi in Carlson's study area is covered with a layer of silt. The average summertime density was 2924 organisms/m², a value intermediate to the different bottom types in the mountain brook. The most abundant organism in the sluggish water of this lowland river was a small mollusk, *Sphaerium transversum,* which reached densities as high as 12,131/m² on one occasion. It was collected at every sampling station on every sampling date. Burrowing worms (Tubificidae) were also abundant in the sluggish, soft-bottomed river, reaching densities as high as 398/m². The only relatively abundant insect larvae were mayfly naiads of the genus *Hexagenia,* which attained a maximum abundance of 884/m². Carlson concluded, "Detritus seems to be the major source of basic food, indicating that the community probably depends heavily on imported material." The comparatively greater abundance of mollusks and oligochaetes in the lowland stream compared with the predominance of insects in the mountain stream is typical of the ecological organization of sluggish and fast streams, respectively.

THE LOTIC BIOME: STRUCTURE AND FUNCTION

To study the structure and function of lotic producers, McIntire et al. (1964) constructed a series of laboratory-housed streams (Fig. 15-4) in which temperature, light intensity, flow rate, and other variables could be controlled but with the biota derived from natural stream water obtained from an adjacent brook in western Oregon. These lotic ecosystems consisted of wooden troughs in which smooth, water-worn rubble was placed to simulate a natural stream bottom. The current velocity was maintained at 24 cm/sec, which renewed the 200 liter total volume at a rate of 2 liters/min. Light intensities up to 8000 lux (lumens/m²) could be achieved by varying the distance from the light banks to the stream surface. Under these conditions an aufwuchs community typical of western Oregon developed. A pennate diatom (*Synedra ulna*) and a filamentous diatom (*Melosira varians*) were the dominant species. At various times during the summer filamentous green algae of the genus *Oedigonium* were abundant. A filamentous blue-green alga (*Phormidium retzii*) also occurred at light intensities of 2000 lux. Among the animals occurring in the streams were midge larvae (species of the genera *Tendipes, Calopsectra, Polypedilum,* and *Brilla*), naiads of several mayfly genera, and various cladocerans and copepods (McIntire and Phinney, 1965).

Gross primary production, respiration, chlorophyll content, and community

423

THE LOTIC
BIOME:
STRUCTURE
AND
FUNCTION

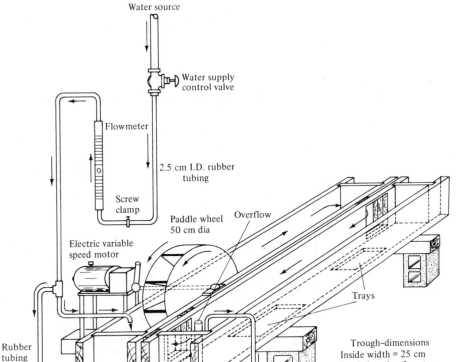

FIGURE 15-4 Diagram of a laboratory stream: water from a stream outside the laboratory enters at the left, is circulated through the troughs by a paddle-wheel, and exits at the right. (After McIntire et al., 1964)

biomass were all higher in ecosystems receiving higher light energy input (McIntire and Phinney, 1965). The export of biomass was determined by straining effluent water through fine mesh sieves, and it was found that export rate varied considerably with time (Fig. 15-5). The biomass export rate was considerably higher in the high-light stream, which was consistent with its greater standing crop biomass. This ecosystem also was subject to more variable rates of biomass export, ranging from only 0.1 gm/day to almost 2.8 gm/day. Biomass export seemed to be facilitated by silt in the incoming brook water, since periods of high incoming turbidity were accompanied with high biomass in the effluent. McIntire and Phinney suggest that this is because of the scouring effect that silt has on the aufwuchs. Later studies indicated that both biomass development and export were promoted by current velocity (McIntire, 1968). At 14 cm/sec biomass was about 100 gm/m^2 and export

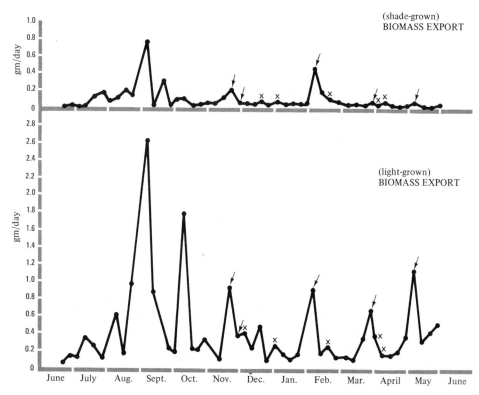

FIGURE 15-5 Seasonal patterns of biomass erosion and export from laboratory streams of the type diagramed in Figure 15-4. The upper stream was grown at low light intensity and the lower stream was grown at high light intensity. The arrows indicate when incoming water was extremely turbid, as a result of high water in the source stream, while the xs indicate slightly turbid conditions. (After McIntire and Phinney, 1965)

averaged 235 mg/day, whereas at 35 cm/sec biomass was about 175 gm/m² and export averaged 399 mg/day. Ecosystem respiration rate also was higher at higher flow rates (McIntire, 1966). Clearly, flow rate, through its effect on ecosystem processes, is a fundamental regulating factor in aquatic ecosystems. Particularly significant is the constant erosion of biomass and its export downstream in the water flow.

Minshall (1967) examined the trophic structure of a stream in Kentucky which flows through a forested area and empties into the Ohio River about 1.5 km from its source. He separated consumer-decomposer food sources into three classes: suspended organic matter, attached organic matter, and leaf material. The average standing crops of these three classes were 0.74 kcal/m³, 15.2 kcal/m³, and 9.5 kcal/m³, respectively. Attached organic matter—consisting of diatoms, a moss, minor algae, and bacteria associated into an aufwuchs—was relatively constant throughout the year, showing no pronounced seasonal variation. The suspended particulate organic matter, in contrast, was maximal in midsummer and midwinter and declined

425

THE LOTIC
BIOME:
STRUCTURE
AND
FUNCTION

at intermediate seasons. The leaf material also showed decided seasonal periodicity, although not in the same phase as the suspended particulate matter. It was maximal during the fall and winter and reached a minimum during the midsummer.

Among the major fauna of the creek were 24 herbivore species, 8 carnivores, and 5 omnivores. Mayflies were essentially grazers and although classified as herbivorous (Fig. 15-6), they, of course, must also consume considerable quantities of bacteria and detritus as they graze on the aufwuchs. Leaf material was important

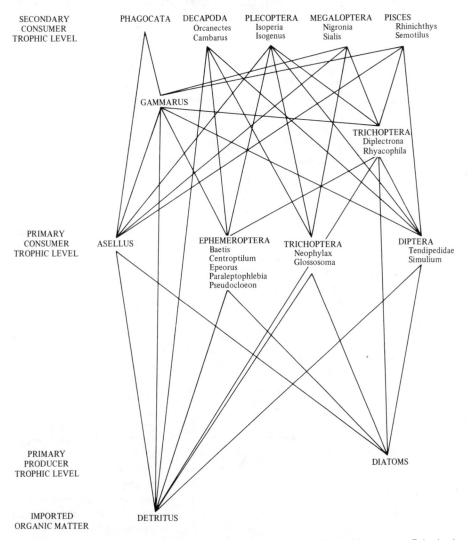

FIGURE 15-6 Trophic structure and energy flow pathways in a Kentucky stream. Principal energy sources are diatoms growing in the stream and leaves (detritus) eroded from a surrounding deciduous forest. (After Minshall, 1967)

in the diets of many species and both scud (*Gammarus minus*) and caddisflies tended to eat almost everything of suitable size in the brook. The scud species was by far the most important consumer, constituting 81 percent of the total numerical standing crop in the brook; its diet was over 90 percent detritus. The largest organisms present were crayfish (*Orconectes rusticus* and *Cambarus tenebrosus*) and the content of leaf material in their diets ranged from 30 to 100 percent. Although the evidence is not conclusive, since algal cells could not be separated from tiny leaf fragments in the attached and suspended particulate organic matter, Minshall concluded that imported organic matter in the form of leaves was by far the most important source of energy in this tiny brook.

THE LOTIC BIOME: THE SIGNIFICANCE OF DRIFT

As the studies of laboratory streams indicate, much of the biomass of lotic producers is subject to erosion by current action which carries the biomass downstream. And the data from the creek in Kentucky indicate that imported organic matter is important in stream energy budgets. Daily drift of invertebrates over a unit area is often several times the standing crop of that area (Waters, 1962). Since most stream organisms are strong swimmers, it is possible that many invertebrates are merely temporarily suspended in the water as a result of movements; so Waters (1965) examined the entry of a mayfly (*Baetis vagans*) and a scud (*Gammarus limnaeus*) into traps over a 24-hour period. First the upstream end was closed so that organisms could enter only from downstream; then the colonization opportunity was reversed. He found that there was no entry from downstream and that all of the individuals were, in fact, drifting downstream, relatively passively, with the current.

Pearson and Franklin (1968) examined eight habitat factors potentially affecting drift and found that population density, temperature, and illumination were particularly important. During the seasonal increase in mayfly population density, the drift of mayfly nymphs was closely related to density (Fig. 15-7). In fact they found that the total density of all invertebrates had a significant relationship to the drift

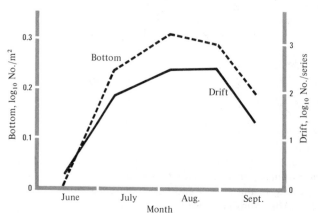

FIGURE 15-7 Seasonal changes in density of a mayfly population on a stream bottom (dotted line, left axis) and their presence in stream drift (solid line, right axis). Presence of the mayfly larvae in drift was closely associated with overall population density. (After Pearson and Franklin, 1968)

rate of the major drifting species. Waters (1961) has suggested that drift results from production in excess of the habitat's capacity to support individuals; the close association between mayfly drift and total invertebrate density suggests that this is probable.

Drift is not entirely passive and shows a pronounced diurnal pattern, reaching a maximum just after sundown. Holt and Waters (1967), utilizing artificial illumination of a small stream in Minnesota, were able to identify an intensity threshold of about 1 lux which initiated the release of drifting organisms from the substrate when intensity was decreasing and caused termination of drift when reached in an increasing intensity trend. It seems likely that light acts as a thigmotaxic (grasping) cue for stream invertebrates, causing them either to release or to grasp the substrate on a diurnal basis. When the release cue occurs in the evening, the drift begins; when the grasping cue occurs in the morning, the drift ends.

Anderson and Lehmkuhl (1968) examined the effect of increased stream flow, resulting from rainfall in a stream's watershed, on drift. They found that rainfall accentuated drift with substantially more animals entering the drift during rainy periods. In the autumn rainy period, although drift rate increased three to four times above normal, there was still a substantial increase in bottom density over the same period. From September 25 to October 22, mean density increased from $334/0.1$ m^2 to $3426/0.1$ m^2. This suggests that the autumn rains, by washing tremendous quantities of fallen leaves into the stream, increased the food supply to the stream invertebrates, resulting in spectacular population growth as well as making it harder for organisms to maintain position in the increased stream flow. Organism size was larger during flood-induced drift in the Oregon stream. In the mayflies (*Baetis* spp.) mean weight of high water drifters was 50 percent higher than during normal stream flow. At normal current velocities the displaced organisms tend to be small, but as velocity increases even large individuals are unable to maintain their positions.

Drift is undoubtedly an important food source to higher order consumers in the lotic environment. It represents a constantly renewing food source for such predators as fish, who merely need to find an advantageous spot in the stream and select food as it passes rather than searching for it. The available prey, we have seen, will be influenced by light, invertebrate density, stream flow, and other factors.

The effect of population density in limiting habitat space is expressed through aggressive encounters in many species. Glass and Bovbjerg (1969), for instance, examined the spacing of caddisfly larvae at different densities and with different types of habitats simulated in the laboratory. In a bowl with no internal habitat heterogeneity, the larvae tended to concentrate at the edge where they were in contact with both the bottom and the wall. The larvae were aggressive, however, including biting and striking one another on contact; this provoked fights that culminated in an avoidance reaction by one of the larvae. Aggressive contacts resulted in a relatively uniform spacing of larvae within the bowl, with very little clumping detectable except at the edge, and even here contact was avoided by spacing. When pebbles were provided in the center of the bowl, aggregation occurred in this refuge. As population density was increased, however, a smaller and smaller percentage of the total population was able to occupy the refuge and the remainder were forced into the marginal homogeneous dish habitat. Glass and Bovbjerg summarized, "Regardless of the population density, no larva was observed to build a tube in physical contact with another. No larva, in any population density, was observed to remain in physical contact with another larva for more than a few seconds."

It seems likely that aversion to contact with other population members must be an important causal mechanism in the correlation between population density and drift. With increasing population size, opportunities for finding a suitable unoccupied space decline, and avoidance reactions upon aggression release individuals into the current; thus drift increases.

Ecosystem character in the lotic biome is influenced primarily by the interrelations among water source, water volume, stream bed volume, and stream bed "fall," the vertical transition per longitudinal transition. The three physical parameters of the stream influence bottom character, drift rate, oxygen content of the water, water temperature, turbulence, and various other factors that regulate species composition of the ecosystem. In addition, since much of the food in a stream is derived from adjacent terrestrial ecosystems via detrital erosion, the latter will influence the character of the lotic ecosystem by regulating properties and supply of food.

THE LENTIC BIOME: GENERAL FEATURES

We considered some of the general features of lotic ecosystems in our earlier discussion of lake succession (Chapter 14). We have seen that lakes also have a flow rate and that the nature of the ecosystem is influenced by the rate of water renewal. Lentic biomes are distinguished from lotic biomes by their much less rapid rate of water renewal. Whereas the flow per volume of pools in even sluggish rivers will be at most a matter of a few months, in lentic ecosystems water renewal rate is of the order of years, sometimes even centuries.

Just as the lotic biome was characterized by pool and rapids ecosystems, the lentic biome can be subdivided into littoral, limnetic, and profundal ecosystems (Fig. 15-8). The littoral type occurs at the edge of lakes and ponds and extends inward from the shoreline to the innermost populations of plants rooted in or attached to the lake or pond bottom. There may be a series of concentric rings of rooted plants from emergent species such as reeds (*Phragmites*) and cattails (*Typha*) at the edge, to floating leafed species such as waterlilies (*Nelumbo*) in deeper water, and submerged species such as waterweed (*Elodea*) and water milfoil (*Myriophyllum*) at the outer edge of the littoral zone. The emergent species are also, of course, marsh species which may occupy large areas where lake succession has resulted in disappearance of open water. Their underwater parts usually will be covered with aufwuchs. The littoral zone contains numerous vertebrate consumers such as frogs and snakes.

The limnetic zone is the euphotic open water, that is, water down to the depth of light penetration. This water contains planktonic producers, particularly diatoms and species of green and blue-green algae. Zooplankton are the primary first-order consumers in the limnetic zone.

The profundal type of ecosystem occurs in those regions below the euphotic zone. This may be relatively minor in ponds, although it represents the major water volume

FIGURE 15-8 An example of
the lentic biome in North
America. Note the herba-
ceous shoreline vegetation
along the left and rear
shores. (American Museum
of Natural History)

429

THE LENTIC
BIOME:
GENERAL
FEATURES

in lakes such as Superior and Baikal which are extremely deep. This zone will be
traversed by a constant rain of detritus from the upper limnetic zone and will have
active decomposer communities in the bottom ooze. Fish fauna will be influenced
by water temperature and successional stage; salmonoid fish such as lake trout
predominate in cold-water oligotrophic lakes; bass, pike, pickerel, and perch in warm
water, mesotrophic to eutrophic lakes; and cyprinoid fish such as carp in warm,
culturally eutrophic lakes. Fish will generally range through the littoral and limnetic
zones and occur less frequently in the profundal.

Because water is such a poor conductor, heat is transferred within water bodies
almost solely by mixing. Clarke (1966) observed that if all the water in a lake 100
meters deep was cooled to 0 degrees and the surface was then heated to 30 degrees,
it would take more than 100 years for a significant change to occur in water temperature at the bottom. Because of the complex relationship between water density and
temperature, heat transfer in large water bodies is accomplished by vertical circulation.

As we have seen when considering eutrophication, there is summer thermal
stratification in many lakes, with an upper warm layer, the epilimnion, a middle
thermocline layer in which temperatures decline abruptly with depth, and a cool
hypolimnion in lake depths. Although the epilimnion tends to circulate as a result
of wind friction over its surface, the denser hypolimnion does not circulate and, as
we have seen, tends to become depleted in oxygen, particularly if there is active
detrital decomposition. In autumn, as air temperatures cool, the epilimnion cools
until it is the same temperature as the hypolimnion, the thermocline disappears,
and the entire lake body is circulated by wind friction. During this *fall overturn,*
profundal water is reoxygenated and limnetic water is enriched with nutrients

released below the thermocline during summer decomposition. As air temperatures continue to cool, ice forms on the surface, the lake temperature becomes uniformly near 4 degrees except for cooler water near the surface, and circulation is again minimal. Finally, during the spring, warming of surface waters results in another period of water circulation, a *spring overturn,* prior to reformation of the summer thermal stratification. Once again nutrients are flushed into the euphotic zone and oxygen is carried into the profundal zone.

Temperate lakes therefore are subject to four distinct seasons triggered by the terrestrial seasons: (a) a stratified season during summer when a thermocline is present and there is an abrupt temperature differential between warm superficial water and cold deep water; (b) an autumn overturn upon cooling of the lake's surface during which nutrients are flushed into the limnetic zone from the profundal; (c) a winter stagnation when lake temperatures are uniformly cold; and (d) a spring overturn in which nutrients are again flushed out of the profundal into the limnetic. This seasonal pattern is somewhat less well developed in both tropical and arctic lakes.

There are certain distinctive lakes, fed by water from mineral springs, in which there is little nutrient exchange between upper and lower regions. These *meromictic* lakes (Hutchinson, 1957b) result from creation of a permanent density gradient by high dissolved salt content. Enrichment by mineral salts creates a permanent and stable density gradient independent of thermal effects coupled to atmospheric temperatures. Since the euphotic zone of meromictic lakes is never enriched by nutrient circulation, they are notoriously unproductive (Russell-Hunter, 1970).

Lakes in some urban areas may be converted into culturally meromictic waters as a result of salt runoff (Bubeck et al., 1971). Numerous cities use salts to facilitate snow and ice removal from streets in wintertime. A bay that receives wastes from Rochester, New York, apparently is being converted into a meromictic body of water as a result of salt enrichment; seasonal overturns are being progressively delayed as salt content builds up in the water depths. In the winter of 1969 to 1970, the drainage basin of the bay received about 1 percent of the United States total salt used for deicing. About half of this drains into the bay and, during the winter, the chloride concentration of the principal feeder creek reached 360 mg/liter compared to 100 mg/liter during summer months. In storm sewer discharges in February a chloride concentration as high as 4000 mg/liter was recorded. From 1910 to 1970 chloride concentration in the bay increased from 12 mg/liter to over 100 mg/liter with a particularly rapid rise from 1950 to the mid-1960s. The winter salt influx has created a substantial density gradient in the bay and, as a result, the period of summer stagnation has been prolonged about a month. The authors concluded, "The changes are not viewed with alarm. The point is that the salt runoff has significantly changed the physical characteristics of the bay and that similar conditions might be expected elsewhere, particularly in heavily salted areas that provide natural traps."

Typical features of the lentic biome are seasonal blooms of populations and seasonal successions in which species replace one another in a regular fashion over an annual period. Holland (1969) examined the seasonal patterns of diatom abundance in Lake Michigan and observed seasonal blooms that differed in extent and timing in offshore and inshore waters (Fig. 15-9). In inshore waters diatom density reached a peak near 4000 cells/ml of water during May and then declined gradually to a minimum in late September. In offshore waters, in contrast, peak diatom densities were reached in late August. These seasonal blooms in primary producers generate a network of population fluctuations that are propagated through the consumers. Although seasonal variations are generally believed to be caused by seasonal fluctuations in supplies of limiting factors such as light and dissolved nutrients, rigid experimental proof of such limitations has been elusive. We saw from Hutchinson's experiments in which Linsley pond water was enriched that primary production may be simultaneously limited by several factors; so simple explanations of seasonal blooms are not persuasive.

In addition to the seasonal blooms of total producer biomass, there is a progressive turnover of populations within the trophic level (Fig. 15-9). Even though total biomass may remain relatively constant, one population replaces another throughout the season. As a result of rapid turnover, productivities of planktonic systems are considerably greater than mere biomass changes suggest. The inverted biomass

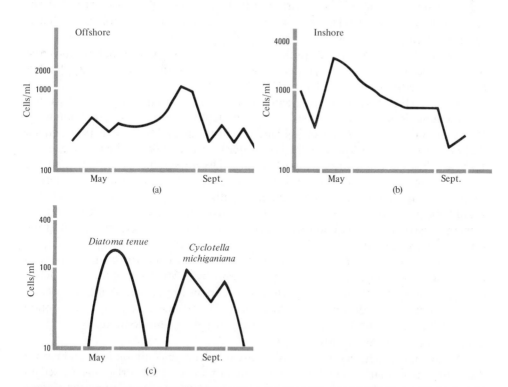

FIGURE 15-9 The seasonal diatom bloom in Lake Michigan in (a) offshore and (b) inshore water, and (c) the change in abundances of two species of diatom over the course of a season indicating the overall curves of (a) and (b) are the sums of many species replacements. (After Holland, 1969)

pyramid of aquatic ecosystems such as we saw earlier when considering energetic organization of ecosystems (Chapter 4) arises out of the rapid turnover of phytoplankton populations and the substantial cropping of these populations by herbivores.

THE LENTIC BIOME: AN EXAMPLE

Picking examples that illustrate biome types is difficult at best and nowhere more difficult than when considering the lentic biome. Lakes, like streams, oceans, forests, and all other ecosystems, are each distinct. They vary in depth, surface, turnover rate of the water, water color, oxygen regime, and so many other attributes that a single abstraction is exceedingly difficult. However, such simplification is valuable for the impression it gives of the general character of the biome.

For our example we will consider Lake 240, one of the Canadian experimental lakes discussed in Chapter 14. Some of its special features that cause it to be atypical are yellowish water caused by a drainage basin occupied by a coniferous forest, generally low nutrient content, weak algal blooms compared with many lakes, and a short summer warm season by virtue of its high latitude (Schindler, 1971; Armstrong and Schindler, 1971). The lake is typical of the biome, however, in that it has well-developed thermal and oxygen stratification, a volume divisible into littoral, limnetic, and profundal zones (although the last is minor), faunal composition characteristic of many north temperate lakes including the Great Lakes (Patalas, 1971), and medium size and depth (Hamilton, 1971).

Lake 240 (Fig. 15-10) has an area of 44.1 ha, drains a land area of 570 ha, has a maximum depth of 13.1 meters, a mean depth of 6.1 meters, a volume of 26.7×10^5 m^3, and a shoreline of 3440 meters (Brunskill and Schindler, 1971). Given the volume and the outflow rate, the lake water is turned over, on the average, every 4.4 years. The lake is frozen over from early December until mid-April and the seasonal overturns occur in late April and mid-October. There is pronounced thermal stratification from June until late September with a 12 degree thermocline between 4 and 8 meter depths during late August and early September. The lake is not deep enough to have a large hypolimnion below the thermocline. Oxygen content at the surface is maximum during midwinter, probably as a result of phytoplankton production under the ice, and by the time of the autumn overturn oxygen has been reduced to zero below 10 meter depth, presumably by decomposer populations active during summer. A similar low-oxygen level occurs in March and April, just before the winter ice cover disappears. Schindler (1971) estimates oxygen consumption during winter as 31 to 34 mg O_2/m^3 of water per day. This compares with 81 for a highly eutrophic lake in Minnesota.

Macroscopic producers (macrophytes) are relatively uncommon in the lake since the bottom is primarily rocky. In sandy and silty areas, quillwort (*Isoetes*), sedges (*Carex*), pipewort (*Eriocaulon*), and bladderwort (*Utricularia*) occur at the water's edge or submerged, according to growth habit (Brunskill and Schindler, 1971). The littoral zone is dominated by epilithic diatoms (Stockner and Armstrong, 1971). This zone extends to a depth of 5 to 6 meters, below which producers do not occur on shore rocks because of poor light penetration. Water temperatures in the littoral vary

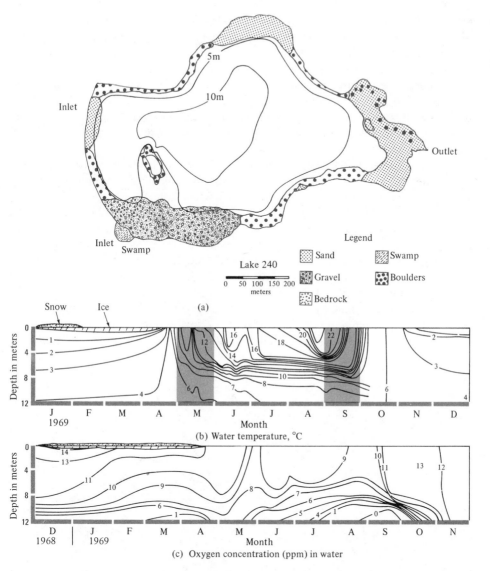

FIGURE 15-10 General features of Lake 240 in the Canadian Experimental Lakes Area. The upper graph is a map of the lake showing bottom types, depth isopleths, inlets, and the outlet. The middle graph is a seasonal map of the relationship between depth and temperature (°C) isopleths. The bottom graph is a seasonal map of the relationship between depth and isopleths of dissolved oxygen concentrations (ppm O_2). (After Brunskill and Schindler, 1971 and Schindler, 1971)

from 4 degrees when the ice melts to 23 degrees in late August. Bottom surfaces in the littoral are covered with a light brown, furlike growth of aufwuchs, the thickness of which varies seasonally from 0.5 to 2 cm. Principal species are the golden algae, *Achnanthes minutissima* and *Fragillaria pinnata*. These species and other golden and yellow-brown algae contribute between 60 and 70 percent of the littoral producer biomass. The rest are a filamentous green and blue-green algae that become more abundant in late summer.

A typical seasonal cycle is present, with diatoms composing over 80 percent of the biomass in April and May to be invaded later by filamentous green algae and desmids (30 to 40 percent of the June-July biomass); these are subsequently invaded by filamentous blue-greens (30 to 40 percent of the mid-August to mid-September biomass), and finally by an autumnal diatom peak (70 to 80 percent in late September to early October).

Biomass and density development (Fig. 15-11) on glass slides suspended at a 1 meter depth indicated that biomass and density both increased rapidly during May and June, whereupon density declined slowly while biomass continued to increase at a somewhat reduced rate. Therefore the littoral zone of Lake 240 is characterized by a rapid early season biomass bloom involving first diatoms and then, more important from the biomass standpoint, filamentous green algae and desmids. These are subsequently replaced by blue-greens and diatoms while biomass addition rate declines. Peak biomass of 12.5 gm/m² was recorded on September 20, just before the autumn overturn.

The principal producers in the limnetic zone were, of course, phytoplankton (Schindler and Holmgren, 1971). The primary taxonomic group present was the golden and yellow-brown algae, with three colonial forms (*Chrysosphaerella, Botryo-*

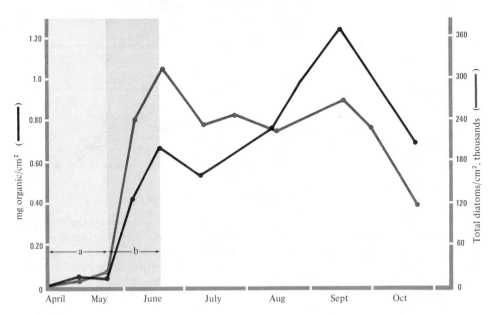

FIGURE 15-11 Changes in biomass of attached algae and diatom density on glass slides suspended in Lake 240 over a growing season. Period a is an initial spring colonization period and b is the period of most rapid algal growth. (After Stockner and Armstrong, 1971)

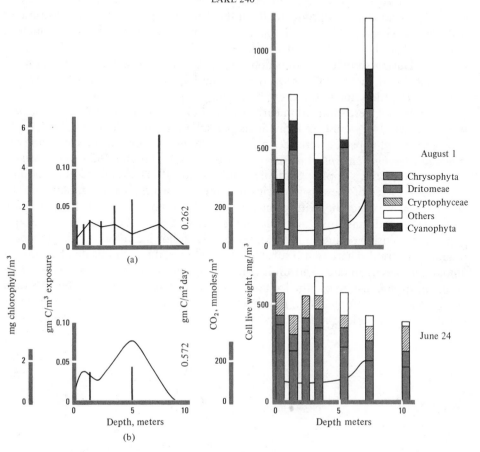

FIGURE 15-12 Phytoplankton productivity (line on left-hand graphs), chlorophyll concentration (bars on left-hand graphs), and abundances of different taxonomic groups (bars on right-hand graphs) in Lake 240 on two dates. (After Schindler and Holmgren, 1971)

coccus, and *Dinobryon*) being most abundant in the June bloom; by August they had been replaced by a dinoflagellate, *Cryptomonas.* There also was a *Euglena* bloom during late spring and early summer.

During late June phytoplankton biomass was 424 mg carbon/m² and photosynthetic rate was 572 mg carbon/m²/day for a turnover time of 0.74 day. By August 1 biomass had increased to 710 mg carbon/m² and net productivity had fallen to 262 mg/m²/day, so turnover time had risen to 2.71 days. In spite of the seasonal increase in turnover time the phytoplankton of the lake still had a high net productivity per unit biomass throughout the summer. In June the biomass was relatively evenly distributed with depth down to 10 meters, but the net productivity reached a decided peak at 5 meters and fell off abruptly below this (Fig. 15-12). By August

1 blue-green algae had become more important and the biomass did not extend as deeply in the lake. Net productivity on this date was relatively evenly distributed with depth, whereas chlorophyll concentration increased with depth to 7 meters, indicating that photosynthetic efficiency per unit chlorophyll was an inverse function of depth. During summer stratification, carbon dioxide concentration tended to be somewhat lower in the epilimnion than in the hypolimnion as a result of photosynthetic depletion above the thermocline and respiratory enrichment below.

Zooplankton were sampled in early May in Lake 240 (Patalas, 1971) and density was extremely low—only six individuals per square centimeter of lake surface. In this sample eight species were recorded, with the most abundant species being the copepods, *Cyclops bicuspidatus* (64 percent of the sample) and *Diaptomus minutus* (31 percent of the sample). Schindler and Nevin (1971) examined the seasonal abundance cycles of these two species in Lake 122, which is very similar to Lake 240. *Cyclops* was dicyclic, with pronounced peaks of abundance in the first larval stage (nauplii) during June and, under the ice, in December (Fig. 15-13). Major population density occurs in this winter period, following a substantial fall egg-production period. There is also substantial larval mortality, as indicated by contrast among nauplii, copepodids, and adults. Maximum reproduction of *Diaptomus minutus* occurs in spring, just before the ice breakup, and mortality is considerably less than in the other major zooplankton. Low but sporadic egg production occurred throughout the summer period, and all surviving organisms during the winter were adults.

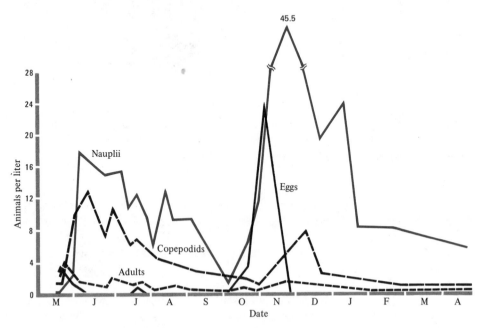

FIGURE 15-13 Seasonal changes in abundances of different stages in the life cycle of a planktonic herbivore, *Cyclops bicuspidatus,* in one of the Canadian Experimental Lakes. (After Schindler and Nevin, 1971)

Both immature stages and adults of this species also exhibited a diurnal vertical migration, moving upward during July from a distinct daytime concentration around 7 meters to a broader peak around 5 meters during the night.

Bottom-dwelling fauna (zoobenthos) of littoral and profundal zones were not sampled separately (Hamilton, 1971). These organisms were sampled in early May, prior to emergence of adults, many of which are aerial, therefore complicating post-emergence population estimates. In May average density in Lake 240 was 1642 individuals/m², of which 89 percent were flies (Diptera) and 7 percent were amphipods. The principal species were both phantom midges (*Chaoborus*), which made up about 45 percent of the total bottom fauna. Another important species was a crustacean, *Pontoporeia affinis*. This species was most abundant in the profundal zone, whereas the dipterans were more important in the littoral zone. Also abundant was the profundal midge, *Phaenopsectra coracina*. In terms of food habits, the dipterans are both grazers and detrital feeders, whereas the crustacean and the midge are primarily detrital feeders. Although no data were reported on the higher order consumers of the experimental lakes, common fish would likely be sunfish, pike, and smallmouth bass.

A food web for Lake 240 would start with colonial golden and yellow-brown phytoplankton, diatoms, and filamentous brown and blue-green algae at the base and end with carnivorous fish. Principal food chains are separated somewhat in the littoral and limnetic, although they tend to converge at the level of the higher order consumers. Limnetic producers are consumed by crawling grazers such as insect larvae which, as we have seen, are also often the most abundant herbivores in lotic ecosystems. What does not occur with comparable importance in lotic systems, except in very sluggish ones, is the planktonic grazing system involving copepods and cladocerans as primary herbivores. Lakes are also more likely to have a higher order consumer such as pike than are most lotic ecosystems in which the highest order consumers will feed primarily on drifting grazers.

THE LENTIC BIOME: STRUCTURE AND FUNCTION

Lentic ecosystems have been fundamentally significant in the development of energy hypotheses in ecology. Among the first attempts to determine a complete energy budget for an ecosystem was Juday's (1940) study of Lake Mendota in Madison, Wisconsin. Much earlier Forbes (1887) had laid the foundation for ecosystem ecology in a historic essay, *The Lake as Microcosm*. Forbes discussed the interrelationships among food chains (which he did not identify as such) and the competition, seasonal cycles, predation, and niche relations among co-occurring species, as well as the balance of abundances in the lake. Although his essay clearly falls within the realm of natural history rather than science, since there is no testing of hypotheses but instead a general discussion of observation, most of the fundamental ideas that characterize "modern" ecology can be recognized. It is, of course, from Juday's energy budget of Lake Mendota that we trace Lindeman's trophic level concept, which has been used throughout ecology to organize feeding relations; and it is to a distinguished limnologist G. E. Hutchinson (1959) who has spent much of his life examining lakes,

437

THE LENTIC
BIOME:
STRUCTURE
AND
FUNCTION

that we owe our conceptual base for understanding the niche. Perhaps it is because the boundaries of lentic ecosystems can be defined with relative ease that they have been favored objects of study in ecology and also have been of fundamental significance in formulating many of the ideas of structure and function examined in previous sections of this book.

More recent studies have examined the balance between producer and herbivore populations in lentic ecosystems. Dickman (1968), for instance, examined the effect that tadpole density had on the aufwuchs community of a small oligotrophic lake in British Columbia. By the first of May the lake is covered with massive clumps of filamentous green algae (*Mougeotia*) which disappear in the subsequent two weeks. The disappearance is coincident with the appearance of tadpoles of the red-legged frog (*Rana aurora*). Plywood cages (3.05 meters on a side) were placed in the lake in 1 meter of water and in some of these frogs deposited eggs. Empty cages served as controls. The first tadpoles appeared on May 12 and at this time the periphyton biomass was 1.6 mg/cm² in all cages. Over the next month, algal biomass plummeted in cages where tadpoles were present (Fig. 15-14), although it stayed constant in control cages. Algal biomass remained low until the tadpoles reached the foreleg stage, whereupon the stomach began to metamorphose and algal feeding stopped. There was some recovery of algal biomass after feeding stopped. In this lake a single herbivore population had a drastic effect on the biomass of the primary producers. This suggests that seasonal oscillations in algal abundance, common in many lakes, may be related in part to consumer abundances.

Whiteside (1970) examined the structure of chydorid cladoceran consumers in a number of Danish lakes. Chydorid community H' was inversely related to lake primary productivity (Fig. 15-15) with

$$H' = 3.661 - 0.004\,P_n \tag{15-1}$$

where H' denotes the chydorid community diversity and P_n net primary productivity

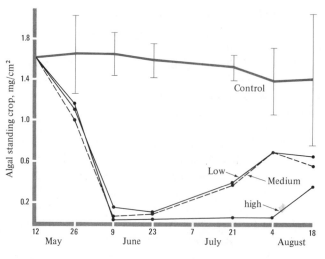

FIGURE 15-14 Effect of tadpole density on algal periphyton biomass: control lacked tadpoles, low density started with 15 tadpoles per 9 m², medium with 24/9 m², and high with 35/9 m². (After Dickman, 1968)

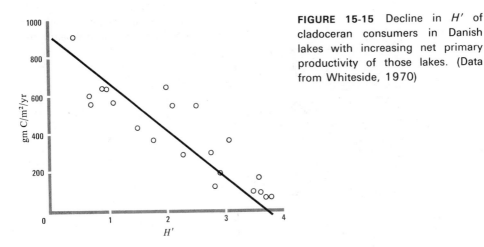

FIGURE 15-15 Decline in *H'* of 439
cladoceran consumers in Danish
lakes with increasing net primary
productivity of those lakes. (Data
from Whiteside, 1970)

HABITAT
DIFFERENTIA-
TION IN
FRESHWATER
BIOMES

in grams carbon per square meter per year. Since, as we saw earlier in our study of lake succession, net primary productivity is highest in eutrophic lakes and lowest in oligotrophic lakes, Whiteside's study suggests that diversity of these detrital grazers declines with lake enrichment.

HABITAT DIFFERENTIATION IN FRESHWATER BIOMES

One of the key problems of ecology to which we keep returning is why the species compositions of ecosystems change. We have just considered many examples of this change in the lentic and lotic biomes. For instance, the yellow-brown algae are important producers in both stream rapids and lake littoral zones, and yet the species compositions of the algal communities are distinctly different. Similarly, crayfish occur in both streams and lakes, but different species are present in the different biomes.

Bovbjerg (1970) has examined the habitat preferences of two crayfish species widely distributed in North America east of the Rocky Mountains: the stream crayfish, *Orconectes virilis,* and the pond crayfish, *O. immunis*. The stream species occupies crevices in the rapids of rocky streams and may dig pits in the banks or under rocks; the pond species burrows in the bottom and builds the "chimneys" well known to people who have walked along pond banks where the species occurs. Relative frequencies of the two species vary along the headwaters of the Little Sioux River in Minnesota and Iowa (Fig. 15-16). Upper reaches of the river often are reduced to a series of ponds during midsummer and *O. immunis* is the exclusive species in this region. In the middle section, the river consists of a series of rapids and pools and here the species are mixed with *O. immunis* primarily in the pools and *O. virilis* in the rapids. The lower gravel- and rock-bottomed region is occupied exclusively by *O. virilis*.

Bovbjerg recognized several adaptations that might account for habitat divergence in the two species: (a) bottom burrowing would be important for survival

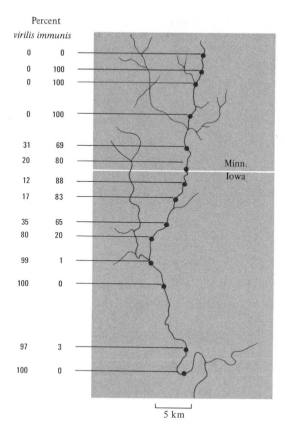

Percent

virilis immunis

0	0
0	100
0	100
0	100
31	69
20	80
12	88
17	83
35	65
80	20
99	1
100	0
97	3
100	0

Minn.
Iowa

5 km

FIGURE 15-16 Distribution of two species of crayfish in the headwater region of the Little Sioux River in Minnesota and Iowa. (After Bovbjerg, 1970)

in temporary ponds; (b) ability to tolerate low oxygen would also be important in the more stagnant pond water; (c) ability to maintain position in a current would be important in flowing waters; and (d) different substrate preferences might result in habitat separation. To examine the first property, he exposed the two species to drying in an artificial pond with a mud bottom. Individuals of *O. immunis* responded to drying by burrowing in the mud; mortality was only 3 percent. *Orconectes virilis* individuals responded to drying by wedging themselves into the drying substrate; mortality was 34 percent. The species also differed substantially in their ability to tolerate low oxygen (Fig. 15-17), with the pond species surviving an average of 12.6 hours under 1 ppm O_2 in water, whereas the stream species average survival time was only 4.2 hours. In the third property, ability to maintain position in flowing water, there was no difference between the species; the pond species, surprisingly, was fully as capable of maintaining its position in a high rate of flow as was the stream species. Finally, Bovbjerg tested the substrate preferences of the two species. A laboratory tank was divided into equal bottom areas of rock, gravel, and muck, and the positions of animals were recorded three times daily for four or five days. This produced 1480 position records per species on 40 individuals per species. The experiments indicated that both species had a pronounced preference for a rocky

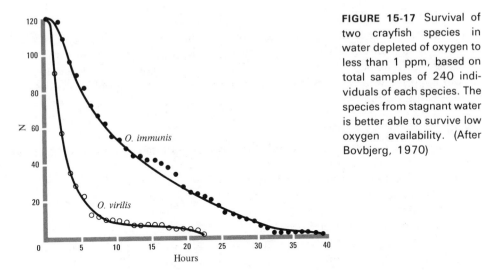

441

HABITAT
DIFFERENTIA-
TION IN
FRESHWATER
BIOMES

FIGURE 15-17 Survival of two crayfish species in water depleted of oxygen to less than 1 ppm, based on total samples of 240 individuals of each species. The species from stagnant water is better able to survive low oxygen availability. (After Bovbjerg, 1970)

bottom (Fig. 15-18). Both distributions differed significantly from random, and the pool species preferred the rocky bottom fully as much as the stream species. Thereupon equal numbers of both species were put in the tank and the effect of interspecific interactions was observed. Under these conditions *O. virilis* was capable of excluding *O. immunis* from the rocky bottoms. In records of aggressive contacts, the former species clearly dominated the latter, particularly over the rocky bottoms.

These experiments on crayfish provide considerable insight into habitat partitioning among closely related species that occur in different habitats. Both species

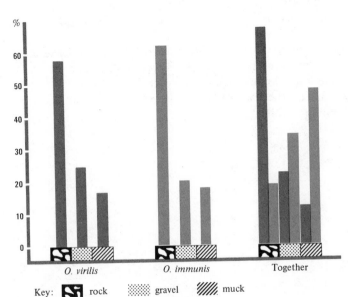

FIGURE 15-18 Habitat preferences of crayfish in an experimental tank divided into equal areas of rock, gravel, and muck bottoms. Data are percent of total occurrences on each substrate. Although both species prefer the rocky bottom, there was competitive displacement of the species occupying muddy bottom areas in nature when both species were added to the tank together. (After Bovbjerg, 1970)

preferred streamlike substrates and both were equally capable of maintaining position in current. The stream species, however, was able to exclude the pond species from its habitat through aggressive dominance. The stream species, in turn, may be excluded from pond habitats by inability to survive low oxygen tensions and lack of appropriate behavioral response to summer drying. Geographic coexistence in the central portions of the river arises out of a balance of appropriate habitats.

GENERAL CONCLUSIONS

1. Freshwater ecosystems can be divided into two major types, distinguished mainly by the volume/flow ratio:
   (a) The lotic biome: streams and rivers
   (b) The lentic biome: lakes and ponds
2. In the lotic biome:
   (a) A critical water velocity occurs at about 50 cm/sec. Above this, the bottom will tend to be stony, oxygen content high, and water temperatures cool. Below it, the bottom will be muddy, oxygen content low, and water temperatures warm.
   (b) Communities within a stream become spatially segregated according to bottom types (that is, stony, muddy, etc.).
   (c) Erosion and downstream export of production will increase with current velocity and many lotic food chains are detritus-based, exploiting energy and materials imported from upstream or adjacent terrestrial ecosystems.
3. In the lentic biome:
   (a) Because heat is transferred in water bodies largely by mixing, temperate lakes undergo seasonal cycles of temperature (and oxygen) stratification.
   (b) There are often conspicuous seasonal blooms of productivity as nutrients flow into the euphotic zone during spring and fall.
   (c) Primary production in offshore areas is dominated by phytoplankton and principal herbivores are zooplankton.
   (d) Seasonal production cycles at various trophic levels will be caused by seasonal patterns of solar radiation, nutrient circulation, oxygen profiles, and predation.
4. Studies of crayfish in the lotic biome suggest that species distribution in a series of ecosystems will depend on the balance between preferred habitat and competitive abilities in different habitats.

marine biomes

16

Problems inherent in ecosystem classification are nowhere more apparent than in a consideration of marine ecosystems. The oceans are the subject of the entire science of oceanography. It is no wonder that a distinct discipline developed around marine study since the oceans cover 361 million km², about 71 percent of the earth's surface. Average depth of the oceans is 3750 meters, with the greatest known depth being 10,750 meters in the Marianas Trench in the Pacific Ocean. Salinity averages around 3 percent, but this varies with depth and geography.

Many oceanic features are strongly influenced by *currents*, moving masses of water resulting from temperature gradients and surface wind friction. Higher solar energy input at the equator produces warm expanding water in the tropics that tends to flow poleward. Winds and the earth's rotation distort the flow so that oceanic waters circulate in a series of cells in the two hemispheres (Fig. 16-1). North of the equator the circulation is mainly in a clockwise direction with westward moving water masses north of the equator and eastward moving masses at about 40° N latitude.

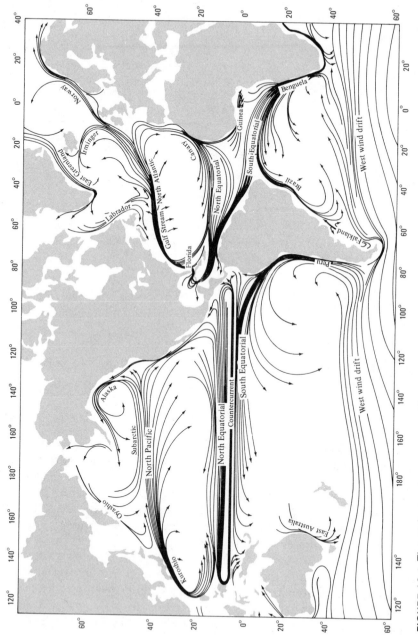

FIGURE 16-1 The major marine currents with current flow rate greatest where the lines are close together. (After Clarke, 1966)

Just south of the equator water masses flow in a westward direction while they move easterly from 40° S to the Antarctic.

A comparatively high evaporation rate in the tropics causes higher salinity in equatorial than in temperate waters. As the saltier water flows poleward and cools, its higher density causes it to sink, thereby mixing upper and lower water.

In contrast to the mixing of lake water which gives rise to seasonal overturns, ocean depths are permanently near 3°C. Seasonal overturns of the complete water volume may occur in shallow coastal areas, but in the open sea such mixing is restricted to upper layers. As Clarke (1966) observed, "If you were a marine fish restricted to 3°C, you could nonetheless travel over more than 60 percent of the globe without being exposed to a significantly higher temperature." The depth of the permanent thermocline varies with latitude and currents, being closer to the surface in temperate areas and deeper in tropical areas.

As a consequence of reduced circulation between deep and superficial water, much of the euphotic zone in the open sea is low in free nutrients. Nutrients immobilized in phytoplankton biomass tend to sink from the euphotic zone to beneath the permanent thermocline. The distribution of marine biomes is significantly influenced by the patterns of nutrient flow into the euphotic zone. The major marine biome-types are (Fig. 16-2): (a) *littoral,* where the ocean and land masses meet along the shoreline, (b) *neritic,* where land masses extend outward as continental shelves; an arbitrary outer limit is a water depth of 200 meters, (c) *upwelling,* where currents force water from deep within the ocean into the euphotic zone, (d) *pelagic,* the vast expanses of open sea, and (e) *coral reefs,* formed where tropical land masses protrude upward into the euphotic zone. Estuaries also are commonly included among marine ecosystems, although they actually are transitional biomes between freshwater and the sea.

The major biome-types in terms of size are the pelagic, occupying about 90 percent of the total oceanic area, and the neritic, about 7.5 percent of the total oceanic area. The remaining 2½ percent or so is divided among littoral, upwelling, and reef

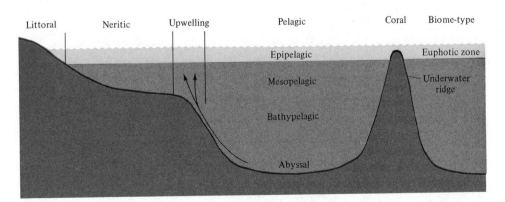

FIGURE 16-2 Generalized diagram of the positions of the marine biomes.

biomes. Although the spatial occurrence of these is limited, they are important ecologically because of comparatively high productivities.

Water is moved in waves and tides in addition to the circulation of water masses in currents. *Waves* are the oscillation of water back and forth without changing position except along the contact zone with the shore where breakers are formed. *Tides* are an alternate rise and fall in water level. They are generated by gravitational effects of the moon and sun, together with the cohesive properties of water molecules. Oceanic bulges occur where the earth is closest to the sun and the moon, and oceanic troughs occur in those portions farthest from the sun and the moon. As the earth rotates, the bulges and troughs are propagated across the oceanic surfaces as tides. Since the moon's gravitational attraction is about two and a quarter times that of the sun, it is more influential in tidal flow. The largest, *spring tides,* occur when the earth, moon, and sun are in line, so that gravitational effects of the celestial bodies are additive. The smallest, *neap tides,* occur when moon and sun form a right angle with the earth in the angle. In this position, gravitational fields are conflicting and tidal effects are minimized. The general period between high tides is 12 hours and 25 minutes, but this is so influenced by local conditions that it can be used only as a rough guide. Rates at which tides move and distances between high and low tides are influenced by local geography; among the highest tides in the world are those in the Bay of Fundy, between New Brunswick and Nova Scotia, where at times they move a vertical distance of 21 meters.

Although the oceans seem monotonous in surface appearance, and they are often monotonous biologically, there still is a vast variation in the ecosystems they encompass. Just as the character of the freshwater biomes is influenced by seasonal variations in light intensity, water temperature, and nutrient supply, so the marine biomes are influenced considerably by these same environmental factors. The great volume of the sea, however, accentuates many ecological distinctions we have previously considered in freshwater.

MARINE BIOMES: SOME GENERAL FEATURES

The transition from shoreline to open water is usually gradual, as it is in lakes. But the vast area of the sea warrants greater distinctions among the changing ecosystems from shore to open water. The organization of marine ecosystems is significantly affected by patterns of nutrient availability associated with spatial position. Inshore areas, the littoral and neritic biomes, are subject to nutrient input from adjoining continental areas. As a result these waters are enriched with both inorganic nutrients and organic detritus compared with upper regions of the pelagic biome. Lower pelagic regions, of course, are enriched by the fall of detritus from above. The most productive marine ecosystems are concentrated along the continental margins and in a few open ocean areas where nutrients from deep within the sea are flushed into the euphotic zone. Except for the littoral and coral reef biomes, where there are firm substrates within the euphotic zone, primary production is dominated by plankton. Estimates of the proportion of marine primary production contributed by phytoplankton are as high as 90 percent (Isaacs, 1969).

Major producers (Fig. 16-3) in the littoral biome are green, brown, and red algae with holdfast organs that attach them to the underlying substrate. Above the seaweed canopy and extending into the neritic biome, diatoms and dinoflagellates are important. Diatoms usually predominate in temperate seas and dinoflagellates in tropical regions.

447

MARINE
BIOMES:
SOME
GENERAL
FEATURES

Coral reef producers consist of dinoflagellates associated with the coral polyps, filamentous green and red algae embedded in the coral matrix, and, often, turtle grass in protected shallow water. In addition to their contribution to primary production, the algae also are instrumental in maintaining reef structure. Goreau and Goreau (1960) found that dinoflagellates facilitated the ability of polyps to synthesize their hard calcareous exoskeleton. Many filamentous algae also precipitate calcium carbonate on their cell walls and Scagel et al. (1965) observed that "corals would probably be reduced to rubble if it were not for the cementing action of the algae."

From the littoral biome outward to the neritic, dinoflagellates become increasingly important in the phytoplankton and the seaweed "forest" disappears as usable sunlight is quenched by the overlying water and phytoplankton. There also is a gradual transition from dinoflagellates with "armor" plates on their outer surfaces in the coastal plankton to "naked" dinoflagellates in the pelagic plankton.

In the open sea the principal producers are *nannoplankton* (nanno = extremely small). It is in this biome that the inverted biomass pyramids discussed in Chapter 4 may predominate. Because of rapid biomass turnover, biomass at the base of the food chain may be somewhat smaller than that at higher levels. In contrast to the tiny producers of the open sea, upwelling areas support the largest marine planktonic producers (Ryther, 1969). Common in upwelling areas are a variety of colonial diatoms, forming both large gelatinous masses and long filaments.

Because littoral water often contains relatively large amounts of detritus, sessile animals are more abundant here than in any other biome, terrestrial or aquatic. Many of these, as well as the sedentary producers, contribute heavily to the littoral and neritic plankton in immature stages of the life cycle. It has been estimated that a single large foliose red alga may liberate 100 million spores into the water upon reproduction (Scagel et al., 1965). In addition many marine animals have planktonic stages in their life cycles. These temporary planktonic stages are called *meroplankton.* They are important in littoral and neritic food chains and also may be abundant around reefs.

Organisms which spend their entire life cycle as plankton are called *holoplankton.* They become increasingly important in pelagic and upwelling biomes where they are the principal constituents of lower levels in the food web. Consumers, producers, and decomposers all may contribute to both meroplankton and holoplankton.

A variety of grazers, such as limpets, may occur in the littoral where they feed on sessile algae. In the neritic biome such zooplankton as copepods and euphasids become important first-level consumers. In the open sea nannoplankton may consist of tiny complex food chains on a single floating particle (Pomeroy and Johannes, 1968), with protozoan consumers associated directly with tiny green algae (microflagellates). Among the longest food chains of any biome are those in the pelagic, where five or more "steps" may be involved from primary producers to such higher order carnivores as tunas, dolphins, killer whales, and squids (Ryther, 1969).

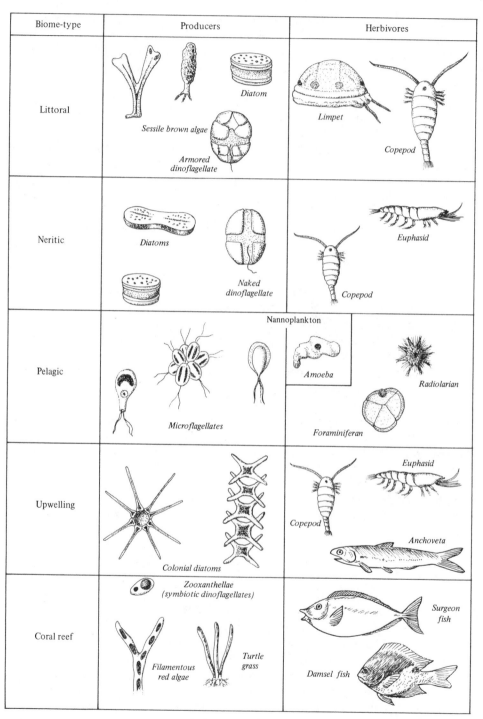

FIGURE 16-3 Some of the conspicuous organisms at different trophic levels in the marine biomes. In the right-hand column are meroplankton, periodically important at various trophic levels in the littoral and neritic biomes and coral polyps, of undefined trophic level.

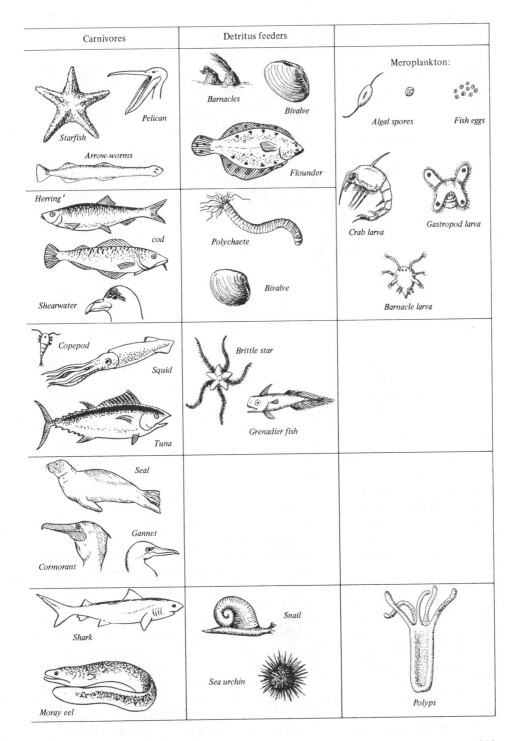

Carnivores	Detritus feeders	
Starfish · Pelican · Arrow-worms · Herring · cod · Shearwater · Copepod · Squid · Tuna · Seal · Gannet · Cormorant · Shark · Moray eel	Barnacles · Bivalve · Flounder · Polychaete · Bivalve · Brittle star · Grenadier fish · Snail · Sea urchin	Meroplankton: Algal spores · Fish eggs · Crab larva · Gastropod larva · Barnacle larva · Polyps

Feeding relationships in marine biomes may be more complex than those in most terrestrial or freshwater biomes (Ryther, 1969; Isaacs, 1969). While there are certain large organisms that have evolved feeding strategies allowing them to exploit phytoplankton directly, many others are at the top of food chains with five or six intermediates. Fish such as sardines, menhaden, and anchovies which filter plankton out of the water, are common in upwellings where large colonial phytoplankton provide a food source that can be efficiently exploited by large herbivores. A typical pelagic food chain would be nannoplankton → microzooplankton → macrozooplankton → fish → larger carnivores; a typical neritic or upwelling food chain would be diatoms → macrozooplankton → fish (Parsons and Lebrausser, 1970). It is exceedingly difficult, however, to classify most consumers according to trophic level since they may take a complex mixture of producers and consumers. In addition, as we shall see later when considering the neritic biome in more detail, the feeding position of a given individual may change radically as it grows.

An important property of aquatic habitats, of course, is the buoyancy of water and its ability to support relatively dense objects. However, plankton are still subject to sinking and they have evolved a variety of adaptations to counteract this tendency. Oil droplets produced as energy-storage vesicles reduce the density of many plankton. Large spines serving to increase the exposed surface are often present, and many plankton have flagella and cilia which help them maintain their position. In the sea flagellated producers tend to be more common where locomotion is required to overcome downward water flow (Isaacs, 1969). Diatoms, which lack efficient locomotion organelles, are more common producers in upwelling areas.

The pelagic biome can be divided vertically into epipelagic, mesopelagic, bathypelagic, and abyssal zones (Fig. 16-2). The epipelagic zone is the shallow upper portion of the sea to the limits of light penetration, perhaps 100 meters at most (Clarke, 1966). The bulk of the pelagic biome lies below this euphotic zone and is traversed constantly by a rain of detritus and sinking plankton from the epipelagic. Bacteria are often associated with the sinking particles, and in the mesopelagic and bathypelagic many mineral nutrients are released that had been immobilized in the biomass of the epipelagic. Detritus reaching the ocean floor provides energy for the largely unknown abyssal ecosystems. This dark environment is inhabited by a variety of decomposers, detritus feeders, and carnivores.

One of the most unusual properties of the abyssal habitat is the great pressure created by the water above it. At a depth of 10,500 meters, for instance, in the Philippine Trench, the pressure is 1050 atmospheres. Even here, however, are found sea anemones, sea cucumbers, and crustaceans (Clarke, 1966). Organisms in the abyssal zone must possess adaptations for existing under very great pressures and must be capable of obtaining food in complete darkness. Because of the permanent thermocline, temperatures in these ecosystems are commonly near 0°C.

Just as lakes were fundamental to formulation of the trophic level concept, studies of mineral circulation in the sea were fundamental to the initial formulations of many of our concepts of nutrient cycling. In 1926 Harvey outlined the pattern of nitrogen circulation in the sea, which involves incorporation of nitrates into plant protein, conversion of plant protein into animal protein subsequent to the feeding

process, release of nitrogen as ammonia from decaying plants and animals, and
conversion of ammonium salts to nitrates. He identified nitrogen and phosphorus
as principal factors limiting energy flow, with iron being of secondary significance.
Silicon also may be an important chemical limiting factor since it is required by
diatoms in which it is a cell wall constituent.

Harvey argued that pelagic nutrient cycles are largely closed, and production
therefore is dependent on the rate at which nutrients are cycled among trophic
classes. He said that "Fertility of any area of the open oceans, not subject to considerable inflows of water from other areas, depends upon three main factors. (a) The
length of time protein formed by phytoplankton remains part of the plant or nourished animal's body. (b) The time which elapses during the decay and formation
of ammonium salts and phosphate from corpses and excreta. To this must be added
the time taken for nitrate-forming bacteria to convert the ammonium into nitrates.
(c) The time which elapses before the reformed nitrate and phosphate again reaches
the upper layers where there is sufficient light for photosynthesis." The high productivities of upwelling areas in comparison with the open sea are primarily the result
of a "chemostat" effect in upwellings where nutrient limitations in the euphotic zone
are minimal because of constant nutrient renewal from the ocean depths.

The following discussion of marine biomes will focus on littoral and neritic
biomes. It is impossible to do justice to the earth's diverse ecosystems as part of
a general consideration of ecology. Readers who wish more information about the
sea should pursue it more directly through the oceanographic literature.

THE MARINE LITTORAL BIOME: TWO EXAMPLES

Although, as we have seen, lakes also have littoral zones, they are neither as extensive
nor as distinctive as the marine littoral region. The littoral biome is characterized
by the physical violence of its inner margin (Fig. 16-4), which is equaled in no other
biome. Organisms are subject to extreme fluctuations of temperature, moisture
supply, light intensity, and other environmental factors as tidal flow surges along
the shore.

The effects of the fluctuation of physical factors and the appropriate adaptations
to them are further defined by the shore substrate. Along rocky shores many organisms possess holdfast organs which allow them to maintain their position in the
pounding surf. There are more sessile animals here than in any other biome. Along
sandy shores, organisms will more often possess adaptations for burrowing into, or
adhering to, the shifting sand. Organisms adapted to rocky shores must be capable
of withstanding physical pounding, whereas organisms adapted to sandy shores must
be capable of surviving in a shifting physical matrix. In both habitats the distribution
of organisms often exhibits a distinct zonation reflecting, in part, capabilities for
withstanding extremes of physical environment. The major volume of the biome,
of course, is not subject to the regular severe environmental fluctuation of the sea
margin.

Stephenson and Stephenson (1949, 1961) surveyed the littoral biome along rocky
coasts in the northern Atlantic and Pacific. The distribution of organisms as well

FIGURE 16-4 The physical violence of the littoral biome's inner margin is matched in few other biomes. (Oregon State Highway Travel Division)

as the species represented tended to be remarkably persistent over large geographic areas. The littoral biome can be divided into three zones: (a) an upper *spray zone,* above the high tide line but subject to moistening by saltwater spray; (b) an *intertidal zone* subject to alternate wetting and drying by the tides; (c) a *subtidal zone* which, although always submerged, is subject to fluctuating water depth and light intensity.

In the spray zone there is an upper band of black, slimy blue-green algae where tidal water rarely flows but where rocks are moistened by saltwater spray (Fig. 16-5). These algae are commonly grazed on by snails of the genus *Littorina.* Immediately below is a barnacle area that extends from the lower portions of the spray zone to the low tide mark. In the upper spray zone barnacles of the genus *Chthalamus* often form a distinctive band below the myxophyceae. In the tidal flow region *Chthalamus* is replaced by barnacles of the genus *Balanus.* We have already seen, when considering Connell's work in Chapter 11, that the distributional patterns of these barnacles are determined by the interaction among competition, predation, and desiccation resistance. The intertidal region on rocky shores is typically covered with a zone of brown and red algae. Particularly abundant are algae of the genus *Fucus,* the alga most often called "seaweed" by those who enjoy walking along oceanic shores. These algae have distinctive floats at the blade tip which hold the blades in a vertical position during high tide. Limpets and other herbivores are also abundant in this zone.

Finally, at the low end of the littoral biome is a forest of the brown alga, *Laminaria.* This region, which is rarely unsubmerged, is often close enough to the tidal margin for the upper blade portions to form a floating jungle at low tide. The area is rich in invertebrates and starfish, many of which flee into the *Laminaria* forest

FIGURE 16-5 A general view of the littoral biome along a rocky coast. Along such coasts the communities change in a fairly definite zonation pattern from the spray zone to areas that are always submerged. (Peter G. Basich)

when the tide goes out. Beyond the *Laminaria* forest, to the limit of light reaching the sea floor, will be red algae. These are particularly abundant in tropical waters (Bold, 1957).

Plankton are abundant in the waters above and around the forests of brown and red algae in the littoral biome. The large algae also may serve as substrates for sessile animals and epiphytic algae (Scagel et al., 1965). Meroplankton are abundant in the outer littoral biome and the larger sedentary algae may contribute significantly to this community. In dense stands of brown algae, up to 3 million flagellated reproductive cells (zoospores) may be produced per liter of water per day through the period from June to September. These zoospores provide an abundant food source for planktonic feeders. In addition, the larval forms of snails, barnacles, sea urchins, coelenterates, and hydroids contribute to a rich meroplanktonic food supply. Diatoms also are major constituents of this planktonic community.

In addition to the three fairly distinct zones in the littoral biome, there are special habitats, such as tidal pools, in which distinct groups of organisms occur.

A different ecosystem-type occurs in the littoral zone over sandy beaches (Fig. 16-6). Pearse et al. (1942) studied the beach littoral zone in North Carolina. Energy supply was derived largely from imported detritus. Although sand grains were

FIGURE 16-6 A general view of the littoral biome along a sandy coast. Communities here will consist largely of organisms which feed by grinding bacteria and algae off the sand grains and by scavenging. (Stock, Boston)

commonly coated with a gelatinous mixture of bacterial cells, algal cells, and particulate organic matter, algae were not abundant and other primary producers were uncommon. Abundant bacterial activity is associated with the detritus, suggesting that the food chains are primarily decomposer based.

As along rocky shores, a series of three zones is recognizable: a supratidal moist zone, an intertidal zone, and a subtidal zone. Principal animals above the tidal flow on the beach studied by Pearse et al. were the ghost crab (*Occypode*) and an amphipod (*Talorchestia*). The intertidal area was dominated by animals that burrow into the sand during low tide. At the upper portions of tide flow a burrowing amphipod (*Haustories*) was common, whereas lower down on the beach it was replaced by a burrowing crustacean (*Emertia*) and a burrowing clam (*Donax*). A variety of nematodes and polychaetes were also found in the intertidal sand. Finally, in the subtidal region were snails (*Terebra*), sand dollars (*Mellita*), portunid crabs (*Arenaeus*), and burrowing crustaceans (*Lepipoda*). Many of these species feed by burrowing through the sand and either ingesting sand particles whole or scraping the surface clean with some type of abrasive feeding organ.

THE LITTORAL BIOME: STRUCTURE AND FUNCTION

Many consumers in the littoral biome are adapted for "filtering" small amounts of food out of environmental debris. Although this feeding method is particularly prevalent in organisms of sandy ecosystems, it is almost as common on rocky shores where such filter feeders as barnacles and sea anemones are abundant.

Marine algae release the bulk of their primary production, directly or indirectly, into the water (Khailov and Burklova, 1969). Marine algae are extremely "leaky" and a significant portion of the energy fixed may be lost directly into the water where it can act as food for microorganisms or other consumers. With gross productivity set equal to 100, about 14.2 percent was expended in respiration, and 37.3 percent was released directly into the water as organic acids, amino acids, and other photosynthetic products. The biomass increment was 48.5 percent of the energy fixed, and 11.2 percent of that was harvested by marine herbivores. At death, tissue lysis released 28 percent into the water and 5.5 percent was converted into soluble organic detritus. Only 3.8 percent of the initial gross productivity remained as insoluble humus. Totaling up the organic matter released into the water immediately on tissue lysis, and on subsequent decomposer activity, 70.8 percent of the marine algal production was converted into dissolved organic matter.

Clearly, detrital feeders around marine algal stands can realize substantial returns from processing the water flowing over these stands. In addition this must represent a considerable export term from the littoral biome into the neritic biome as tidal waters flow outward into offshore currents.

Barnacles are among the most conspicuous of the sessile littoral consumers. They both feed and breathe by driving water through an interior space and over gills. Food particles are trapped by tiny appendages and transferred into the mouth. They can distinguish food particles from inert debris, and the latter are usually rejected (Crisp, 1964). Barnacle density on artificial substrates can be controlled by their requirement for a pitted attachment surface as their meroplanktonic larvae become sessile. Crisp and his associates examined the effects of population density and food supply rate on *Balanus*. Flat surfaces containing different pit densities were placed in three habitats: (a) a rapidly flowing tidal area where food supply would be high, (b) an intertidal area with a slower flow rate and therefore less abundant food, and (c) the inside of boxes with a few holes drilled in the sides that were positioned in the slower flow area with, of course, food supply rate further restricted by the boxes.

Average weight of an individual *Balanus* was a negative function of density and a positive function of food supply rate (Fig. 16-7). With rapidly flowing water passing over them, barnacles at low densities reached weights of up to 700 mg after 8 months growth. Although population biomass was much higher in the rapidly flowing water, it declined at extremely low population densities in which the large size per individual could not compensate for the small number of individuals. It was found that the effect of food supply on fecundity was even more pronounced than the effect on population biomass. Whereas biomass per unit area in the rapidly flowing water was about three times that in the intermediate position, fecundity per unit area was almost ten times higher in the former location. These experiments indicate that population size in this littoral species is dependent on the density of attachment sites and the rate of water flow over it, presumably as the latter influences food supply per individual.

Other littoral consumers are less passive feeders than barnacles. Among these are the herbivorous echinoderms, sea urchins. Paine and Vadas (1969) examined the effect of a sea urchin (*Stronglylocentrotus*) on density of littoral algae. Over a period of three years, they removed sea urchins from plots and observed the effect of this on the algal flora. Algal biomass increased considerably when the herbivores were removed. In addition, many algal species invaded soon after sea urchins were

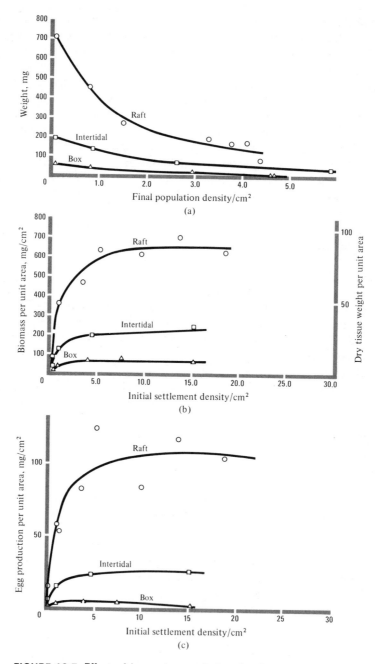

FIGURE 16-7 Effect of barnacle population density and food supply rate (highest on the raft, lowest in the box) on several population parameters: (a) average weight of a barnacle, (b) population biomass density, and (c) fecundity. (After Crisp, 1964)

excluded; that is, the sea urchin population was reducing diversity of the algal community by cropping some species into extinction. Eventually, however, the sites came to be dominated by brown algal species capable of forming a canopy that shaded out shorter species. At the end of three years the intertidal zone was dominated by *Hedophyllum* with *Laminaria* dominant in the subtidal. These experiments indicated that sea urchins cropped the larger species preferentially, thereby opening up the algal canopy and allowing shorter species to penetrate the community. The early increases in diversity were short-lived because of the pronounced dominance of taller brown algae.

Similarly, Southward (1964) found that removal of limpets from littoral plots resulted in an initial increase in diversity that was rapidly reversed as *Fucus* dominated the site. *Fucus* is much more abundant on sheltered than on wave-beaten shores and it was long believed that the alga was limited by wave action. Limpets, however, are much more abundant on open shores and Southward's experiments indicated that *Fucus* abundance was limited on open shores by limpet cropping.

In the sandy littoral ecosystem, feeding takes the form of either scraping adherent organic matter off sand grains or digesting the organic matter off ingested sand grains (Boaden, 1964). Bacterial concentrations of 50,000 to 500,000 cells per milliliter of sand have been reported, and diatom densities of up to 20,000 cells per milliliter of sand are known. Organisms that exploit these food sources are largely omnivorous and most of them probably do not discriminate among algae, bacteria, and detrital organic matter as they feed.

THE NERITIC BIOME: AN EXAMPLE

One of the most thoroughly studied neritic ecosystems occurs in the North Sea, that portion of the Atlantic Ocean bordered by Britain on the west and Norway, Denmark, Germany, the Netherlands, and Belgium on the east. Although the deepest point known is about 2680 meters, much of it is shallow enough to be considered continental shelf.

Intensive study of this area has been catalyzed by the fact that a major European food fish, the herring, is a consumer there. The major food chain in the ecosystem includes herring, sand-eels, a copepod (*Calanus*) which is a major herring food, a few flagellates and diatoms (Fig. 16-8). Other important primary consumers include meroplankton larvae of barnacles, molluscs, and tunicates, as well as copepods. Carnivores include arrow-worm, sand-eels, several cladocerans, and an amphipod.

Young herring feed primarily on smaller copepods and meroplankton, whereas the adult herring feed primarily on *Calanus* and sand-eels. Problems in applying the trophic level concept are graphically demonstrated by the food web in this neritic ecosystem. Herring, although nominally a higher order consumer, feed at the third to fifth trophic levels, consuming everything from herbivores to second-order carnivores.

Feeding is not merely a random process, for herring and copepods (Harvey, 1937) both clearly select prey items. Size and shape appear to be principal cues in feeding by both consumers and, as the food web indicates, growing herring switch from

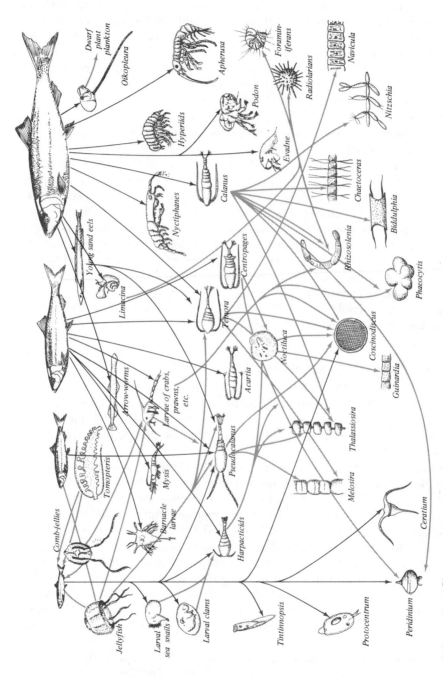

FIGURE 16-8 Changes in the feeding habits of herring as they grow from a length of 1 cm on the left to greater than 13 cm on the right. Major food chains are indicated by dark-colored arrows, and other organisms important in the food chains supporting herring are indicated by light-colored arrows. (After Wells, Huxley, and Wells, 1939)

smaller to larger copepods. This example of the neritic biome is limited to a small
portion of the food web and although the *Calanus*-containing food web may represent 90 percent of the herbivore-based energy flow in the North Sea, the example
is limited by our lack of knowledge of overall trophic structure.

One characteristic feature of marine biomes in temperate latitudes, as in the lentic
biome, is the spring algal bloom. The relationship between this bloom and limiting
factors is as obscure in oceanic waters as it is in freshwaters. Harvey's statement
of the importance of phosphate and nitrate in limiting algal productivity, mentioned
earlier, has led to several studies of the relationship between the spring bloom and
nutrient levels in the water. We saw from Edmondson's studies (Chapter 14) that
depletion of these two nutrients was associated with the spring bloom in Lake
Washington. Considerable work by Riley and his associates (Riley, 1963) identified
several environmental factors important in marine algal blooms including radiant
energy supply, water transparency, water temperature, depth to the thermocline,
phosphate concentration, and intensity of zooplankton grazing.

Studies by Cushing and his associates (Cushing, 1964) in the North Sea indicate
that decline of the spring bloom is closely related to overconsumption by *Calanus*.
A *Calanus* patch about 50 to 80 km in width was observed from late March to early
June. Over this area the vast majority of the algal consumption was accomplished
by *Calanus*. Algal abundance, relatively insignificant until early April when the
bloom began, reached a peak in late April (Fig. 16-9). This was followed by a
precipitous decline in algal abundance, associated with a rise in the *Calanus* population. The relationship between these two curves suggests such classical predator-prey
cycles as those discussed in Chapter 10.

Although algal biomass declined abruptly early in May, primary productivity
continued to climb to a peak in June. During this period the algal population was
consumed at a rate that equaled production and maintained biomass steady state.
Therefore the bulk of solar energy flowed into *Calanus*. Total algal mortality was

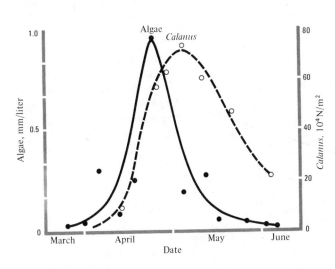

FIGURE 16-9 The spring blooms of primary producers (algae) and a principal herbivore (*Calanus*) in the North Sea. Note that the herbivore peak lags somewhat behind the peak of its major food source. (After Cushing, 1964)

calculated from the relationship between biomass at one sampling period, biomass at the next sampling period, and productivity. From the relationship between algal mortality and growth of the *Calanus* population, the proportion of consumed algae that was wasted could be calculated. The argument here is fairly complicated, however; perhaps an outline will prove helpful: (a) algae incorporate energy at some measurable rate, P_n; (b) this energy may be incorporated into algal biomass, ΔB; (c) the algal biomass may increase with time; (d) it may be harvested by herbivores (primarily *Calanus*); or (e) it may be lost through death. Thus we can write

$$\Delta B = P_n - H - D$$

where ΔB is change in algal biomass, P_n is net productivity, H is consumption by *Calanus,* and D is loss by death and sinking into lower ecosystems. The data indicate that $\Delta B = 0$ after a brief biomass increase. Since P_n can be measured directly, the problem of estimating H resolves to separating it from loss through death. To do this, *Calanus* biomass was plotted against algal mortality, and the intercept was taken as D (Fig. 16-10); that is, algal mortality at zero *Calanus* biomass must be loss through death. There are several tenuous assumptions involved in this sequence of reasoning—for instance, that nonconsumptive algal mortality would be a constant over this period. That the model was only a partial description of population dynamics is demonstrated by the scatter around the line of Figure 16-10. Nevertheless it does provide one of the few estimates of energy flux through natural neritic populations.

Finally, by calculating the rate at which biomass accrued to the *Calanus* population and the respiratory costs associated with this biomass, consumptive wastage could be calculated as

$$W = 100 \frac{G + R}{H}$$

where W is percentage of algal biomass eaten that was wasted, G is biomass increment to the *Calanus,* R is respiratory cost associated with this biomass, and H is

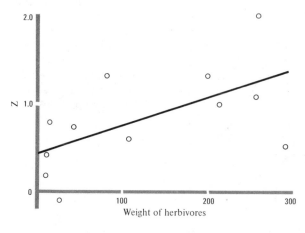

FIGURE 16-10 Increase in total mortality of algae (Z) with increase in herbivore biomass. The Y-intercept is nonconsumptive algal mortality. (After Cushing, 1964)

consumption by *Calanus*. Over the period of the algal bloom, consumption was 370 percent of that required by growth and respiration. In the period following the bloom from late May to early June this value declined to 26 percent. In other words *Calanus* wasted much of the algal net productivity through excess consumption during the period of the bloom, and then encountered insufficient algal net productivity for growth and maintenance requirements. Such negative energetic economy must be accompanied by mortality; *Calanus* biomass declined 30 percent during the last two weeks of study.

Whereas the herbivorous copepod wastes much of the algal biomass, studies with radioactive tracers indicate that most of the phosphorus in the algae may be removed as algal biomass is being rapidly processed (Marshall and Orr, 1964). *Calanus* therefore may process algae rapidly during the period of the spring bloom, crop the algae back to a low biomass level, and depend on stored nutrients for subsequent growth while cropping the remaining algal biomass at a high rate.

Overconsumption releases large quantities of fecal pellets into the sea that settle into lower regions as an energy supply for detrital feeders. Algal destruction by herbivores is probably particularly important in preparing particle sizes suitable for bacterial action, rather than the large, siliceous bodies characteristic of dead, but uneaten, diatoms.

Just as there are blooms in phytoplankton and zooplankton populations, so there are periodic occurrences of high reproduction in animals higher on the food chain. One of the most thoroughly studied examples was tabulated from the Norwegian spring herring catches (Fig. 16-11). Herring first begin appearing in the harvest when they are about three or four years of age. The percentage composition of various age classes was tabulated for fishery catches from 1908 to 1925. By extrapolating backward, then, it was possible to determine in which years reproduction and subsequent survival were highest. Over the 18-year period, three years made contributions to fishery yield that were significantly greater than average, namely 1904, 1913, and 1918. In fact, the 1904-age class dominated the herring population for ten years, and remained important for almost fifteen. Although there was fairly low average year-to-year reproduction and survival rate in herring, there were certain reproductive periods, at intervals of around six years, in which reproduction and survival were particularly high.

THE CORAL REEF BIOME: A BRIEF GLANCE

It has already been pointed out that the name "coral" reef is something of a misnomer since algae may be as important as coelenterates in reef formation. There are two principal types of such reefs: *atolls* which are ringlike reefs enclosing a lagoon and *barrier reefs* which are narrow coral ridges relatively close to a shoreline but separated from it by a lagoon. Atolls occur in tropical and subtropical waters where submarine volcanoes or ridges extend into the euphotic zone. The largest and most famous barrier reef is off the northeastern coast of Australia.

The algae within coral polyps are nonmotile dinoflagellates (Freudenthal, 1962) which, however, have flagellated zoospores. Although experiments with radioactive

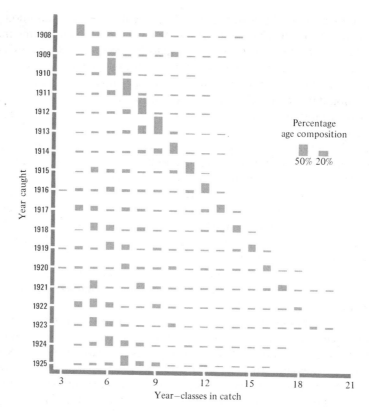

FIGURE 16-11 Relationship between age of a herring when caught and the year in which it was caught. From these data, the year in which herring were born can be extrapolated. For instance, the major catch from 1908 to 1918 was of fish born in 1904. (After Russell-Hunter, 1970 and Lea, 1930)

tracers indicate that organic compounds are transferred between dinoflagellates and polyps (Muscatine and Hand, 1958), the nutritional significance of such transfer is not clear. In addition to transfer from their symbionts, polyps also feed on zooplankton swimming over the reef.

Studies of polyp nutrition reveal that although they rapidly digest protein and animal carbohydrate (glycogen) they cannot digest plant structural carbohydrates (Yonge and Nicholls, 1931). As a result the dinoflagellates are indigestible. It is likely that digestion of zooplankton captured by the polyps releases nitrogen, phosphorus, and sulfur in forms which are utilized by dinoflagellates. Experiments with radio-phosphorus (Pomeroy and Kuenzler, 1969) indicate that there may be a relatively closed cycle between polyps and algae which keeps this element from being lost to the water.

We already considered the complex structure of the reef (Chapter 4), indicating that filamentous red and green algae are embedded within the coral matrix. There are often large turtle grass (*Thallasia*) flats in sandy areas within the lagoon (Wiens,

1962). These dense grass stands may contribute significantly to total primary production of the ecosystem, but some studies indicate that they are curiously devoid of consumers (Banner, 1952).

In general, reefs support quite diverse communities in comparison with adjacent marine areas. The rich spatial diversity of the reef creates a variety of microhabitats supporting species which feed directly on the reef components or on associated food sources (Fig. 16-12). There is considerable conflicting evidence about the energetic

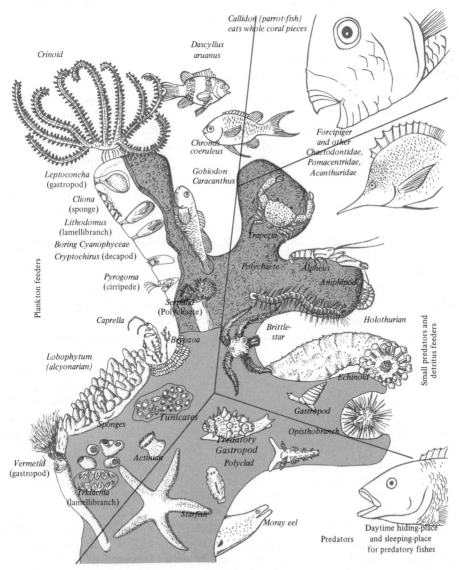

FIGURE 16-12 Some of the feeding and spatial relationships among organisms in reefs around the Maldive Islands. (After Friedrich, 1969 and Gerlach, 1960)

self-sufficiency of coral ecosystems, however. Odum and Odum (1955) concluded that an atoll which they studied had a net primary productivity of 26 gm/m^2/day. Of this, 24.4 gm/m^2/day was exported or consumed as respiration. The balance between input and output led them to conclude that the reef was "a true ecological climax or open steady-state system." Gordon and Kelley (1962), on the other hand, found that a Hawaiian reef which they examined was not self-supporting, but received significant energy input from adjacent areas. It is possible that consumer and decomposer populations in atolls surrounded by pelagic ecosystems may be more closely tied to yield from *in situ* primary production than those of reefs receiving detrital input from adjacent neritic ecosystems.

THE UPWELLING BIOME:
DIATOMS, FISH, BIRDS, AND PEOPLE

The largest marine catches in commercial fisheries (Holt, 1969) are the Peruvian anchoveta (10.5 million metric tons/yr), Atlantic herring (3.8), Atlantic cod (3.1), mackerel (2), Alaskan walleye pollack (1.7), South African pilchard (1.1), oysters (0.83), squid (0.75), shrimps and prawns (0.69), and clams and cockles (0.48). By and large these species are most abundant either in neritic or upwelling biomes. Over 80 percent of all fish caught are taken from water depths of less than 200 meters (Ricker, 1969). By far the most important species is the Peruvian anchoveta (*Engraulus ringens*) which contributes over 40 percent of the total tonnage among the ten major fisheries species. This yield comes from an upwelling about 60 km by 1000 km, an area about half that of the state of Pennsylvania. The total catch of all fish by all United States fishermen in 1969 was about 20 percent of the yield from the Peruvian anchoveta fishery alone (Paulik, 1971).

The base of the food chain in this area is primarily colonial diatoms (Fig. 16-3). Net primary production values as high as 10 gm carbon/m^2/day have been observed, although the annual average is about a tenth of this (Paulik, 1971). Shrimplike euphasids, as well as copepods, are important zooplanktonic consumers of phytoplankton.

Much of the primary production, however, is harvested directly by adult anchoveta which have modified gills allowing them to feed on the colonial diatoms. These fish spawn in huge schools in the upwelling. Although spawning is spread over the entire year, a major reproductive period is in September and a less pronounced peak occurs in April or May. The eggs have negligible quantities of stored nutrients, allowing newly hatched fry only a limited ability to search for food. If there are not copepod larvae or other small plankton in the immediate vicinity, the fry are doomed. Eggs and larvae also are subject to substantial predation by euphasids and adult anchoveta. Cannibalism is probably an important stabilizing mechanism in the anchoveta population. If reproductive output is high and adults are abundant, egg and larval cannibalism will reduce recruitment to the adult population, and at the same time yield energy to maintain that population. If, however, the adult population is small, reduced cannibalism will result in increased recruitment of fry into adult classes. The fish mature at one year of age and reproductive output is substantial, with two-year-old females sometimes producing over 20,000 eggs apiece. Growth rate is rapid from 5 to 25 months of age, with average individual biomass increasing from 2 to 25 gm over this period.

Predation on fish, in addition to cannibalism, is high (Paulik, 1971). A bird fauna adapted to harvesting anchoveta has developed along the Peruvian coast. The major bird predators are cormorants (60 to 80 percent of the total bird community), gannets (18 to 35 percent), and pelicans (2 to 5 percent). All have throat pouches and feed almost exclusively on anchoveta. Ninety-five percent of the cormorant diet and about 80 percent of the diet of the other species consists of this fish. Feeding occurs primarily at dawn when the fish are abundant in the upper 10 meters of water. Because these birds are so dependent on a single food source, they occasionally are subject to extremely high mortality. Every six or eight years the upwelling is submerged beneath a sheet of warm, low salinity surface water from the north. This water keeps the anchoveta from feeding near the surface and up to 70 percent of the birds may starve. Young and diseased birds suffer almost complete mortality.

The bird community normally harvests about 4 million metric tons of anchoveta each year (Ryther, 1969). There also is heavy predation by tuna, squid, sea lions, and a variety of lesser predators which consume a quantity equivalent to the bird communities. By far the major predator species in recent years, however, is man. Most of the immense harvest is converted into fish meal. The first fish meal factory in Peru was built in 1950 (Paulik, 1971) and by 1961 Peru was the world's leading fishery nation in tonnage landed (Borgstrom, 1971). Processing factories were built largely with capital from western Europe and the United States, and over 90 percent of the meal is consumed by the livestock industries of these areas. Chickens, dairy cattle, and pigs are the principal consumers of the rich protein source. Borgstrom estimated that the fish protein exported from the Peruvian current during 1963 to 1965 was equivalent to the South American continent's entire production of meat protein.

Recent export of fish meal from the Peruvian upwelling marks the second time that products of this ecosystem have been important to western European diets. The fish-eating birds produce droppings at a rate of about 45 gm/bird/day (Paulik, 1971). Since roosts are along the coastal desert, the droppings are not subject to significant erosion or leaching and accumulate in vast deposits of guano. As Russell-Hunter (1970) observed, "These deposits were shipped to western Europe, largely from 1840 to 1880 and, used as fertilizers, helped support the more intensive agriculture which nourished the expanding and more urbanized populations of the industrial revolution."

Upwellings are one of the few "natural" ecosystems capable of yielding a high quality food source in large quantities. Because relatively large fish feed directly on the primary producers, the loss from trophic level transfer is minimized, much as it is in agriculture by employing ruminants as "first-level" consumers. There is considerable debate, however, about the ability of these upwellings to support such intensive fishing on a sustained basis. Because of their high reproductive output, continuous reproduction, and high growth rate, anchoveta are an efficient source for a predator to depend on. It should be clear from preceding sections of the book, however, that it is not impossible for predators to overconsume their prey. In fact, there is precedent for such overconsumption, by man, in pelagic biomes.

THE PELAGIC BIOME:
A BRIEF GLANCE AND A REQUIEM FOR WHALES

Most of the open sea, we have pointed out, is relatively unproductive. Calculations of the food supply potentially available to copepods (Marshal and Orr, 1964) indicated that although neritic waters usually contain sufficient food to support them, food density in the pelagic biome is generally insufficient. And, in fact, copepods are extremely rare here (Pomeroy and Johannes, 1968).

Much of pelagic primary production is by microflagellates associated with organic particles. These nannoplankton particles are often a complex little ecosystem with protists on their surface functioning as primary consumers. Because of the tiny primary producers, food chains leading to larger carnivores are often several steps long. Seasonal plankton blooms are less pronounced in pelagic biomes and are generally lacking entirely in tropical waters.

The constant rain of detritus from above tends to be immobilized in benthic pelagic ecosystems by the permanent thermocline. Nitrate and phosphate concentrations in tropical surface water average less than 1 percent of peak concentrations in typical temperate neritic ecosystems (Russell-Hunter, 1970). Highest pelagic concentrations of these elements will occur in dense cold water near the sea floor, far from the euphotic zone and hence unavailable to the food chain base.

As we have seen, there are exceptions to the low nutrient content of oceanic waters beyond the continental shelves, where upwelling forces benthic water into the euphotic, and in reefs where efficient internal cycling tends to keep nutrients from being lost. There also are certain pelagic ecosystems toward the polar margins, where the pelagic euphotic zone is not depauperate. Arctic and antarctic convergences are formed by mixing of cold, lower salinity water melted from the ice caps with warm, higher salinity tropical water along the interfaces of oceanic current cells (see Fig. 16-1). Although there may be a relatively constant thermal gradient, there is no thermocline. While upwellings represent, in a sense, "holes" punched in the thermocline by water currents, convergences are a destruction of the thermocline over relatively large pelagic areas by water mixing.

Productivity may be relatively high in the polar convergences during their respective summers, but continuous nights of two months or so during winter result in decidedly seasonal biological activity. Among the most pronounced manifestations of this seasonality is the migration of the Arctic tern. This bird, which feeds on fish and zooplankton abundant during the convergence's summer seasons, migrates from the Arctic to the Antarctic and back over a 12-month period, timing its arrivals to coincide with seasonal food abundance.

Although we said earlier that dinoflagellates tend to be more important producers in tropical waters, they are also one of the major components in polar convergences. In the Arctic, nannoplankton also are abundant and in the Antarctic diatoms are important. Major herbivores in these ecosystems are krill, shrimplike members of

the genus *Euphasia,* which often occur in massive populations several kilometers
in extent. The average biomass of these animals during the summer bloom averages
100 to 200 mg per cubic meter of water in the North Pacific and 750 to 1500 mg/m^3
in the Antarctic (Nemoto, 1970).

THE PELAGIC
BIOME: A
BRIEF GLANCE
AND A
REQUIEM
FOR WHALES

Remarkable food chains have evolved to exploit the summer blooms of polar
convergences. One of the shortest food chains known (Mackintosh, 1965) occurs in
the Antarctic: phytoplankton → herbivorous euphasids → blue and fin whales. The
blue whale is the largest animal alive, with individuals reaching a length of 100
feet and a biomass of 150 tons. The effect of man on the species is one of the most
vivid examples known of a predator drastically overexploiting its prey. In less than
30 years, from the 1937–1938 to the 1966–1967 whaling seasons, the number of blue
whales killed dropped from 15,035 to 70 (Watt, 1968; Russell-Hunter, 1970). An
even more precipitous population decline occurred in the other carnivore in the food
chain, the fin whale. Over a six-year period in the early 1960s, the number of fin
whales killed plummeted from 20,000 to 2,000 (Russell-Hunter, 1970).

Whales are members of the mammalian order Cetacea which includes two
suborders, the toothed whales and the baleen, or whalebone, whales. Because there
is only one birth annually per female, they are particularly susceptible to overpreda-
tion. The only toothed whale which has been heavily exploited is the sperm whale,
which was the base of New England's nineteenth century whaling industry. Large-
scale whale predation was organized by the Dutch in Spitsbergen following Henry
Hudson's observation of their abundance nearby in 1807. Later, New Bedford,
Massachusetts became the world's greatest whaling port. From 1835 to 1871, an
estimated 163,400 sperm whales were killed by New England hunters (Watt, 1968).
The decreasing profit margin resulting from depleted sperm whale populations and
use of petroleum distillates for lamps finally led to death of the industry in the late
nineteenth century.

Modern whaling, based on baleen whales, and "regulated" by the International
Whaling Commission expanded enormously after the Second World War. This
industry is dominated by Japan, Norway, and the U.S.S.R. which together accounted
for almost 70 percent of all whales killed in the 1960 to 1961 season (FAO, 1965).
Kill quotas are set in "blue whale units." Employing an increasingly sophisticated
technology of whale finding, killing, and processing, the industry harvested an
average of around 15,000 blue whale units in each season from 1947–1948 to 1961–
1962. In a single year, however, the harvest dropped 25 percent, from 15,252 in 1961–
1962 to 11,306 in 1962–1963. Throughout the 15-year period, there were a variety
of lines of evidence indicating that whale populations were being seriously over-
exploited (Watt, 1968): (a) blue whales were a decreasing proportion of the catch,
(b) the proportion of pregnant females dropped precipitously, (c) mean length of
whales killed declined, (d) survivorship curves indicated that mature adults were less
abundant. By 1964 the world's estimated blue whale population was between 600
and 2000. In the 1966 to 1967 season, 70 were killed. Russell-Hunter observed that
"Mankind may have already hunted into extinction the largest animal species which
has ever existed on the planet Earth." This, of course, was in spite of the fact that
scientists had been warning for years that whales were being drastically overhunted,

the industry was "regulated" by a modern social bureaucracy (the IWC), the experiment had already been done (on sperm whales), and there was no overwhelming need for whale products.

Baleen whales are capable of exploiting krill efficiently because they evolved an efficient morphology and behavior for sieving zooplankton out of water masses. Nemoto described two major feedings strategies. Swallowing whales (including both blue and fin whales) swallow water as they swim and then expel it through whalebones which retain the krill. Skimming whales swim through the water with their mouth open and filter outgoing water through the whalebones. Although it is clear that food must be present in relatively large concentrations to support such strategies, the large size and thick blubber accumulations (about 20 percent of body weight) indicate that there also has been selection for low rates of basal metabolism. Because of the seasonality of krill abundance, many whales are migratory, moving between upwellings, neritic areas, or equatorial convergences during winter and polar convergences during summer.

Within convergences, there is a certain amount of spatial segregation. Blue whales used to feed at higher latitudes than fin whales in the Antarctic. One of the best evidences that the competitive exclusion principle operates in nature was provided by expansion of the fin whales' geographic distribution upon extermination of blue whales. Now that the fin whale also is being exterminated, sei whales are expanding into the same area.

From estimates of whale biomass supported in the Antarctic convergence prior to hunting, Nemoto (1970) calculated that they consumed about 772,800 tons of krill per day over a 100 day feeding season. This amounts to 77.3 million tons a year. Because of massive depletion of whales, much of the krill population now remains unharvested. The U.S.S.R. has begun to examine the ability of man to move downward on the trophic ladder, now that he has eliminated the major carnivore of these ecosystems. To date, attempts to convert krill into a meal have not been commercially viable. But perhaps with increasing human populations and exhaustion of other food sources, man may begin to eat fish meal from upwellings and krill meal from polar convergences.

GENERAL CONCLUSIONS

1. The distributions of the five major marine biomes are strongly influenced by oceanic currents and proximity of the euphotic zone to land masses:
   (a) In the littoral biome, along shorelines, food chains contain sedentary producers, marine grazers, and filter-feeding detritus consumers.
   (b) In the neritic biome, above the continental shelves, planktonic producers and first-order consumers support large filter-feeding fish populations.
   (c) In the pelagic biome, or open sea, the food chain base is extremely small plankton and food chains may be many steps long except in areas where the thermocline is destroyed by converging currents.
   (d) In upwelling ecosystems, primary productivity is dominated by colonial phytoplankton and large filter-feeding fish may exploit these directly in short food chains.

(e) In coral reefs, the symbiotic relationship between coelenterates and algae, together with great habitat heterogeneity, create complex trophic relationships.

2. Temperate oceanic areas are often subject to pronounced spring blooms of productivity associated with warming temperatures and high concentrations of inorganic nutrients in the euphotic zone. This bloom is often propagated through the ecosystem as waves of population abundance at progressively higher trophic levels.

3. More constant nutrient and energy fluxes are maintained in tropical oceans compared to the decided seasonality of temperate seas. Stable internal circulation systems are important in generating high productivity in tropical seas.

terrestrial biomes:
desert, tundra, prairie, and savanna

17

We have classified terrestrial ecosystems into seven biome-types: desert, tundra, prairie, savanna, temperate coniferous forest, temperate deciduous forest, and tropical forest (Fig. 17-1). This chapter concentrates on the first four of these; specific features of forest biome-types are considered in the following chapter.

The biomes are distributed in a general east-west direction on most continents. Tropical forest, for instance, is concentrated in a band along the equator between approximately 20°N and 20°S latitudes. Similarly, the occurrence of temperate coniferous forest is concentrated between 50°N and 70°N latitude. There are, however, important exceptions to this east-west distribution pattern. For instance, the temperate coniferous forest biome in North America extends down the Cascade-Sierra and the Rocky Mountain chains as a north-south belt to around 35°N; and the savanna biome in eastern Africa has a tremendous north-south range. The distribution of terrestrial biomes reflects, to a significant degree, the worldwide pattern of precipitation which arises out of the global movement of atmospheric air masses.

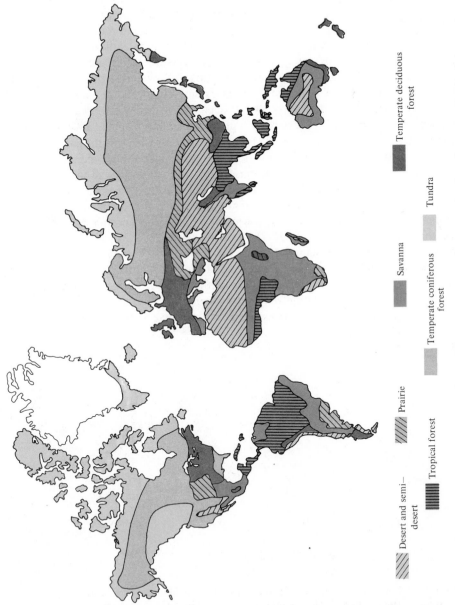

FIGURE 17-1 The earth's major terrestrial biomes. Bare areas are either large bodies of water or arctic-alpine regions largely devoid of organisms.

Temperate deciduous forest

Savanna

Temperate coniferous forest

Tundra

Prairie

Desert and semi—desert

Tropical forest

CHAPTER 17
terrestrial
biomes:
desert,
tundra,
prairie,
and savanna

Climate is a regional pattern of meteorological conditions including temperature, rainfall, and radiation. Global variation in climate arises out of nonuniform distribution of incoming radiation from the sun (C. G. Rossby, 1941). Because of the low angle at which the sun's rays approach polar surfaces, radiation input to these surfaces per unit time is considerably lower than it is in equatorial regions. Regions of warm ascending air at the equator are created by high radiation input. Air is a fluid, of course, and regions of lower surface pressure are created by ascending air at the equator. Cold polar air therefore tends to flow across the earth's surface into lower density regions at the equator. The rising equatorial air cools and begins to fall, at the same time flowing poleward over the surface equatorward flow. This transport of equatorial air poleward at high altitudes and of polar air equatorward near the earth's surface diminishes the temperature contrast across the earth's surface compared to that which would result from radiation inequalities alone.

In both Northern and Southern Hemispheres, three atmospheric circulation cells are formed because the high altitude air flowing poleward becomes sufficiently cooled at about 30°N and 30°S for some of it to sink downward to the earth's surface. A branch of this sinking air spreads toward the poles, and another branch spreads back toward the equator (Fig. 17-2). At these latitudes surface air tends to be motionless; these bands of stagnant air are called the *horse latitudes.* North and south of the horse latitudes, surface air flows toward the poles until it contacts the polar fronts at about 60°N and 60°S. Because of the earth's thermal balance and the resulting atmospheric circulation patterns, barometric pressure at the surface drops from the poles to 60° latitude, rises again to the horse latitudes, and then drops

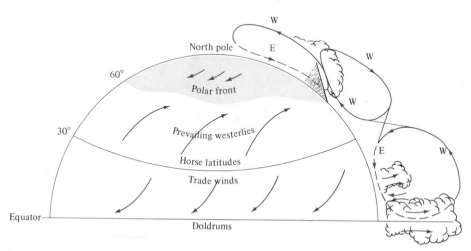

FIGURE 17-2 Generalized diagram of the pattern of atmospheric circulation for the Northern Hemisphere. Heavy precipitation bands occur at the equator and the polar front. Surface winds are distorted from straight north-south movement by the earth's rotation and declining circumference toward the pole. The Southern Hemisphere is a mirror image of this pattern. (After Rossby, 1941)

from the horse latitudes to the equator. At high altitudes in the atmosphere, pressure declines constantly from the equator to the poles.

These general zones of northward and southward flowing air are modified by three factors: (a) the rotation of the earth, (b) its decreasing radius away from the equator, and (c) the effects of continents and oceans on heat exchange between the atmosphere and the earth.

First, because the radius of the earth declines toward the poles, air flowing poleward at a given speed moves faster and faster, relative to the earth's surface, as it gets farther from the equator. In other words, surface air from the horse latitudes to the polar front tends to blow from southwest to northeast in the Northern Hemisphere and from northeast to southwest in the Southern Hemisphere. In contrast, air from the horse latitudes to the equator loses velocity relative to the earth's surface, so the wind blows from northeast to southwest north of the equator and the opposite direction—that is, southwest to northeast—south of the equator. Also, the apparent speeds of these winds depend on their speed relative to the speed of the rotation of the earth. Air rotating less rapidly than the earth will appear to blow from the east in both hemispheres. Similarly, air rotating more rapidly will appear as a west wind in both hemispheres.

The second important factor modifying the zonal circulation pattern is the distribution of large water and land masses on the earth's surface. These modifiers are more important in the Northern Hemisphere, where the major continental masses are concentrated, than in the Southern Hemisphere. Because of the high latent heat of water, as we saw when considering aquatic biomes (Chapter 15), large bodies of water act as heat sinks and radiators absorbing and releasing relatively large quantities of heat to the atmosphere without undergoing large temperature changes. During the winter, cold arctic temperatures become displaced southward over the North American and Asian continents (Fig. 17-3). In addition, snowcover on the interiors of the land masses acts as a reflector, returning the sun's radiation back toward space. The buildup of cold air in the continental interiors produces large high pressure air masses which stagnate for a time and then flow outward toward the oceans where they contact warm, moist, oceanic air. The warm air is forced upward over the denser polar front, moisture condenses out of the cooling tropical air, and precipitation occurs. In the summer the continental masses lose their snowcover and are generally warmer than adjacent oceanic areas so that isotherms are displaced northward over the continents (Fig. 17-3). The oceans become high pressure areas and the continental interiors vary between high and low pressure, relative to the oceanic air, as they heat up during the day and cool down during the night. As a consequence of pronounced diurnal variation in continental temperatures, the global air circulation pattern of winter exemplified in the repeated equatorward flow of polar air masses becomes less pronounced. Weather becomes more strongly influenced by local factors, particularly elevation.

Zonal atmospheric circulation is an important influence on the distribution of terrestrial biomes, just as water currents are important regulators of marine biomes. In equatorial regions the rate of atmospheric ascension and cooling reaches a peak shortly after the sun has passed overhead. This results in afternoon thundershowers

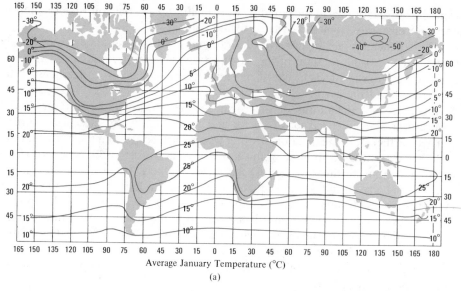

Average January Temperature (°C)

(a)

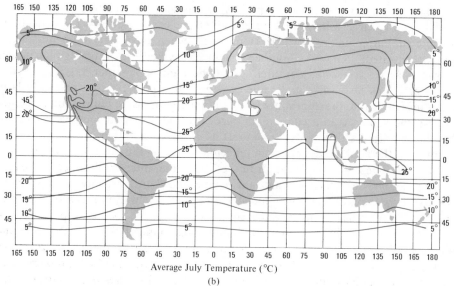

Average July Temperature (°C)

(b)

FIGURE 17-3 Temperature patterns on the earth's surface: (a) January isotherms, (b) July isotherms. In the Northern Hemisphere, January isotherms are displaced southward over the continents, while July isotherms are displaced northward. Again, the Southern Hemisphere is the converse of this pattern. (After Rossby, 1941)

associated with tropical regions. The air flowing poleward loses moisture until it becomes cool enough to sink downward above the horse latitudes. As it sinks it warms and is therefore able to hold more water. This results in low rainfall areas over the horse latitudes which are associated with the earth's major deserts. Finally, atmospheric cooling near the polar fronts at 60 degrees generates other high rainfall regions (Fig. 17-4).

Global precipitation patterns are clearly related to global biome distribution (compare Figs. 17-1 and 17-4). The cooling of warm, moisture-laden tropical oceanic air as it rises and passes over South America, Africa, and the Malayan Archipelago generates rainfall in excess of 200 cm/yr. It is in these regions of high rainfall that the tropical rainforest biome occurs. The world's major deserts occur near the Tropics of Cancer and Capricorn (that is, the horse latitudes) where annual rainfalls of less than 10 cm are common. Deserts also occur on the continental sides of mountain ranges where oceanic air loses its moisture content as it moves up over the mountain range; this generates a low rainfall region, known as a *rain shadow,* on the inland side of the mountain range (Fig. 17-5). Although we generally associate deserts with the warm temperatures characteristic of the horse latitudes, it is possible for deserts to occupy relatively cool climates in rain shadows.

THE DESERT BIOME: GENERAL FEATURES AND EXAMPLES

Deserts occur in regions where annual rainfall averages less than 25 cm and are particularly affected by the balance between rainfall and evaporative potential of the air. Because they are characterized by a large excess of potential evaporation over rainfall, available water tends to become saline since dissolved salts are not removed from the soil by leaching or runoff. Under desert conditions organisms are under strong natural selection to conserve the limited water supply.

By far the largest desert area is centered around 20°N latitude from the Atlantic Coast of Africa to Central Asia. This Saharan-Arabian-Gobi desert complex covers over 10 million km², or about 8 percent of the nonpolar land mass. Over much of this area, particularly in the central portions, rainfall may occur only every few years, and the vegetation is reduced to scattered shrubs and creeping plants in protected depressions (Polunin, 1960). Because of the high evaporation, with humidities as low as 5 percent in the daytime, depressions are often quite saline, so even those habitats in which moisture may occasionally be present are very rigorous. The principal vegetation in this desert includes various shrub *Acacia,* succulents (*Euphorbia*), and tamarisk trees (*Tamarix*), and the most conspicuous consumers are various locusts, which undergo periodic population outbreaks that are often associated with vegetative production following rainfall, and such rodents as hamsters and gerbils. Another exceedingly harsh desert region occurs along the west coast of South America. In Peru and Chile a band of land 100 km wide adjacent to the Pacific Ocean receives virtually no rainfall.

In North America, at about the same latitudes as the Sahara desert complex, are the Sonoran and Chihuahuan deserts of Mexico and western United States (Fig. 17-6). Whereas rainfall in the great deserts of Africa, Asia, Australia, and South

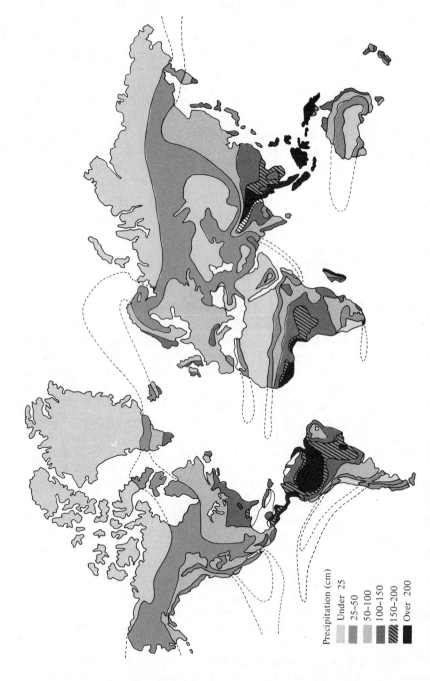

FIGURE 17-4 Annual precipitation patterns over the earth. Note that areas with similar amounts of precipitation may have different seasonal distributions, with resulting effects upon ecosystem organization. (After Rossby, 1941)

Precipitation (cm)

Under 25
25–50
50–100
100–150
150–200
Over 200

477

THE DESERT
BIOME:
GENERAL
FEATURES
AND
EXAMPLES

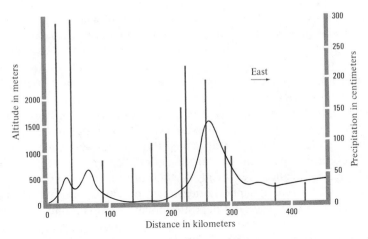

FIGURE 17-5 A map of precipitation (bars) and land mass altitude from the Pacific Ocean (on the left) inland across the Olympic and Cascade Mountains of Washington. Highest precipitation occurs on the seaward mountain slopes while the inward slopes are in a rain shadow. (After Daubenmire, 1959)

America is typically less than 10 cm annually, these American deserts receive about 25 cm/yr of rain and so are often called *near-deserts* (Polunin, 1960; Riley and Young, 1966). Principal plants are shrubs such as creosote bush (*Larrea tridentata*), the succulent saquaro cactus (*Cereus giganteus*), and various prickly pears (*Opuntia* spp.). Important herbivores include kangaroo rats (*Dipodomys* spp.), black-tailed jack rabbits (*Lepus californicus*), and grasshopper mice (*Onychomys torridus*). Insects are also important consumers and among the most abundant carnivores are the insectivorous birds such as the cactus wren (*Campylorhynchus brunneicapillum*) and bats of various species. Reptiles, particularly lizards, are also common.

FIGURE 17-6 General view of a shrub desert in North America. Note the large areas of bare soil between the shrubs. (U.S. Forest Service)

478

CHAPTER 17
terrestrial
biomes:
desert,
tundra,
prairie,
and savanna

The hot desert complex in which creosote bush is abundant merges to the north with a cold desert complex that is characterized by sagebrush (*Artemisia tridentata*). This colder region occurs in the Great Basin rain shadow that extends from eastern Washington southward into southern Nevada. Although many of the species differ in the two regions, major consumers in the colder area include insects, kangaroo rats, ground squirrels, jack rabbits, bats, and a variety of birds.

Although the desert biome is caused primarily by the shortage of water, it is shaped topographically by occasional excessive rainfall. Rains in desert regions are characterized by their violence as well as their infrequency and when combined with the sparse vegetative cover, they result in periodic severe erosion, producing the washed-out channels called *arroyos* in North America and *wadis* in the mid-East. Desert soil is often stony and poorly developed on slopes, erosion having carried away the smaller soil particles and deposited them as alluvial fans on valley floors. This spatial differentiation in the desert biome often produces a pronounced "patchiness" in community structure with distinct communities occurring on slopes and in valleys.

ADAPTATION TO WATER SHORTAGE

Adaptations of desert organisms, whether plants or animals, to the shortage of water have taken two forms: (a) avoidance of drought and (b) conservation of water. Generally all successful desert occupants have combined these two mechanisms into efficient strategies for coping with the environment. Many plants, for instance, grow rapidly in response to available moisture, with subsequent periods of inactivity when moisture is in short supply. Went (1955) found that the common ephemeral annuals, which often bloom abundantly only every several years on the Mohave desert, produce seeds that will germinate only if moisture supply is sufficient to sustain flowering. Rapid seed production is followed by persistence as dormant seeds until later rains. Went found that the seeds contained water-soluble germination inhibitors that could be washed out of the seeds only with relatively heavy dousings, generally about the equivalent of a 10 to 20 mm rainfall. Desert perennials, in contrast, are always present as vegetative individuals but often go through pronounced seasonal cycles of leaf production keyed to rainfall (Orshan, 1963). Such shrubs will have a minimum of leaves until there is sufficient rainfall, then will rapidly produce leaves which may be present for a few weeks before they are shed. In other words the seasonal cycle of leaf production and leaf drop in deserts is coordinated with moisture supply rather than with temperature.

Many desert plants, of course, are succulent and as such can store large quantities of water for use during periods of drought. Some succulents possess a mechanism for assimilating carbon dioxide during the dark; that is, they store relatively unstable photosynthetic intermediates during the day with the stomates closed to prevent water loss, then stabilize the photosynthetic intermediates during the night by fixing carbon dioxide when water loss through open stomates will be minimized (Neales et al., 1968). In a typical plant unadapted to desert conditions photosynthesis and transpiration are both high during the day and carbon dioxide is lost by nighttime respiration (Fig. 17-7). In a desert succulent, in contrast, the cycle is reversed, with carbon uptake being maximized during the dark period and falling to near zero when illuminated. As a consequence, water lost per unit photosynthesis is considerably lower in the desert plant. The ratio of water loss to carbon dioxide assimilated over a 24-hour period was 166 for the sunflower and only 47 for the desert agave. Certain plants therefore have become adapted to water shortage by inverting the stomatal rhythm, keeping the stomates closed during the day but storing the light energy in a form that can be used during the dark so as to assimilate carbon when water loss through open stomates will be minimized.

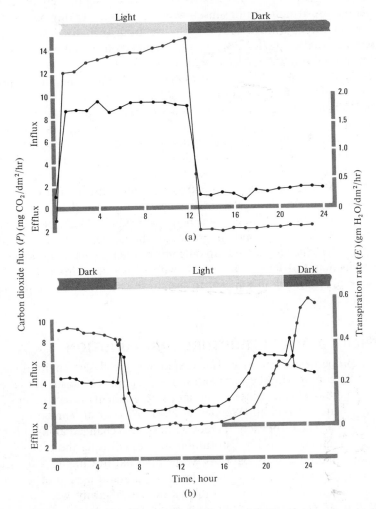

FIGURE 17-7 Diurnal patterns of photosynthesis (colored line) and transpiration (black line) in (a) sunflower, a plant adapted to high moisture habitats, and (b) a desert succulent. The desert species has an inverted stomatal rhythm, with the stomates open during the dark when evaporative stress is low. (After Neales et al., 1968)

480

CHAPTER 17
terrestrial
biomes:
desert,
tundra,
prairie,
and savanna

The ability of animals to cope with the desert's water shortage depends on various combinations of drought avoidance and reductions in water consumption. Most desert mammals show a pronounced daily activity pattern, being active at night and inactive during the day, and many live in burrows where they may encounter relatively mild temperatures during the day. The kangaroo rat, for instance, is active only at night. In addition, it produces urine and feces with an extremely low water content, has a relatively low rate of water loss from lung surfaces as it breathes, and is able to meet its water requirements from water produced during metabolism (Schmidt-Nielsen, 1964). Similar mechanisms for conserving water are present in lizards. Although many lizards are active during the day, their principal activity periods are at sunrise and sunset, with the animals seeking out shade during the hot part of the day. In periods of extreme drought many desert animals will go into semihibernation, with a reduced rate of metabolism and with water requirements met through the metabolism of fat deposits.

One of the most striking adaptations to the desert biome is manifested in periodic population eruptions of the desert locust (*Schistecera gregaria*), which was among the plagues inflicted on Egypt during the Jewish captivity. Maturation of the gonads is controlled by aromatic terpenoid compounds in the desert shrubs eaten by locusts (Carlisle et al., 1965). The concentration of these compounds is highest in dormant buds just as the seasonal rains begin, which is when egg laying takes place. When early rains are heavy enough to produce abundant forage, egg laying is stimulated and the locust population may reach extremely high densities. Because of decided spatial variation in rainfall, locusts often reach locally high densities in islands of rich vegetative growth surrounded by a poor food supply. Occasionally when such outbreaks exceed the food supply, migration of the horde begins and a "plague of locusts" originates. Such a plague is an outcome of the interaction between synchronized reproduction and heavy early rains with subsequent drought, producing a local pulse of plant growth insufficient to sustain the herbivore population that has been generated.

THE DESERT BIOME: STRUCTURE AND FUNCTION

Just as the properties of the organisms exploiting the desert environment reflect adaptations for coping with the short and sporadic supply of moisture, so the overall structure of the biome is modified along moisture gradients. Creosote bush density was linearly related to rainfall (Woodell et al., 1969) over an annual range of 4 to 30 cm (Fig. 17-8). The dispersion of creosote bush also changed along this moisture gradient. Randomness of shrub distribution was tested using the ratio of variance to mean from quadrat density data. As we pointed out in Chapter 4 on ecosystem structure a ratio greater than one indicates clumping and a ratio of less than one indicates even spacing. The average ratio in the desert shrubs was 1.75 for stands with rainfall of 15 cm or greater, but only 0.588 for stands with rainfall of 10 cm or less. In other words creosote bush tended to be clumped in the higher rainfall regions of its range and evenly spaced in low rainfall regions. Woodell et al. argue,

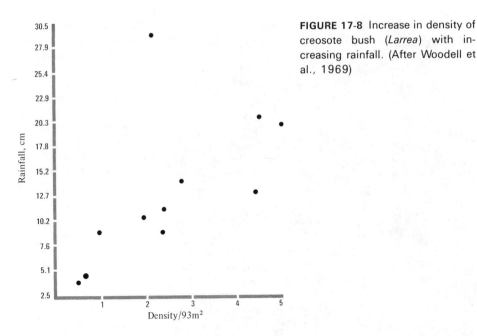

481

THE DESERT
BIOME:
STRUCTURE
AND
FUNCTION

FIGURE 17-8 Increase in density of creosote bush (*Larrea*) with increasing rainfall. (After Woodell et al., 1969)

"It is likely that the regular spacing of *Larrea* in many sites of low rainfall is the result of root competition for available water."

Characteristic plants of the hot desert in North America are a variety of large columnar cacti. Typical of these is the saguaro that occurs abundantly in Arizona and in Sonora, Mexico. This species often is decidedly nonrandom in distribution, with individuals tending to cluster around shrubs and trees which they may eventually outlive (Fig. 17-9). Turner et al. (1966) examined factors associated with such "nurse plants" which might be important in saguaro establishment and found that survival of saguaro seedlings differed markedly in sunlight and under shaded conditions (Fig. 17-10). Survivorship curves showed that the cactus seedlings were unable to survive in direct sunlight but had comparatively high survival rates when shaded. There was also evidence that soils underneath nurse plants were somewhat higher in nitrates and phosphates than soils in the open desert, and the growth rate of seedlings when shaded and irrigated was somewhat higher in the soil from under nurse plants. This suggests that the shrubs and trees, many of which are deeprooted, may concentrate nutrients from a large soil volume into leaves which later enrich the underlying superficial soil layers from litter fall. The combination of shading and higher nutrient status therefore enhances survival of seedling saguaros considerably. Clumping in ecosystems often may arise out of the formation of locally favorable establishment sites by biological modification of the physical environment.

In certain locations saguaro has failed to reproduce since around 1900 (Niering et al., 1963). Around 1880 large numbers of cattle were taken into Arizona and a drought from 1891 to 1893 as well as widespread destruction of grazing lands and adjacent land areas from overgrazing and trampling resulted in huge cattle die-offs.

482

CHAPTER 17
terrestrial
biomes:
desert,
tundra,
prairie,
and savanna

FIGURE 17-9 Saguaros clustered around a palo verde tree which established a microhabitat suitable for survival of saguaro seedlings. Note the severe rodent damage to the saguaros. A wood-rat den occurs in the clump, and a spiral rodent runway inside the second stem from the right has broken through to the exterior. (Niering et al., 1963)

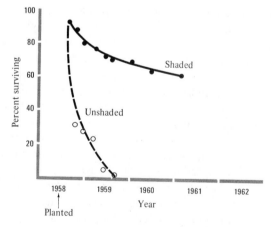

FIGURE 17-10 Survivorship curves for shaded and unshaded saguaro seedlings. (Data from Turner et al., 1966)

Niering et al. found that saguaro in grazed areas reproduced much more efficiently on rocky slopes than on the alluvial catch basins in arroyos, although the overall population density and individual size were larger in the arroyos. The inability of cattle to range over rocky slopes apparently reduced mortality from trampling. In addition, saguaros are eaten with relish by many rodents; in particular, wood rats (*Neotoma albigula*) are often devastating to saguaro (Fig. 17-9), tunneling inside the stem and thereby increasing the individual plant's susceptibility to a fatal bacterial disease that rarely afflicts noninjured cacti. Wood rats are much more abundant on grazed areas; grazing also increases the abundance of jack rabbits which browse upon nurse plants, often killing them and thereby eliminating future establishment areas.

Niering et al. concluded that declining reproduction of saguaro in many areas arose out of amplification of grazing effects through the ecosystem as determined by the effect that substrate had in regulating grazing intensity. Successful reproduction of this species depends on the presence of nurse plants to provide appropriate establishment microhabitats and on protection from predation by herbivores. The single change of introducing cattle into the desert caused indirect modifications in the grazed ecosystems toward eliminating the two conditions essential for saguaro reproduction.

Herbivores may be important in regulating reproduction of desert plants. Chew and Chew (1970) examined energy flow from shrub species into mammalian consumers and found that although herbivores consumed only 1.95 percent of the total aboveground net primary productivity, they consumed 86.5 percent of the total seed production. The principal species in their study included a seed eater (Merriam's kangaroo rat, *Dipodomys merriami*), a browser (black-tailed jack rabbit, *Lepus californicus*), and an insectivore (southern grasshopper mouse, *Onychomys torridus*). In terms of total energy flow through the small mammal community, these species contributed 55 percent, 22 percent, and 6.5 percent of the total, respectively. The other ten species occurring on the site shared the remaining 16.5 percent among them.

The Chews found that although the kangaroo rat sustained a relatively constant biomass over the year and yielded very little energy to higher trophic levels, some of the rarer species suffered substantial mortality with resulting oscillations in abundance. They deduced from this that the kangaroo rat was important in accelerating the recycling of chemicals in the ecosystem by returning the large quantities of nutrients in the seeds into the detrital cycle, thereby allowing plant production to be sustained from year to year. The less abundant species, they argued, are more important in supporting upper trophic levels and sustaining total energy flux through the food web.

THE TUNDRA BIOME: GENERAL FEATURES

The tundra is a treeless biome-type occurring from the forest limit to the ice caps in polar regions and from the forest tree line limit to glaciers in mountain ranges (see Fig. 17-1). The biome is largely restricted to the Northern Hemisphere because

484

CHAPTER 17
terrestrial
biomes:
desert,
tundra,
prairie,
and savanna

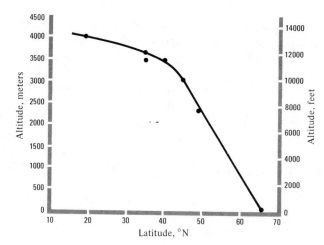

FIGURE 17-11 Decline in the altitude at which timberline occurs at progressively more northern latitudes. (After Bamberg and Major, 1968)

its land masses are considerably closer to the pole than are the continents in the Southern Hemisphere. The transitional zone between tundra and coniferous forest lies farther north along the western margins of the continents and sweeps progressively southward toward the eastern continental margins, as modified, of course, by the occurrence of alpine tundra along mountain chains. The general northwest-southeast sweep of tundra arises out of the eastward wind circulation at these latitudes which brings oceanic air into the interior of the continental land masses from the west. The milder oceanic winds are generally "quenched" as they move across continental interiors in North America and Eurasia.

The climate of the tundra is characterized by long, cold, harsh winters, and short, cool, mild summers. Frost is likely at any time of year, and the soil volume accessible to plant roots is limited by the occurrence of permafrost, a layer of permanently frozen soil. Physiological drought is common because although water is abundant, it is generally in the form of ice, thus making it unavailable to organisms. The distance down to permafrost decreases northward across the biome, and it is a less pervasive component of alpine tundras than of arctic tundras. The tundra occurs at successively higher altitudes along mountain ranges in the more southern latitudes (Fig. 17-11) until it finally disappears near the equator. In this topographic transition there is a general decline in the importance of soil freezing and long, frigid winter temperatures but an increase in solar radiation and in wind. Wind speed is often high in alpine situations so its desiccating effects may be particularly important there (Bliss, 1962). In fact, the strongest surface wind ever recorded, 372 km/hr, was on Mount Washington, New Hampshire. Precipitation also is usually higher in alpine tundras, which often have a snow pack several meters deep. Such snow accumulation is rare in the arctic.

A characteristic geological process in tundra that affects the distribution of the biome's biotic components is frost heaving (Eyre, 1963). As water freezes and thaws, its volume increases or decreases, respectively, by 8 to 9 percent. Because of the presence of permafrost and low evaporation, arctic tundra sites tend to be moist.

Freezing and thawing of a large water volume per given soil volume result in pronounced heaving and moving of the soil profile as the seasons change. As a result of different magnitudes of soil movement according to the amount of water present in the soil, there is a gradual sorting of soil particles according to size (Fig. 17-12). Tundra soil becomes highly patterned in a variety of forms that reflects the land slope and severity of frost heaving. Vegetation becomes patterned according to the underlying soil, so there are often poor areas where large stones are deposited and an area of relatively dense vegetative cover where soil particles are smaller.

Important plants include sedges, grasses, dwarf willows, and foliose lichens (Fig. 17-13). The last are particularly important because they are a principal food source of two of the arctic tundra's few large herbivores, caribou in North America and reindeer in Eurasia. Other herbivores include voles, lemmings, musk oxen, and hares. There are relatively few insect species, but those present, such as blackflies and mosquitoes, are often abundant during the brief summer. Migratory birds are also conspicuous during the summer. Carnivores include the arctic fox, wolves near the contact zone with forest, and a variety of raptors, including the snowy owl. Alpine tundras often serve as summer grazing grounds for bighorn sheep, elk, and deer that winter at lower elevations, and the mountain goat is a common herbivore on stony slopes in the alpine areas as are pikas and marmots. Lichens are generally a less abundant constituent of the alpine vegetation, but sedges, dwarf willows, and grasses are common.

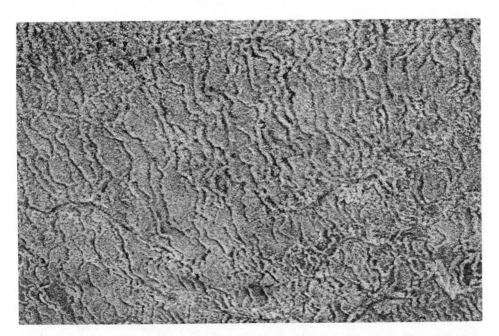

FIGURE 17-12 Repeated freezing and thawing of surface soils underlain by permafrost often result in patterned ground visible in this tundra landscape. (Steve McCutcheon)

486

CHAPTER 17
terrestrial
biomes:
desert,
tundra,
prairie,
and savanna

FIGURE 17-13 General view of the tundra biome in the Arctic. (Steve McCutcheon)

THE TUNDRA BIOME: STRUCTURE AND FUNCTION

The general decline in diversity from tropical to arctic regions reaches its terrestrial culmination in the tundra where ecosystems are comparatively poor in species and food-web diversification is low. In the vicinity of Point Barrow, Alaska, for instance, almost all of the plant biomass occurs in three species, and there is only a single herbivorous vertebrate, the brown lemming (*Lemmus trimucronatus*) (Schultz, 1964). The growing season is only two months and because of permafrost, the layer of soil accessible to plant roots is less than 40 cm deep. Precipitation is only 10 cm/yr, which places it within the range of desert precipitation. Evaporative stress, however, is comparatively low and permafrost plus a short frost-free period keep the soil profile relatively moist.

This is an area of herbivore cycles, for the brown lemmings oscillate in abundance with about four-year periodicity and peak densities of 150 to 200 lemmings per hectare. In the discussion of predator-prey relationships (Chapter 10) we saw that such predators as snowy owls and jaegers change their distribution from year to year to concentrate in areas of high lemming density. In addition to predator oscillation in a given ecosystem, lemming oscillations are propagated downward in the food chain to the producers. Thompson (1955) examined net primary productivity in areas where lemmings were excluded by fencing as compared with areas freely grazed by the lemmings. He found that lemming grazing had a severe effect on standing crop, reducing it by almost 50 percent during a year of peak lemming abundance, but not affecting it when the cycle was at its nadir (Fig. 17-14). During periods of lemming population peaks, the vegetation actually may be uprooted by the animals (Schultz, 1964).

Schultz (1964) also observed that oscillations in the biotic component of the tundra biome are reflected in the abiotic component. Cycles of vegetation abundance, in particular, result in cycles of permafrost depth as the amount of sunlight reaching the soil fluctuates inversely with vegetative cover. When the vegetation is depleted, the permafrost tends to melt to a lower depth. As the vegetation cover becomes more dense with a lemming decline, the soil cools and the permafrost migrates upward. Schultz hypothesizes that changes in soil volume and in nutrient flow from the soil to the vegetation volume may be a part of the ecosystem cycle in this tundra. A high lemming population, by destroying and eating a large portion of the vegetation, would accelerate the flow of available nutrients into the soil profile. Decline of the lemming population would result in a gradual recovery of the vegetation which would draw on the enlarged soil nutrient pool with a resulting enrichment of foliage nutrient content. Assuming that lemming reproduction is affected by nutrient content of the foliage, the attainment of appropriate nutrient levels in the plants would enhance lemming reproduction, and this would set the stage for another outbreak. The hypothesis of a self-propagating cycle involving soil-plant-lemming is an intriguing one, but there are few data available to test it.

Not all tundra ecosystems are subject to the oscillations occurring at Point Barrow. Carl (1971) examined a population of arctic ground squirrels (*Spermophilus undulatus*) along the Alaska coast somewhat west of Point Barrow. Although there were annual cycles of total abundance as young were born each spring, the number of breeding adults each spring was relatively constant. This stability was a product of breeding territories held by males. In the spring there were three classes of males: territory holders, floaters, and refugees. The first class mated with the females, which emerge from hibernation later than the males. The floaters ranged over the territories held by the first class of males, but did not hold permanent territories and did not reproduce. Refugees were driven out of the population center into marginal sites where they also failed to breed. The two subpopulations established burrows in marginal lowland sites and were subject to almost total mortality during periodic

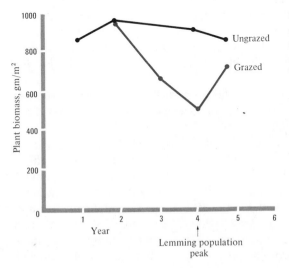

FIGURE 17-14 Effect of lemming grazing on plant biomass in a tundra ecosystem at Point Barrow, Alaska, during a lemming cycle. (Data from Thompson, 1955)

488

CHAPTER 17
terrestrial
biomes:
desert,
tundra,
prairie,
and savanna

floods. Foxes were also instrumental in eliminating population members in excess of territorial carrying capacity. During the summer 82 percent of the fox feces collected contained ground squirrel remains.

Prehibernation territories are established in the fall and this results in further decimation of the population. Grizzly bears often prey on the squirrels during this period and since the bears dig the animals out of their burrows, this predation is destructive of suitable habitat. Finally, squirrels forced into hibernation burrows in the lowland appear to be subject to 100 percent mortality during spring floodings. As a result of annual reproduction the size of the population fluctuates approximately fourfold on an annual cycle. During the establishment of breeding territories in the spring and hibernation territories in the fall, however, "surplus animals are driven from the colonies and enter the refugee population where they are subject to heavy predation by foxes and bears"; and heavy mortality from burrow flooding also acts as a leveling agent.

As the two examples above indicate, certain tundra ecosystems are subject to periodic population outbreaks, whereas other populations may be relatively stable in ecosystems where constantly effective feedback mechanisms operate. Where these mechanisms are lacking, oscillations are propagated throughout the ecosystem, and they are not, of course, restricted to the tundra biome (Pitelka, 1964). If they involve coupled oscillations of the type that occur in the Point Barrow ecosystem, assigning cause and effect becomes exceedingly difficult if, in fact, it is possible at all. A more important question than the "cause" of such oscillations may be what properties of such oscillating ecosystems distinguish them from the more constant ecosystems such as the tundra in which ground squirrels rather than lemmings are the principal herbivore.

THE TUNDRA BIOME: ADAPTATION TO COLD

A pervasive selective factor in the tundra environment is the extreme cold. As in the adaptation of desert organisms to water shortage, adaptation of tundra organisms to severe winter cold has taken two forms: avoidance or tolerance. Relatively few heterotherms occur in the arctic, and those that do have life cycles that can be completed in the short summer period with the winters spent in a quiescent life cycle stage.

Mooney and Billings (1961) examined the properties of alpine sorrel (*Oxyria digyna*), a tundra plant distributed widely in both arctic and alpine locations. They found that plants from alpine populations at more southern latitudes than tundra plants produced more flowers but fewer rhizomes, indicating natural selection for vegetative reproduction in arctic tundra and for sexual reproduction in the alpine tundra. Alpine populations had a higher rate of net photosynthesis, and were light saturated at higher intensities than tundra plants (Fig. 17-15). They also survived high night temperatures somewhat better than arctic samples. And, finally, respiration rates were consistently higher in arctic plants. The studies of Mooney and Billings indicate that although the same species often occur in tundra and alpine sites, they may be differentiated into ecotypic populations by a variety of distinctive

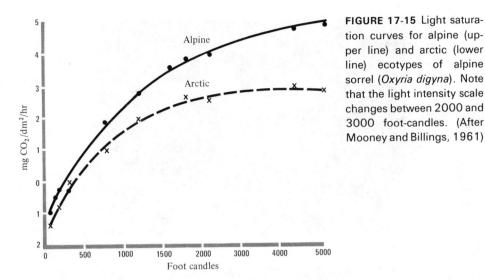

FIGURE 17-15 Light saturation curves for alpine (upper line) and arctic (lower line) ecotypes of alpine sorrel (*Oxyria digyna*). Note that the light intensity scale changes between 2000 and 3000 foot-candles. (After Mooney and Billings, 1961)

adaptive responses to environmental factors distinguishing alpine and arctic environments.

Homeothermic animals in the tundra biome possess a variety of adaptations keyed to conservation of heat. Many large herbivores, such as deer and elk, are summer transients in the alpine tundra, migrating upward in the summer to graze and browse and migrating downward during the winter into different biomes at lower elevations. The marmot, a permanent resident in alpine tundra, hibernates during the winter and the pika, another permanent resident, stores plant materials as "haystacks" which are consumed during the winter (Fig. 17-16).

Two "rules" have been proposed to describe adaptive responses of homeotherms to the latitudinal gradient which terminates in the tundra. These are Bergmann's rule and Allen's rule (Clarke, 1966). Bergmann's rule states that body size within a species tends to increase toward cooler climates. Allen's rule states that the limbs of animals become shorter in colder climates. Both of these mechanisms represent reductions in the surface-volume ratio, resulting in greater heat conservation in cold climate animals. These rules are widely violated, however, and Scholander (1955) has pointed out that external insulation in the form of feathers and hair may be more important as an adaptive mechanism than changes in body size and proportions. The thick cover of insulating fur on arctic mammals undoubtedly represents a consequence of strong selection for minimizing heat loss in a cold habitat.

In addition, there are organisms such as the arctic fox which go through seasonal color phases (Fig. 17-17), being in a dark color phase during the summer and developing a white coat for the winter period. An important prey species, the varying hare, also goes through such annual color phases. The importance of color camouflage to both predator and prey has resulted in natural selection for a coat color change at the spring and fall shedding periods which produce appropriate coloring for reducing visibility in fairly uniform habitats.

490

CHAPTER 17
terrestrial
biomes:
desert,
tundra,
prairie,
and savanna

FIGURE 17-16 The pika (*Ochotona* sp.) stores cured plants for food during the alpine's long winter. (*Top:* Lee Rue from National Audubon Society. *Bottom:* W. L. Miller from National Audubon Society)

THE PRAIRIE BIOME: GENERAL FEATURES

The prairie biome, more than any other collection of ecosystems, is a thing of the past. No other biome has been more completely reshaped by the mechanization of man than the prairie biome. Most of the earth's regions designated as prairie on the general biome map (Fig. 17-1) have been converted into cropland or grazing land from which the major native grasses have disappeared. The great grain belts

FIGURE 17-17 Many tundra animals, like the Arctic fox (*Alopex lagopus*), undergo seasonal color changes conferring lower visibility against the landscape. (*Top:* Gordon Smith from National Audubon Society. *Bottom:* Lee Rue from National Audubon Society)

of North America and Eurasia have left the prairie biome in the form of isolated tracts from a few hectares to a few hundred square kilometers in area scattered within a huge expanse of maize, wheat, milo, soybeans, rye, flax, and barley. Even pasture land in the prairie biome now consists largely of alien plant species. For several hundred thousand years, however, prior to the explosion of man's European population in the eighteenth and nineteenth centuries, the prairie biome occupied the heart of North America and some areas just inland on the Pacific Coast and occurred as a long narrow band across the center of Eurasia. Prairies of varying sizes occur on all the continents. Certain genera of grasses occur throughout most of the biome-type, including needle grass (*Stipa*), fescue (*Festuca*), bluestem (*Andropogon*), wire grass (*Aristida*), and others.

492

CHAPTER 17
terrestrial
biomes:
desert,
tundra,
prairie,
and savanna

The general impression conveyed by the prairie biome is of an unbroken sea of grass waving from horizon to horizon; Wells (1965), however, has shown that the North American prairie was a matrix within which there were woodlands of pine, juniper, and oak in areas of topographic relief (Fig. 17-18). Rainfall declines substantially in an east to west gradient across the North American continent and the character of the prairie biome changed with the rainfall. In the eastern portions of the biome, the major plant species were big bluestem (*Andropogon gerardi*), switchgrass (*Panicum virgatum*), Indian grass (*Sorghastrum nutans*), and bluegrass (*Poa* spp.). Further west, little bluestem (*A. scoparius*), needle and thread (*Stipa comata*), and wheat grass (*Agropyron smithii*) occurred. They were replaced in what is now the Central Plains states from North Dakota to Texas with grama grasses (*Bouteloua*) and, finally, along the margins of the Rocky Mountains such short grasses as buffalo grass (*Buchloe dactyloides*) were found. A similar gradient occurred in the Eurasian prairies, or steppes as they are called, with tall grasses along the northwestern border and progressively shorter grasses toward the south and east. From these Eurasian prairies came many of man's most important grains, including wheat and rye, as well as such domesticated animals as horses, sheep, and cattle (Russell, 1969).

The most extensive area of prairie biome was in North America (Fig. 17-19). In addition to the grasses already mentioned, a wide variety of dicotyledenous plants occurred in the grassland. The most conspicuous herbivore, of course, was the bison (*Bison bison*), which ranged from northern Canada to Mexico in huge herds totaling perhaps 75,000,000 animals (Watt, 1968) at the time that European man began to invade the prairie biome in the nineteenth century. By 1888 only 88 bison still lived within United States boundaries. Occurring from the short grass regions westward into the desert biomes was the pronghorn (*Antilocapra americana*), and elk (*Cervus canadensis*) were relatively common in the woodland-grassland areas within the prairie matrix. Other conspicuous mammalian herbivores were prairie dogs (*Cynomys* spp.), jack rabbits (*Lepus* spp.), and ground squirrels (*Citellus* spp.). Arthropod consumers presumably were also abundant, but there are few records to document this. At any rate, arthropods, particularly grasshoppers, are now extremely abundant in the biome where they reach plaguelike abundances in drought years. Important carnivores included small mammal predators such as badgers (*Taxidea taxus*), coyotes (*Canis latrans*), and blackfooted ferrets (*Mustela nigripes*). Larger predators, following the bison herds, included gray and red wolves (*Canis lupus* and *C. niger,* respectively) and the cougar (*Felis concolor*). Many of the native consumers, of course, have accompanied their biome into near extinction, being replaced by cattle and man. In the Canadian prairie provinces, the domesticated livestock population (as cattle units, with five sheep and 0.8 horse per cow) went through a semilogistic growth pattern during the late nineteenth and early twentieth centuries (Fig. 17-20). Since the 1920s, this herbivore population has oscillated from a low of 3.5 million to a maximum near 6 million. Although much of the prairie biome has been cultivated, many marginal areas have been grazed by cattle, displacing the native herbivores.

We saw from our consideration of the desert biome that overgrazing by cattle

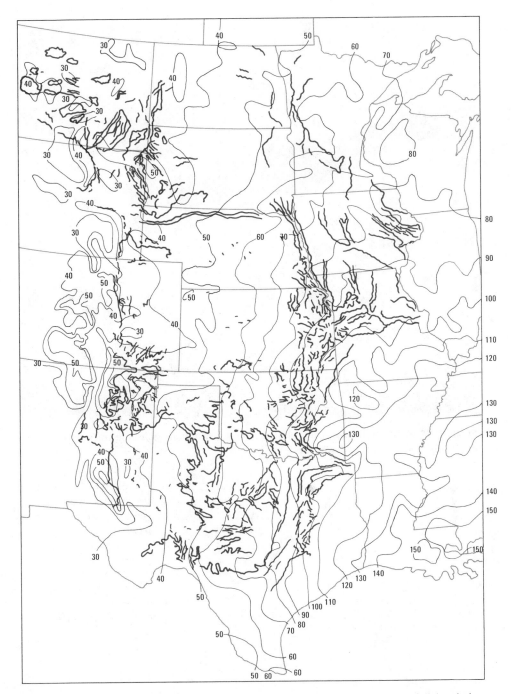

FIGURE 17-18 Map of the prairie biome in the United States showing rainfall isopleths (cm/yr) and woodlands occurring in upland locations (colored lines). (After Wells, 1965)

494

CHAPTER 17
terrestrial
biomes:
desert,
tundra,
prairie,
and savanna

FIGURE 17-19 General view of the prairie biome in Montana. Note the trees on the rocky escarpment in the background. (De Wys)

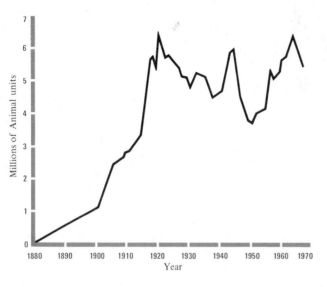

FIGURE 17-20 Livestock population in the Canadian prairie provinces during colonization of the biome by European man. Numbers of beef cattle, horses, and sheep are given in animal units (one unit equals 1.0 cow, 0.8 horse, or 5.0 sheep). (After Lodge et al., 1971)

495

THE PRAIRIE
BIOME:
STRUCTURE
AND
FUNCTION

can have a devastating effect on certain populations. There is considerable evidence, in addition, to indicate that the desert biome has expanded into the short-grass prairie as a result of the destruction of the latter by overgrazing. Buffington and Herbel (1965) compiled a history of ecosystem change in southern New Mexico using surveyors' notes for early records and direct data for recent information. In this prairie-to-desert shrub contact area they found that the prairie biome had disappeared during a hundred years of grazing, being now completely replaced by desert shrubs (Fig. 17-21). Over half of the area had been prairie in 1858 with *Bouteloua* and *Hilaria* abundant. By 1963 there were no shrub-free areas. Many of these desert shrubs have seeds that are consumed, but only partially digested, by cattle. Enough pass through the digestive tract for cattle to function as efficient dispersal agents, particularly since the seeds are deposited in a high-nutrient microenvironment— cow feces. Buffington and Herbel concluded that "Seed dispersal, accompanied by heavy grazing and periodic droughts, appeared to be the major factor effecting the rapid increase of shrubs."

Just as the desert biome climate can be characterized by the long periods of drought relieved by occasional scanty rainfall, the prairie biome can be characterized by regular precipitation oscillations. Records from Brandon, Manitoba, in a midgrass area, and Medicine Hat, Alberta, in a short-grass area, document the variation in precipitation from year to year (Fig. 17-22). Rainfall at Brandon over a 70-year period varied from a low of 30 cm to a high of 56 cm. At the more arid Medicine Hat location the range was from 25 to 46 cm. These records also indicate that although precipitation patterns are broadly regional, cycles of drought may be quite different in different parts of the biome. While Brandon was suffering its most severe drought of the period during the second decade of records, Medicine Hat was in one of its wettest periods. Because of the pronounced oscillations in moisture supply, the carrying capacity of the prairie will vary from year to year in a given place and from place to place in a given year. So long as the principal consumers can migrate as did the bison, overall energy flow for the biome may remain relatively constant. Conversion of the biome to a fenced rangeland, however, produced locally severe overgrazing during periods of drought. In marginal areas like those examined by Buffington and Herbel, such overgrazing may result in permanent eradication of the biome.

THE PRAIRIE BIOME: STRUCTURE AND FUNCTION

The species composition of the prairie biome changes with the rainfall gradient, and so does its productivity. In a tall-grass prairie where rainfall averages near 100 cm annually, Kucera et al. (1967) reported a net aboveground primary productivity of 482 to 570 gm/m^2/yr over a three-year period. In a midgrass prairie where rainfall averages near 50 cm, Hadley and Buccos (1967) reported net aboveground primary productivity of 338 gm/m^2/yr.

The effects of burning on the prairie biome have been a controversial subject for decades. There have been extensive experiments examining the effect of burning on prairie yield. Kucera et al. (1967) found that burning enhanced yield of a tall-grass

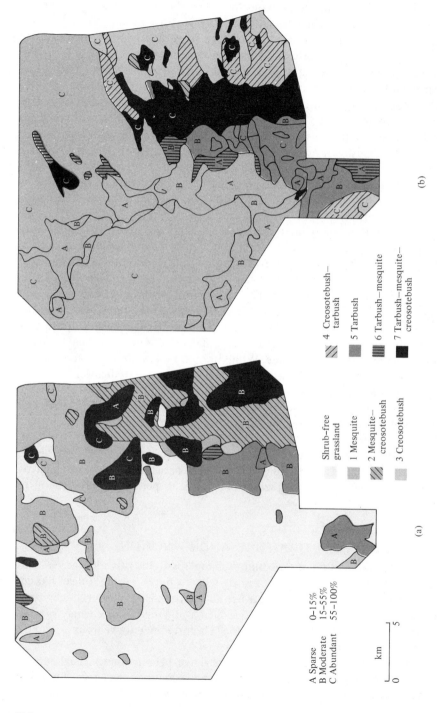

A Sparse   0–15%
B Moderate 15–55%
C Abundant 55–100%

km
0   5

Shrub-free grassland

1 Mesquite

2 Mesquite—creosotebush

3 Creosotebush

4 Creosotebush—tarbush

5 Tarbush

6 Tarbush—mesquite

7 Tarbush—mesquite—creosotebush

(a)

(b)

FIGURE 17-21 Maps showing the invasion of grasslands by desert shrubs between (a) 1858 and (b) 1963. (After Buffington and Herbel, 1965)

497

THE PRAIRIE
BIOME:
STRUCTURE
AND
FUNCTION

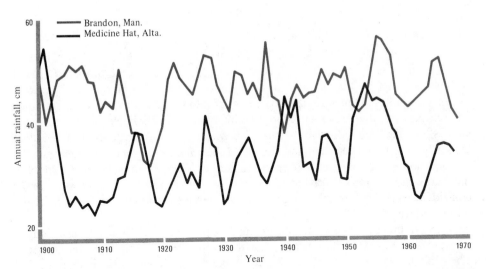

FIGURE 17-22 Patterns of annual precipitation at two locations in the Canadian prairie provinces showing the erratic nature of rainfall in the biome. (After Lodge et al., 1971)

prairie during years of high rainfall but had a negligible effect during a dry year. In our discussion on primary productivity (Chapter 5) we saw that Weaver and Rowland (1952) were able to increase net producer productivity by removing litter from prairie plots. They suggested that a natural mulch inhibited growth by acting as an insulating layer to keep soil temperatures cold later in the spring. Burning would eliminate this effect and, in addition, might destroy toxic organic compounds in the litter and release mineral nutrients. Several experiments in short- or midgrass prairies suggest, however, that burning may have an inhibitory effect on them (Anderson, 1953; Dix, 1960). It is likely that in such prairies a natural mulch would conserve moisture in the soil, thereby enhancing productivity. The effect of burning therefore will undoubtedly depend on the important limiting factors in the ecosystem and the effect that burning has on the balance of these limiting factors.

Another important effect that fire may have had on the prairie biome during prehistoric times was preservation of the biome from invasion by trees. As Figure 17-18 indicated, woodlands were well distributed throughout the prairie biome. Wells (1965) said, "There is no range of climate in the vast grassland province of the central plains of North America which can be described as too arid for all species of trees native to the region." Various species of trees and shrubs are potentially capable of invading the grasslands and it is misleading to suggest that the sole factor controlling such a distribution in this, or any other biome, is a single climatic complex. Prairie plants are primarily cryptophytes or hemicryptophytes, with their dormant buds buried beneath the soil surface. Trees and shrubs, of course, have their buds in an aerial position. Prairies are an almost perfect ecosystem for the propagation of fires during the fall since the dry aboveground portions form a closely intermixed,

498

CHAPTER 17
terrestrial
biomes:
desert,
tundra,
prairie,
and savanna

tinder-dry matrix. It is likely that autumnal fires were a common occurrence in the prairie, started by lightning and by hunting Indians. A prairie fire in 1910 in Nebraska developed a 200-km front in a single day (Wells, 1965). An early student of the prairies, C. O. Sauer (1952), has written, "I grew up in the timbered upland peninsula formed by the junction of the Missouri and Mississippi rivers. The prairie began a few miles to the north and extended far into Iowa. The broad rolling uplands were prairie . . . the stream-cut slopes below them were timbered; river and creek valleys and flanking ridges were tree covered. . . . From grandparents I heard of the early days when people dared not build their houses beyond the shelter of the wooded slopes, until the plow stopped the autumnal prairie fires. In later field work in Illinois, in the Ozarks, in Kentucky, I met parallel conditions of vegetation limits coincident with break in relief. I gave up the search for climatic explanation of the humid prairies."

Prairies, then, are grasslands occurring generally in geographic regions of inter-mediate and highly variable rainfall, with comparatively high evaporative stress. Their extent was determined by the combination of climate, broad uplands, and frequent fires. Such fires would kill the aerial buds of invading woody plants while sparing the buried buds of the prairie species. The latter phenomenon is likely to be more important toward the periphery of the biome where it contacted forest, savanna, and desert biomes. The eastern margins of the North American prairie were probably defined by a climate sufficiently moist to reduce the fire frequency enough for trees to overtop the grass, thereby reducing the tinder-dry layer produced every fall. Similarly, a reduction in cover by recurrent drought along the southwestern margins reduced the litter available for fire propagation, thus allowing replacement by desert shrubs. Along the western margins, the great Rocky Mountain escarpment provided a combination of rocky, open soils (breaking up the grass cover as do the numerous escarpments within the prairie matrix), colder temperatures, and frequent midafternoon showers during the late summer and fall.

Even more speculative than the role of fire in the prairie biome is the effect of grazing. It is astonishing perhaps that we know so little about this process in natural ecosystems, since it is a principal source of high-quality protein in our managed ecosystems. There were two types of herbivores in the grasslands of North America: migratory large herbivores and sedentary small herbivores. The migratory bison herds apparently ebbed and flowed with the seasons, migrating north and south to good grass and also moving out into forests along the Great Lakes and Rocky Mountains, perhaps following grassy clearings in the forests. This migration resulted in locally heavy grass defoliation for short periods of time. A number of studies have indicated that grasses recover quickly from such defoliation (Alcock, 1964). A mixture of wheat, rye, and oats was sown in the fall and grazed the following spring by dairy cattle to a 6-cm stubble from an initial height of 25 cm. This grazing removed almost all of the leaf blades. After an initial lag subsequent to grazing, the relative growth rate of the grazed cereals reached a high level (Fig. 17-23). As a consequence leaf area ratio of the previously grazed stand soon exceeded that of the ungrazed stand. In general, grazing studies indicate that periodic defoliation, even when fairly severe, has little effect on producer structure and function. The

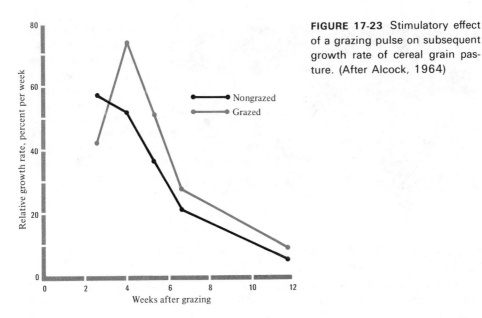

FIGURE 17-23 Stimulatory effect
of a grazing pulse on subsequent
growth rate of cereal grain pasture. (After Alcock, 1964)

499

THE PRAIRIE
BIOME:
STRUCTURE
AND
FUNCTION

Nongrazed
Grazed

effect will be regulated, however, by defoliation frequency, the stage of the plant
life cycle at which defoliation occurs, and general vigor of the plant.

Studies of grazing on native North American prairies have shown that the most
abundant plant species were not adapted to sustained heavy grazing (Voight and
Weaver, 1951). In midgrass prairies near Lincoln, Nebraska, the major plants in
lightly grazed plots were big and little bluestems. In more heavily grazed pastures
they were replaced by Kentucky bluegrass (*Poa pratensis*) and grama grass. Both
of the latter species are low-growing species that occur deep in the canopy of the
native grasslands. Heavy grazing opened the upper canopy and grazing cattle were
much less efficient in feeding on the low-growing bluegrass and grama. Heavily
grazed prairies were invaded by annual dropseed grasses (*Sporobolus* spp.), which
are generally much less palatable than the bluestems. Bluestems made up 49.8
percent of the cover in lightly grazed pastures, but only 1.9 percent in heavily grazed
pastures. The annual dropseeds were absent in lightly grazed pastures, but contributed 23 percent of the cover in heavily grazed pastures.

The large migratory herbivores probably had a minimal effect on the vegetation.
The sedentary herbivores, however, and particularly prairie dogs, could often have
a locally devastating effect (Fig. 17-24). In modern prairies preserved as museum
pieces in tiny public land fragments, the land in the vicinity of prairie dog colonies
is often largely denuded of vegetation. This denudation apparently represents both
the effects of locally high herbivore density and a behavioral response by prairie
dogs to reduce predator cover within the colony area. Prairie dogs were subject to
predation by rattlesnakes, badgers, hawks, owls, coyotes, and ferrets. Reduction of
cover within the colony area reduced the ability of these predators to approach the
burrows undetected.

500

CHAPTER 17
terrestrial
biomes:
desert,
tundra,
prairie,
and savanna

FIGURE 17-24 Prairie dogs (*Cynomys* spp.) build dikes around the burrow entrance that serve to keep heavy rainfall from flooding the burrow. Destruction of vegetation near the burrow also gives a better view of many predators. (Arthur H. Bilsten from National Audubon Society)

THE PRAIRIE BIOME: PERPETUATION OF A SPECIES COMBINATION IN SPACE

The geographic distribution of different producer species combinations within the prairie biome is in the form of bands with an extensive north-south spread but a narrow east-west spread. Short grass prairie, for instance, with buffalo grass and blue grama grass as major species extends from Canada to Mexico but with an east-west width of about 3 hundred kilometers. This suggests that species differentiation is required to provide sufficient genetic divergence to compete efficiently along a moisture gradient; it is, however, merely the relative abundances that change along this east-west gradient. Most of the major prairie species are found throughout the biome, with short grasses being most abundant in the west where tall grasses are restricted to special microhabitats, and tall grasses being most abundant in the east where short grasses are restricted to special microhabitats. The north-south bands, then, are primarily bands of shifting relative abundances, as defined by gradients of available moisture.

McMillan (1959a) inquired into the mechanisms defining the ability of species to occupy such a diverse array of climates by observing the phenological responses of geographically diverse populations under uniform garden conditions. When grown in a garden at Lincoln, Nebraska, populations of little bluestem (Fig. 17-25) and many other species showed earliest flowering in northern and western populations and progressively later flowering in southern and eastern populations. All populations began spring growth near the same date, so these gradients of flowering time represent gradients of developmental time, northern populations having a very rapid developmental speed and southern populations developing flowering organs at a much more leisurely pace. Not all species, however, showed rapid development in northern populations. In June grass (*Koeleria cristata*), for instance, populations from diverse sites flowered synchronously. A consequence of this was that the span of flowering period in the entire community was considerably greater in southern samples than in northern samples (Fig. 17-26). In northern ecosystems there was strong selection for rapid development, a necessity if seed is to mature during the short growing season. In southern ecosystems there was strong selection for temporal

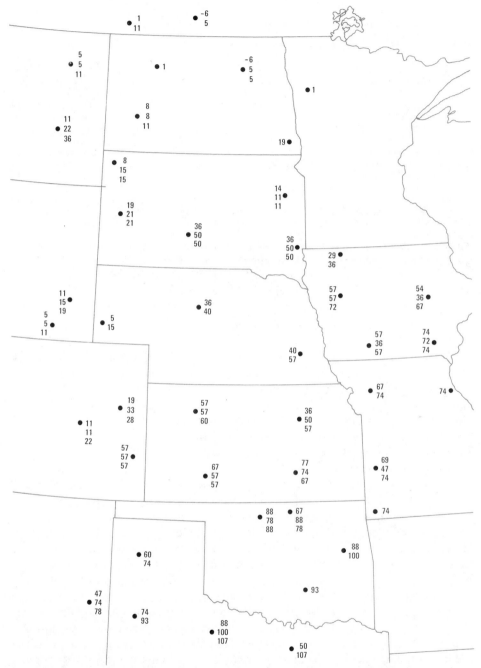

FIGURE 17-25 Flowering time of little bluestem (*Andropogon scoparius*) population samples from different geographic locations when grown in a transplant garden in Lincoln, Nebraska. For each individual, the number of days before (-) or after June 15 flowering first occurred is indicated. Rate of development to flowering decreases substantially from south to north. (After McMillan, 1959a)

502

CHAPTER 17
terrestrial
biomes:
desert,
tundra,
prairie,
and savanna

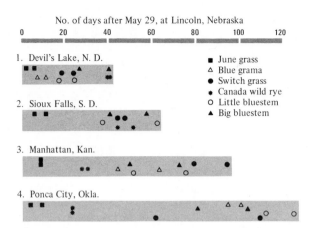

FIGURE 17-26 Relationship between flowering time of prairie grasses in Lincoln, Nebraska, transplant garden and their geographic origin. The period over which community flowering was sustained almost tripled from North Dakota samples to Oklahoma samples. (After McMillan, 1959a)

differentiation of flowering time, thus spreading resource demands on moisture and nutrient pools over a longer time period.

By programming photoperiod and temperature sequences from climatic records at sites from North Dakota to Mexico, McMillan (1965) further examined the interaction among flowering requirements, native climate, and climate within which population samples were grown. North Dakota clones of blue grama (*B. gracilis*) flowered 10 weeks after the last frost in North Dakota (the equivalent of July 6 in North Dakota); they flowered after 7 weeks in the Nebraska climate (June 4 in Nebraska), and 4 weeks in central Texas (April 6). They failed to flower after 20 weeks in the Mexico City climate. Mexico City clones, in contrast, failed to flower in the North Dakota climate, flowered after 17 weeks in Nebraska (August 13), 12 weeks in central Texas (June 1), and 20 weeks in Mexico City (June 10). These experiments indicate that developmental speed of a northern population may be telescoped considerably by moving the population southward. Conversely, the developmental period of the southern population sample was decreased somewhat when moved northward. Neither population flowered at the opposite climatic extreme from their origin, and in climates in which both populations flowered, the northern population did so much more rapidly. More significantly, however, the North Dakota population flowered on July 6 in its native climate, and the Mexico City population flowered on June 10 in its native climate. So although the intrinsic developmental speed to flowering is much faster in the northern population, its last killing frost date is so late that it actually flowers later than the Mexico City population in nature.

Olmstead (1944) has shown that ecotypes of grama grass differ in the photoperiod requirements to induce flowering. Northern populations required photoperiods in excess of 15 hours for floral organs to be formed, whereas southern populations required only 13 hours. This suggests that northern populations would be incapable of flowering at southern sites because of the insufficient photoperiod.

These studies of prairie plants indicate that the perpetuation of a certain species combination through a biome is dependent on selection for ecotypic variants adapted to local combinations of photoperiod and temperature. The patterns of changing relative abundances across the moisture gradient suggests that there has been little, if any, ecotypic differentiation in response to changing water relations.

THE SAVANNA BIOME: GENERAL FEATURES

The savanna biome is among the largest of the earth's biotic systems, occupying huge areas in Africa, South America, and Australia (Fig. 17-1). A savanna is a combination ecosystem with trees forming a scattered open canopy within a grassland matrix; it occurs in tropical or subtropical sites, where temperatures are typically warm all year. The fact that treelessness of the prairie biome cannot be explained by a fire hypothesis alone is demonstrated by the regular occurrence of fire in savannas (Bourliere and Hadley, 1970). The grass cover in savannas tends to be more open, however, allowing occasional openings for tree seedlings to become established. The growing season in the prairie biome is defined primarily by temperature, except on the prairie margins where the biome grades into the desert or savanna biomes. In the savanna biome, as in the desert, a principal controlling factor is seasonal rainfall. Annual precipitation in the savanna biome averages between 50 and 150 cm (Fig. 17-4) and it is decidedly seasonal. This weather pattern is defined by a latitudinal expanse from 5° to 20°N and S latitudes in which the varying insolation of a "tilted" earth generates periods of heavy rainfall alternating with drought. Toward the lower end of the rainfall range the savanna becomes a thorn scrub ecosystem that grades into desert shrublands. Toward the upper end of the range the savanna biome grades into tropical forest.

Extensive areas of the South American savanna biome occur on land relatively impervious to water so that the soil alternates between being waterlogged during the wet season and extremely dry during the dry season. Palms (*Mauritia* spp., *Capernicia* spp.) are common trees in these savannas. The Australian savannas, which have largely disappeared under exploitation by technological man, were characterized by open stands of eucalyptus trees (*Eucalyptus* spp.) with a grass and shrub understory. Principal trees in the North American savanna, which begins in Texas and sweeps southward through Mexico, are mesquite (*Prosopis* spp.) and several species of acacia (*Acacia* spp.).

When we refer to savanna, we generally think of Africa, both because much of the continent is covered by savanna and because the earth's most extensive and thoroughly studied savannas are there (Fig. 17-27). The trees characteristic of the African savanna are acacias and species of *Brachystegia, Combretum,* and *Julbernardia;* the grass cover ranges from wire grasses (*Aristida* spp.) in the drier regions to bluestems (*Andropogon* spp.) and elephant grasses (*Pennisetum* spp.) in the wetter regions. Wire grasses, occurring along the desert margins of the savanna, form a cover only a few centimeters high; elephant grasses occur as flanks of the tropical

504

CHAPTER 17
terrestrial
biomes:
desert,
tundra,
prairie,
and savanna

FIGURE 17-27 General view of the savanna biome in East Africa. (Eric Hosking from National Audubon Society)

forest biome and may form a dense growth 3 to 4 meters high (Eyre, 1963). Tree abundance varies from scattered individuals to fairly dense stands and, as in the prairie biome, escarpments may be covered almost completely by woody plants.

Certainly among the most characteristic features of the African savanna is the wide variety of grazing ungulates that it supports. Lamprey (cited in Bourliere and Hadley, 1970) has identified five classes of ungulates in Tanzania, separated by their migratory behavior: (a) giraffe, Grant's gazelle, and hartebeest are resident species that rarely move out of a local home range; (b) impala disperse in the dry season and form herds during the wet season; (c) dik-dik, waterbuck, warthog, and rhinoceros become dispersed in the rainy season; (d) eland and wildebeest immigrate into the area during the dry season; and (e) buffalo, elephant, and zebra undergo seasonal migrations through the area following the forage. On the plains of East Africa there is a vast flow of the entire animal fauna as they follow the distinctly seasonal forage and water availabilities in the area (Anderson and Talbot, 1965). The migratory properties of the ungulate community appear to be a function of climatic seasonality in a given region (see Chapter 10).

Reproduction as well as migratory behavior varies with the rainy season, and toward the arid margins of the biome, where there is a well-developed dry season, reproduction is synchronized with the 90- to 120-day wet season (Dasmann and Mossman, 1962). In less seasonal areas, however, ungulate reproduction may be spread throughout the year (Bourliere, cited in Bourliere and Hadley, 1970).

In addition to its herbivore populations, the savanna biome is famous for its major predators—lions, cheetahs, leopards, Cape hunting dogs, and hyenas. These carnivores, of course, follow the huge game herds across the biome and are differentiated by the prey that they seek. Lions kill all the antelope up to such large species as the eland, whereas, at the opposite end of the spectrum, hyenas are largely scavengers, eating the remains of carcasses killed by one of the larger predators.

505

THE SAVANNA
BIOME:
STRUCTURE
AND
FUNCTION

THE SAVANNA BIOME: STRUCTURE AND FUNCTION

The net primary productivity in the savanna varies considerably in the gradient from thorn shrub and wire grass to acacia and elephant grass ecosystems. Productivity estimates range from 37 to 800 gm/m²/yr or, on a growing season basis, 0.67 to 8.12 gm/m²/day (Bourliere and Hadley, 1970). These estimates, however, are confined to the herbaceous ground layer and there are few data on the contribution of trees to primary productivity, although they are the major food source for such browsers as giraffes and elephants. Hopkins (1968) measured the leaf area ratios of herb and tree strata during their peak and recorded values of 3.71 for grasses, 0.85 for nongrass herbs, and 1.46 for trees; these represent 62 percent, 14 percent, and 24 percent, respectively, of the plant cover. Many of the trees are deciduous and some, notably *Acacia albida* which is abundant in the northern savanna region of Africa, grow out of synchrony with the rainfall. This latter species produces leaves during the dry season and loses its leaves at the onset of the rains. The leaf fall enhances herb productivity underneath the trees (Radwanski and Wickens, 1967), probably as a result of nutrient enrichment, suggesting that trees may serve as nutrient pumps in the savanna, taking up nutrients from deep within the soil by virtue of their deep root systems and enriching surface layers by leaf fall.

Even trees that do not grow asynchronously with the grasses may enhance grass productivity under their crown since leaf decay during the dry season in negligible (Hopkins, 1966). There is also twig and limb fall from trees, which is processed by termites prior to soil microbial action. Because termites are more active during the dry season, the initial stages of wood decomposition are accomplished then, providing a rich substrate for soil microbial colonization during the rainy season.

The savanna contains a variety of nonungulate consumers and Bourliere and Hadley (1970) tabulated biomass data on a variety of consumer classes for savanna near the biomes' northwestern border in Africa (Table 17-1). Arthropods, which peak

Table 17-1 Fresh weight biomasses of several minor consumer classes in African savanna

Organism-type	Average biomass (kg/ha)	Peak biomass (kg/ha) and month of peak
Herbivorous arthropods	4.4	8.1 (November)
Carnivorous arthropods	2.1	3.1 (November/December)
Seed-eating birds	0.5	0.7 (July)
Fruit-eating birds	0.21	0.39 (July)
Insect-eating birds	0.31	0.41 (February)
Other carnivorous birds	0.08	0.14 (February)
Rodents	0.46	0.98 (October)
Insectivorous mammals	0.024	0.06 (November)
Amphibians	0.103	0.196 (May/June)
Lizards	0.066	0.106 (August)
Snakes	0.149	0.254 (November)

Bourliere and Hadley, 1970

506

CHAPTER 17
terrestrial
biomes:
desert,
tundra,
prairie,
and savanna

at the beginning of the dry season, are by far the principal nonungulate consumers in this area. The rainy season in the area is from April to October, with a decided peak in April and birds that feed on plant reproductive structures are most abundant during the middle of the rainy season, as are amphibians. Insectivorous mammals peak at the same time as their food source, but insect-feeding birds were most abundant when insect populations were near their nadir. This suggests that these carnivores must depend on an insect source that is most abundant during the dry season when insect abundance overall is low.

THE SAVANNA BIOME: NICHE STRUCTURE IN A COMPLEX COMMUNITY

One of the most puzzling aspects of the savanna biome in Africa is the variety of large herbivores that exploits its plant populations. The variety of antelopes alone almost defies imagination: wildebeests, blesbok, duikers, dik-diks, oribis, klip-springer, waterbucks, gazelles, springbuck, sable antelope, oryxs, gerenuk, kudus, and elands, to say nothing of zebras and other nonantelope grazers. How is it possible for such a vast array of large herbivores to coexist? Compare this with the North American prairie, supporting only three large herbivores, or the South American savannas and prairies that are largely devoid of large herbivores.

Our earlier consideration of niche structure suggests that several mechanisms might operate to separate the niche dimensions of these grazers. Among these might be a temporal separation in grazing. Vesey-Fitzgerald (1960) observed that there was a seasonal cycle of foraging on the Serengeti with Thomson's gazelle following wildebeest, which had followed zebra and topi. In other words the initial grazers in an area were zebra and topi, and there was a regular progression of appearance and replacement in the two subsequent grazers. Gwynne and Bell (1968) examined the stomach contents of zebra and wildebeest at the end of the wet season and found that they differed in the grass parts taken while grazing. The wildebeest took more leaf material while the zebra took more stem. In samples of stomach contents of herbivores at the end of the dry season, several species could be differentiated by their feeding habits. In a comparison of stomach contents of four species, zebras took equal parts of stem and sheath with no leaf blade; the topi took predominately sheath, but with more blade than the zebra; the wildebeest took more blade than any of the other three species; and the Thomson's gazelle showed a decided preference for sheath material. The different proportions of plant parts taken reflect the time sequence in which the animals graze an area of grass as well as the grazing method. Considering the nutritional properties of different plant tissues, Gwynne and Bell indicated that wildebeest was obtaining a higher protein and soluble carbohydrate yield than the other species. The poorest diet, by these criteria, was that of the zebras. They also pointed out that Thomson's gazelle is a low-level browser, eating the tips off low-hanging tree limbs and shrubs, so that it may be differentiated from the other three species by this behavior.

507

THE SAVANNA
BIOME:
NICHE
STRUCTURE
IN A
COMPLEX
COMMUNITY

Table 17-2 Tabulated index values of niche breadth on the habitat and time dimensions for the larger animals of the Mkomazi Reserve

Niche breadth on four habitat types ($1 \leq x \leq 4$)		Niche breadth on three seasons ($1 \leq x \leq 3$)	
Rhinoceros	(3.42)	Eland	(2.988)
Eland	(3.31)	Gerenuk	(2.975)
Wart hog	(3.22)	Reedbuck	(2.974)
Giraffe	(3.11)	Giraffe	(2.973)
Elephant	(3.09)	Wart hog	(2.953)
Gerenuk	(2.92)	Rhino	(2.947)
Dik-dik	(2.86)	Ostrich	(2.934)
Hartebeest	(2.79)	Gazelle	(2.933)
Gazelle	(2.54)	Kudu	(2.908)
Zebra	(2.52)	Hartebeest	(2.898)
Ostrich	(2.46)	Elephant	(2.895)
Kudu	(2.45)	Oryx	(2.867)
Impala	(2.36)	Impala	(2.806)
Oryx	(2.31)	Dik-dik	(2.679)
Steinbok	(2.10)	Steinbok	(2.568)
Waterbuck	(2.03)	Zebra	(2.509)
Klipspringer	(1.98)	Buffalo	(1.897)
Reedbuck	(1.97)	Waterbuck	(1.708)
Wildebeest	(1.75)	Wildebeest	(1.592)
Buffalo	(1.74)	Duiker	(1.000)
Duiker	(1.63)	Klipspringer	(1.000)
Bushbuck	(1.22)	Bushbuck	(1.000)
Lion	(1.90)	Lion	(1.763)
Jackal	(1.64)	Jackal	(1.000)
Hunting dog	(1.00)	Hunting dog	(1.000)
Hyena	(1.00)	Hyena	(1.000)

Harris, 1972

Harris (1972) examined the niches of 22 species of large herbivores in northern Tanzania, differentiating two niche dimensions: the type of habitat (mountain, tall-grass savanna, short-grass savanna, and shrub savanna) and the seasonal periodicity (wet season, dry season, and transitional periods). Animals were counted along ten transects (a total of 7500 km) divided into 37 habitat specifications according to the proportions of the four major habitat types represented in the sample. About 70 percent of the area was shrub savanna, so sampling time was not random for the area but was identical for each habitat-type. From the information index formula (H'), he calculated a habitat niche breadth and a temporal niche breadth for each of 22 herbivore species and 4 carnivore species. By taking e^H, a range of habitat niche breadths from one to four and of temporal niche breadths from one to three is obtained. He found that rhinoceros and eland ranged through the greatest variety of habitats (Table 17-2). Through all of the seasons most of the species

508

CHAPTER 17
terrestrial
biomes:
desert,
tundra,
prairie,
and savanna

occurred somewhere in the area except for duiker, klipspringer, and bushbuck, which appeared only in the wet season. Carnivores tended to be both more seasonal and more restricted in habitat range than most herbivores. Harris also calculated a measure of niche overlap (Horn, 1966) based on the proportion of a species count that occurred in a given habitat type. This, of course, does not account for temporal patterns like those reported by Vesey-Fitzgerald, but it does give an indication of the degree of spatial overlap between the species. A two-factor niche overlap matrix would require comparing the co-occurrences of species in a given habitat type at a given time, necessitating combining the spatial niche overlap matrix (Fig. 17-28) with 33 monthly records. The highest spatial overlap was between impala and hartebeest. These two species tended to be associated in a complex that also included ostrich, steinbok, giraffe, and gazelle. Low index values indicate spatial separation of the species realized niches. Wildebeest, for instance, was quite separate from klipspringer, dik-dik, gerenuk, and kudu. It is also interesting that the hunting dogs show a much lower index value with herbivores than the other three carnivore species. It has been reported frequently that this is the only African carnivore from which prey tend to flee on sight rather than only on pursuit. The other three carnivores, in contrast, are found in spatial coexistence with herbivores fairly frequently.

Taken together these studies of niche diversification in the African herbivore populations suggest a variety of isolating mechanisms that separate the ecological demands the species place on the ecosystem. The species, however, are rather tightly packed along the habitat dimensions, suggesting substantial food competition among the diverse herbivores.

GENERAL CONCLUSIONS

1. Global patterns of climate are important determinants of the distribution of terrestrial biomes:
   (a) Deserts occur in areas of low rainfall and high evaporative stress. Principal producers are often shrubs, succulents, and drought-resistant grasses. Rodents, insects, reptiles, and birds are important consumers. Mechanisms of drought avoidance and water conservation are important adaptations at all trophic levels.
   (b) Tundra occurs in areas with short frost-free periods and cold temperatures. Principal producers are often sedges, lichens, and dwarf shrubs. Insect larvae, mammals, and birds are important consumers. Population fluctuations may be propagated through the trophic levels and even into the ecosystem's abiotic component.
   (c) Prairies occur in areas of moderate and variable rainfall subject to periodic but not sustained drought. Principal producers are grasses and the trophic structure is based on grazers.
   (d) Savannas are intermediate ecosystems with trees scattered in a grassland matrix.

	hartebeest	impala	ostrich	steinbok	giraffe	gazelle	wart hog	dik dik	eland	elephant	gerenuk	zebra	oryx	kudu	rhinoceros	reedbuck	wildebeest	waterbuck	buffalo	duiker	klipspringer	bushbuck	jackal	lion	hyena	hunting dog
hartebeest	1																									
impala	.86	1																								
ostrich	.84	.82	1																							
steinbok	.80	.80	.72	1																						
giraffe	.85	.81	.83	.72	1																					
gazelle	.83	.77	.83	.75	.75	1																				
wart hog	.76	.77	.68	.68	.76	.62	1																			
dik dik	.72	.64	.67	.59	.74	.66	.61	1																		
eland	.68	.69	.62	.51	.71	.68	.46	.54	1																	
elephant	.65	.59	.58	.50	.73	.63	.50	.56	.78	1																
gerenuk	.80	.60	.62	.59	.63	.63	.55	.66	.55	.57	1															
zebra	.72	.60	.62	.53	.69	.79	.52	.54	.69	.74	.50	1														
oryx	.61	.51	.61	.57	.67	.68	.61	.69	.57	.60	.58	.78	1													
kudu	.69	.63	.53	.29	.69	.50	.42	.62	.53	.61	.49	.43	.49	1												
rhinoceros	.74	.66	.52	.41	.67	.53	.45	.35	.51	.69	.35	.50	.34	.40	1											
reedbuck	.77	.75	.54	.48	.66	.53	.46	.33	.39	.52	.34	.38	.26	.40	.71	1										
wildebeest	.70	.64	.44	.27	.53	.51	.20	.04	.47	.63	.05	.49	.12	.04	.46	.43	1									
waterbuck	.64	.57	.40	.29	.48	.39	.32	.34	.42	.66	.33	.48	.34	.34	.45	.34	.35	1								
buffalo	.54	.48	.35	.29	.42	.35	.26	.23	.40	.52	.22	.44	.24	.27	.57	.51	.74	.44	1							
duiker	.53	.45	.38	.32	.46	.35	.37	.28	.36	.67	.27	.41	.34	.26	.46	.30	.20	.61	.38	1						
klipspringer	.60	.59	.07	.14	.31	.25	.22	.23	.28	.39	.24	.20	.41	.55	.14	.13	.04	.38	.19	.23	1					
bushbuck	.40	.23	.18	.12	.26	.12	.18	.10	.21	.51	.05	.24	.10	.15	.47	.25	.11	.54	.30	.67	.13	1				
jackal	.84	.82	.70	.60	.73	.76	.59	.43	.65	.68	.53	.74	.62	.29	.39	.29	.32	.51	.51	.40	.01	.01	1			
lion	.78	.74	.53	.50	.56	.60	.44	.35	.53	.60	.41	.60	.49	.31	.36	.41	.40	.48	.53	.38	.39	.20	.66	1		
hyena	.67	.73	.61	.57	.60	.58	.59	.54	.27	.35	.57	.11	.41	.47	.44	.55	.00	.19	.13	.26	.12	.09	.13	.22	1	
hunting dog	.51	.16	.14	.10	.14	.49	.05	.03	.14	.57	.05	.64	.48	.06	.27	.07	.16	.08	.16	.02	.00	.01	.07	.12	.00	1

FIGURE 17-28 A niche overlap matrix, based on spatial occurrence, for 22 herbivores and 4 carnivores in the African savanna. Among the herbivores, overlap ranged from 86 percent between hartebeest and impala to 4 percent between oryx and wildebeest and between klipspringer and wildebeest. Hunting dogs had a particularly low frequency of co-occurrence with other species in the community. (After Harris, 1972)

510

CHAPTER 17
terrestrial
biomes:
desert,
tundra,
prairie,
and savanna

2. Population fluctuations in biomes with distinctly seasonal climates may be triggered by particularly high levels of an important limiting factor and lack of strong K-responsive mechanisms in the populations.

3. Local ecosystems in a biome consist of distinct ecotypes although the species composition may be quite similar.

4. Diversity in different ecosystems is a function of the variety of ways in which the energy and chemical resources may be exploited.

terrestrial biomes: forests

18

There are three main types of trees which predominate in the vegetation over broad geographic areas: evergreen conifers, deciduous broadleafs, and evergreen broadleafs. Although we have divided the world's forested areas into three general biome-types, these, like other biomes, may best be regarded as ecosystem spectra, grading into one another over broad areas and constantly changing in composition and ecological properties. In general, evergreen coniferous forests are centered in the Northern Hemisphere, deciduous broadleaf forests occur in the middle latitudes, and evergreen broadleaf forests are restricted to high rainfall regions on each side of the equator. Along their margins these forests grade into one another or into sparse woodlands, savanna, tundra, or prairie.

THE CONIFEROUS FOREST BIOME: GENERAL FEATURES

Two huge expanses of coniferous trees sweep across the northern portions of Eurasia and North America, southward through the Northern Hemisphere's great mountain chains and the level plains of the southeastern United States. This is the coniferous forest biome with outlying islands occurring as far south as northern Africa and Nicaragua. The similarity of producer life form in this broad area should not suggest, however, a unifying environmental explanation for the biomes. For although the northern expanse is clearly related to the cold moist climate, the pine forests of the southeastern United States are influenced strongly by fire and soils.

The northern expanse is the boreal forest (Fig. 18-1), which opens into the sparser taiga (Fig. 18-2) as it approaches the tundra biome. Principal tree species in the North American boreal forest are white and black spruce (*Picea glauca* and *P. mariana*) with balsam and subalpine fir (*Abies balsamea* and *A. lasiocarpa*), all of which are evergreen species. Also abundant may be larch (*Larix decidua*), a deciduous conifer, and a variety of species of birch (*Betula* spp.), the latter likely to occur in areas that have been burned, logged, or subject to "blow-down" from wind. Southward in the Rocky and Cascade-Sierra mountain ranges, important tree species are Douglas fir (*Pseudotsuga menzesii*), ponderosa pine (*Pinus ponderosa*), and white fir (*Abies concolor*). Along the Pacific coastal mountains, the coniferous forest changes from Sitka spruce (*Picea sitchensis*) in northern regions to California's great redwood (*Sequoia sempervirens*) forests in the southern regions.

In the central region of the continent, around the Great Lakes, the boreal forest

FIGURE 18-1 A general view of the coniferous forest biome in Russia. (Novosti from Sovfoto)

513

THE
CONIFEROUS
FOREST
BIOME:
GENERAL
FEATURES

FIGURE 18-2 Tree density becomes sparse near the coniferous forest biome's transition to tundra at the northern tree-line. (Steve McCutcheon)

formerly merged into the lake forest in which white pine (*P. strobus*), red pine (*P. resinosa*), and eastern hemlock (*Tsuga canadensis*) were abundant. This forest, however, was almost totally destroyed by logging during the last two decades of the nineteenth century (Eyre, 1963), leaving huge expanses of shrubby wasteland through much of Michigan, Wisconsin, and Minnesota. Along the mountain ranges of eastern North America the coniferous forest biome skips southward as isolated stands of spruce-fir forest at high elevations buried within the matrix of the deciduous forest biome. In the southeastern United States, however, there are large coniferous forests in which several species of pines predominate, particularly longleaf pine (*P. palustris*), loblolly pine (*P. taeda*), pitch pine (*P. rigida*), and shortleaf pine (*P. echinata*). This extension of the coniferous forest biome occurs as "pine barrens" along the Atlantic Coast from Long Island southward and occupies much of the interior farther south where it occurs on the regions of sandy soil abundant in the area.

Principal large herbivores in the biome are two species of deer, the white-tail deer (*Odocoileus virginianus*) in the east and the mule deer (*O. hemionus*) in the west. The moose (*Alces americana*) is also common throughout the boreal forest, and the elk (*Cervus canadensis*) occurs in the western mountain forest. The deer and moose in particular, however, inhabit mainly disturbed and successional areas within and at the margins of the forest. The most important consumers in a mature forest include the snowshoe hare (*Lepus americanus*), a variety of rodents, particularly red squirrels (*Tamiasciurus hudsonicus*), chickarees (*T. douglasi*), and birds, including jays (Corvidae), common raven (*Corvus corax*), chickadees (Paridae), nuthatches

(Sittidae), and a variety of warblers (Parulidae). These birds, of course, may feed at several trophic levels, depending on season and habits.

The most important large carnivore was the timber wolf (*Canis lupus*), although the cougar was important in western mountain forests and in the southeastern pine forests. Both of these preyed primarily on the big ungulates, but now are extremely rare except in remote areas. The lynx (*Lynx canadensis*) ranges through the boreal forest along with the red fox (*Vulpes fulva*), which also extends out into adjacent biomes. Arboreal predators that feed upon the squirrel populations in the canopy include martens (*Martes americana*) and fishers (*M. pennanti*). Several weasels are also common predators. A characteristic omnivore in the biome is the black bear (*Euarctos americanus*), and the grizzly bear (*Ursus horribilis*) once was common in western regions of the biome.

The boreal forest of Eurasia, which begins in Scandanavia and extends across the continent onto the northern Japanese islands, is much the same as the North American boreal forests with only species being different. Once again there is a forest of spruce and fir, with pines common in certain areas. From west to east, for instance, Norway spruce (*P. abies*) is replaced in turn by Siberian spruce (*P. sibirica*) in the center of the biome, and *P. ajanensis* on Sakhalin and Hokkaido Islands. Just as there are pine forests on the sandy soils of the southeastern United States, so there are pine forests on the sandy soils of the central Volga River basin (Riley and Young, 1966). And just as ponderosa and lodgepole pine extend down the western North American mountain ranges, so Scotch pine (*P. sylvestris*) extends southward across the European mountain chains into the Spanish Sierra Nevada. Although the species are different, important consumers include squirrels, deer, moose, jays, warblers, bears, wolves, martens, and the European equivalent of the fisher, the sable (*M. zibellina*). Most of the carnivores and large herbivores have long been gone from the mountain extensions of the Eurasian biome, although deer parks are maintained by states and wealthy individuals. And where large predators such as wolves still occur near the northern limits of the biome, they are the subject of concerted efforts at extermination, similar to such attempts in northern Canada and Alaska.

It should also be pointed out that coniferous trees are subject to predation by a wide variety of beetles, lepidopterans, and other insects and that these are preyed upon in turn by wasps, shrews, bats, and other insectivores. Also common, particularly in clearings or edge areas, are such small mammals as voles and, in North America, chipmunks.

This biome has been subject to considerable disruption by man through extensive logging and burning, but it still remains largely unpopulated. Southern extensions, particularly around the Mediterranean Sea, have been destroyed, and areas of the lake forest that were clear-cut during the 1880s and 1890s appear to have been directionally modified to a shrubby type of community. But still, man's influence on the biome has consisted of harvesting its primary producers for export into human societies that are centered in other biomes. This contrasts markedly with the extended human usage of the prairie biome.

515
THE
CONIFEROUS
FOREST
BIOME:
STRUCTURE
AND
FUNCTION

THE CONIFEROUS FOREST BIOME: STRUCTURE AND FUNCTION

Although the major coniferous forest species are evergreen, they go through activity cycles that reflect the pronounced seasonality of the climates throughout the biome. Bourdeau (1959) examined the seasonal physiological activity of spruce trees in a plantation in New York state and found that both photosynthesis and respiration varied seasonally (Fig. 18-3). Net photosynthesis at 2000 foot-candles of light and 25°C was zero or near zero during December through February and reached a peak rate during October. In other words the net productivity potential varied seasonally even when actual assimilation measurements were made at a standard combination of light and temperature. During the winter the spruces had low photosynthetic potential. Respiration at 25 degrees in the dark was much less seasonal than the net productivity, but still showed a gradual decline into midwinter and then an increase through spring.

To determine how important temperature variations might be in regulating the seasonal course of net productivity, Bourdeau grew hemlock clones in the greenhouse and outdoors and compared their photosynthetic patterns. Plants grown outdoors went through the typical seasonal cycle of photosynthetic efficiency, with net productivity ability minimal during midwinter. Plants maintained in warm conditions

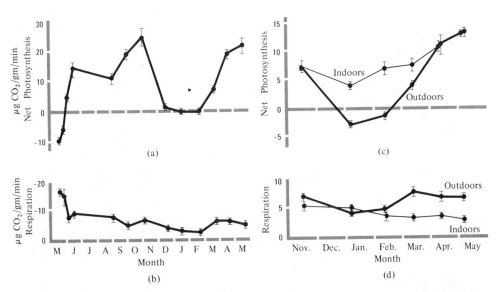

FIGURE 18-3 Seasonal changes of conifer photosynthesis at 2000 ft-candles and dark respiration at 25 degrees assay temperatures: (a) P_n of spruce seedlings growing outside, (b) R of spruce seedlings growing outside, (c) P_n of hemlock seedlings growing either outside or in a greenhouse, and (d) R of hemlocks growing inside or outside. The pronounced seasonal rhythms of photosynthesis and respiration are at least partially independent of seasonal temperature changes, as indicated by their appearance in greenhouse-grown plants. (After Bourdeau, 1959.)

also showed a seasonal cycle of photosynthesis, but it was much less pronounced than in the plants grown outdoors. This suggests that the plants were either responding to the annual cycle of light intensity, with a pronounced thermal regulation of response, or there is an intrinsic rhythm of photosynthetic potential in conifers that is synchronized with the growing season.

Larsen (1965) examined the annual growth rates of black spruce along a latitudinal gradient in the boreal forest from the taiga southward into lake forest. The radial growth increment was calculated from cores taken from 10 to 20 randomly selected individuals in 55 stands from Wisconsin to latitudinal timberline in the Northwest territories (Fig. 18-4). Although there was a decrease in average productivity from south to north, minimum productivity was not very different. Change in the average represented a higher upper extreme toward the south. The productivity in the Southern sites was quite variable, ranging up to four times greater than the productivity in the stands adjacent to the tundra. Southern stands, however, were considerably more variable in growth, and minima were well within the range of timberline sites. Boreal forest sites near the lower latitudinal limit of this forest are much more variable in their yield than the uniformly poorly yielding marginal sites. This suggests that sites on the margins of a biome will provide resource-poor niches, whereas interior sites may vary from poor to good.

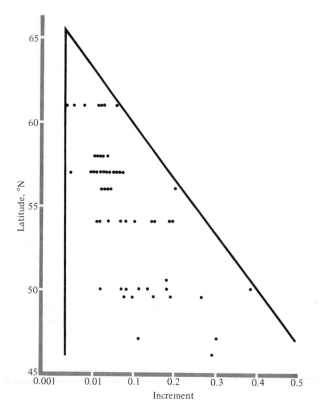

FIGURE 18-4 Relative growth rates of black spruce (*Picea mariana*) populations (each dot is the average of 10 to 20 trees) in randomly selected stands at different latitudes. The growth increment was estimated for the total volume of a cylinder by multiplying basal area (at breast height) by height, with this product divided by tree age. Northern populations had a uniformly low growth rate, while southern stands ranged from low to high, presumably reflecting site quality. (After Larsen, 1965.)

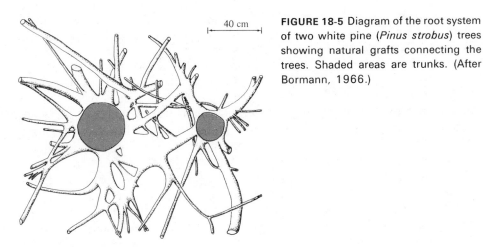

40 cm

FIGURE 18-5 Diagram of the root system of two white pine (*Pinus strobus*) trees showing natural grafts connecting the trees. Shaded areas are trunks. (After Bormann, 1966.)

517

THE CONIFEROUS FOREST BIOME: STRUCTURE AND FUNCTION

Separate plants are usually regarded as separate functional entities, but it has been shown that there may be interconnection among trees in the same area as a result of root grafting (Bormann, 1966). Excavation of root systems in white pine forests has revealed extensive root grafting among adjacent trees (Fig. 18-5). Members of a grafted system often undergo conspicuously different development as they age, with one becoming dominant and others being gradually suppressed; the suppressed tree may finally be converted into a living snag as its top dies, but its root system survives by virtue of its graft into another living top. After the crown of the suppressed tree dies, its root system may be "appropriated" by the still living companion, with translocation occurring between the crown and the captured root system. Death of a crown is primarily determined by competition with adjacent trees in the canopy. As one crown is suppressed, however, organic compounds may actively move from the active crown into the suppressed crown, thereby delaying the ultimate competitive outcome. This permits the smaller trees to survive longer than they could in the absence of grafting.

In the interior of undisturbed coniferous forest stands, the ungulates usually associated with forests are relatively rare and the principal herbivores are seed eaters, including squirrels, birds, and numerous insects with larvae that consume conifer seeds. Cone production is typically highly erratic, often varying from almost complete failure to a "bumper crop" in consecutive years (Lowry, 1967).

Cone crop is not a cyclical phenomenon but varies randomly with time, the intervals between maxima in a given stand varying from one to over five years (Cole, 1954b). Cone production is usually much less variable in lodgepole pine than within other coniferous species (Smith, 1970). Where this species is abundant, as on the eastern slopes of the Cascade Mountains, it may suffer heavy seed predation during years of poor cone production by other species. In areas where lodgepole is rare, as on the Cascade's western slope, squirrels and other seed eaters are forced into marginal habitats in which they harvest alder and maple seeds. On the Cascade eastern slope lodgepoles have evolved a cone-type (called *serotinous* cones) in which the cone scales are tightly appressed to one another. Natural selection for serotinous cones occurs in frequently burned areas since they retain the seeds under an insulating layer that opens up on heating to release the seeds. Lodgepole, then, is a coniferous biome successional species, adapted to colonizing burned areas.

Because of the tough closed cone structure, squirrels must expend more effort getting seeds out than is required when feeding on the open cones of the Douglas fir. Red squirrels occurring in areas with large lodgepole stands (that is, on the eastern slopes) have proportionately larger temporal jaw muscles than those occurring outside of lodgepole areas (Smith, 1970). Smith found the ratio ($\times 10^4$) of temporal muscle weight to body weight was 4.78 away from lodgepole and 6.30 in lodgepole stands ($t = 3.34$; $P < 0.01$). This suggests a degree of evolution or acclimation in squirrel populations in response to the necessity for utilizing closed cones as an alternate food source in poor cone years. Thus heavy squirrel predation during poor cone years could constitute further selection on the pine population for a harder, more resistant cone structure, thereby constituting negative feedback onto the squirrel population. This example from the coniferous biome suggests that coevolution among organisms in the same ecosystem is a complex process, influenced by such independent variables as fire frequency.

Relatively little is known about the effect of seed predation on tree reproduction in the mature forest. Since the canopy is closed except for occasional wind throws or other natural disturbances, however, we can deduce that seedlings have high mortality from competition if not from predation. Logging, of course, opens up forest areas for reproduction and several studies have examined the importance of predation in limiting the establishment of trees in logged areas.

On a clear-cut area in Oregon's Cascade Mountains, all the trees were harvested on 62 ha (25 acre) in the spring of 1959 and the area was burned the following fall (Gashwiller, 1970). The major seed predators on the site were deermice, although there were also chipmunks, voles, and seasonally abundant seed-eating birds. Only 10 percent of the Douglas fir seeds survived to produce a one-year-old seedling. Heaviest predation (41 percent) was by deermice; birds and chipmunks ate 24 percent, and the remaining 25 percent mortality was accounted for by invertebrate predators, diseases, and nonviable seeds. In western hemlock, 22 percent of the seeds were eaten by deermice, only 3 percent were eaten by birds and chipmunks, but miscellaneous factors, particularly nonviable seed, accounted for 53 percent of the seed. Conifer seed crop in the area averaged 180,000/ha. Seed eaters, however, reduced the viable yield to less than 25,000/ha. Clearly a substantial fraction of the tree reproductive effort was channeled into seed-eating herbivores.

THE CONIFEROUS FOREST BIOME: UNGULATES AND THEIR PREDATORS

The principal large herbivores in the coniferous forest are ungulates, particularly deer. Abundance of the two North American deer species has increased substantially since colonization of the continent by technological man converted much of the formerly closed forest into disturbed successional sites favored by deer. In addition, deer predators have been virtually exterminated. A principal predator, the timber wolf, has been reduced to probably less than 100 individuals in the United States excluding Alaska. As a consequence of reduced predation and a greatly expanded favorable habitat, many deer populations have gone through population outbreaks followed by large die-offs.

An overabundant deer population can have a devastating effect on vegetation in its browsing area. In Minnesota's Itasca State Park, whitetail deer were protected from predation for about 50 years (Ross et al., 1970), during which period the deer herd reached a density of up to 29/km². In this forest the heavy browsing caused such severe destruction that seedlings of dogwood, willow, alder, white pine, balsam fir, and other species could not survive. In a 1 ha area enclosed with deer-proof fence, there were 84 red pine seedlings and 1740 white pine seedlings. In a similar area outside the exclosure, there were no red pine seedlings and only 12 white pine seedlings per hectare. Finally, following several winter die-offs of deer from starvation, hunters were allowed into the park to function as predators. Continued hunting has reduced the deer population considerably, but it continues to have a significant effect on producers. Average height of seedlings inside and outside the exclosure after ten years of hunting were 1.19 and 0.28 meter for white pine, 0.48 and 0.23 meter for red pine, 0.71 and 0.25 meter for balsam fir, and 1.75 and 0.15 meter for white spruce. In addition to their effect on the tree populations, the deer also reduced shrub densities considerably.

Both food quality and food quantity have an effect on the vigor of deer populations. Transferring whitetail deer out of the Missouri Ozarks onto more fertile grassland soils resulted in 28 percent heavier individuals after five years (Steen, 1955). Adult male mule deer from a logged area of Oregon were 40 percent larger than males from mature closed canopy forest (Einarsen, 1964).

As we indicated earlier when discussing the tundra biome, alpine tundra is subject to periodic grazing by ungulates moving upward during the summer. Johnson et al. (1968) found that alpine grasses contained twice as much phosphorus and 50 percent more protein per unit weight than similar grasses in adjacent lowland areas. This suggests that the upward migration of ungulates may be a response to better quality forage in alpine areas. Deer also appear to select higher quality forage from the available food. Swift (1948) found that deer selected plants with a higher phosphorus, calcium, and fat content than was average for plants in the field where they were grazing.

Food quality and quantity are important in regulating deer reproduction. Cheatum and Severinghaus (1950) examined the corpora lutea of adult does and found that the count varied from 1.11 on poor ranges to 1.97 on good browsing areas. Viable embryos varied from 1.06 on the poor range to 1.71 on the good range. In addition to this direct feedback of food supply onto fecundity, other studies have indicated that age of first reproduction may be affected by range quality. When food supply was poor, only 5 percent of the females bred successfully during their first year, whereas up to 45 percent were successful first-year breeders on good rangelands (Severinghaus and Tanck, 1964). Verme (1962) fed captive does food of different nutritional quality and found that reproductive failure increased from 7 to 90 percent as diet quality declined. The principal cause of reproductive failure in does with poor diet was immediate post-parturition death of weakened fawns. More than a third of the fawns from poorly fed does died within two days of birth. Principal mechanisms of this death were poor milk yield from the doe, weak fawns, fawns too small to reach the teats, and does not allowing their fawns to suckle.

A number of studies suggest that heavy mortality is concentrated in males (Klein, 1970). In Alaska, for instance, 22 percent of the females shot by hunters were over five years of age, whereas only 7 percent of the males were that old. Even during the period when mortality is normally slight, from $1\frac{1}{2}$ to $5\frac{1}{2}$ years of age, the ratio of dying males to dying females was 1.67 to 1. In spite of these mechanisms for regulating population size, North American deer frequently outstrip their food resources in areas where population is low. Klein (1970) has said that, "North American deer . . . are adapted to early successional stages of vegetation, which are of a transitory nature [and] appear not to have well-developed self-regulatory mechanisms and are normally characterized by wide population fluctuations." North American deer, in other words, are colonizing species, subject to previous r-selection (see Chapter 9) by the transitory nature of their favored habitats. They now find that habitat-type considerably expanded over the last century and, as a result, have undergone population expansion that often outstrips the resource supply in the habitat.

Timber wolves were undoubtedly an important factor regulating deer populations in the native forest. The summer food supply of wolves in the boreal forest of Ontario's Algonquin Park was primarily whitetailed deer (Pimlott, 1967). Eighty percent of the summer wolf feces contained deer remains. The next most abundant food source was moose which comprised 8 percent of the food supply. Of the two ungulates taken, 71 percent of the deer were fawns and 88 percent of the moose were calves. Other summer food sources, contributing only a minor portion of the diet, were beaver, snowshoe hare, muskrat, marmot, porcupine, raccoon, and mice. During the winter, feces were impossible to collect in sufficient quantity to constitute a sample, but deer kills were common in the park. Age distribution of kills indicated that wolves kill predominately very young or very old deer. Of the kills 58 percent were deer five years of age or older and fawns constituted 17 percent, so less than one-quarter of the deer killed were in the prime age classes of $1\frac{1}{2}$ to 5 years.

Pimlott (1967) developed a hypothetical feeding balance to indicate the relationship between deer and wolves in which a deer density of $10/25$ km² would be required to support a wolf density of $1/25$ km² (Table 18-1). The total deer kill would be $177/250$ km² during the winter and $190/250$ km² during the summer for a fairly balanced distribution of numbers taken. The winter period would be twice as long as the summer, however, so somewhat larger deer would be taken during the winter months.

Considerable disagreement surrounds the importance of wolves in regulating population size of their principal prey, the large ungulate species. Studies of the moose population in Isle Royale National Park in Michigan suggest that wolves are important in regulating the population of this ungulate (Jordan et al., 1967) primarily by killing old animals. In whitetail deer herds in Wisconsin (Thompson, 1952)

521

THE
CONIFEROUS
FOREST
BIOME:
UNGULATES
AND THEIR
PREDATORS

TABLE 18-1 Calculation of number of deer required
to support a wolf population of one per 25 km²

Basic Assumptions	
Size of area	250 km²
Wolf population	10
Gross food consumption by wolves (avg. wt. 60 lb)	
Oct.–May	3.8 kg/day
June–Sept.	3.3 kg/day
Wastage	20%
Species other than deer—winter	10%
summer	20%

Age-composition and weight of deer killed		
Winter—fawns	30%	36 kg
adults	70%	68 kg
Summer—fawns	80%	18 kg
adults	20%	68 kg

Total kill of deer—winter	177
summer	190
	367 deer

Density of 10 deer/25 km² with productivity of 37% is
required to support one wolf per 25 km²

Pimlott, 1967

in which population size expanded rapidly from logging and burning in the 1930s, wolves were incapable of regulating deer populations; population outbreaks that occurred in areas where wolves were present were similar to those that occurred where wolves were absent. Wolves are relatively broad feeders and will switch prey according to abundance; in various areas they may prey primarily on deer, elk, bison, mountain sheep, caribou, moose, or a variety of smaller game.

Unraveling the ecological significance of predation to ecosystem processes in systems that have been subjected to substantial modification is extremely difficult. As Pimlott (1967) has written:

Contemporary biologists often have a distorted viewpoint about the interrelationships of ungulates and their predators. We live in an age when there is a great imbalance in the environments inhabited by many of the ungulates. In the case of deer and moose, the environmental changes . . . have been favorable and populations are probably higher than they have ever been. Under such circumstances it is not much wonder that we have been inclined to argue that predators do not act as important limiting factors on deer and moose populations. I doubt, however, that it was a very common condition prior to intensive human impact on the environment. In other words, I consider that adaptations between many of the ungulates, particularly those of the forest, and their predators probably evolved in relatively stable environments that could not support prey populations of high density.

It seems likely that in the native forest average deer density was much lower, and occurrence tended to be concentrated in localized areas of disturbances. Average wolf density was probably somewhat higher, and their diet was probably somewhat more diverse than it is in this day of abundant ungulates. Localized deer outbreaks would undoubtedly serve as predation loci for wolf packs, rapidly reducing such local outbreaks. In the present, however, ungulate populations are so large and widespread and wolf populations are so small and localized that it seems unrealistic to expect these conditions to reflect the undisturbed situation with any precision.

THE DECIDUOUS FOREST BIOME: GENERAL FEATURES

The temperature deciduous forest biome of the Northern Hemisphere occupies the northeastern United States and sweeps across Europe as a broad band from the Atlantic in the west to the Ural Mountains in the east. Then it disappears into the steppes and deserts of central Asia only to reappear again around the Yellow Sea and in northeastern China. Southward into central China it used to merge with a broadleaf evergreen forest complex where evergreen oaks (*Quercus*) were common. Other extensive broadleaf nondeciduous forests at one time occurred in Australia where *Eucalyptus* was abundant, in the Andes where the Southern Hemisphere conifers (*Araucaria* and *Podocarpus*) were mixed with evergreen broadleaf trees, and in southern Africa along the Atlantic Coast. Most of these mixed forests, which are neither coniferous nor deciduous but a mixture of diverse growth forms, have long since disappeared under the axe and the plow. Similarly, the great temperate deciduous forest biome of Eurasia has largely disappeared with the development of agricultural man. Among the characteristic trees of the biome in Eurasia were the European beech (*Fagus sylvatica*) which was found from the British Isles to Russia. Also common were a variety of deciduous oaks (*Quercus*), ash (*Fraxinus*), chestnut (*Castanea*), birch (*Betula*), elm (*Ulmus*), and others (Fig. 18-6).

Among the large herbivores in the Eurasian deciduous forest are roe deer (*Capreolus capreolus*) and red deer (*Cervis claphus*); also common are a wide variety of voles, mice, squirrels, and birds, including one of Europe's favorite game birds, the partridge. The omnivorous wild boar (*Sus scrofa*) once roamed throughout the biome, as did the European bison or wisent (*Bison bonasus*). The latter species was widely distributed throughout Europe, but by the Middle Ages had shrunk to scattered herds pushed into the coniferous forests of the mountains by agricultural man. Finally in 1930 there were 30 European bison, all confined to zoos. Since that time a free-ranging herd of about 100 has been reestablished in a Polish forest preserve. The story of the European bison, largely an exploiter of forest edges and clearings, is not unlike the story of the plains-roaming American bison, two once abundant animals saved from extinction at the last moment.

Among the large predators in the deciduous forest biome in Eurasia were the European wildcat (*Felis silvestris*), the European lynx (*Lynx lynx*), wolves, the

FIGURE 18-6 General view of the deciduous forest biome in France. (P. Berger from National Audubon Society)

523

THE
DECIDUOUS
FOREST
BIOME:
GENERAL
FEATURES

common red fox (*Vulpes fulva*), and a variety of owls, particularly the tawny owl (*Strix aluco*). Most of the smaller predators, such as owls and the foxes, are relatively common in forest preserves and other areas where they are protected from hunting. But the wolf and lynx are now extinct in the deciduous forest, and wildcats are extremely uncommon.

In North America the deciduous forest is much like it is in Eurasia. Whitetail deer (*Odocoileus virginianus*) are the most common large herbivore and such small herbivores as voles, mice, squirrels, and the eastern chipmunk (*Tamias striatus*) are found throughout. The large carnivores have disappeared here also, but both red and gray foxes (*V. fulva* and *Urocyon cincreoargenteus*) are still present and there are occasional black bears (*Euarctos americanus*) in more remote regions of the northeastern United States. There are a number of species common to the deciduous forest that also occur in other biomes, such as the red fox, whitetail deer, Eastern cottontail (*Sylvilagus floridanus*), longtail weasel (*Mustela frenata*), Virginia opossum (*Didelphis virginiana*), and raccoon (*Procyon lotor*).

The deciduous forest changes gradually from northeast to southwest in North America (Braun, 1950). In the northeast the principal forest trees are maples (*Acer*) and beech. Toward the south, prior to extinction of the American chestnut, the maple-beech community was gradually replaced with oak-chestnut forests. Toward the west it was replaced by oak-hickory (*Carya*) forests. These forests stretch as a band along the prairie-forest contact zone from Indiana southwestward into Texas. The oak-hickory permutation of the North American deciduous forest extends westward into the Great Plains as far as Nebraska and Kansas where it occupies sheltered slopes and occurs as scattered woodlands within the prairie matrix. Deciduous trees were always found throughout the prairie biome along streams, with the principal species in these riparian woodlands being willow (*Salix*) and cottonwood (*Populus deltoides*).

The heart of the deciduous forest biome in North America contained a large number of tree genera common to the Eurasian deciduous forest. Principal among these, in addition to beech and oak, were ash, elm, birch, and chestnut, the last being an almost extinct species, killed by chestnut blight introduced from Asia in 1904. Within 35 years this parasitic fungus had obliterated the unresistant American chestnut. A similar phenomenon is now occurring in elms as a result of Dutch elm disease.

The deciduous forest biome in North America, like the biome in Eurasia, has been heavily cultivated. In North America this cultivation began in the seventeenth century and reached a peak in the nineteenth century. With the opening of the fertile lands of the prairie biome to large-scale farming, much farmland in the deciduous forest biome has been abandoned and allowed to return gradually to deciduous forest (Fig. 18-7) as succession proceeds. The farmlands of Europe have, of course, existed for centuries with a comparatively high productivity in spite of lower innate fertility of forest soils in comparison with grassland soils. The topography of Europe is considerably more level than that of the American deciduous forest land and, as a consequence, erosion is a much less serious problem. American deciduous forest occurs on the old, eroded mountains that stretch from Maine to Georgia to Arkansas,

FIGURE 18-7 Throughout the forest biomes in North America, areas once farmed are slowly returning to forest after being abandoned. (U.S. Forest Service)

the Adirondacks, the Appalachians, and the Ozarks. This steep, rocky ground stands in sharp contrast to the gently rolling hills and broad river plains of northern Europe. Europeans soon found that the type of agriculture they were used to practicing could not be sustained on North American forest soils and farms were progressively abandoned in the westward search for a viable agricultural base; it finally culminated in the maize belt of the eastern prairies and the wheat belt of the western prairies.

525

THE
DECIDUOUS
FOREST
BIOME:
STRUCTURE
AND
FUNCTION

THE DECIDUOUS FOREST BIOME: STRUCTURE AND FUNCTION

All ecosystems exhibit some degree of seasonality. We have discussed the functional organization of the savanna biome around alternating wet and dry seasons in a generally warm climate; we also considered the severely restricted growing period in the tundra biome. Nowhere is seasonality more conspicuous, however, than in the deciduous forest biome with the alternating periods of leafiness and bareness that are its distinguishing property. Leafing out of the canopy during the spring results in a marked diminution of the amount of sunlight reaching the forest floor (Fig. 18-8). After reaching a minimum just after leafing is complete, there is a consistent, though small, increase in ground level light intensity as the leaves age, and then when leaves fall in autumn and there is a rapid increase in the proportion of direct sunlight that reaches the forest floor.

Because of the shading effect of limbs, branches, and trunks, the forest floor never receives sunlight in intensities equivalent to open areas. As a consequence, there has been selection for ephemeral species in the understory—species that grow rapidly during the early spring, flower before the trees leaf out, and then mature fruit rapidly or very slowly through the summer shade. In addition many plant populations may have differentiated into low light adapted forms under the forest canopy. In a series of elegant experiments, Björkman and Holmgren (1963) examined the photosynthetic properties of goldenrod populations originating in woodlands and in open fields when grown under identical conditions. Net photosynthesis at light saturation in

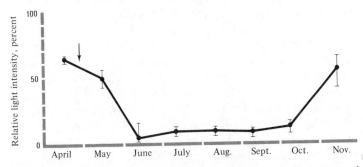

FIGURE 18-8 Seasonal changes in light intensity (as a percentage of full sunlight) under the canopy of a beech forest in Japan. The period of leafing out is indicated by an arrow and the vertical lines are ranges of observed variation. (After Nomoto, 1964.)

clones from the shade of a Danish beech forest was 18.27 mg CO_2/dm^2 of leaf surface per hour. In clones from a sunny meadow in southern Sweden, the net productivity at light saturation was 25.4. The values were significantly different ($t = 5.14$ for $P < 0.01$ with $df = 4$), indicating that the plants originating under shady conditions were much less efficient under high light intensities.

They also examined the initial slope of net productivity on light as a measure of the efficiency of utilization of low light intensities. This slope was 2.97 in shady origin plants and 2.35 in sunny origin plants ($t = 9.50$; $P < 0.01$), indicating that plants adapted to growing under a forest canopy were better able to use low-intensity light. The effect that growing the plants under different light intensities had on the net productivity at those intensities was studied as a measure of the abilities of the plants to acclimate to different light conditions. Net productivity of plants from a shady habitat grown in high light was 206 percent of the rate of comparable plants grown at low intensities. In other words the photosynthetic apparatus had a greater light-trapping ability in these shade-adapted plants when they were grown at higher light intensities. In the sun-adapted plants, the net productivity at high light was 343 percent of the rate at low light. This was a significantly greater stimulation than shown by the shade-origin goldenrod ($t = 7.21$; $P < 0.01$); so the goldenrod ecotype adapted for growing in the shade of beech canopy was more efficient at utilizing low light intensities, but had lower net productivity at light saturation and was not as capable of acclimating to high light as the meadow ecotype.

In addition to the seasonality of leaf production and internal light climate there often is pronounced seasonality in the chemical properties of plant parts. In oak woodlands, for instance, foliage palatability to most herbivores is highest just after the oaks leaf out since they soon begin accumulating complex organic molecules called *tannins* that discourage herbivores and are inhibitors of certain digestive enzymes. As oak leaves age, tannin content may increase to 5 percent of leaf dry weight and water content also declines (Feeny and Bostock, 1968; Feeny, 1968).

Larvae of a leaf feeder, the winter moth, showed poor growth with 1 percent tannins in the diet and pupation size declined. This moth therefore has evolved a life cycle keyed to larval feeding upon young, low-tannin oak leaves (Feeny, 1970; Varley, 1970). Adults emerge from pupae in November and lay eggs that overwinter until hatching in April as the oak leaves emerge. The larvae complete their development in slightly over a month and then pupate just as tannin concentrations are beginning to increase substantially; the summer and fall are spent in the pupal stage, thereby avoiding the poor food supply of mature oak leaves. A number of other invertebrate herbivores have evolved similar strategies for avoiding oak leaves except during early spring. One moth has larval stages that feed on young oak leaves and then migrate onto understory species in May (Varley, 1970).

Respiration in woody portions of trees is keyed to cyclic primary production. In oak trees on Long Island, Woodwell and Botkin (1970) found that stem respiration continued at a low, but nonzero, rate throughout the winter and increased twenty- to thirtyfold as the trees leafed out in the spring (Fig. 18-9). Stem respiration peaked in June, and then declined steadily throughout the growing season. Unit leaf photosynthesis showed a much less homogeneous pattern, varying from quite high to quite

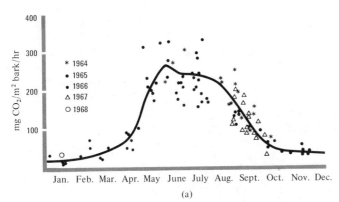

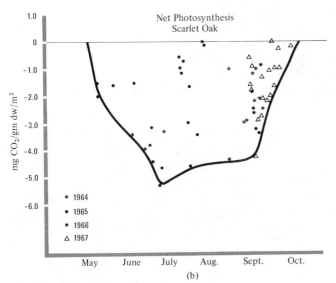

527

THE
DECIDUOUS
FOREST
BIOME:
STRUCTURE
AND
FUNCTION

FIGURE 18-9 Seasonal patterns of (a) stem respiration and (b) leaf net photosynthesis of scarlet oak (*Quercus coccinea*) trees on Long Island for four annual cycles. Each point is the 24-hour mean of approximately 50 measurements. Although there is a definite annual cycle of stem respiration, net photosynthesis is much less consistent. (After Woodwell and Botkin, 1970.)

low at any given time during the growing season. Although maximum unit leaf rates were recorded in July, there certainly was a less consistent pattern of average net productivity than of average stem respiration.

Considerable work has been done on the foliage area and biomass geometry of deciduous woodlands and, in general, both leaf area index and leaf biomass tend to be concentrated in space. In a Japanese birch stand (Satoo, 1970) most of the leaf surface and most of the leaf biomass were concentrated at a height of 16 meters above the ground (Fig. 18-10). Trunk biomass, in contrast, was concentrated near the ground, whereas branch biomass was concentrated just below the canopy. The latter data suggest that most of the leaves tend to occur on branch tips with the bulk of the branch remaining leafless. A deciduous tree canopy is often in the form of an inverted bowl with the leaves around the outside surface, in contrast to the cone-shaped canopy of coniferous trees.

The total mass of leaves produced may be more important in defining production

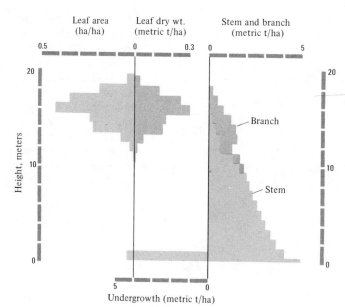

FIGURE 18-10 Vertical profiles of leaf area (*left*), leaf biomass (*middle*), and wood biomasses (*right*) in a Japanese birch (*Betula platyphylla*) stand. (After Satoo, 1970.)

than net productivity per unit leaf area. Satoo found that net production of birch trees was much more closely related to the total leaf area produced than to the net productivity per unit leaf (Fig. 18-11). Net assimilation rate per unit leaf (NAR) varied considerably, as it did in the Long Island oaks, but was not closely related to total yield; trees with a high leaf area, however, invariably had high primary net productivity. So a large tree with a sizable canopy need not have a particularly high photosynthetic rate per unit leaf area to contribute conspicuously to forest

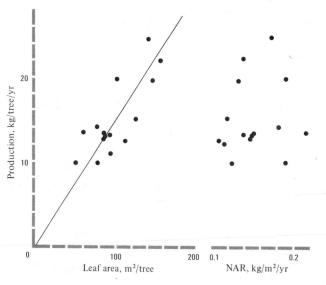

FIGURE 18-11 Relationship of net productivity of birch trees (*Betula maximowicziana*) in Japan in relation to total leaf area per tree (left-hand graph) and P_n per unit leaf area (NAR, right-hand graph). The productivity of a given tree is much more dependent upon total leaf area than on photosynthesis per unit leaf area. (After Satoo, 1970.)

production. On the other hand, trees with a low leaf area may compensate for this, in part, by a high NAR.

One of the most intensive examinations of matter and energy flow through the producer component of an ecosystem has been on an oak-beech-hornbeam forest in the Netherlands (Duvigneaud and Denaeyer-De Smet, 1970). The oldest trees on the site are around 75 years old and up to 23 meters high. Density averaged 1500 tree stems/ha, basal area was 21 m^2/ha, and leaf area index was 6.8. Standing crops and flux of both energy (Fig. 18-12) and principal mineral elements (Fig. 18-13) have been determined. Total biomass of the stand was 156 metric tons/ha, of which almost 80 percent was aboveground. The dead organic matter in the soil, 125 tons/ha, was equal to the aboveground biomass, 121 tons/ha.

Although oaks dominated the stand in terms of biomass, they contributed much less to leaf area index than either hornbeam or beech. Leaves of the latter species are much lighter per unit area than oak leaves, and this was reflected in the very similar values for leaf biomass among the three species.

A quite different impression of the forest arises from comparing the net productivity for the principal species (Fig. 18-12b). In terms of energy flow hornbeam was clearly dominant, contributing over 40 percent of the total tree net productivity. Oak and beech, somewhat less productive than hornbeam, contributed equivalent amounts to the primary energy flux. The understory, that is, shrubs and ground flora, was also about twice as important in energy flux as it was in standing crop, contributing less than 3 percent to the latter but over 6 percent to net productivity. The significant energy storage capacity of a woody community is evident from the fact that less than 40 percent of the annual net productivity was returned to the soil as litter.

The importance of the biotic compartment as a nutrient storehouse in forests is documented further by data on the principal mineral elements (Fig. 18-13a). There was five times more potassium in the plants than in the exchangeable soil pool, and the amount of magnesium in the plants was almost equal to the available soil pool. The soil contained undetectable amounts of exchangeable nitrogen and phosphorus. Oaks were generally less important in the nutrient pools than their biomass would indicate, but they were still the most important tree, followed by hornbeam, with beech being the least important of the three major species in the biotic nutrient pools. The tendency for trees to immobilize nutrients in their biomasses is documented by the patterns of mineral flux through the forest (Fig. 18-13b). About 40 percent of the total nutrients taken up during the growing season were not returned to the habitat but became incorporated into the forest's biotic compartment.

In our examination of some of the properties of the coniferous forest biome we considered evidence that trees suffer high reproductive mortality from herbivorous predators. Similar data are available from the deciduous forest biome. In an English oakwood (Shaw, 1968) success of acorns in producing year old seedlings was less than 0.6 percent. Although acorn viability was over 80 percent during the entire period, few of these acorns survived predators to produce a year old seedling. Subsequent studies with exclosures (Shaw, 1968) indicated that small mammals were the principal acorn harvesters; seedling production inside small mammal exclosures

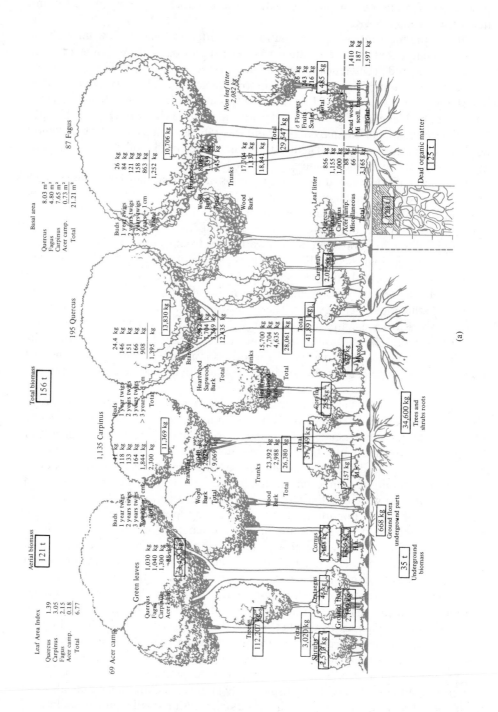

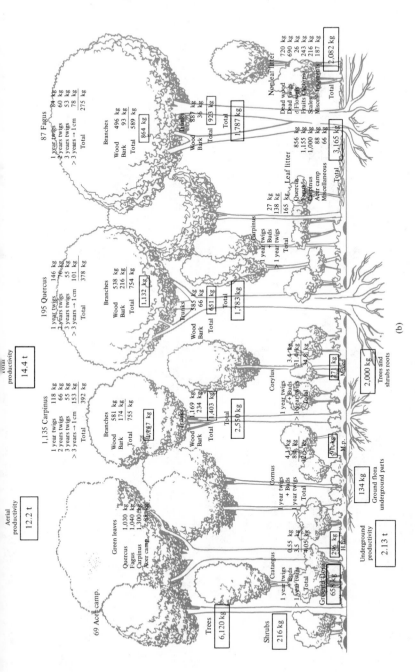

FIGURE 18-12 Energy balances for a European deciduous forest: (a) biomasses (kg/ha) of important trees and understory species; (b) annual energy fluxes (kg/ha/yr) in the same populations. Although extremely complicated, as is any ecosystem, these figures will repay careful study. For instance, among the trees, oaks (*Quercus*) contribute 37 percent of the biomass, 21 percent of the leaf area, and 29 percent of the productivity; hornbeams (*Carpinus*) contribute 34 percent of the biomass, 45 percent of the leaf area, but only 29 percent of the productivity; beeches (*Fagus*) contribute 26 percent of the biomass, 32 percent of the leaf area, and 42 percent of the productivity; and maples (*Acer*) contribute only 3 percent of the biomass and leaf area, and make no significant contribution to productivity. (After Duvigneaud and Denaeyer-DeSmet, 1970.)

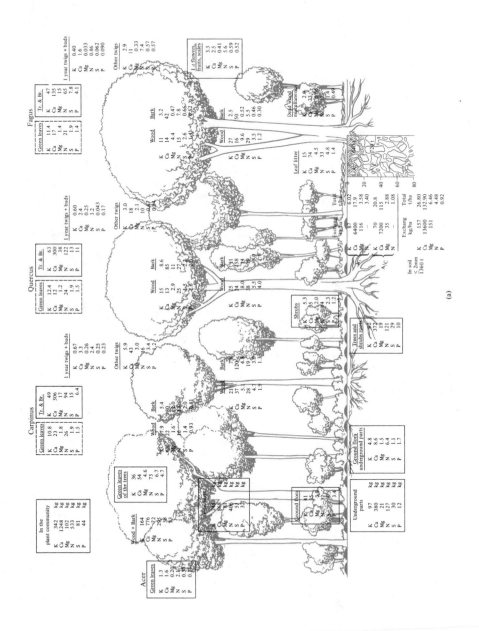

(a)

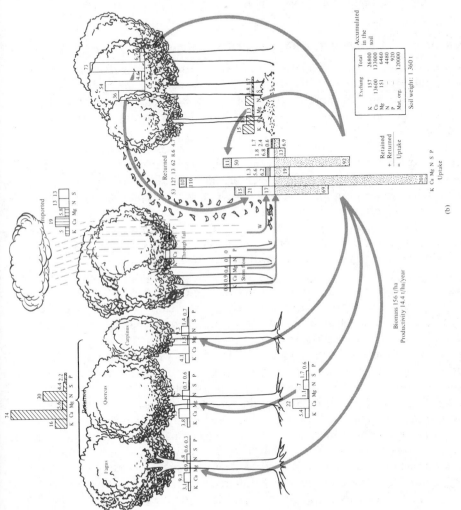

FIGURE 18-13 Balance of principal nutrient elements in the European deciduous forest of Figure 18-12: (a) nutrient pool sizes in kg/ha, and (b) nutrient flux rates in kg/ha/yr. Note that these data can be used, as those of Figure 18-12, to calculate turnover times and other parameters useful in estimating population function in ecosystem dynamics. (After Duvigneaud and Denaeyer-DeSmet, 1970.)

533

was 70 to 80 times that in the unprotected controls. Acorn germination success varied considerably according to position on the soil. Those on the surface had a relatively high germination percentage, but were so vulnerable to predators that mortality was essentially 100 percent. Buried acorns, in contrast, were less vulnerable to predation but germinated poorly. The most favorable place for acorns was above the soil surface but buried under a layer of litter; this protected the acorns somewhat from predation while allowing adequate germination compared to buried seeds.

THE DECIDUOUS FOREST BIOME: ECOLOGICAL INTEGRATION

In our earlier consideration of the desert biome we discussed the triggering of desert locust eruptions by volatile compounds that were produced by the opening buds of desert shrubs. An equally intriguing example of a complex adaptive system that keys insect reproduction to the presence of an important food plant is available from the deciduous forest biome.

For a number of years scientists had attempted to breed polyphemus moths under laboratory conditions without success, even though mating often occurs in caged individuals placed outdoors. Pondering the observation that the moths bred particularly well when placed within the foliage of certain larval food plants (Ludwig and Anderson, 1942), Riddiford and Williams (1967) placed red oak seedlings in a cage with moths and thereby induced mating. With no red oak leaves present 30 females failed completely in subsequent mating tests, whereas success rate was 33 percent with leaves in the cages and 29 percent with red oak near the cages. A variety of other leaves known to be acceptable food—maple, birch, chestnut, horse chestnut, elm, hickory, and beech—were also tested but none induced mating.

Since the moths mated if red oak was merely in the vicinity, they deduced that some volatile compound released by the leaves must induce mating. A series of ingenious experiments followed that were designed to determine the mechanism of the phenomenon. In one series only males or only females were exposed to red oak leaves prior to, but not during, mating. A six-hour exposure of males did not generate mating behavior, but a six-hour exposure of females resulted in 50 percent mating success, which indicated that females were the receptor organisms. Since antennae are known to be important receptor organs, they removed the female antennae 24 hours before exposure; this completely eliminated the mating response to oak leaves.

The series of experiments indicated that mating in polyphemus moths arises out of the following sequence of events: (a) red oak leaves produce a volatile compound; (b) this compound is received by the female antennae; (c) reception induces production of a male sex attractant by the female; (d) males receive the attractant signal with their antennae; and (e) mating proceeds. In another interesting observation Riddiford and Williams found that 24 hours of female pretreatment could be negated by an intervening 30-minute period in the absence of red oak leaves.

The complexity of this evolutionary mechanism almost defies imagination. To ensure that mating will not occur in the absence of suitable larval food sources, polyphemus moths have evolved a mating inducer keyed to leaves of one of the most important of those food plants. This example provides graphic evidence of the complexity of the integrative mechanisms sometimes involved in such seemingly straightforward ecosystem processes as energy flow. The magnitudes of energy flow pathways in ecosystems arise out of a complex system of adaptations such as those of the polyphemus moth and the earlier discussed antiherbivore mechanism of tannins in oak leaves. Clearly a mechanistic understanding of all the factors determining flow magnitudes is far away.

535

THE
TROPICAL
FOREST
BIOME:
GENERAL
FEATURES

THE TROPICAL FOREST BIOME: GENERAL FEATURES

The tropical forest biome occurs as three distinct bands lying between 20°N and S: (a) in South and Central America, (b) in Africa, and (c) in Southeast Asia, the East Indies, and the Malaya Archipelago (Fig. 17-1). The American tropical forest is centered in the Amazon Basin but spreads northward into the Central American lowlands as far as southern Mexico. The largest single area of tropical forest occurs in the Amazon River basin (Richards, 1952). The African tropical forest runs along the Gulf of Guinea and extends westward into the heart of the continent in the drainage basin of the Congo River. The Indo-Malayan tropical forest extends from Ceylon and western India through the lowlands of Southeast Asia and the Malaya Archipelago to New Guinea.

The tropical forest biome occurs in regions of high rainfall, generally exceeding 200 cm annually and always exceeding 150 cm/yr. The heart of the biome is the tropical rainforest, although this grades, along its edges, into a band of deciduous or semideciduous forest that eventually merges into savanna or thorn scrub (Fig. 18-14). The rainforest is composed of tall, evergreen broadleaf trees. There is less seasonality than in temperate climates, rainfall is high, temperatures are high and relatively constant, photoperiods are short and uniform through the year, and humidity is high. Mean annual temperatures are around 25°C, and both diurnal and seasonal variations are slight. Relative humidity is often 80 percent or more and the combination of high temperatures, high humidity, and regular heavy rainfall make the climate oppressive to some visitors from temperate latitudes. In peripheral areas in which a pronounced dry season occurs, the tropical forest develops seasonal leaf fall, with the forest taking on the appearance during the dry season of a temperate deciduous forest during winter. These deciduous border forests are particularly prevalent in the Indo-Malayan region where they are called *monsoon* forests. The distinctly seasonal rainfall generated by the southwest monsoons over the biome's eastern Asian regions is associated with a more seasonal forest than in the biome's center. Teak (*Tectona grandis*) is the best known tree found in these forests, although it is less abundant than pyinkade (*Xylia xylocarpa*), a legume.

It is impossible to list the most common trees in the tropical forest biome since such a large number of plant species occur in the vegetation (Fig. 18-15). In the deciduous forest, for instance, we saw that beech, oaks, and several other genera were represented throughout the biome. This cannot be said for the tropical forest. The number of tree species present in a limited area is huge; for instance, a hectare typically will contain 50 to 70 different tree species (Richards, 1952) compared to 10 to 20 in an equivalent area in a deciduous forest in the northeastern United States.

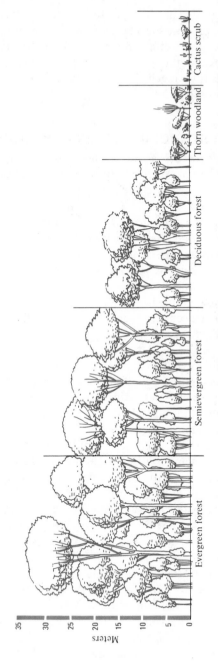

FIGURE 18-14 Gradient of change in tropical ecosystems from evergreen rainforest in the lowlands of Trinidad to the desert biome in neighboring Venezuela. (After Richards, 1952 and Beard, 1944.)

FIGURE 18-15 General view of a tropical rainforest in Central America. (American Museum of Natural History)

537

THE TROPICAL FOREST BIOME: GENERAL FEATURES

Predominance of a single species in tropical forest is largely limited to such special areas as riverbanks or unique soil types. There is often, however, a tendency for a single plant family to be particularly prevalent in the forest. In South America members of the Leguiminosae form a conspicuous part of the total flora. Similarly, African forest tends to have large numbers of members of the family Meliaceae, and members of the Dipterocarpaceae are common in Indo-Malayan forests.

Large herbivores, common in savanna and deciduous forest, are uncommon in the tropical forest biome. The largest are two species of tapir (*Tapiridae*), one in South America and one in Asia, and Africa's okapi (*Ocapia johnstoni*). Most herbivores, of course, are arboreal and among the most conspicuous and abundant are primates. Among these are a large number of different monkey species in African and Asian tropical forest, including India's sacred monkeys, the langurs (*Presbytis* spp.), and an abundant African group, the guenons (*Cercopithecus* spp.) and the apes: the gorilla (*Gorilla gorilla*) and chimpanzee (*Pan troglodytes*) occurring in Africa, and the orangutang (*Pongo pygmaeus*) and gibbons (*Hylobates* spp.) in Asia.

Apes are absent in the American tropical forest, but monkeys, distinct from those in Africa and Asia, are common. Among the most conspicuous are spider monkeys (*Ateles geoffroyi*), howler monkeys (*Alouatta* spp.), marmosets (Callithricidae), capuchins (*Cebus* spp.), oukaris (*Cacajoa* spp.), and squirrel monkeys (*Saimiri sciureus*). Fruit-eating bats are a common constituent of the tropical forest fauna.

There are also some large rodents among South American forest herbivores, particularly the capybara (*Hydrochoreus* spp.), paca (*Cuniculus* spp.), and agouti (*Dasyprocta* spp.). The latter species, of course, are terrestrial, living on the forest floor. Among the arboreal mammals, tree sloths (Bradypodidae) are strikingly adapted to life in the trees. Those animals spend their lifetimes in an inverted position in the canopy and fur of one species often becomes coated with an algal encrustation that blends into the forest foliage.

Large predators are relatively uncommon in the tropical forest, since grazing ungulates are absent. Leopards (*Felis pardus*) range through the edges of the African and Asian forests, and medium-sized predatory mammals include the serval (*F. serval*). In the South American forest, jaguars (*F. onca*), ocelots (*F. pardalis*), margay cats (*F. wiedi*), jaguarundis (*F. eyra*), small-eared dogs (*Atclocynus microtis*), and bush dogs (*Speothos venaticus*) range through the biome. Also common is the omnivorous coati (*Nasua nasua*). The variety of fairly large predators in the American forest probably developed because of the presence there of several large rodents (cabybaras, pacas). Insect-feeding bats are also important predators in tropical forest, suggesting that the canopy supports an abundant insect fauna.

Insects, both herbivorous and carnivorous, are common in the tropical forest, which holds the distinction of containing some of the largest insects known. Tropical rhinocerous beetles are among the heaviest insects known, cockroaches are gigantic in the forest, and walking sticks are longer here than anywhere else. When a tropical forest insect predator captures a prey, he often has a comparatively large energy packet compared to the prey of temperate insectivores. Ants and mosquitoes also are common in tropical forests.

Birds feed at a variety of trophic levels and among the more conspicuous in American tropical forests are parrots (Psittacidae); these, like monkeys, are characteristic for the biome. Toucans and tinamous (Tinamidae) are also important herbivores. Common insectivores include antbirds (Formicariidae) and puffbirds (Bucconidae). Also conspicuous among the bird fauna are the omnivorous trogons (Trogonidae).

Many bird species in tropical forest are conspicuous for their bright and unusually colored plumage. Among the most unusual are New Guinea's birds-of-paradise, of which there are several species that occur at distinct altitudes in the mountainous terrain of the island. Under severe hunting pressure, first from the aboriginal population which used the plumage for ceremonial purposes and then during the twentieth century when the plumes were used for hat manufacture in Europe, many of these species have been exterminated, or nearly so. The birds-of-paradise exemplify one of the most conspicuous properties of the tropical forest biome that distinguishes it from other biomes—high species diversity in local areas. The great variety of similar, but distinct, species in small areas extends from the producers through the consumers. A typical hectare of tropical forest will contain more species of plants, insects, birds, and other organisms than a typical hectare in any other biome.

Finally, one general feature of the tropical forest biome that was mentioned earlier but should be reemphasized here is the low soil fertility. As Richards (1952) has pointed out, "A soil bearing magnificent rainforest may prove to be far from

539

THE
TROPICAL
FOREST
BIOME:
STRUCTURE
AND
FUNCTION

fertile when the land is cleared and cultivated. In a mature soil the capital of plant nutrients is mainly locked up in the living vegetation and the humus layer, between which a very nearly closed cycle is set up. The resources of the parent rock are only necessary in order to make good losses due to drainage."

THE TROPICAL FOREST BIOME: STRUCTURE AND FUNCTION

Although the tropical forest is evergreen, with leaf fall occurring throughout the season, it surpasses all other forests in annual litter production (Fig. 18-16). Changes in growing season, moisture supply, and temperatures generate different litter productions in the tropical forest biome, the deciduous forest biome, and the coniferous forest biome.

In contrast to the uniform timing of leaf production and leaf fall selected by the temperate climates, tropical forest trees are conspicuously polymorphic in leaf production (Richards, 1952). As a general rule the old leaves fall simultaneously with production of a new leaf crop, so the trees are bare for only a few days. In some species leaves may be produced and shed constantly and thus the tree is never bare of leaves; in others, particularly toward the seasonal margins of the biome, the species may be bare for a period of several weeks, although the bare periods are less synchronous among species than they are in the temperate deciduous forest. The proportion of trees that undergo periodic bare periods is higher among upper-canopy species than among lower-canopy species. Beard (1942) reported percentages of upper- and lower-canopy species with bare periods as 24 percent and 0 percent in the heart of the Trinidad rainforest, 33 percent and 10 percent in an intermediate forest, and 50 percent and 10 percent in a marginal forest. The percentage of upper-

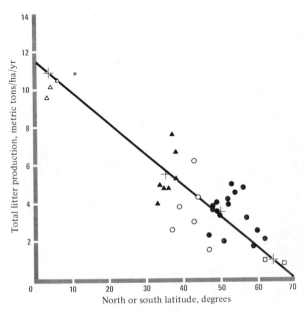

FIGURE 18-16 Decline of annual litter production of forests with latitude. Open triangles are primarily tropical rainforests, solid triangles are largely deciduous forest, circles are largely coniferous forest in North America (open circles) and Europe (closed circles), and squares are arctic-alpine coniferous forest. The line is fitted to means of the climatic zones with the means indicated by crosses. (After Bray and Gorham, 1964; data indicated by asterisk from H. T. Odum, 1970.)

and lower-canopy individuals that shed their leaves for a period was 6 percent and 0 percent, 16.5 percent and 10 percent, and 66 percent and 25 percent for the three types of tropical forest.

In addition to the constant and nonsynchronous pattern of leaf fall and production, reproduction among tropical forest trees also is comparatively evenly spaced throughout the year (Baker and Baker, 1936). Although certain species may flower and produce fruits for only a month or two out of the year, these periodically flowering species are usually so spaced that their flowering and fruiting times are nonoverlapping. Other species, however, may flower and fruit almost continuously.

There is a wide variety of relationship between flowering and vegetative activity. Some trees have fruit that ripens just as the leaves fall, others always flower when bare of leaves, and flowering in some species occurs only on bare branches so that part of the tree will be green and vegetative while another part is bare and reproductive (Richards, 1952).

We saw that in the deciduous forest there was a distinct region of major leaf mass and just below this a distinct region of maximum branch mass. Biomass profiles from Puerto Rico (H. T. Odum, 1970) indicate that spatial clumping may be absent in some tropical forests (Fig. 18-17). Although there was a tendency for clumps of

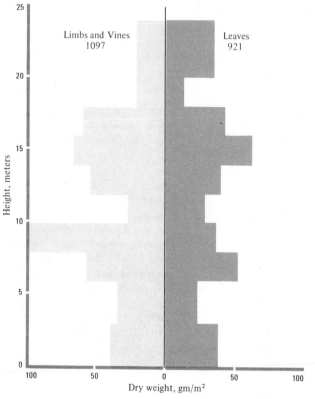

FIGURE 18-17 Vertical profile of leaf and woody biomasses in a Puerto Rican rainforest. Compare with Figure 18-10. (After H. T. Odum, 1970.)

branches to appear at 15 meters and 9 meters, leaves were quite uniformly distributed throughout the ecosystem's aerial depth. In a Costa Rican tropical forest (Stephens and Waggoner, 1970) a more stratified canopy has been reported with a lower layer extending from the ground to 5 meters, and an upper layer from 25 to 35 meters above the ground.

541

THE
TROPICAL
FOREST
BIOME:
STRUCTURE
AND
FUNCTION

One of the most intensively studied tropical forest ecosystems is at the Puerto Rico Nuclear Center (H. T. Odum, 1970). The energy budget for the forest in kilocalories per square meter per day was 3830 from incoming sunlight with a gross productivity of 131.0 partitioned between respiration of 116 and a net productivity of 15.2. Of the respiration, 60 percent was root respiration, 33 percent was leaf respiration, and the remainder was partitioned among trunks, branches, and fruits. The net productivity measured from biomass increments was 16.31 kcal/m²/day, which is not far from the value above that was measured by gas analysis. Biomass increments were partitioned among leaf fall (5.8 kcal/m²/day), branch fall (6.68 kcal/m²/day), root death (2.3 kcal/m²/day), wood accumulation (0.72 kcal/m²/day), fruit fall (0.7 kcal/m²/day), and leaf leaching losses by rain drip (0.03 kcal/m²/day). Since the only cumulative biomass addition to the forest was through wood growth, 3.8 percent of net productivity was retained by the forest, the remainder passing through decomposer and grazer food chains.

Although tropical forests are quite productive on an annual basis, and much of this production is returned to the soil as litter, there is considerable evidence to indicate that short-term productivity may be somewhat higher in temperature ecosystems. Lemon et al. (1970) compared carbon dioxide assimilation of Costa Rican rainforest with temperate ecosystem producers and found that the latter had net productivity rates up to ten times those of the Costa Rican forest. Okali (1971), in contrast, compared growth rates of four tropical tree seedlings with a temperate herbaceous plant, sunflower, and found that one of the species had a growth rate fully as high as the sunflower. The data of Lemon et al. on annual net production yielded 3.5 metric tons/ha, which is clearly far below the Dutch hornbeam-oak-beech forest examined earlier with an annual production of 14.4 metric tons/ha (Duvigneaud and Denaeyer-DeSmet, 1970). However, data on the Puerto Rican forest (H. T. Odum, 1970) convert to 55.5 metric tons/ha/year, which is clearly far above either the Costa Rican forest or the temperate deciduous forest.

These data suggest that either (a) there is a tremendous range of productivity potential within the tropical forest biome, with some ecosystems extremely productive and others showing very meager production or (b) good repeatable data on tropical forest productivity are not yet available. The former explanation seems more likely. We saw earlier when examining the coniferous forest biome that productivity of a major species, black spruce, ranged from very high to very low toward the southern limit of its range, whereas at the northern limit productivity was uniformly low. Although the extrapolation is a large one, it seems not unreasonable to assume that a similar phenomenon may operate on a broad basis. Tropical sites may sustain extremely high productivity when the whole environmental complex is nearly optimized. But locally limiting conditions such as interruption of nutrient cycling by an extremely high leaching rate, locally shallow soil, or a variety of other conditions could provide an infertile site, thereby limiting productivity in spite of generally favorable climatic conditions.

We have tended to emphasize the importance of climate in regulating the distribution of biomes, and the correlation between climatic patterns and biome distribution is apparent. It should always be borne in mind, however, that local conditions of soil and microclimate can modify regional climatic impact to a considerable extent. We have seen that woodlands are common within the prairie biome whenever rivers supply sufficient moisture for riverine species or where geological relief produces natural firebreaks, thus deterring the ability of autumnal prairie fires to kill tree seedlings. Similarly, there may be considerable variation in productivity of different ecosystems within a biome, depending on local variations in important limiting factors. We may expect the significance of these locally limiting factors to become more important as overall climatic patterns become less and less limiting to growth. It seems likely, therefore, that the large variations reported in net primary productivity of tropical forests represent real differences rather than merely methodological differences.

Little is known about the structure and function of tropical forest animal communities. It has long been known that there tend to be two distinct animal communities, one of the canopy and one of the forest floor (Allee, 1926). In American tropical forest, for instance, there is a monkey and fruit-eating bird community in the canopy and a rodent-bird-coati-predatory cat community on the forest floor. However, quantitative data of the type we have examined for African ungulates, or for wolves and their ungulate prey, are almost nonexistent for tropical forest consumers. Little can be said about energy flow, niche organization, and functional role in the ecosystem.

One of the puzzling problems in plant-animal interactions in the tropical rainforest arises out of the great distances, a kilometer or more, between individuals of the same species. The sporadic flowering of individuals raises the question of how such widely dispersed plants are fertilized. Baker (1960) has argued that many tropical plants may be inbreeding, particularly species in the mature forest. Such a breeding system would allow reproduction in the absence of another flowering individual in the vicinity. In addition inbreeding is a key mechanism that can give rise to the widespread speciation prevalent in the tropical flora as newly differentiated individuals reproduce by self-pollination.

Many tropical tree species, however, are known to be outcrossing, and Janzen (1971) has presented evidence that some bees are capable of flying great distances in the tropical forest. Twelve marked bees released as pairs 1, 2, 3, 4, 5, and 6 km from their nests at 11:15 A.M. had all returned by 3 P.M. In subsequent experiments marked bees were able to return from as far as 30 km from their nests. A bee released 30 km from its nest returned 65 minutes later with a full pollen load. Similarly, another bee released 14 km from its nest returned in 47 minutes with a full pollen load. Janzen concluded that "The existence of a complex of large bees, hummingbirds, and sphinx moths may provide reliable outcrossing over distances far greater than would be expected if one were only to consider the pollinators of temperate zones, such as wind and small bees."

One of the more interesting plant-animal interactions in tropical sites is that between a successional genus of tree (*Cecropia*) and ants (*Azteca*) (Janzen, 1969a). The ants live in large hollow internodes in the tree and tend aphids from which they derive "honeydew," a sugary secretion of aphids. The principal food sources of the ants are tiny enlargements growing from the trees' petiole bases; these are harvested as a larval food source. A third *Cecropia* trait which is beneficial to the ants is the presence of thin internodal wall regions that allow them to penetrate into the stem interior. We might well ask what selective benefit the ants confer on occupied trees that would lead to natural selection on the tree for the complex of traits that encourage ant occupancy.

Based on Barnwell's (1967) suggestion that the ants may reduce competition between *Cecropia* and climbing vines by chewing the latter, Janzen performed a series of experiments in which vines were twined over ant-occupied *Cecropia*. These vines are common occupants of successional tropical forest sites in which they may overgrow *Cecropia* (Fig. 18-18) that have not been colonized by ants. In Janzen's experiment, one week after the vines were introduced to ant-containing *Cecropia*, 8 percent had blow off, 75 percent had died from ant chewing, 8 percent showed evidence of being chewed upon, and only 10 percent showed no evidence of ant damage. The tendency for ants to chew heavily on vines growing over their host plant provides that plant with an efficient competitive mechanism against a type of plant competition that is extremely difficult to "defend" against.

FIGURE 18-18 On the left is a *Cecropia* seedling lacking an ant colony and, therefore, vine-covered. On the right is an ant-occupied and vine-free seedling. (D. Janzen)

Although we have already referred to the greater diversity of tropical communities, it seems appropriate when terminating our discussion of biomes to return once again to the latitudinal pattern of diversity since it reaches its culmination in the tropical forest biome. There are several manifestations of this diversity. MacArthur (1969) presented a map of the number of breeding bird species per geographic quadrant from the arctic tundra to the tropical forest and demonstrated the association between biome-type and bird species variety (Fig. 18-19). Tundra bird communities typically contain less than 40 bird species. There is a fairly abrupt transition between the tundra and coniferous forest biomes, with the latter having about twice as many species per unit area. Similar species densities occur in the deciduous forest and prairie biomes, but diversity increases substantially to between 150 and 250 species per geographic quadrant in the desert and savanna biomes. The most abrupt discontinuity, however, is clearly seen in the tropical rainforest of lower Middle or Central America, where species density increases to 500 or 600 species per quadrant. Although there are general latitudinal gradients evident within biomes, as, for instance, the gradual increase in species number down the mountain extensions of the coniferous forest biome in the western United States, there are also discontinuities associated with transitions between biomes.

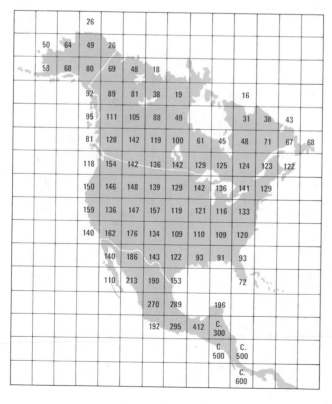

FIGURE 18-19 The numbers of land bird species breeding in geographic quadrants of equal size in different parts of North and Central America showing the increase in diversity from north to south. (After MacArthur, 1969 and MacArthur and Wilson, 1967.)

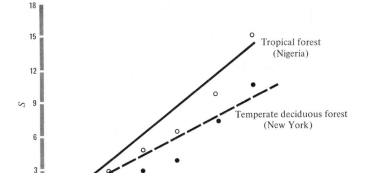

FIGURE 18-20 Relationship between number of individuals sampled and number of species in sample from tropical rainforest and temperate deciduous forest. The rainforest data are from Richards' (1952) transect of a Nigerian site and the temperate deciduous forest are unpublished data of McNaughton.

The bird data represent a broad-scale change in the variety of species occurring in different biomes. In our earlier consideration of succession we observed that the number of species in a sample often is a straight line when plotted against the log of number of individuals sampled. This suggests the possibility that the greater species density of tropical sites may arise out of greater numbers of individuals per unit area. As a test of this, we can plot the species-individual curve for tropical forest and temperate deciduous forest trees (Fig. 18-20). For a tropical forest sample,

$$S = -1.60 + 8.03 \log N \qquad (18\text{-}1)$$

whereas for a temperate deciduous forest sample,

$$S = -0.37 + 4.86 \log N \qquad (18\text{-}2)$$

The slopes of these two lines are significantly different ($t = 3.23$ for $P < 0.02$), indicating that species are added much more rapidly to a tropical forest sample as individuals are added. This indicates that the greater diversity per unit area in the tropical forest biome arises out of a greater variety of species per number of individuals, rather than merely out of greater individual density. We shall deal with the meaning of such a pattern, and mechanisms which might generate it, in the following chapter.

BIOMES: A BRIEF CONCLUSION

As the preceeding chapters suggest, the diversity of the world's ecosystem-types is considerable. Different types of organisms occur in different habitat-types. As a consequence, energy and materials flow patterns vary in different ecosystems. As

TABLE 1 Summary comparison of the areas and productivities of the earth's major ecosystem-types.

Biome-type	Areal extent 10⁶km²	Average caloric value kcal/gm	Biome Gross production 10¹²kg/yr	Biome Gross production 10¹⁶kcal/yr	Biome Net production 10¹²kg/yr	Biome Net production 10¹⁶kcal/yr	Individual ecosystems Gross production gm/m²/yr	Individual ecosystems Gross production kcal/m²/yr	Individual ecosystems Net production gm/m²/yr	Individual ecosystems Net production kcal/m²/yr
Tropical forest	20	4.2	133	55.9	40.0	16.8	6,700	28,140	2,000	8,400
Temperate deciduous forest	6	4.6	26	12.0	7.8	3.6	4,300	19,780	1,300	5,980
Temperate coniferous forest	10	4.7	27	12.7	8.0	3.8	2,700	12,690	800	3,760
Savanna	15	4.0	18	7.2	10.5	4.2	1,200	4,800	700	2,800
Prairie	25	4.0	13	5.2	7.5	3.0	500	2,000	300	1,200
Desert	25	4.0	3	1.2	1.8	0.7	120	480	70	280
Cultivated land	15	4.3	16	6.9	9.8	4.2	1,100	4,730	650	2,795
Tundra	10	4.7	3	1.4	2.0	0.9	300	1,410	200	940
Permanent snow and ice	15	0.0	0	0.0	0.0	0.0	0	0	0	0
Lentic and lotic*	7	4.0	2	8.0	1.4	0.6	300	1,200	200	800
Total terrestrial and freshwater	148		241	110.5	88.8	37.8				
Pelagic	332	5.0	69	34.5	41.5	20.7	200	1,000	125	625
Other marine biomes	27	5.0	16	8.0	9.5	4.7	600	3,000	350	1,750
Estuaries	2	5.0	7	3.5	4.0	2.0	3,300	16,500	2,000	10,000
Total marine	361		92	46.0	55.0	27.4				
Biosphere	509		333	156.5	143.8	65.2				

Golley, 1972

a conclusion to our discussion of biomes, a consideration of their areal extent and net primary productivity may provide insight into their relative importance in the earth's ecological energy budget. Golley (1971) assembled data comparing different biomes. In terms of its areal extent over the earth's surface, the pelagic biome is by far the most extensive (Table 1), occupying over 65 percent of the biosphere. The productivity of this biome per unit area is so meager, however, that it contributes less than 30 percent of the annual energy flux, a figure comparable to tropical forest. The most productive ecosystems on a per unit area basis are those in the three forest biomes. Intermediate efficiences occur in prairie and savanna.

Golley also tabulated data from a variety of sources estimating the percentage of annual net primary production which moved upward in the food web into herbivores (Table 2). These data suggest that energy flow per unit incorporation is minimal in forests and maximal in grasslands. The ability of trees to incorporate much of their energy in the form of support structure represents a major food source not yet effectively penetrated by herbivores. On the other hand, of course, it is likely that penetration of this adaptive mode would result in destruction of the earth's forests. At any rate the transfer of energy upward through the food chains is proportionately greater in herbaceous ecosystems than in woody ecosystems. It should be apparent from the variation in estimates of Table 2, as well as from our compromises in summarizing the variability over huge geographic areas into a single abstraction, that the world's ecosystems are almost infinitely variable, both from place to place and from time to time. As we have seen, however, this variability may be organized through a consideration of primary producer structure, feeding strategies, and spatial repetition of certain types of organisms. The following chapter concentrates on the latter facet of ecology.

TABLE 2 Summary comparison of the percentage of net primary production in different types of ecosystems which is consumed by herbivores.

Ecosystem	Organic input consumed (percent)
Tropical forest	8.5
Tropical forest	7
Temperate forest	1.5–2.5
Temperate forest	1.5
Temperate forest	3.4–9.2
Temperate forest	40
Meadow	25–30
Tanganyika grassland	28
Indian grassland	49
Uganda grassland	60
Cultivated potatoes	12–20
Spring	38
Spring	24
Ocean, Georges Bank	5

Golley, 1972

1. (*Continued from Chapter 17*)
    (e) Coniferous forests occur at high latitudes and altitudes where rainfall is high relative to evaporative stress or on soils, such as sand, with a high moisture delivering capacity.
    (f) Deciduous forests occur in mild, moist climates with moderately long frost free-periods and predictable rainfall.
    (g) Tropical forests occur in a wide band along the equator and range from evergreen rainforest where rainfall is abundant to seasonally deciduous forests where there is a pronounced dry season.
2. Birds, mammals, and insects are important consumers at all trophic levels in the forest biomes. Seed eating is a particularly common herbivore activity which often has a devastating effect on tree seed crops.
3. Within a biome, productivity of its characteristic populations may be uniformly low near the biome's margins but may range from high to low, according to site quality, within the biome.
4. Expansion of browsing ungulate populations, often kept low by predation in undisturbed forests, can have substantial deleterious effects on ecosystem function.
5. Seasonality of the major producers in deciduous forests results in striking micro-habitat variation in the forest. Understory species, both consumers and producers, have evolved adaptive mechanisms in response to the special microhabitats.
6. Much of the nutrient and energy content of forest ecosystems is immobilized in the wood of major producers.
7. Chemical substances present in producers may act alternately as both repellants of some herbivores and attractants of others. These chemicals are important integrators of the ecosystem's populations.
8. Although the tropical rainforest is evergreen, there is a constant turnover of plant leaves and other nonpersistant biomass, with phenological cycles of various species out of phase with one another.
9. Species diversity is greater in the tropical rainforest than in any other biome.

the nature of ecological communities 19

The community concept is among the oldest in ecology. Our consideration of the earth's biome-types revealed that certain species combinations are repeated over broad geographic areas. In this chapter we shall examine this phenomenon more critically to determine (a) the degree to which species associations repeat and (b) the mechanisms responsible for species association.

In 1844 Edward Forbes, a professor of botany at London's Kings College, studied the distribution of marine animals in the Aegean Sea. He concluded that "There are eight well marked regions of depth distinguished from each other by the associations of species they include" (Kormondy, 1965). Later, a German biologist, Karl Mobius (1877), observed that "An oyster bed is not inhabited by oysters alone, but also by other animals. Every oyster bed is a community of living beings, where the sum of species and individuals mutually limited and selected have continued in possession of a certain definite territory. Space and food are necessary as the first requisites of every social community." Following these lines of reasoning, a Danish botanist, Eugene Warming (1909), pointed out that "Certain species group themselves into natural associations, that is to say, into communities which we meet with more or less frequently and which exhibit the same combination of growth forms. Species that form a community must either make approximately the same demands on the environment, or one species present must be dependent for its existence upon another species."

Mobius coined the word *biocoenosis* (from Greek roots meaning life with something in common) to refer to all of the organisms in an ecosystem—which we have referred to in previous chapters as the ecosystem's biotic component. We think it is more useful to restrict the concept of community to groups of populations with similar ecological functions. As Daubenmire (1968) observed, "No biologist could possibly be competent enough in the taxonomy, life histories, and ecology of all the plants and animals of an average community to make a thorough study of them as a unit." In fact in recent ecology books (Whittaker, 1970b; Boughey, 1971a; Odum, 1971) the concept is rarely applied beyond a group of organisms with similar ecological properties. Thus these books separate plants, fish, birds, and so on, in most of their community discussions. As defined in Chapter 1 and applied throughout the book, we have utilized the operational definition of communities as groups of species with similar life requirements, referring to all the communities in an ecosystem as its biotic component. It seems more realistic to treat plants, herbivores, and bacteria as members of different communities since their ecological requirements are not alike but differ markedly. That is, a *community* is one or more populations with similar resource demands co-occurring in time and space. A typical ecosystem, then, consists of several communities related to one another through the trophic structure. Populations in the same community are related to one another by the fact that they may compete for the same resources. Although communities usually consist of several populations, certain habitats may support a community consisting of a single population. For instance, marshes or herbicide-treated agricultural fields are often monospecific communities.

The problem of how populations in a community are functionally related is one of the most difficult in ecology. Because considering this complex subject presupposes knowledge of the organizing principles of ecology and of the character of specific communities, we have delayed presenting it until we had discussed in detail both the concepts of ecosystem organization and the properties of biomes. In this chapter we apply these observations to communities in an attempt to understand (a) how distinct are species associations in different areas and (b) what defines the variety and relative abundances of species in a community.

In almost every chapter we have referred to the idea of equilibrium. This chapter is particularly concerned with the equilibria between co-occurring populations with similar resource demands. Mobius observed in studying the oyster bank that "If favorable temperature makes the species more fruitful, since there is neither room nor food enough for the maturing of all germs, the sum of individuals in the community soon returns to its former mean. The surplus which nature has produced by the augmentation of one of the forces is thus destroyed by a combination of all forces, and the equilibrium is soon restored."

Perhaps the most significant distinction between "natural" ecosystems and those shaped by man is that the former generate both spatial and temporal equilibria from the energy of the sun. Man, in contrast, depends on ecosystems whose persistence is dependent on constant energy inputs from fossil fuels and other forms of stored energy. Our study of succession indicated that species diversity is often associated with ecosystem persistence, and the chapters on populations (Part 3) suggested several mechanisms that may be involved in producing species diversity. Although we are far from a complete understanding of ecosystem equilibria, such understanding is essential if man is to reduce his dependence on nonpersistent ecosystems. Community study may contribute to this knowledge by defining the nature of persistence and diversity among co-occurring populations.

SPATIAL SCALE AND SPECIES ASSOCIATION

As Mobius observed, organisms occupy space. Space either defines resources in the sense that a certain space contains an array of resources in certain concentrations or it may itself be a resource, as we saw when considering competition among barnacles. Therefore at the simplest level we can argue that individuals co-occur because they occupy different spaces. However, as Erickson (1945) pointed out, we can consider the space within which a species occurs at several different levels (Fig. 19-1). Leather flower, for instance, is a plant that occurs within a geographic space around the confluence of Missouri, Kansas, and Nebraska, although populations are not uniformly distributed throughout this range. Erickson recognized a hierarchy of distributional subdivisions consisting of (a) species range, (b) concentration regions, (c) clusters of populations within the regions, (d) individual populations within the clusters, and, finally, (e) aggregates of individuals within the populations. At the final level of resolution we find different individuals occupying different spaces and therefore exploiting different resources; but the difficulty still remains of defining the resource structure allowing an individual of a certain species to occupy this space.

We have seen that competition is an ecological mechanism by which the most efficient genotypes are sorted out of a species combination. In the limited habitat provided by Gause (1934) for *Paramecium aurelia* and *P. caudatum*, the former always won. Similarly, in the flour beetles competing at different combinations of humidity and temperature, one species always won, albeit not always the same species. So for leather flower we may pose two contrasting questions: (a) why it does not solidly occupy its entire range to the exclusion of all other species and (b) why it occurs at all. Since this plant requires the same basic resources as all other plants within its range, we may well ask why it does not exclude them by virtue of being a better competitor or, alternatively, why they do not exclude it if it is a poorer competitor. One of the fallacies in such an extreme statement of the problem of co-occurrence resides in the premise that leather flower "requires the same basic resources as all other plants." Clearly this is true in that all plants require the same basic nutrient elements and energy source; however, to say that all plants have the same fundamental niche is a considerable oversimplification. In fact, we know this is not so. Leather flower, we may assume, co-occurs with individuals of other species because the spatial organization of factors defining the niches generates alternating

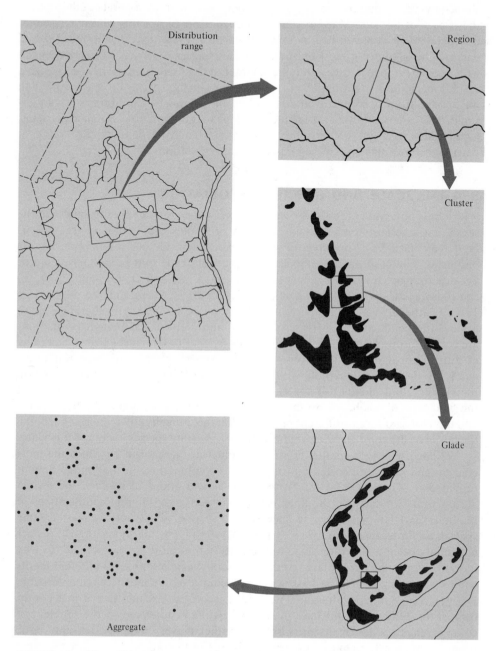

FIGURE 19-1 Maps illustrating the organization of the distribution range of leather flower (*Clematis fremontii*) into a hierarchy of subdivisions from a marcoscale showing the species range to a microscale mapping separate individuals in a forest glade. (After Erickson, 1945.)

spaces, some in which leather flower may effectively exclude competitors and some in which it is excluded by them.

Agriculture has relied on two principal ecological strategies throughout its existence: (a) expansion of the realized niches of cultivated species through elimination of competitors and (b) expansion of the fundamental niches through selective breeding. Our objective in a study of the community is understanding how realized niches are organized in nature, as determined both by the competitive efficiencies of species and the spatial and the temporal distribution of resources.

Kershaw (1964) examined the relationship among abundances of two grass species, fescue and bentgrass, that are common in grasslands in the British Isles. A transect was laid out along a geological transition between soils derived from chalk and flint. Bentgrass was more abundant on chalk soils and fescue on flint soils, so abundances of the two species were negatively related (Fig. 19-2). Within the grasslands these two species represent a balanced ecological organization of the community in response to soil. There is a trade-off between fescue and bentgrass according to the balance of environmental properties. At the chalk extreme of the transect bentgrass is favored, at the opposite end fescue is favored, and in between the abundances are balanced according to the balance between chalk and flint.

Given the patterns of alternating abundances accompanying environmental gradients, the question is how these communities may be delimited. At one end of the transect studied by Kershaw there was one type of grassland and at the other end there was a somewhat different grassland. In practice, communities tend to be defined pragmatically by the relative abundances of the most important species. We saw this when discussing biomes in which, for instance, plant communities in the prairie biome of the central United States were characterized by a transition from a big bluestem, switchgrass combination in the moist eastern extreme to a blue grama, buffalo grass combination in the arid western extreme.

The geographic distribution of the principal species of a certain community is usually much wider than the particular combination of species abundances that

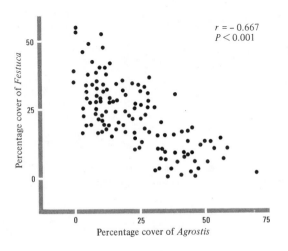

FIGURE 19-2 Negative relationship between abundances of two grass species in grasslands in Britain. (After Kershaw, 1964.)

$r = -0.667$
$P < 0.001$

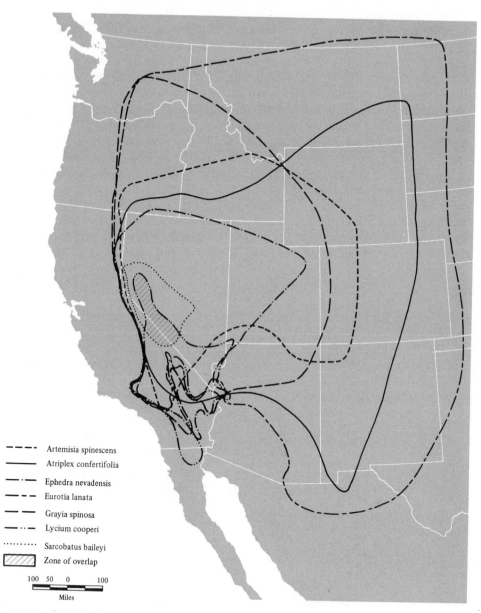

– – –	Artemisia spinescens
——	Atriplex confertifolia
–·–·–	Ephedra nevadensis
– – –	Eurotia lanata
— —	Grayia spinosa
–··–··–	Lycium cooperi
··········	Sarcobatus baileyi
▨	Zone of overlap

100 50 0   100
Miles

FIGURE 19-3 Distribution ranges of several plant species important in the shadscale desert community. The species ranges extend far beyond the areas where they define a particular type of ecosystem. (After Billings, 1949.)

define that community. Billings (1949) mapped the distributions of seven desert biome species that form a transitional community type between the hot and cold deserts (Fig. 19-3). These species, which are important in Nevada's shadscale desert, prove that the area in which a species is most abundant may not be related in a straightforward manner to its distribution range. The overlap region in which they define a particular type of community occurs on the western margin of all their ranges, and two of the species are distributed as far east as Nebraska and the Dakotas, where they are insignificant components of the vegetation.

The question of scale and of what constitutes a common spatial arrangement of populations admits of no objective resolution. The distribution of no single important species defines the shadscale zone, for example, and the continuous change in abundances of fescue and bentgrass allows no definite demarcation between the different grasslands. One technique that was widely used for several years to delimit communities objectively was the species-area curve. If the cumulative numbers of species are recorded in an expanding sampling area, a familiar form of curve results (Fig. 19-4). It has often been argued that the area where this curve begins to level off represents the minimal area that would include the community's representative species combination. We have seen enough curves of this sort, however, to recognize that this is merely a curve of the form

$$S = a + b \log A$$

where S is number of species in the sample and A is the sampling area (Gleason, 1925). Since density tends toward a constant for most communities, the species-area curve usually is a manifestation of the species-individual curve we examined repeatedly in the preceding chapters. The question of what constitutes a common spatial arrangement therefore is a somewhat arbitrary one. If we were interested in sampling an objectively defined ecological unit we would not, for instance, sample across the boundary between a forest and an old field. This boundary, however, or *ecotone* as such contact zones are called, is just as much a common spatial arrangement,

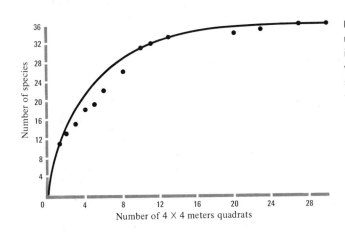

FIGURE 19-4 Increase in the number of species recorded in a community sample with an increase in the area sampled. Data are from a forest understory community in North Carolina. (After Oosting, 1958.)

or co-occurrence in time and space, as the middle of the forest or the middle of the field. Clearly the placement and nature of our sample will depend on the sort of hypothesis we are testing (Mills, 1969).

CONTINUITY AND DISCONTINUITY IN COMMUNITIES

One hypothesis about community organization is that the landscape is organized into a series of more or less homogeneous communities separated by fairly abrupt ecotones. The question of whether such abrupt areas can be delimited objectively, however, has been a point of contention for decades (Gleason, 1926; Clements, 1936; Whittaker, 1953; McIntosh, 1958; Daubenmire, 1966). Two basic approaches to testing for discontinuities have involved: (a) examining the relationships among the relative abundances of species populations in a large number of stands and (b) examining changes in the species abundances along environmental gradients. The former approach poses the question: Are there groups of species whose abundances are associated, that is, where one is important are they all important, and where one is rare are they all rare? The other approach poses the question: Are species distributed along environmental gradients in a coordinated fashion?

Bray and Curtis (1957) utilized the data of Brown and Curtis (1952) on upland deciduous forests in Wisconsin to test for distinct species associations by the first approach. Density and trunk basal area of trees and frequency of herbs and shrubs in 59 forest stands were employed. A coefficient of community similarity (Czekanowski, 1913; Gleason, 1920) was used to compare the stands. The coefficient is (Table 19-1)

$$C = \frac{\Sigma(2m_i)}{\Sigma(a_i + b_i)} \tag{19-1}$$

where a_i is the abundance of species i in community a, b_i is the abundance of species i in community b, and m_i is the minimum value for the species (that is, either a or b, whichever is smaller) and values are summed for all species in the communities. If the stand's species compositions and species abundances are exactly the same,

TABLE 19-1 Example of the calculation of a community coefficient for two hypothetical communities

| | Species abundance (gm/m²) | |
Species	Community a	Community b
1	50	20
2	100	5
3	30	30
4	3	0
5	0	80

$$C = \frac{2\,(20 + 5 + 30 + 0 + 0)}{(50 + 20 + 100 + 5 + 30 + 30 + 3 + 80)} = 110/318 = 0.346$$

then $a = b = m$ and $C = 1$. For stands with identical species composition, C will decrease as a and b become less similar; and, of course, two stands with no species in common would have a C of zero.

To compare all 59 stands Bray and Curtis had to calculate a total of 1711 Cs. The two stands that generated the C closest to zero were chosen as the end points of a graphical axis and the remaining 57 stands were separated from them by calculating $1/C$; that is, one of the two most dissimilar stands was taken as the end point of a graphical axis, and the array of 58 $1/C$ was calculated to place each remaining stand on the axis. A stand that had a coefficient of community of 0.2 with the reference stand would have a graphical distance of $1/0.2 = 5$. When all of the stands had been placed on the first axis, the two stands from the midrange with the most different Cs were selected and a second axis was calculated. This placed each of the stands in a two-dimensional array. Finally, a third axis was established and each stand was placed within this three-dimensional matrix.

The Bray and Curtis approach represents a three-dimensional ordination in which the stands are ordered by virtue of their similarity of species composition and relative abundances. Species could now be placed within the ordination and patterns of coordinated distribution examined (Fig. 19-5). Although species exhibited definite centers of distribution at which they were relatively abundant, Bray and Curtis concluded that there was a continuous change in species abundances throughout the ordination and that no distinct groups of species could be delimited.

The three axes were related to several environmental factors. Soil maturity and fertility were associated with one axis, drainage characteristics with another, and the degree of selective logging to which the stand had been subjected was related to the third axis. Bray and Curtis concluded that, "Each species has a separate area of location, and within this area its distribution is interspersed to varying degrees with other species distributions so that there is a continuous change in stand composition from any part of the ordination to any other part." This conclusion, that

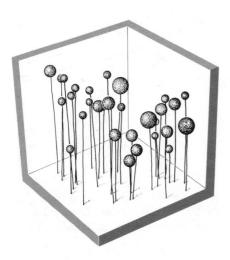

FIGURE 19-5 Distribution of red oak (*Quercus borealis*) in an upland forest continuum of southern Wisconsin. The size of the balls is proportional to the basal area of the species in the stands where it occurred. (After Curtis, 1959.)

communities are continuously changing species collections with no definite gaps between them, has come to be known as the *continuum hypothesis.*

There were two important difficulties with Bray and Curtis' approach to organizing community information: (a) there was a mathematical bias in the placement of stands on the axes and (b) the ordination was not related in a straightforward way to the distribution of environmental factors. Environmental determinants could only be deduced secondarily by correlation of axes with environment after the fact. The mathematical bias is apparent when we realize that 1/0.1 and 1/0.2 are five units apart, whereas 1/0.8 and 1/0.9 are only 0.14 units apart, even though the *C*s are the same distance apart. Since Bray and Curtis there have been several refinements for organizing continua of vegetation (Hall, 1970; Norris and Barkham, 1970).

Another approach to identifying discontinuities between communities relates community structure to the environment by sampling a series of communities along a habitat gradient. *Gradient analysis* (Whittaker, 1952) usually has involved sampling along intuitive habitat gradients at different elevations or on different hillside exposures. Exposure gradients are interpreted as transitions of moisture and temperature (ravines being moist and cool, whereas exposed hilltops are dry and warm). Altitudinal gradients, of course, represent patterns of change in temperature as well as insolation.

In a search for distinct demarcations between different communities, Whittaker (1953) examined the distribution of plant morphological types along an exposure gradient in the Smoky Mountains of the southeastern United States (Fig. 19-6). The percentage of deciduous trees in the forest declined continuously from the cool, moist habitats to the warm, dry habitats. Coniferous trees showed exactly the opposite distribution. These forests represented an essentially continuous replacement of one

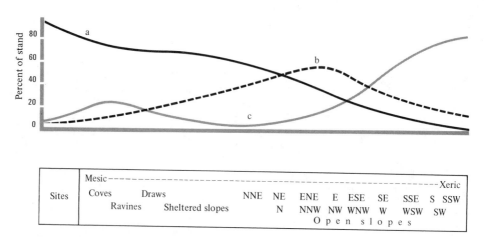

FIGURE 19-6 Distribution of different growth forms in relation to an intuitive moisture gradient at low elevations of the Great Smoky Mountains in Tennessee: (a) deciduous broadleaved trees, (b) semisclerophyllous deciduous trees, and (c) coniferous trees. (After Whittaker, 1953.)

type of growth form by another with no abrupt changes. Broadleafed plants with leathery leaves occupied an intermediate portion of the habitat gradient in which neither deciduous nor coniferous trees were predominant.

Whittaker (1956) examined mixed deciduous-coniferous forest communities along moisture and altitudinal gradients in the Smoky Mountains. The general form of species distributions was a bell-shaped curve (Fig. 19-7). Each species had a center of distribution at which it reached its maximum relative abundance and from which it tapered gradually to disappearance. Species with their maximum importance at the end of the gradient, of course, had truncated curves. The general impression gained from this gradient analysis is that of a system of interacting populations with, "The balance among populations [shifting] with change in environment so that the vegetation is a pattern of populations corresponding to the pattern of environmental gradients" (Whittaker, 1953).

The two techniques of community analysis just considered—biological ordination and habitat ordination—represent somewhat different approaches to organizing community information. Both are based on the premise that the most realistic technique for discovering whether there are objectively definable groups of species is to sample a number of stands and then look for correlated distributions of species abundances within those stands. Biological ordination is based on ordering the stands according to their biological similarity, whereas habitat ordination is based on ordering the stands according to apparent habitat similarity.

A monumental attempt to refine the habitat approach was undertaken by Loucks (1962). He sampled 63 mixed deciduous-coniferous forest stands in New Brunswick and developed the most comprehensive approach yet attempted to organize habitat data into functional environmental terms. He attempted to isolate a number of habitat properties that might be important in defining niches of forest plant species. These factors included (a) the distance to the soil water table, soil water-holding capacity, and water run-off and accumulation on the site, which were combined into a moisture index; (b) humus type, soil depth, silt and clay contents of the soil, organic composition of the soil, depth of decomposition in the soil, and nutrient import to the site from run-off, which were combined in a nutrient index; and (c) insolation, nighttime temperature, heat load as a modifier of daytime air temperature, and moisture content of the soil (representing a latent heat sink in the soil), which were combined into a local climate index. Loucks called his indices *scalars*, since they involved placing each stand on a scale of moisture regime, a scale of nutrient supply, and a scale of local climate.

The technique for synthesizing the scalars was complex and involved a number of judgments about how organisms respond to different environmental variables. For instance, the proportion of small soil particles (silt and clay) in the soil is a measure of nutrient-holding capacity since nutrients are held on particle surfaces, and surface per unit volume increases dramatically as particles in the volume are subdivided. Loucks reasoned that a given change in small particle percentage would be more important if small particles were rare than if they were abundant. In other words he felt that a small change in nutrient-holding capacity is more important when this capacity is low than when it is high; he therefore used a log scale in

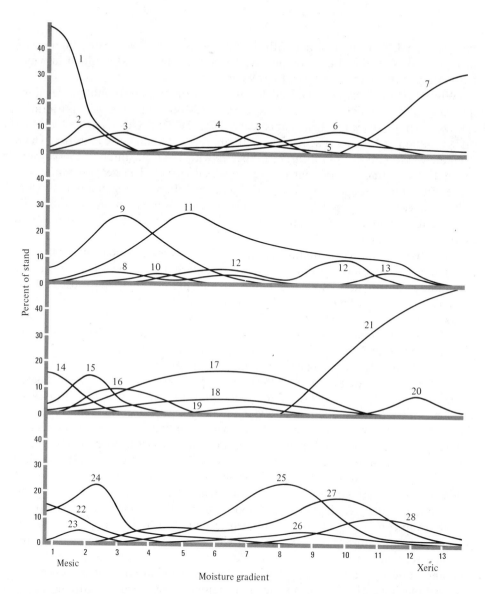

FIGURE 19-7 Distribution of 28 plant species in relation to an intuitive moisture gradient (as in Fig. 19-7) in the Great Smoky Mountains. Relative abundances are percentage contributions to density of tree stems greater than 1 cm in diameter. Communities intergrade continuously from those dominated by snowdrop-tree (*Halesia monticola*) in ravines to pine-dominated forests on dry slopes. Species are coded as: 1, *Halesia monticola*; 2, *Acer saccharum*; 3, *Hamamelis virginiana*; 4, *Carya tomentosa*; 5, *Nyssa sylvatica*; 6, *Pinus strobus*; 7, *P. rigida*; 8, *Quercus borealis*; 9, *Tsuga canadensis*; 10, *Fagus grandifolia*; 11, *Acer rubrum*; 12, *Qu. alba*; 13, *P. echinata*; 14, *Aesculus octandra*; 15, *Betula allegheniensis*; 16, *B. lenta*; 17, *Cornus florida*; 18, *Carya glabra*; 19, *C. ovalis*; 20, *Qu. marilandica*; 21, *P. virginiana*; 22, *Tilia heterophylla*; 23, *Cladrastis lutea*; 24, *Liriodendron tulipifera*; 25, *Qu. prinus*; 26, *Qu. velutina*; 27, *Oxydendrum arboreum*; 28, *Qu. coccinea*. (After Whittaker, 1965.)

defining the importance of silt and clay in the nutrient scalar. Loucks made a concerted effort to combine a series of independent habitat properties into scalars reflecting his three niche dimension complexes.

Functional significance of the scalars was documented by the correlation of both community and species properties with them. Community richness, for instance, was related to the nutrient scalar in a fashion reminiscent of the species-individual curve (Fig. 19-8). At low scalar values, that is, on poor soils, richness was low, and at high scalar values richness was high. Similarly, relative growth of an important species in these forests, balsam fir, was positively correlated with the index. Annual net productivity estimated as basal area increment was small relative to present basal area on poor soils, whereas on good soils the relative basal area increment increased to nearly 5 percent of the basal area present at the beginning of the year. It should be apparent, however, that the scalars are not completely independent of the communities. We have already seen abundant evidence, particularly in Chapter 13, that organisms modify the habitat they colonize.

Whereas gradient analysis involves an intuitive environmental ordering based on a variety of habitat properties associated with elevation, moisture, soil type, and other factors, the scalar technique is an attempt to separate important factors constituting niche dimensions and then recombine them into a synthetic measurement scale. To examine the relationship of a given species along one dimension, the range of the other dimensions must be specified. For instance, if we examine the distribution of sugar maple in the matrix formed by the three scalars (Fig. 19-9), we find that its center of distribution is at the confluence of warm climate, low moisture, and intermediate nutrient supply. The species has a definite distribution center in this environmental combination and then drops off sharply outside it. Balsam fir, in contrast, has a distribution center similar to sugar maple, but its importance drops off much less abruptly from this center. A definite environmental interaction between local climate and moisture is suggested for fir. It is found under

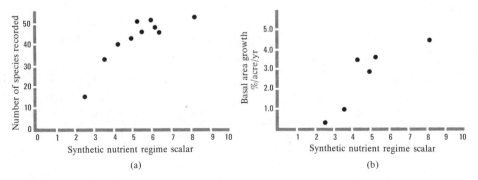

FIGURE 19-8 Relationship of two forest stand properties to soil nutrients, from low nutrient supply on the left to high nutrient supply at the right end of the scalar: (a) number of species present, (b) relative growth rate. All stands are in the moisture scalar range of 4.8 to 6.0 since the patterns may vary for different nutrient-moisture combinations. (After Loucks, 1962.)

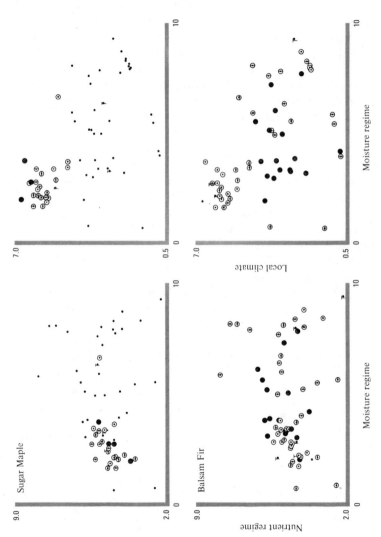

FIGURE 19-9 Distribution of two tree species within a three-dimensional forest ordination based on scalars of moisture, nutrients, and local climate. Distributions of the trees within the space defined by these axes constitute realized niches of the species in the forests studied. To visualize the space, mentally move the left-hand graph up above the right-hand one, then tip the top up to form the floor and one side of a box. The species "positions" are then floated over the floor (local climate and moisture regime axes) according to the height on the nutrient and moisture coordinates as in Figure 19-6. Relative abundances in a given stand are coded as follows: dot = less than 5 percent, open circle = 5 to 25 percent, circle with horizontal bar = 26 to 50 percent, circle with vertical bar = 51 to 75 percent, and circle with cross = 76 to 100 percent. Sugar maple is important only in relatively dry forests with soils of intermediate fertility and a decidedly warm microclimate. Balsam fir ranges over a much wider variety of environments, with relative abundance greatest toward the center of the environmental space. (After Loucks, 1962.)

warm conditions if moisture is low and cool conditions if moisture is high, just the converse of what we might deduce to be the compensatory relationship between these factors. An additional environmental interaction affecting balsam fir is indicated by the relationship among species abundance, nutrient supply, and moisture supply: When moisture is low, the species is found only on relatively poor soils, but if moisture is high, it occurs over a range of soils from relatively poor to relatively fertile.

The scalar data on these two species suggest a complex answer to the question of whether there are discrete community groups. Clearly, a community defined by an abundance of sugar maple would be relatively discrete in New Brunswick since this species drops off in importance quite abruptly outside of its distribution center; and whereas balsam fir drops off abruptly if moisture is low, it is widespread across other environmental indices if moisture is abundant. Although there were definite gradients of forest composition associated with the scalars, once again distinct species associations could only be defined within specified regions of the gradients.

Using plant communities as a model of community organization, "It is apparent to many ecologists that under some conditions vegetation forms a continuum, under other conditions it forms discrete communities, and that most vegetation is somewhere between. The question is: What factors determine the relative continuity of the vegetation?" (Beals, 1969). Beals hypothesized that the relative continuity of communities is a function of the relative continuity of the habitat. He tested this hypothesis by examining desert shrub stands in Ethiopia. He analyzed densities along altitudinal gradients, one with a slope of 63.5 meters/km and another with a slope of 3 meters/km. The vertical transition per unit horizontal distance was more than 20 times greater along the steep gradient. The community coefficient (Table 19-1) was used to compare a given sample with adjacent samples on the elevational gradient. A dissimilarity index was calculated as $1 - C$, thereby obviating the problem created by reciprocals. The possible range of dissimilarity is from 1 for samples with no species in common to 0 for samples with the same abundances of the same species.

Beals found that the average value of $1 - C$ was 0.41 on the steep slope and 0.37 on the gradual slope. Of more interest, however, was the strikingly different variabilities of the indices on the two slopes (Fig. 19-10). Variance of the dissimilarity index was significantly greater ($F = 5.85$; $P < 0.001$) on the steep slope, indicating much more abrupt changes in community structure for a given change in altitude. Along the gentle slope there was a gradual and relatively continuous change in community structure. Along the steep slope, regions of minor change were often separated by areas of extremely rapid change.

The preceding analyses of community structure suggest that communities may change abruptly or continuously, according to the degree of environmental discontinuity. Clearly there are relatively distinct combinations of species abundances, but these combinations represent "nodes" on a continuously changing network. An important objective of ecology is to organize this network in such a way that our understanding of ecological function is increased.

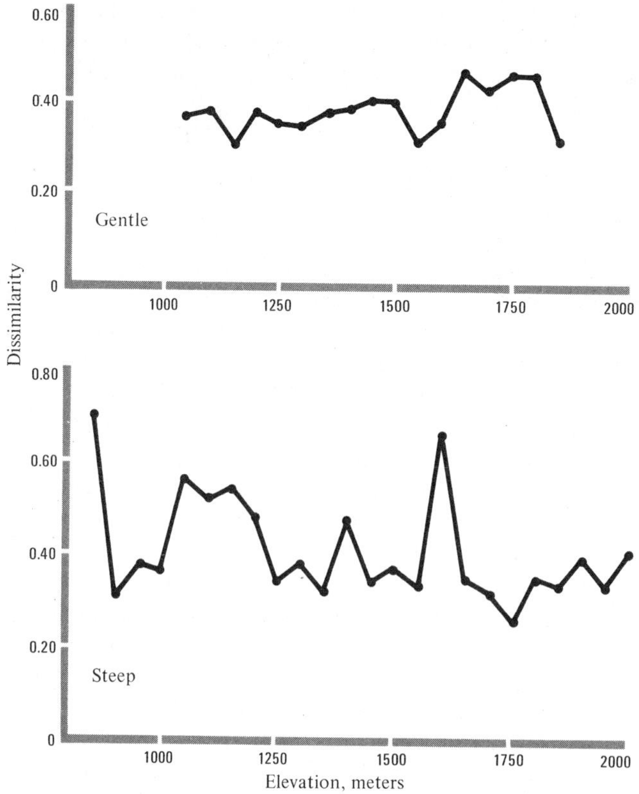

FIGURE 19-10 Dissimilarity indices of adjacent community samples over a similar altitudinal gradient along a gentle slope and a steep slope. High values, indicating rapid community changes, are more common on the steep slope. (After Beals, 1969.)

COMMUNITY ORGANIZATION: A SPECIFIC EXAMPLE

Certain California grasslands constitute a useful ecosystem for examining the properties of a community in dynamic equilibrium. They occur in uncultivated areas throughout the Central Valley and in various places along the Coast Ranges and Sierra Nevada foothills (Biswell, 1956). Introduced species have largely replaced native species in them, most of the important plants being from the Mediterranean region. Many of these are annuals and so the grasslands represent a dynamic response to prevailing habitat conditions, with the community partially reconstructed *de novo* each year from a reservoir of seeds present in the soil (Major and Pyott, 1966). In a sense these grasslands are reminiscent of the lentic and neritic biomes' planktonic communities that are reconstituted in a cyclic fashion each year. Cycles in the California grasslands are generated by a distinct summer dry period, June through October, when they are dry and brown alternating with winter rains, December through April, when they are lush and green.

McNaughton (1968b) examined the organization of these grasslands on a Coast Range foothill, Jasper Ridge, maintained by Stanford University as a scientific preserve. Eight stands that were studied and sampled through a growing season were designed to provide both continuous habitat change and habitat discontinuity:

Exposure gradients from a northeast exposure to a southwest exposure were sampled on both sandstone-derived and serpentine-derived soils. The former soil is characterized by relatively large soil particles with a resulting poor moisture and nutrient-holding capacity. The latter is characterized by numerous large rocks and rocky outcrops which create variable soil depths and drainage properties. In addition serpentine soil has a unique chemical composition (very high magnesium concentrations and a low calcium level) which is toxic to many plant species (Cooper, 1922; Whittaker, 1954).

The poor nutritional properties of the serpentine soil were reflected in a much lower peak biomass (in May at the beginning of the dry season) than on the sandstone soil (Fig. 19-11). In addition biomass on serpentine was independent of the moisture gradient, whereas it fell off conspicuously on the dry end of the sandstone continuum. Productivity was conspicuously moisture dependent on the sandstone soils, falling to a level below that of the serpentine community on southwest slopes. An examination of community diversity, as measured by the number of species recorded in a sample of 100 individuals, also indicated a tendency for grasslands on different substrates to converge on the dry end of the exposure gradient.

In spite of a tendency toward similarity of many community properties, a comparison of the relative abundances of species in the different stands indicated a substantial dissimilarity between sandstone and serpentine grasslands (Table 19-2). Stiff brome (*Bromus rigidus*), for instance, was restricted to sandstone soils in which it made up as much as 67.8 percent of the peak standing crop. And needlegrass (*Stipa pulchra*) was found to be much more important on serpentine, in which it contributed as much as 41 percent of the peak biomass. In addition 9 of the 15 species present on sandstone were absent on serpentine, and 17 of the 23 species on serpentine were absent on sandstone. Thus we can pose the same question as to the similarity or difference of these grasslands that was posed earlier for forests and shrublands. Once again we find that although they are similar, differences also exist. The floristic compositions are relatively distinct, for example, a majority of

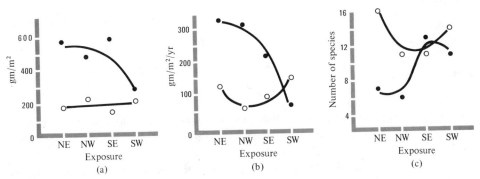

FIGURE 19-11 Changes in California grassland properties along an intuitive moisture gradient from moist (NE) to dry (SW) slopes on two different soil-types (solid circles = sandstone; open circles = serpentine): (a) peak biomass, (b) net primary productivity, and (c) number of species present. (After McNaughton, 1968b.)

TABLE 19-2 Relative importance of plant species on two soil types and four exposures in the Jasper Ridge grasslands—expressed as percentage of peak standing crop

Species	Sandstone				Serpentine			
	North-east	North-west	South-east	South-west	North-east	North-west	South-east	South-west
	Percent of peak biomass							
Avena fatua	44.4		8.6					
Medicago hispida	0.7		3.7					
Avena barbata			1.7					
Torilis nodosa		1.8	0.9					
Centaurea melitensis			2.0					
Festuca megalura				0.2				
Lolium multiflorum	7.0	7.4		3.5				
Erodium botrys	4.4	0.8	5.9	8.3				
Bromus rigidus	35.4	67.8	42.6	22.6				
Bromus mollis	7.0	21.1	21.3	38.1	7.9	19.6	27.3	53.2
Clarkia purpurea	1.1	1.1	4.7	1.2	3.3	1.2	2.7	
Hemizonia luzulaefolia			1.2	8.3	11.8			5.9
Eriastrum abramsii				3.0				0.8
Lotus subpinnatus			0.2	3.7	4.0	6.5	3.6	14.4
Stipa pulchra				16.4	40.7	41.0	26.4	10.5
Eschscholzia californica					11.8	11.6	13.4	1.3
Festuca grayii					1.1	1.2	10.2	1.4
Plantago erecta					2.0	4.1	5.6	1.8
Melica californica					8.0	11.2		
Linanthus androsaceus					1.7	0.45		
Brodiaea laxa					0.35	3.0		
Calochortus venustus					3.3			1.0
Agroseris heterophylla					0.31			2.4
Festuca dertonensis					0.30		2.1	
Achillea millefolium					1.4			
Polypogon monospeliensis					0.20			
Trifolium tridentatum						0.10		
Brodiaea pulchella							4.0	
Sitanion jubatum							3.6	
Lomatium utriculatum								6.5
Madia gracilis							1.0	1.6
Poa scabrella								0.6

McNaughton, 1968b

the species on one soil type being absent from the other. Sandstone stands generally have a greater peak biomass and a greater productivity than the serpentine stands, and yet these converge on the dry slopes.

We are still faced with the problem of organizing this network in such a way that our understanding of ecological function is increased. One objective approach is through the community coefficients based on species contributions to peak biomass in different stands. For the eight stands there are 28 different Cs. Using absolute abundances (grams per square meter) rather than relative abundances (percent of total grams per square meter), so that C will reflect the pattern of biomass change documented in Figure 19-11 as well as the distribution of stand biomasses among species, a matrix of Cs was generated that could be arranged in a consistent fashion reflecting the functional similarities of the different stands (Table 19-3). The two most similar stands were the northeast and northwest stands of serpentine, and C for this comparison is in the lower right-hand corner of the community matrix. The least similar stands were the northeast exposure on sandstone and the northeast exposure on serpentine, and C for this comparison is in the lower left-hand corner of the matrix.

The matrix can be divided into three regions: (a) an upper left triangular region in which sandstone stands are compared, (b) a lower right triangular region in which serpentine stands are compared, and (c) a lower left rectangular region in which grasslands on different substrates are compared. From this matrix we can see that grasslands on serpentine and sandstone soils tend to be distinct, and we can test this objectively by doing a t test on the means of within-soil-type Cs and between-soil-type Cs. The average of within-soil Cs was 0.531 and the average of between-soil Cs was 0.226. These values are significantly different ($t = 5.32$; $P < 0.001$). The matrix, however, is a form of continuum, with the northeast slopes at the extremes and the southwest slopes in the middle. The northeast stands were more similar to other stands on the same substrate than to stands on the other substrate ($t = 4.91$; $P < 0.001$). At the middle of the continuum, however, this between-substrate difference disappeared. For instance, the average C for southwest stands compared with other stands on the same substrate was 0.429. Comparing these same slopes with stands on the opposite substrate gave an average C of 0.357. These values are not different ($t = 1.00$; $P = $ n.s.).

A *vector* is defined as a directed magnitude, and the familiar example of a vector is velocity or force from physics. By directed analyses through the community matrix, we can formulate community vectors describing the functional change in communi-

TABLE 19-3 Matrix of community coefficients (biomass based) for California grassland communities

	$\dfrac{\text{SN}}{\text{NE}}$	$\dfrac{\text{SN}}{\text{NW}}$	$\dfrac{\text{SN}}{\text{SE}}$	$\dfrac{\text{SN}}{\text{SW}}$	$\dfrac{\text{SR}}{\text{SW}}$	$\dfrac{\text{SR}}{\text{SE}}$	$\dfrac{\text{SR}}{\text{NW}}$
SN–NE							
NW	0.550						
SE	0.556	0.740					
SW	0.342	0.473	0.509				
SR–SW	0.093	0.259	0.335	0.592			
SE	0.116	0.187	0.152	0.437	0.539		
NW	0.099	0.167	0.162	0.398	0.410	0.670	
NE	0.060	0.073	0.090	0.385	0.306	0.533	0.746

SN = stands on sandstone soil.
SR = stands on serpentine soil; other symbols are exposures from northeast to southwest along an intuitive gradient of decreasing moisture supply and increasing temperature.
McNaughton, unpublished

ties along different habitat gradients. We have been doing that intuitively in a simplified form in the previous analyses. Starting at the lower left-hand corner and directing vectors outward from that point, we see that C increases in all directions; also, as we have proven, there are abrupt transitions in the vectors if we go straight up or directly to the right. If we move diagonally across the matrix, however, the discontinuities become less and less abrupt. Finally, a diagonal vector from upper left to lower right is more or less of constant magnitude for its entire length.

This community matrix provides an objective mechanism for ordering components of the community network and is clearly related to the community properties (Fig. 19-12) considered earlier. We saw that peak biomass, productivity, and species diversity tended to converge in the grasslands, and replotting them against the ordination derived from the matrix indicates that the ordination effectively separated extremes in the community network. In addition many individual species show the familiar bell-shaped curve when plotted against this ordination, even though the intuitive environmental heterogeneity is much greater than in simple exposure gradients. In fact, as ordered by the diagonal matrix vector, there are no abrupt discontinuities in the grasslands. Assuming that community organization reflects environmental organization, we may deduce that matrix extremes are determined by the environments, but testing this hypothesis would be difficult. Both ends of the gradient are cool and moist and perhaps the most we can say at this point is that substrate differences are maximized on cool, moist sites.

Intuition suggests that one extreme of a habitat gradient in these grasslands would be the southwest slope on serpentine where the inherent nutritional limitations of this soil type are combined with limitations imposed by low available moisture. The community matrix, however, suggests that this site is intermediate in the grasslands. If abundances of species in the different habitats are indicative of environmental properties, then we have learned something about the grasslands that contradicts our intuition, that is, that northeast sandstone and northeast serpentine sites are extremes of an environmental gradient. In addition, and even less apparent in the absence of the matrix analysis, the expression of the moisture gradient on serpentine was the converse of its expression on the sandstone and was exactly the opposite of our initial assumption that NE $\rightarrow$ SW represented the appropriate stand ordering, regardless of substrate. This analysis provides an important dimension to our previous statement that communities are both distinct and indistinct from one another. By allowing us to order them in a manner that not only demonstrates the ways in which they are distinct and indistinct, it also suggests functional relationships that were not previously apparent.

ABUNDANCES OF SPECIES POPULATIONS IN THE COMMUNITY

The matrix of community coefficients allowed the introduction of some order into the pattern of population change on the landscape. Another complex community problem is how resources are divided among different species in a community. This has been stated historically in somewhat different terms as the abundance problem

569

ABUNDANCES
OF SPECIES
POPULATIONS
IN THE
COMMUNITY

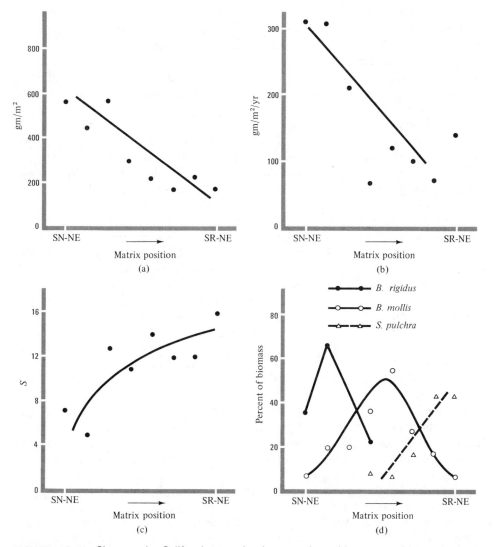

FIGURE 19-12 Changes in California grassland properties with communities organized according to similarity from the matrix of community coefficients of Table 19–3: (a) peak biomass, (b) net productivity, (c) number of species present, and (d) relative abundances of three species important in different stands. (Data from McNaughton, 1968b and unpublished.)

(Preston, 1948; MacArthur, 1957), the dominance problem (McNaughton and Wolf, 1970), the species-area problem (Gleason, 1925), the species-individual problem (Williams, 1964), and various combinations of these. We have already considered the problem implicitly, particularly in our discussions of competition (Chapter 11) and succession (Chapter 13) when we examined how the number of species and their comparative abundances are related. The problem consists of two separate, but related, parts: (a) how resources are divided among different species and (b) what the dividing of these resources tells us of the relationships among species in nature. Resolution of the first part of the problem has proven so elusive that the second part has barely been considered. In particular, the question of dividing *resources* has been only an implicit part of studies of how individuals are distributed among species.

The problem of species abundances in a sample is intricately related to the problem of species diversity, that is, how many species are present in a sample. Consider, for instance, the species-individual curves shown in Figure 18-20, comparing the rate at which species are added for a given addition of individuals from tropical and temperate forests. The fact that there are many more species in the tropical forest for a given number of individuals proves that the average number of individuals per species must be smaller in this community than in the temperate community. In addition, in Chapter 4, when considering the general properties of ecosystems, we observed from Raunkaier's law of frequencies that most species present in a community tend to be rare. In other words there are a few species with many individuals and many species with a few individuals.

Preston (1948, 1962) has shown that for very large samples it is possible to give satisfactory empirical solutions to the relationship among species abundances. These very large samples, which Preston refers to as "universes," generally must exceed 40,000 individuals (Preston, 1957). A community of organisms presupposes co-occurrence in time and space and it is hard to justify the expansion of this idea to a sample this large. In fact a universe is generally defined as a closed or independent system, which certainly does not apply to a community. A universe, therefore, is a level of organization intermediate between the community and the biosphere, including enough individuals to constitute a relatively accurate sample of all the organisms of a given ecological type or at a given trophic level. Preston (1962) says, "For our purposes, most samples are to be regarded as small if the number of species involved is less that about a hundred."

In such a large sample it has been shown repeatedly that the numbers of species in abundance classes are normally distributed when plotted against log to the base 2 of those classes (Fig. 19-13). In this distribution the number of individuals assignable to a given species is tabulated. The range of numbers of individuals is then divided into classes according to log to the base 2 so that the upper limit of a given class is double the upper limit of the next lower class. The smallest number of individuals that can be counted, of course, is one. Therefore the smallest class includes all those species for which one or two individuals were counted. The next class includes all those species with two to four individuals (borderline species are divided equally among the two classes). The next class includes species with four to eight individuals, and so on, until a class is generated that includes the most abundant species. When the abundance data for a universe are ordered according to these "octaves" of abundance, the numbers of species in the octaves are normally distributed; so for a very large sample most species are neither common nor rare, but of intermediate abundance by the standards of the lognormal. This approach

571

ABUNDANCES
OF SPECIES
POPULATIONS
IN THE
COMMUNITY

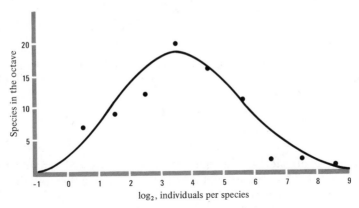

FIGURE 19-13 Lognormal organization of abundances of 86 bird species observed during the breeding seasons around Westerville, Ohio, over a 10-year period. The *Y*-axis is the number of species with a given abundance, and the *X*-axis is the $\log_2$ of abundance as classes. (After Preston, 1962.)

spreads the least abundant class of the Raunkaier law into a series of classes, indicating that there are a few very common species, a few very rare species, and many species of low to intermediate abundance. The lognormal distribution has been widely applied to abundance samples and has proven reasonably accurate in describing the large samples for which Preston prescribed it.

Interpretation of the lognormal in functional terms has been obscure at best. The closest that Preston (1962) has come to a mechanistic interpretation of the curve is an intuitive argument, "If . . . fitness is distributed normally, numbers of individuals will be distributed lognormally." Preston then continues by saying, "I myself distrust the argument. . . ." No biological explanation of the series has been forthcoming in the interim. It should be pointed out that this is not a "random" distribution in the same sense that digits occur randomly in a random number table. The lognormal is random only in the sense that variation in class size is randomly distributed around the mean class. This is somewhat different from saying that abundances are distributed randomly among species.

The latter argument was first proposed by MacArthur (1957). In an influential paper he presented one of the first attempts to reason to an ecological conclusion based on a few explicitly defined assumptions that held promise of biological validity. Ecology, by and large, has been an empirical science in which hypotheses are developed by inductive reasoning from the appearance of nature. MacArthur, however, applied deductive reasoning to biological properties of species to arrive at a logical model of how abundances might be distributed among co-occurring species. We considered this model briefly in Chapter 4 and more detailed consideration of MacArthur's arguments is now appropriate. He first assumed that the logistic growth equation was a reasonable, although perhaps not complete description of the way populations grow. If we let $N_{i(O)}$ be the size of the *i*th population at time *O*, we can ask what the size of the population will be at time *t*, that is, after some time interval dt has passed. Let

$$r = dN/N \, dt$$

Recall from equation (9-4) in our discussion of population growth that this is the reproductive output per individual member of the population. The population size at time t is

$$\log N_{i(t)} = \log N_{i(0)} + \int_0^t r \, dt \qquad (19\text{-}2)$$

where the last expression in the right term is the realized reproductive output integrated over the time period. MacArthur said that either (a) the integral is important compared with initial population size or (b) the integral is unimportant compared with initial population size. In the former case there is a marked change in population size over the time period; in the latter case the population size is relatively constant.

The first proposition is likely to apply only to species that are colonizing unexploited habitats, for instance, early successional species or annual plants and animals that expand each year into an unfilled niche during the time period, dt. The second proposition, in contrast, will apply primarily to species in equilibrium communities in which there are no drastic changes in niche structure on the time scale of dt. MacArthur (1960) utilized Williams' bird censuses in a woodland near Cleveland, Ohio, to estimate the nature of the change in population size in a relatively stable ecosystem (Fig. 19-14). Over the 18-year study period populations tended to increase if they were small and to decrease if they were large, with "smallness" and "largeness" being defined by some apparent equilibrium value (about 20 birds in the ovenbird example of Fig. 19-14).

The latter type of species MacArthur called *equilibrium* species and he asked further how the abundances might be distributed among co-occurring equilibrium species. He argued that there are two possible relationships among relative abundances of these populations: (a) the total resources available to the community are constant and fully exploited so that an increase in one species must be accompanied by a decrease in other species or (b) the total resources are not fully exploited and population abundances are independent in the sense that an increase in one species need not be accompanied by a compensating population decline. If we now rank the species from least (1) to most (S) abundant, we can calculate the expected abundances of each of the S species. Assuming that niche sizes evolve randomly, in the first or nonoverlapping niche model, abundance of the rth species is

$$A_r = \frac{E}{S} \sum_{i=1}^r \frac{1}{S - i + 1} \qquad (19\text{-}3)$$

where E is the total abundance (or resource content) of the community, S is the number of species in the community, and i is the number of intervals from the rarest species to the rth species. For the second, or overlapping niche model, abundance

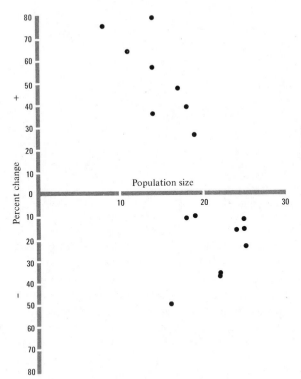

573

ABUNDANCES
OF SPECIES
POPULATIONS
IN THE
COMMUNITY

FIGURE 19-14 Changes in abundance within an ovenbird (*Seiurus aurocapillus*) population in a forest near Cleveland, Ohio, over an 18-year period in relation to population size. Upper half of the Y-axis is increases, lower half is decreases. When population density was high, it tended to be smaller the succeeding year; when density was low, it tended to be higher the next year. (After MacArthur, 1960.)

of the *r*th species is (Pielou and Arnason, 1966)

$$A_r = E\left(\frac{S - r}{S - r + \frac{1}{2}}\right) \qquad (19\text{-}4)$$

where *r* is rank of the species from rarest (1) to most abundant (*S*). These are called the *random niche* hypotheses and they describe two different organizations of limiting factors in an ecosystem. In the first model the limiting resources are saturated with exploiting organisms and a population may increase its size only by eliminating individuals of another species. In the second model limiting resources are not saturated and a species may expand its size without eliminating individuals of another species.

Reasoning from the logistic growth model, assuming that resource exploitation ability will be randomly distributed among species and that equilibrium species may be limited either by competitive efficiency on a completely exploited resource or by their ability to exploit an incompletely exploited resource, MacArthur developed two alterative hypotheses of species relative abundance in a community. The accuracy of these models of community structure, however, can well be questioned (Fig. 19-15). Many ecologists have compared the random niche distributions with real community samples and by and large they have found that these models show a tendency to underestimate the most abundant species and overestimate the least

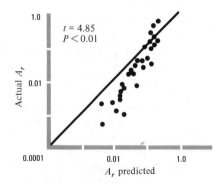
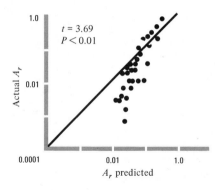

FIGURE 19-15 Failure of nonoverlapping (*left*) and overlapping (*right*) random niche models to predict abundances of bird species in wintertime flocks. (Data from Morse, 1970.)

abundant species (Pielou and Arnason, 1966). The degree of departure from reality is somewhat greater for the second model than for the first. Given the frequency of adequate fits out of the number of communities compared, it would seem that these fits are more a product of chance than any real relationship to the models; MacArthur (1966) has indicated that he no longer believes the models are ecologically realistic. We have discussed them here, however, because they represent one of the finest examples of applying deductive reasoning to ecological problems.

This reasoning appeared to supply two alternative models of community organization, with defined functional assumptions delimiting them, and both were amenable to testing. Although they have not proven realistic (Cohen, 1969), they provoked much careful community study and were instrumental in relating population growth, a principal concern of animal ecologists, to community structure, a principal concern of plant ecologists. There are those who would undoubtedly argue that the models produced more heat than light in ecology, but our belief is that the heat generated led to a healthy fusion that warrants considering them as an example of basically wrong hypotheses that generated productive scientific activity. Few hypotheses are correct, but stimulating ones at least increase our knowledge of what is not correct.

SAMPLE SCALE AND RELATIVE ABUNDANCE

Hairston and Byers (1954) found that the arrangement of relative abundances in a sample depended on sample size, and we have already mentioned that Preston restricted application of the lognormal to "universes." Conversely MacArthur found that small samples fit his random niche models better than large samples. Hairston (1959) performed a thorough analysis of the effect of arthropod sample size on fit to the first MacArthur model and found that the model underestimated the most abundant species and overestimated the least abundant species. In other words common species were more common and rare species were more rare than a random partitioning of niche space among populations would suggest.

McNaughton and Wolf (1970) examined the ability of the lognormal series to predict abundances of plant species in pooled samples from a variety of habitat-types on different exposures or of different successional ages. The pooled samples fit the lognormal quite well, but when these were again subdivided into samples of co-occurring populations, there was a consistent tendency for the lognormal to underestimate the number of rare populations (Fig. 19-16). For the 26 plant communities

$$\log F/F_r = 0.481 - 0.471 \log \mu \qquad (19\text{-}5)$$

where F is the actual frequency of populations in an abundance octave, F_r is the frequency predicted by the lognormal distribution, and μ is the upper limit of the abundance octave in percent of total community abundance. From this equation $\mu = 10.5$ percent when $F = F_r$. In other words there were fewer populations than predicted with relative abundances greater than 10.5 percent, and many more populations than predicted with relative abundances of less than 10.5 percent.

We may return for a moment to the meaning of the lognormal and see what it tells us about the organization of nature. Since a species' niche has a carrying capacity that will be defined by the number of individuals the niche supports, we can deduce that the lognormal is a measure of the distribution of carrying capacities of complete species' niches; that is, we can conclude that niche-carrying capacities in a universe are distributed lognormally. Data on individual numbers, however, are an indirect estimate of niche resource capacity. Whittaker (1965) measured net productivity of plant populations in three deciduous-coniferous mixed forests in the Smoky Mountains (Fig. 19-17). Energy flux into these plant populations was distributed according to the lognormal, providing more direct evidence that carrying capacities of species' niches tend to be distributed lognormally.

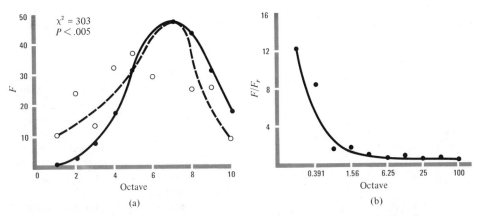

FIGURE 19-16 Inability of the lognormal distribution to predict relative abundances in plant community samples: (a) relationship between abundances predicted by the lognormal equation (filled circles) and those observed in 26 community samples (open circles); (b) relationship between the ratio of actual (F) and lognormal-predicted (F_r) abundances and the upper limit of a relative abundance octave. There are fewer common species in communities than the model predicts, and many more rare species. (After McNaughton and Wolf, 1970.)

log normal dist seen here

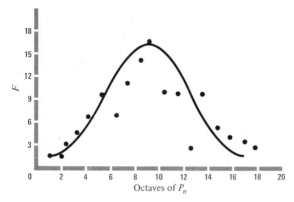

FIGURE 19-17 Fit of combined net productivity data of plant populations in three mixed forests in the Great Smoky Mountains to the lognormal equation. The lower limit of the smallest octave is 0.001 gm/m²/yr and the upper limit of the largest octave is 524 gm/m²/yr. (Data from Whittaker, 1965.)

Although the lognormal distribution provides us with a description of a species' niche-carrying capacities summed for all its occurrences, it does not provide information about the most rigorous ecological interactions, those among populations in the same community (McNaughton and Wolf, 1970). Many of the species in a universe may be so remote from one another, either in space or in time, that they never co-occur. In the California grasslands, for instance, *Avena fatua* and *Stipa pulchra* never co-occur on the same slope, and yet both are abundant on certain slopes (Table 19-2). The departure of progressively smaller samples from the lognormal suggests that species with a high abundance on one site will tend to occur on other sites, but that species with a low abundance will be limited in the number of sites on which they occur. The California grasslands tend to support this hypothesis although there is considerable scatter in the data (Fig. 19-18). Generally, species with a maximum biomass of 10 gm/m² or more were found on four or more slopes, whereas species with a maximum biomass of less than 10 gm/m² were found on less than four slopes. This suggests that the major species tend to repeat from stand to stand, whereas the rare species tend to be different in different stands.

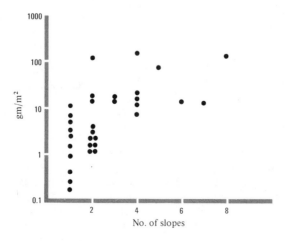

FIGURE 19-18 Relationship between the number of slopes on which a plant species occurred in California grasslands and the maximum biomass of that species on any one slope. While species with a small peak biomass are relatively narrowly restricted, species with high peak biomasses range from common to relatively restricted. Data from McNaughton (unpublished).

Whittaker (1965) pointed out that the lognormal distribution has a characteristic sigmoid form when plotted on what he called a "dominance-diversity" curve. In this graphical presentation of relative abundances Y is population abundance on a log scale and X is the rank of the species from most abundant (1) to least abundant (S). Note that this is the reverse of the ordering used to calculate expected abundances from the random niche hypotheses. There are two extreme forms of dominance-diversity curve (Fig. 19-19): (a) the sigmoid lognormal and (b) a simple straight line. The straight line is a consequence of a constant proportional decrease in importance from most to least important species. This form of curve is called a *geometric series* and it has been applied in ecology primarily by a number of Japanese ecologists (Numata et al., 1953; Shinozaki and Urata, 1953; Nobuhara and Numata, 1954) and Whittaker (1965). The two forms may be distinguished by the central plateau of the lognormal where there are a number of species (those in the modal and adjacent octaves) that have similar abundances. In the geometric series there are no transition regions but only a constantly decaying curve.

The functional interpretation commonly given to the geometric series is the niche preemption hypothesis (Whittaker, 1965). If the community has a total energy content or energy flux E, and the most abundant species appropriates a proportion of this, p, then its abundance is pE. According to this hypothesis the second species takes the same proportion of what is left $p(E - pE)$, and so on, until the proportion remaining is insufficient to support a population. From this, abundance of the rth species becomes

$$A_r = E\left[p(1-p)^{r-1}\right] \tag{19-6}$$

In other words if the first species contributed one-third of the total community energy content, the second species would contribute $\frac{1}{3}(1 - \frac{1}{3}) = \frac{1}{3}(\frac{2}{3}) = \frac{2}{9}$; the third species would contribute $\frac{1}{3}(1 - \frac{1}{3} - \frac{2}{9}) = \frac{1}{3}(\frac{4}{9}) = \frac{4}{27}$.

The hypothesis is based on the assumption that the contribution to E is dependent on resource utilization of the population. The resources remaining after the rth

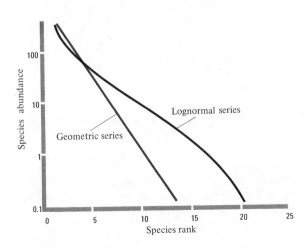

FIGURE 19-19 Two abundance hypotheses for a sample of 20 species and 1000 individuals (or gm/m² or gm/m²/yr, or another measure of abundance). Species are ranked from most abundant (1) to least abundant (20). The rare species are much rarer in the geometric series, extending far below the lower limit of the X-axis here (see Fig. 4-11). (After Whittaker, 1965.)

population is added to the community are $1 - (1 - p)^r$ where the total resource pool is set equal to one. The series is terminated, therefore, when the resource fraction remaining is too small to support an additional species.

When all the biomass data from the California grasslands were pooled and plotted according to the dominance-diversity form, they were organized according to the lognormal (Fig. 19-20). As the individual stands were progressively separated from the universe, however, there was progressively poorer fit to the lognormal and a closer and closer approximation to the geometric series. The total serpentine and total sandstone samples, separable in the community matrix by their average community coefficients, were a poorer fit than the total sample to the lognormal, with the series "plateau" less pronounced on sandstone than on serpentine. Finally, although there were some distinct groups along the lines, each of the individual slopes approximated a geometric series. The approach to this series was a negative function of the number of species in the sample. A universe, then, is many communities, unrelated by interactions involving a common resource pool.

This analysis suggests that the geometric series and lognormal series represent extremes of a sample-size-dependent continuum of relative abundances. The geometric series describes the partitioning of realized niche space among co-occurring populations. The lognormal series describes the partitioning of realized niche space among species. The latter is a consequence of the evolution of diversity in the species along the niche parameters that it exploits; that is, species that have a high abundance in a given community must combine this with ability to range through a large number of communities to be abundant in a universe. The geometric series is a consequence of the frequency of occurrence of certain niche parameter combinations

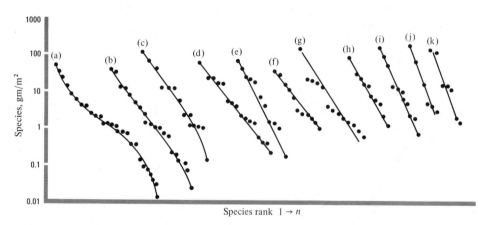

FIGURE 19-20 Data from California grasslands indicating that averaging a group of geometric series generates a lognormal distribution. Samples are arbitrarily spaced along the X-axis to separate them: (a) species averages for all slopes on both soils, (b) all serpentine slopes, (c) all sandstone slopes, (d) serpentine NE, (e) serpentine NW, (f) serpentine SE, (g) serpentine SW, (h) sandstone SW, (i) sandstone SE, (j) sandstone NW, and (k) sandstone NE. (Data from McNaughton, 1968b and unpublished.)

in a given locality. The fact that combining geometric series gives rise to a lognormal suggests that (a) few species combine wide occurrence with high abundances, (b) few species combine limited occurrence with rarity, most species (c) occur with medium abundances in a medium number of communities, or (d) have high abundances in a few communities, or (e) have low abundances in a large number of communities.

In a variable habitat any of these adaptive strategies is likely to be successful, except for (b), which combines rarity with limited occurrence. Such species would appear to be perched in an extremely precarious ecological position, subject to extinction on a minor niche perturbation. Species with this adaptive strategy are likely to occur only in habitats that are exceedingly stable where they will not be eliminated by the vagaries of environmental change. We will return to this question later, but first let us examine a somewhat different, but related, question.

Hairston (1959) examined the relationship between spatial distribution of individual microarthropods and the abundance of their populations in a community. He used a coefficient to measure the degree of clumping among individuals (Fisher, Corbet, and Williams, 1943; Bliss and Fisher, 1953). The coefficient approaches infinity for a random distribution of individuals in space and zero for a contagious distribution. Hairston found that the rarest species in the community either tended to be randomly dispersed in the community or had a highly contagious distribution (Fig. 19-21). This suggests that there are two types of rare niches in communities: (a) those where the confluence of appropriate niche parameters is randomly dispersed in space and (b) those where the confluence of appropriate niche parameters is congregated in a limited area. A population with the former niche distribution pattern is likely to have a low probability of reproductive success, due to the distance

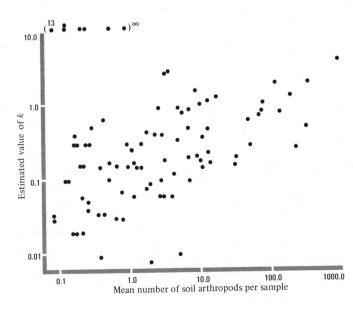

FIGURE 19-21 Data from arthropod communities showing the tendency for rare species to consist of individuals either highly clumped in space (low value of k) or essentially randomly distributed ($k = \infty$). (After Hairston, 1959.)

between population members, but a high probability of surviving local extinction. A population with concentrated niche distribution, in contrast, is likely to have a high reproductive success, but a higher probability of extinction from local habitat fluctuation than the randomly dispersed population. Assuming that the probability of such a fluctuation occurring at any point is fairly constant for different points in the community, contagion would appear to be a safer strategy than dispersion, if a species depends on sexual reproduction for perpetuation in the community.

SPECIES DIVERSITY: ITS DEFINITION AND CAUSES

Diversity is a state of variety, of difference among members of a collection. A population, for instance, may be diverse in its age structure (that is, the developmental state of population members) or its genetic structure (that is, individual members may be different from one another, or may possess a wide variety of alleles, or both) or in various other properties. As usually employed in ecology, however, diversity refers implicitly to the ecological variety of the individual members of a community. To assess ecological variety we would have to assess the genetic diversity of all members of a sample that is designated a "community." It is generally assumed that species populations in a local area represent distinct ecological units. Community diversity therefore may be measured by how individuals are distributed among species. We have been doing this throughout the book by measuring the slope of S on $\log N$, although there are myriad other approaches to defining diversity (Whittaker, 1965; Pianka, 1966; McIntosh, 1968; Hurlburt, 1971).

If we sample a single individual, the diversity of our sample is zero, and the relative abundance of that species is one. If we sample two individuals, the diversity may be either zero or $2/\log 2$, and the relative abundance of species may be either 1 or 0.5, depending on whether the second individual sampled is of the same species as the first. We have seen that small samples will approach the geometric series and that progressively larger samples will converge on the lognormal. We must then ask why some communities have a greater slope of S on $\log N$ than others and why the greater slope suggests a greater departure from the geometric series and approach to the lognormal.

Pianka (1966) inquired into the meaning of diversity differences among communities. We have seen that species diversity increases with successional age, with less variable climate, and so on. Pianka pointed out that a variety of alternative hypotheses had been proposed to explain these gradients:

1. Temperate regions are considered to be impoverished due to recent glaciations and other disturbances.
2. Diversity arises out of a general increase in environmental complexity.
3. Niches are "smaller" in more diverse communities.
4. Predators (and/or parasites) . . . hold down individual prey populations enough to lower the level of competition in more diverse communities.
5. Stable habitats allow the evolution of finer specializations.
6. Greater production results in greater diversity.

581

SPECIES
DIVERSITY:
ITS
DEFINITION
AND
CAUSES

Although Pianka developed these arguments, as did the original proponents he cited, mainly in terms of tropical versus temperate diversity, there is no reason not to expand them into more general expressions as we have done. Data are available from various sources to test each of these hypotheses.

Data to test the proposition that glaciation diminishes diversity, separated from the temperate-tropical comparison, are available from areas in Wisconsin. The southeastern portion of the state was free of glaciation throughout the repeated periods when the rest of the state was covered with ice sheets. Curtis (1959) presented diversity data on communities restricted to these areas. The average species density (number of species per community sample) was 53 for nonglaciated areas and 45 for glaciated areas. These values are not different by Student's t ($t = 1.41$ for $P = $ n.s. with $df = 19$). Neither was the total number of species recorded in a community type different in the two areas. This number was 209 for the nonglaciated area and 239 for the glaciated area ($t = 1.008$; $P = $ n.s.). Although we have no direct data on the numbers of individuals involved in these samples, the fact that the non-glaciated area is principally prairie and deciduous forest whereas the glaciated area is primarily coniferous forest suggests that individual density would be higher or similar in the glaciated area. The data suggest acceptance of the null hypothesis.

The second hypothesis, that diversity is a function of environmental complexity, appears almost circular. Since niches are defined by the occurrence of species within them, an increase in species diversity provides de facto evidence of increasing environmental diversity. There is other evidence, however, that shows habitat complexity as a generator of species diversity. We saw several examples of this in Chapter 4 when we considered the mechanisms of distribution patterns. For instance, in the two experiments by Harper and his associates on (a) the segregation of buttercup habitats in a field as a result of drainage patterns in the field and (b) seedling establishment indicating that different species became established under different soil preparation treatments, we saw evidence that habitat complexity increased community diversity. We also saw evidence, when discussing succession, which indicated that diversity of bird species was a function of the diversity of the vertical structure of tree foliage. Finally, Zinke (1962) has shown that individual trees modify the soil properties around their bases (Fig. 19-22), thereby creating resource shells in the form of complex gradients of soil properties. The presence of such resource shells around individuals may be an important mechanism for generating a confluence of environmental parameters that could define the niche dimensions of other species. All of this evidence, derived from independent definitions of species niches, suggests that environmental complexity is an important mechanism of species diversity.

The third hypothesis, that species in diverse communities may have "smaller" niches than those in species poor communities, was tested by McNaughton and Wolf (1970). They measured niche size as the variance in the relative abundances of species across habitat gradients. Niche size was defined as

$$W = \left[\frac{\Sigma(y_p \cdot p^2) - (\Sigma y_p \cdot p)^2 / \Sigma y}{\Sigma y} \right]^{1/2} \tag{19-7}$$

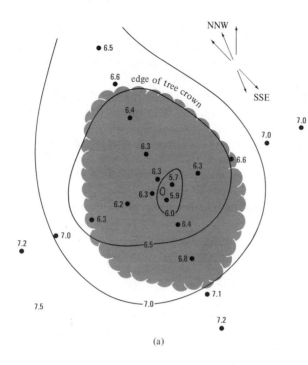

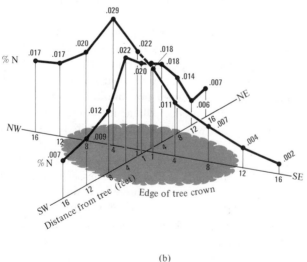

FIGURE 19-22 Patterns of (a) soil pH and (b) soil nitrogen around a single individual of lodgepole pine. Prevailing wind directions are indicated by proportional arrow lengths in the upper right-hand corner of (a). This ability of organisms to modify the habitat in their vicinity may be an important component of the establishment of complex habitat gradients creating conditions suitable for colonization by a species not previously present in the community. (After Zinke, 1962.)

where p was the position of the community on the environmental gradient, y_p was the relative abundance of the species in that community, Σy was the total relative abundance of the species for all its occurrence, and W was species niche size. This index approached the number of positions on the gradient for a species that was uniformly distributed over the gradient, and approached zero for a species that was concentrated on a limited portion of the gradient. For any community, then, there was a collection of species present and to each of these species a niche size could be assigned that depended on the ability of the species to colonize an environmental gradient partially independent of that community.

583

SPECIES
DIVERSITY:
ITS
DEFINITION
AND
CAUSES

McNaughton and Wolf found that mean niche width of the species in a community was an inverse function of the number of species in that community (Fig. 19-23). Although the average niche width of trees was somewhat greater than that of grasses or shrubs, the decline in niche width with increase in diversity was similar in both types of communities. This provides support for the hypothesis that species in diverse communities are more restricted in their ability to colonize a range of habitats than species in less diverse communities. In other words they have smaller niches.

The fourth hypothesis is that predation reduces the level of competition enough to allow coexistence of a greater variety of species (see Chapter 11). Harper (1969) summarized a series of six papers published in 1933 by a British agronomist, Martin Jones, on the effect of grazing on pastures. Using both newly seeded simple species mixtures and complex native grasslands, Jones established that there was no simple relationship between herbivore grazing intensity and plant species diversity. Instead the effect depended on palatability of the major species on the dominance-diversity curve. If the principal species were the most palatable in the community, grazing increased diversity. If, however, the rare species were more palatable, grazing diminished diversity. As Harper concluded, "A complex balance of species . . . is sensitive to rather precise control by the action of the grazing animal." Brooks and Dodson (1965) examined the effect of predation on zooplankton in a series of Connecticut lakes into some of which a marine fish, the alewife, had been able to migrate and some which were free of alewives. They found that alewives, because of their feeding strategy, preferentially took the large zooplankton that had dominated the lakes before the influx of alewives. Predation on these larger-sized species allowed a number of previously insignificant components of the community to

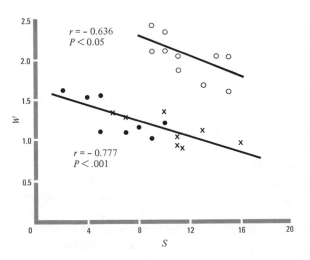

FIGURE 19-23 Decline in the average niche size ($\bar{W}$) of species in plant communities as diversity increases. Samples are from desert shrubs (•), grasslands (x), and mixed forest (o). (After McNaughton and Wolf, 1970.)

increase in abundance, with a general increase in diversity. Overall the data indicate that predation influences diversity, but not in the simple way suggested by the original hypothesis.

The fifth hypothesis, that habitat stability or predictability allows the evolution of niche specialization, is supported in a general way by the latitudinal clines in species diversity discussed at the end of Chapter 18. The data on the glaciated and nonglaciated areas of Wisconsin, however, contradict this hypothesis since the repeatedly glaciated portions of the state are no less diverse than those portions of the state that have never been glaciated. In addition the hypothesis seems to be contradicted by data indicating that habitat heterogeneity is a powerful mechanism for generating species diversity. Given that habitats can be defined by temporal variation (as we saw from the diurnal spacing of *Oenothera* pollinators in the example of Chapter 2), a certain amount of temporal instability might facilitate diversity as much as diminish it.

The final hypothesis, that increased productivity generates increased diversity, is both supported and contradicted by the evidence. For instance, we have seen several examples of a decline in the net productivity at a trophic level with increase in diversity at that level. In the California grasslands, for example, there was an inverse relationship between net primary productivity and plant diversity. In our consideration of the lentic biome (Chapter 15), we also saw that net primary productivity of Danish lakes was inversely correlated with diversity of cladoceran consumers, suggesting that high net primary productivity diminished diversity of communities at the next higher trophic level. In contrast to this we saw in Chapter 13 an example of a perturbation experiment in which enhancing net productivity at the primary trophic level increased both diversity and net productivity at higher trophic levels, although producer diversity remained constant. This single experiment both supports and contradicts the hypothesis. Our interpretation of these examples is that increasing net productivity does not have a simple effect on diversity or an intrinsic relationship to its effects, probably depending rather on other more complex mechanisms.

In summary, it appears that the most important mechanisms generating species diversity are habitat complexity and decreased niche size. Increasing specialization of invading species in a diversifying system, through subdividing the niche parameters, would allow species diversity to increase more rapidly than a simple accretion of new niches might. We have seen that niche size diminishes with diversity and we can further ask if the sum of the niche sizes also increases, suggesting an addition of new niches in addition to a greater subdivision of total niche space. McNaughton and Wolf (1970) examined this question for a series of communities and found that the sum of the species niche sizes (ΣW) for a stand was a positive function of stand diversity for shrub and grass stands but was independent of diversity for forests (Fig. 19-24). This suggests that an accumulation of new niches was a more important mechanism of diversification in the shrub and grasslands than in the forest.

An intensive study of how niches may be added and subdivided has been conducted by Waloff (1968) and his associates. They have been studying the insect communities on a common bush in the British Isles—Scotch broom. Of 31 primarily herbivorous insects found on this single plant species, there are two species that feed

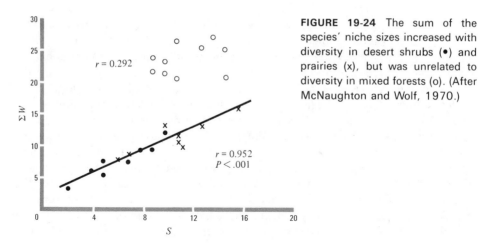

FIGURE 19-24 The sum of the
species' niche sizes increased with
diversity in desert shrubs (•) and
prairies (x), but was unrelated to
diversity in mixed forests (o). (After
McNaughton and Wolf, 1970.)

585

RELATIVE
ABUNDANCE
AND THE
NICHE

in the seed pod, three that feed on tissues of the pod cavity perimeter, six species
that produce galls, four species that feed directly on the foliage, five species that
feed on plant juices, four species of stem miners, two species that feed on the bark,
and five species that mix predation with juice feeding. So within the 31 species there
is considerable spatial differentiation of habitat and feeding activity; but this raises
the question, for instance, of how it is possible for two species to co-occur in the
seed pod. Waloff said, "Larvae within the pods have different habits. *Contarinia
pulchripes* are found early in the year in green and unripe pods. They reduce the
insides of the pods, including the seeds, to a brown mass. The larvae of *Clinodiplosis
sarothamni* are fungus feeders and are found later in the year, when the broom pods
have hardened."

RELATIVE ABUNDANCE AND THE NICHE

Most of the preceding evidence indicates that the relative abundances of species
are a reflection of the different carrying capacities of their niches. It seems likely
that all species are specialists for a certain combination of environmental factors
and if this combination is widespread on a site, the species will be abundant there.
The most reasonable description of the distribution of carrying capacities in a local
area is the geometric series. Mixing several of these series, as we have seen, produces
the lognormal distribution which is, in fact, a measure of the distribution of species
niche carrying capacities in many communities. The explanation for why local niches
should be distributed according to the geometric series is obscure. The niche preemption hypothesis, although stated above, is difficult to interpret functionally. It
implies that co-occurring populations line up at a resource trough and each takes
a certain proportion of what is in the trough when its turn comes. We can think
of no ecological justification for such a hypothesis and assume instead that the series
in some way describes the relative abundances of resource specializations in nature.

There is some evidence to support the hypothesis that all species are specialists and that the most abundant species have specialized on an abundant environmental complex, thereby forcing other species into peripheral specializations. Connell (1961), as we have seen (Chapter 11), has shown that the most abundant barnacle in marine communities of the Scottish coast is specialized for occupation of space below the high tide line. Above this line it is replaced by another species capable of living under periodic desiccation. The interesting point is that the desiccation-adapted and less abundant barnacle, which has a narrower realized niche, is actually capable of occupying innundated sites if the other species is removed. In other words the most abundant barnacle is a true dominant species in the sense that it occupies niche space potentially occupied by other species in the community (McNaughton and Wolf, 1970). The most abundant species, however, cannot occupy sites above the tide line, even if competition is eliminated; that is, the dominant species has sacrificed a portion of its "potential" fundamental niche by specializing on space occupation below the tide line. The other species has sacrificed a portion of its realized niche by maintaining sufficient genetic capacity to colonize both habitats.

Similar evidence is available from the marsh plants, cattails. The broad-leaved cattail (*Typha latifolia*) is a widely distributed species and is much more abundant than another species, the narrow-leaved cattail (*T. angustifolia*), except in areas where saline habitats are common (Smith, 1967). The latter species is largely restricted to saline habitats. From this we would conclude that the narrow-leaved cattail is a saline habitat specialist, which would be wrong. McMillan (1959b) has shown that whereas the narrow-leaved cattail, the presumed specialist, can occupy both high and low salt conditions, the broad-leaved cattail can occupy only freshwater. Once again the species with the largest fundamental niche is forced into a subordinate distribution by maintaining sufficient flexibility to occupy both habitats.

STABILITY AND DIVERSITY IN COMMUNITIES

We considered earlier the patterns of succession that give rise to species combinations which are increasingly persistent through time. We also examined the resistance of some of the early successional communities to perturbations in the environment (Chapter 13). From the limited data that are available, we concluded that there is no clear relationship between stability and diversity, although the argument has been widely circulated in both the technical and popular ecological literature that diversity generates stability. Community stability can be evaluated in at least three different ways: (a) persistence of certain species combinations through time, (b) persistence of certain relative abundance relations through time, and (c) ability of the community to maintain its functional organization following an external perturbation.

We saw, in the succession chapter, that early successional communities are directionally modified through the effects that they have on site properties. Although the later communities are less rapidly modified by their effect on site properties, they may be directionally modified by such extrinsic variables as climate. We also saw, in the biome chapters, that certain species combinations, with similar relative abundance relations among the species, tend to occur over broad geographical areas.

The balance among co-occuring species may be drastically affected by environ-

mental changes. Weaver and his colleagues, for instance, observed a replacement
of bluestem-dominated prairie by grasslands typical of more arid environments
during severe drought (Weaver, 1954). He observed that "One of the most outstanding phenomena of the dry climatic cycle was the destruction of a portion of
one plant association and its replacement by a more xeric one. This change occurred
very gradually over a period of seven years." In Kansas prairies prior to drought,
little bluestem contributed 50 to 85 percent of the cover and big bluestem contributed
15 to 50 percent. Short grasses, such as blue grama, buffalo grass, and wheatgrass,
abundant in more arid western prairies, were insignificant in the vegetation. Within
three years of drought, little bluestem declined by 95 percent and big bluestem by
15 percent. Both were replaced by the short grasses. After rainfall returned to normal,
the tall grasses again became the predominant species. Although producer biomass
plummeted during the drought, the degree of disruption was certainly less than it
would have been if the tall grass prairie had not contained a reservoir of short grass
species. This suggests an important functional role for rare species in partially
stabilizing the community's functional properties in a varying environment. While
the relative abundances of prairie grasses were modified considerably by drought,
species poor agricultural systems in drought areas were completely devastated (Fig.
19-25). What is not known, however, is whether less diverse prairie sites, lacking
as extensive a reservoir of arid climate grasses, were more severely affected than
those studied by Weaver.

It is apparent that the adjustment of relative abundances in a community may
stabilize functional properties. Response to environmental variation, however, will
depend on the type of variation. In the succession chapter, for instance, we saw that
plant communities of early successional stages in the northeastern United States were
more sensitive to a nutrient enrichment than later stages (Table 13-6). Insects, in
contrast, were less stable to perturbation in the more diverse communities (Table
13-7). In addition, a climax deciduous forest community in the same area clearly
would be more susceptible to destruction by fire than would the early successional
stands. While fire would be a noncatastrophic perturbation to a stand in which the
major plant species were perennial grasses and herbs, it might completely destroy
the forest. An appropriate test of stability relations of communities of different
growth form requires, of necessity, noncatastrophic perturbation of both.

Another aspect of stability is the degree to which populations tend to undergo
irruptions in different types of communities. We have discussed in numerous places
the irruptions of lemmings in the tundra biome, ungulates in the temperate forest
biomes, and locusts in the desert biome. Similar irruptions seem to be rare in the
tropical forest biome, particularly in the rainforest. Lack of pronounced population
oscillations in these ecosystems may be less a property of the inherent food web
diversity, since many of the species are highly specialized in feeding habits, than
of the relative lack of seasonality. Of course, the tropical biome has not been as
carefully examined as the other biomes and irruptions may be more common there
than we suspect.

An important component of food web diversity may be the length of time
consumers have had to respond evolutionarily to a given food source and, in partic-

FIGURE 19-25 Two outcomes of drought in the North American Great Plains: agricultural ecosystems were totally devastated, whereas native ecosystems, although often undergoing shifts in species abundance, were much less severely disrupted. (*Top:* USDA; *bottom:* U.S. Forest Service.)

ular, the abundance of that food source through evolutionary time. Southwood (1961) examined the relationship between diversity of insect communities on different tree species in Britain and the abundance of the tree in Quaternary geological records. He found a clear positive relationship between Quaternary abundance and the number of species in insect communities on the trees (Fig. 19-26). A large number of insects fed on oaks, willows, and birches, which were abundant during the Quaternary, while many fewer species fed on species introduced into Britain recently. In spite of the high diversity of the oak insect communities, however, certain of their associated insects undergo irruptions so severe that the oaks are completely defoliated (Fig. 19-27). This suggests that there is no clear relationship between the number of co-occurring species in a community and the temporal stability of those species populations.

The problem of ecosystem stability, and its relationship to the properties of communities in that ecosystem, is of considerable practical significance. Man presently depends on agricultural ecosystems with extremely low species diversity. The development of industrial agriculture, particularly in the United States since the Second World War, has resulted in the use of weed and insect poisons, resulting in low diversity ecosystems over huge geographic areas. The eastern prairie biome has been converted into extensive areas where almost the sole plants are maize and soybeans. The western prairie biome has been converted into thousands of square miles of wheat. Extensive applications of herbicides have been used to reduce interspecific plant competition to an absolute minimum, while other biocides have reduced animal populations. A more accurate knowledge of the role of rare species in community function, and the relationship of diversity to population stability should allow the design of an agricultural system less dependent on the energy of fossil fuels for its maintenance. Although the preceeding consideration of the nature of ecological communities indicates how complicated the problems are, it also shows

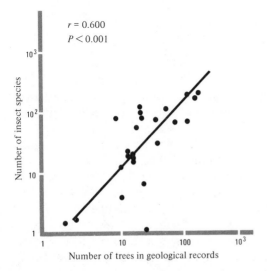

$r = 0.600$
$P < 0.001$

Number of insect species

Number of trees in geological records

FIGURE 19-26 Relationship between the number of insect species found on various tree species and the abundance of those tree species in geological deposits of the Quaternary. (After Southwood, 1961.)

FIGURE 19-27 Oak trees in Great Britain severely defoliated by herbivorous insects, in spite of the high diversity of insect communities on the oaks. Note that sycamore, in foreground, which harbors about 5 percent as many insect species as oaks (Southwood, 1961), is much less severely defoliated. (Paul Feeny.)

that much has been learned over the last few years. This recent knowledge has helped clarify the problems that must be examined in the future.

GENERAL CONCLUSIONS

1. A community is one or more populations with similar resource demands co-occurring in time and space.
2. Species composition changes continuously in space, with the same combination of species in the same abundance relation rarely, if ever, repeated in different areas.
3. The degree of discontinuity between communities will be a function of habitat similarity. Relatively abrupt transitions may occur where the habitat changes abruptly, while gradual shifts will occur if the habitat changes gradually.
4. In a local area the relative abundances of species tend to be distributed according to the geometric series. Mixing many community samples, and hence geometric series, gives rise to a lognormal distribution of abundance. These series suggest that few species combine either high abundances in a large number of communities or low abundances in a few communities. At present there is no strong

evidence suggesting the functional mechanisms that give rise to the geometric series.

5. High species diversity in a community seems to be primarily a function of habitat heterogeneity on the site and invasion by species with small niches, that is, species that specialize on relatively rare resource combinations.

SOME HYPOTHESES FOR TESTING

Two hypotheses about community organization seem particularly important and amenable to testing: (a) that diversity is generated primarily by habitat heterogeneity of a site, and (b) that diversity is a mechanism which stabilizes community function. We have seen several tests of the latter idea at various points in the book and the data are generally conflicting. It seems important to provide carefully controlled conditions for community perturbations of various types. For instance, it would be interesting to introduce grazers or predators into communities of different diversity and to see whether more diverse communities were more resistant to disruption by the introductions. Alternatively, it would be interesting to selectively remove species from ecosystems with different diversities and to see whether more diverse food webs were less sensitive to such removals.

With reference to the hypothesis that diversity of a community is a consequence of habitat heterogeneity, it would be possible to create microhabitats in communities of different diversity (for instance, by creating localized moisture and nutrient gradients) and observe whether invasion by new species was more common in a species poor system than in a species rich system.

Finally, one of the gaps in our understanding of community organization is the absence of a mechanistic explanation of the geometric series which describes abundances among co-occurring species in a local area. We have seen that the lognormal distribution arises out of mixing many geometric series. But what accounts for the geometric distributions of abundances in a local area? Why should niche carrying capacities in a community be distributed in such a fashion?

PART 5

CONCLUSIONS

The two final chapters offer different types of ecological conclusions. Chapter 20 is a look at man, placed in the context of the principles developed in preceding parts. Chapter 21 considers ecology as a scientific endeavor, and consists of the authors' personal views of what constitute summarizing generalizations of ecology and problems that are most important from the standpoint of creating a predictive theory of ecology. Both chapters therefore tend to be somewhat speculative, but for different reasons. About the ecological development of mankind we know much, but projections from that development vary radically. We present those in Chapter 20 for what they are—the interpretation of man's nature, history, and likely future as seen by two practicing ecologists. Similarly, the conclusions of Chapter 21 are those of two professional ecologists looking at their field. Just as no two persons are likely to view man in quite the same light, so our conclusions about the science must reflect, in part, our own interests.

a brief ecological history of mankind

Our objective in this chapter is to place man in the context of the ecological principles developed in preceding chapters. First, we examine the general outline of human evolution; then we consider the properties of modern human populations which define our ecological present and suggest our ecological future.

Like any discussion of the past much of what we say is an approximate reconstruction realizing that there will be errors in detail. For, after all, there are few hard data, in the sense that data were utilized to develop ecological principles. "Population samples" in the study of human physical evolution are fragmentary at best, often consisting of merely a few bone fragments. Even more tenuous are reconstructions of the social organization of early man, which are deduced largely from comparative study of the present behavior of ourselves and other species translated into the presumed environments of early humans.

Concepts relating to the present ecology of man, particularly pollution and population growth phenomena, have been considered previously where they related to ecological principles being discussed. It is our belief, however, that a thorough

596

CHAPTER 20
a brief
ecological
history of
mankind

understanding of modern man's ecological situation can be most clearly viewed from a vantage point including his evolutionary origins. We present in this chapter what we believe are the most logical arguments about our past and future. Others will disagree. Since experiments are impossible the disagreements are largely of inter-pretation. As first premises we believe (a) with Lorenz (1966) and Tinbergen (1968) that in spite of our tremendous capacity for learning, significant aspects of our behavior are still genetically determined and (b) with Murdoch (1959) and Vadya (1961) that culture is adaptive in the sense of providing people with the means of adjustment to the environment.

In Chapter 1 we said that man is a successful species. We defined this success by the ability of a population to maintain or increase its size. Man's nature arises out of his descent from populations that possessed Darwinian fitness, traits which allowed them to persist while co-occurring populations became extinct or more limited in size. It should be remembered, however, that success is a relative term. Individuals that are less fit in a given niche may be unsuccessful at competition there. If, however, they can respond to competitive exclusion via directional selection, they may penetrate a new adaptive mode and thereby become more successful, in terms of ultimate abundance and persistence, than a "winning adversary" in the previous niche. Whether the population will become more abundant, of course, depends on the carrying capacity of the new niche.

We can consider a sort of niche field defined by habitat properties and population properties (Fig. 20-1) in which carrying capacities may vary as determined by both the resource structure of the ecosystem and the adaptive structure of the population. Imagine a population at point A that grows to the carrying capacity of the niche it occupies. Individuals within that niche will now be under severe intraspecific competition and this will tend to drive the population toward genetic diversification

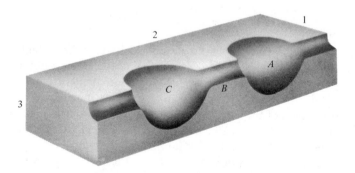

FIGURE 20-1 Cross-sectional diagram of an evolutionary field in which axes 1 and 2 are genotypic combinations and axis 3 is the carrying capacity of the niche, as defined by the environment plus the genotype. Populations which evolved by natural selection from the right-hand margin to point A can evolve further only down the narrow channel (B) connecting with new niche C.

as a mechanism for decreasing the intensity of that competition; that is, those individuals with gene combinations that allow them to exploit "peripheral" portions of the niche may be favored in reproduction by the greater distance to K for that character combination. The carrying capacity is essentially zero, however, for most of the possible gene combinations, either as a result of more efficient competitors in other populations or because of a lack of sufficient resources. The population therefore will tend to "spill" down the channel B with an exploitable carrying capacity. Reproduction of individuals in the center of the adaptive mode will be limited by the efficiency of their gene products within this mode. Reproduction of individuals at the periphery of the niche, however, will be limited also by the frequency with which their offspring represent gene combinations capable of moving further down the channel. In other words the rate of evolutionary progress in the channel will be limited primarily by the rate at which mutation and recombination produce new adaptive forms, rather than just more efficient members of an existing adaptive form.

Along the evolutionary channel a new character combination C may evolve to exploit a resource combination in the ecosystem that is currently unexploited but is capable of supporting a large number of individuals. Because the carrying capacity is zero outside the channel, individuals within it will have a relatively low reproductive success. Individuals arriving by mutational and recombinational chance at C have a tremendous reproductive potential since they are in an essentially unlimited environment. We would expect that initially there would be strong selection for increasing r_m within the population since the maximum reproductive potential feasible for a regulating population in the channel now may be the major factor limiting the population at C.

Although a niche is defined both by the genetic properties of exploiting individuals and by the resource structure of the ecosystem, it is apparent that there may be "empty niches" in the sense of unexploited resource combinations in many ecosystems. Explosive expansions of introduced pests indicate that such unexploited or less efficiently exploited resources may be common, lacking only the appropriate evolutionary or migrational event. For the past 50 million years or so primate evolution has been a process of defining such unexploited or less efficiently exploited resource combinations.

PRIMATE ORIGINS

Primates are a mammalian order that include lemurs, marmosets, monkeys, apes, and man (Simons, 1972). They evolved from mammalian insectivores at least 60 million years ago. This evolution was related to the appearance of the tropical forest 120 million years ago (Moreau, 1966). Present in this vegetation was a wide variety of "potential niches" associated with arboreal herbivory. As we have seen from our consideration of the tropical forest biome, these niches came to be occupied by birds, insects, and, to a significant degree, by apes and monkeys.

Although both apes and monkeys feed in the forest canopy, the way they exploit food sources is quite different. Monkeys run along the tops of branches and use

598

CHAPTER 20
a brief
ecological
history of
mankind

them as a system of intersecting roadways. Tree-dwelling apes brachiate; that is, they swing from branch to branch, thereby hanging beneath the branches. As a consequence of these different means of locomotion, apes and monkeys encountered somewhat different foods and evolved different feeding habits (Grand, 1972), although both occur in the forest canopy.

Life in the forest canopy was made possible by a series of mutations and recombinations that allowed their possessors to exploit food sources unavailable to their ancestors. Primary among these was the evolution of grasping feet which allowed early primates to move off the ground and into the branches above. Greater precision of movement in this hazardous and variable physical habitat generated selection for increased visual acuity and particularly for binocular vision. Color vision was an important component of visual evolution because it provided additional depth and perspective cues. There also was strong natural selection for expansion of the cerebral cortex, that portion of the brain responsible for coordinating visual cues and limb movement. In terms of reproductive characteristics the new arboreal life style gradually selected for reduced litter size and somewhat larger offspring size so they were capable of at least hanging onto the mother to avoid a lethal fall from the canopy. Increasing cerebral and visual complexity also placed a selective advantage on lengthening the gestation period so that a reasonable development of brain tissue could be accomplished. Except for strong selection for what we would call "reflexes," there was a general diminution of selection for inherited behavior and a lengthening of infancy which allowed a larger portion of other behavior to be learned. Reduced litter size and lengthened infancy also placed a premium on longevity. This period of evolutionary development was accompanied by genetic diversification as certain adaptive modes became filled and evolution was channeled by the tropical forest habitat structure down a varying system of genetic and environmental fields such as those illustrated in Figure 20-1.

By 30 to 40 million years ago, perhaps 30 million years after the appearance of the first primates, the evolutionary process had channeled primates into three separate adaptive strategies: (a) the prosimians which include present-day lemurs and tarsiers, (b) monkeys, and (c) apes. The prosimians, judging from their living relatives, were predominately nocturnal, often insect feeding, and were the product of strong selection for tremendous visual acuity, which allowed them to stalk prey during the night in the forest canopy. The feeding strategy consisted of a glacially slow approach along limbs and twigs to unsuspecting prey. The prosimians' almost imperceptible stalking and great visual acuity handicapped the prey in the predator-prey game.

Monkeys had evolved along quite a different route, being selected for flawless and rapid "foraging" in a precarious physical system. Unlike in the largely solitary prosimians, there also was selection for group formation in monkeys, perhaps as an antipredator device and probably also important in the behavioral learning associated with the vigorous arboreal feeding strategy. Because running over branches was likely to be particularly precarious in the dark, these primates have largely diurnal activity patterns. Members of this group, such as the baboons, eventually moved out of the trees and into a terrestrial life style. Group formation was probably of even greater selective advantage in this habitat with its many potential predators.

Finally, apes were channeled into yet a third niche type, that of the canopy interior, in which hanging fruits and leaves are less accessible to the surface-grazing monkeys. Brachiating primates, the apes, were able to exploit these food sources by moving through the interior branching system, swinging along underneath the canopy.

The prosimians are particularly distinct in their feeding strategy and reflect a basically conservative response to the tropical forest vegetation. Maintaining the insectivorous feeding habits of their immediate ground-dwelling ancestors, they merely were selected for extension of this niche into a new type of physical matrix. Primates less successful at this feeding strategy were forced into two alternate evolutionary channels: (a) surface foraging on the superficial forest canopy and (b) interior foraging on the lower surface of the canopy. Monkeys and apes partitioned the tropical forest canopy spatially. This partitioning was allowed by mutation and recombination which provided variation within primate populations. The habitat structure of the forest, in turn, defined natural selection through resource supplies capable of supporting a new type of individual. Perhaps man's origin from this evolutionary matrix is reflected in most elementary school playgrounds, where "jungle gyms" reconstruct man's former habitat for his children's amusement.

HUMAN ORIGINS

We have seen from the pollen records discussed in Chapter 14 that biomes are subject to constant change and oscillation as habitat conditions fluctuate. During the period of primate evolution considered previously, the tropical forest went through alternating periods of expansion and contraction associated with world-wide climatic oscillations. Through most of this period the climate was somewhat warmer and more rainy than it is now, but periodic ice ages were followed by droughts that caused the tropical forest biome to expand and contract (Moreau, 1966). About 30 million years ago Africa's tropical forest extended far outside its present range, extending northeastward into what is now southern Egypt. At this time the water level of the seas was considerably higher than at present and the rainforest contacted the Mediterranean shores in this area.

Alternate expansion and contraction of the forest created fluctuations in the carrying capacities associated with primate niches. In addition an intermediate type of biome, the savanna, ebbed and flowed across the continent with the climate. This biome currently is widely distributed in Africa, as we have already seen. Expansion and contraction of the savanna and forest biomes, respectively, generated selective advantage within primates for a return whence they came, that is, for primates capable of exploiting energy sources on the ground. Both the monkeys and the apes were able to respond to this selection, culminating in baboons from the former group and man from the latter. Perhaps prosimians were unable to respond because mutations never arose which allowed them to compete efficiently with other small terrestrial mammals or to escape terrestrial predators. From baboons and man we

600

CHAPTER 20
a brief
ecological
history of
mankind

can deduce that an important component of their selective response was social behavior and group formation.

Fossil remains of early semi-terrestrial apes have been found in geological deposits in southern Egypt that date from 30 to 40 million years ago (Simons, 1964). Still probably a good brachiator, these apes occurred along the forest margin and were capable of walking upright, at least for short periods; such upright carriage was of selective significance because it allowed them to see over the grass canopy as they moved between groups of trees (Oakley, 1961). Since they were probably hair-covered (Hockett and Ascher, 1964), it is likely that they moved out of the forest only during early morning or late evening hours when foraging in bright sunlight would not overheat an animal basically adapted for existence in dense shade (Boughey, 1971b). The genus *Proconsul,* one of these early ground apes, was widespread in Africa between 20 and 30 million years ago, but is unknown elsewhere, although a number of other ground apes ranged through Europe and Asia. Other genera of gound apes were contemporaneous with *Proconsul* in Africa, but they probably were not within the evolutionary lineage leading to humans. During this period of climatic flux there was strong selection along the forest border for apes capable of moving out of the trees and into the adjoining savanna. A large number of ape species occurred in the forest. As the forest area contracted those most fit for tree habitat exploitation were likely to exclude competitively less fit genotypes (Fig. 20-2) from their shrinking niche. A contraction in the carrying capacity of forest niches with forest disappearance would place a selective premium on genetic events allowing exploitation of the expanding savanna habitat. The character of these genetic events probably differed from community to community, thereby producing a proliferation of ground ape ecotypes, the most efficient of which would be sorted out in further expansion by forest that diminished the carrying capacities of savanna niches.

From the structure of man we know that important properties ultimately preserved from the earliest ground apes are (Boughey, 1971b) (a) erect body carriage

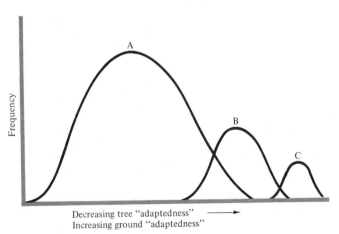

FIGURE 20-2 Diagram of phenotypic frequencies in three populations showing varying degrees of adaptation to arboreal living. Population C, less adapted to exploiting tree habitats but more adapted to ground living, would be crowded out of a shrinking forest and would be under strong selection for increasing efficiency on the ground.

arising out of a reorganization of the hind limbs, pelvic girdle, and spine; (b) manipulative ability arising out of opposable inside digits on the forelimb and a flexible shoulder structure; (c) daylight activity pattern arising out of visual capabilities; (d) behavioral learning arising out of brain structure, small litter size, long gestation period, and extended infancy; (e) cooperative activity and a complex social system arising out of brain structure; (f) an omnivorous diet arising out of dentition and dietary requirements.

THE APPEARANCE OF HUMANS

From the ground apes of 30 million years ago to the appearance of *Homo sapiens* 0.25 million years ago (Fig. 20-3) there were numerous intermediate forms, evolutionary dead ends, and considerable reorganization of the ground ape group. This reorganization represents the interaction between genetic events and ecosystem resource structure. A genetic event allowing more effective exploitation of that resource structure would be preserved and radiated throughout interbreeding populations. Traits that diminished the effectiveness of resource exploitation, that is, fitness, would be eliminated from the population. It must be borne in mind that the resource structure was not a constant; the evolution of ground apes was carried out in a context of biome fluctuation which exposed them to disruptive selection. Certain phenotypes were favored at certain times and in certain places, whereas other phenotypes were favored at other times and places.

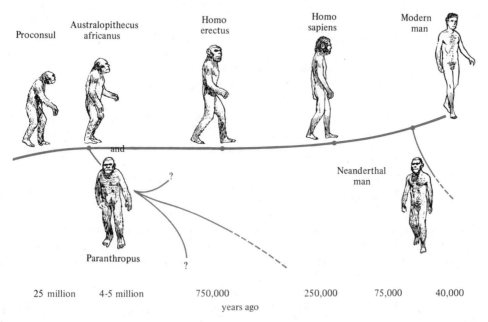

FIGURE 20-3 Presumed evolution of man in Africa and Eurasia from about 15 million years ago to the present.

602

CHAPTER 20
a brief
ecological
history of
mankind

Disruptive selection increases the genetic diversity of the population and therefore increases its ability to respond to subsequent directional selection (see Chapter 3). Man's origins are buried in the genetic diversity of the ground ape populations which began to be exposed to stringent directional selection because of the increasingly prolonged dry spells that culminated in the present distribution of biomes. As we have seen, the progressive conversion of the East African highlands from tropical forest to larger and larger areas of savanna, beginning 6 to 8 million years ago, shrank the carrying capacities of niches exploited by ground apes dependent on forest species for a substantial portion of their diet. This generated strong selection for alternate food sources. In the African savanna there are two abundant food sources: grass and grazing ungulates (see Chapter 17). Grass is a food source not likely to be exploited by populations whose dentition had been selected for chewing soft fruits and leaves rather than the abrasive mineral-laden leaves of grasses, and whose digestive systems had not been selected for processing huge quantities of low-quality food.

By 4 to 5 million years ago a variety of *hominoids* (early humans) called australopithecines had become abundant in the expanding savannas and some of these began using stones as digging and, perhaps, hunting implements (Robinson, 1963). As the savanna continued to spread over East Africa, australopithecines diverged into two distinct forms, perhaps ecotypes of the same gene pool but generally given generic status in the confusing system of fossil ape classification (Simons et al., 1969). One of these ape populations (*Paranthropus*) had evolved dentition more suitable for exploiting plant food sources and thus it probably was most common along the forest border and lake margins where it could find sufficient fruits, rootstocks, and other soft plant organs to support successful populations. The other (*Australopithecus africanus*) had a somewhat larger brain size and was probably an extensive tool user (Dart, 1953).

The *A. africanus* population was midway in a continuum of rapid brain-size expansion that began around 3 million years ago and was completed by 1 million years ago (Pfeiffer, 1969). During this period cranial capacity increased from an average of 500 cm^3 in the australopithecines to 975 cm^3 in *Homo erectus* (Campbell, 1966), a widespread species 750,000 years ago. The period of rapid evolution of brain size was associated with a number of other changes converting early man into formidable hunters.

As the climate became progressively drier, there was probably increasing selection for populations which preyed on game animals during the lengthening dry season (Robinson, 1963). Not only was there an increase in the abundance of prey bones in sites associated with hominoids, the size of prey species also increased substantially (Howell, 1966), suggesting an increase in ability to kill larger species. Such an adaptive strategy opened a new niche, with a substantial carrying capacity, exploiting the world's abundant herds of grazing mammals. Although there is no way population size can be accurately estimated at this early date, extrapolating backward from recent estimates suggests that there were perhaps 125,000 hominoids 1 million years ago (Deevey, 1960).

The genetic diversity of apes, generated by disruptive selection in a spatially and temporally variable habitat, provided the genetic raw materials for a subsequent

period of rapid directional selection as the forest shrank and the savanna expanded. The efficient forest competitors retreated with their shrinking niche and excluded the less efficient ground apes from it. Thus ground and forest apes diverged into two population types, one restricted in abundance and primarily herbivorous and one becoming progressively more abundant and more carnivorous.

Leakey (1960) gives an interesting description of small mammal hunting techniques which suggests why there was strong selection for increased brain size in the hunting apes.

> When you see a hare, it runs straight away and you run straight after it. It has its ears back as it goes, but not all the way back. The ears move all the way back when it's about to dodge, a sharp right or a sharp left. Now if you're right-handed, you always dash to the right, anticipating a dodge to the right. That means the odds are fifty-fifty, and you should catch half the hares you chase right off. If you've guessed correctly, the hare runs directly at you and you can scoop it up like fielding a fast grounder.

It is no wonder that there was strong selection for increasing brain size as primates returned to the ground since this required that their superb reflexes, evolved in the trees, be preserved, and that memory, association, logic, and judgment be grafted onto them.

A further difference between the evolving hominoid brain and the brain of arboreal primates was an increase in the internal interconnection of brain areas (Geschwind, 1964). The brain of modern man contains a nerve bundle at the junction of the visual, auditory, and touch centers which suggests a switching system for interrelating sense impressions. This center is particularly important in the association of visual and auditory impulses that are essential to communication with a complex auditory language.

Hockett and Ascher (1964) suggest that formal language, rather than a system of ill-defined vocal calls, probably began developing about 2 million years ago. Organizing sounds into a consistent communication system would have produced an increasingly efficient mechanism for organizing social activity in the communal hominoids. The ability of some of these populations to become progressively better hunters is undoubtedly associated with the evolution of a language capability. The relationship between an expanding niche and brain evolution suggests a positive feedback system involving brain size and the activity patterns necessary in successful big-game hunting.

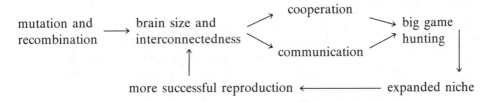

That is, those hunters with traits such as larger and more complex brains, allowing

604

CHAPTER 20
a brief
ecological
history of
mankind

more effective hunting, had an increased reproductive success, therefore shifting the population's mean brain size and structure. Such an evolutionary amplification system is, of course, limited by the rate at which genetic changes can be accomplished and finally by the environmental carrying capacity.

A variety of other traits in addition to brain properties distinguished *H. erectus* from his evolutionary predecessors. Increasing hunting activity suggests, for instance, that a decline in body hair allowed him to forage more efficiently during the heat of the day. He was a relatively talented toolmaker, shaping bones and stones into choppers, cutters, clubs, and spears (Clark, 1963; Robbins, 1972). He also possessed some traits which decreased his fitness. As the hairy coating diminished, he was likely to be chilled by the cold nights of the African plateau.

As hominoids evolving into *H. erectus* migrated outward from the African savanna, there was selection for decreasing skin pigmentation (Loomis, 1967). Dark skin increased fitness in a tropical habitat by screening ultraviolet radiation, thereby reducing the rate of synthesis of vitamin D. A light-skinned, hairless ape spending much of its time in bright sun would be subject to synthesis of levels of vitamin D that could result in elevated blood phosphorus and calcium levels, ultimately producing kidney stones, certainly fatal in a premedical population. There was therefore strong selection for dark pigmentation accompanying the evolution of hairlessness among the ground apes. With migration out of tropical areas, however, there was new selection for decreased pigmentation, increasing vitamin D synthesis, and thereby decreasing the susceptibility to rickets, the bone disease caused by vitamin D deficiency.

By 250,000 years ago the dispersing *H. erectus* had evolved into *H. sapiens* with a total population probably near 1 million (Deevey, 1960). The differences between the two species are minor: a further increase in cranial size, somewhat higher skulls, flatter faces, and reduced brow ridges. There may have been more or less parallel evolution of *H. erectus* populations into *H. sapiens* populations over a broad geographic area (Coon, 1963), although the exact nature of the evolutionary transition is widely disputed (Birdsell, 1963). At any rate, three semiisolated populations were formed as various migration barriers involving alternating periods of glaciation, rising sea levels, and desert formation tended to separate the African, Asian, and European populations to varying degrees. Sporadic genetic and cultural exchange probably took place during transitional climatic periods when favorable habitats connected the populations for a few thousand years before they were isolated again.

SOCIAL INNOVATION

Man's history is a history of social-technological changes which represent adaptive responses to the habitat. Among the first, as we have seen, was tool use, which facilitated the ability of ground apes to hunt. Group hunting ability represented a positive feedback system in evolution since any trait increasing hunting ability would increase fitness, thereby increasing the frequency of that trait in the population. A positive feedback system is inherently unstable, however, and hunting *H. sapiens* is a graphic example of this instability. Fire was first used by Asian ecotypes of *H. erectus* about 400,000 years ago, but there is no evidence that its use spread into Africa prior to population differentiation into *H. sapiens*. Fire was an important adaptive device in marginal climatic areas where it allowed populations to survive the winter; but African populations employed it as an efficient method of increasing food yield from the savanna ecosystem. These populations used fire to drive game animals into swamps and other topographic traps where they could be slaughtered more easily.

At about the same time, hafted weapons became widespread and the combination of spears and fire made man a devastating hunter (Martin, 1967). The extinction rate of large mammals increased dramatically during this period and about 40 percent of the game genera in Africa were eliminated in 30,000 years. Even more dramatic was the coincidence between the migration of *H. sapiens* into North America and the extinction of large mammals there (Martin, 1966), with about 71 percent of the genera being exterminated in a brief period following the appearance of technologically advanced human predators. Although there is considerable controversy about man's exact importance in these extinctions (Martin and Wright, 1967), it seems likely that this new predator was an important element.

Trivers (1971) has pointed out that the evolution of altruism, that is, behavior of benefit to others but not of any apparent immediate benefit to the individual(s) exhibiting the behavior, was an important component of the ecological success of humans. Selection for altruistic behavior is dependent on long lifetime, low population dispersal rate, and a degree of mutual interdependence among population members. These conditions define the circumstances under which an altruistic act may increase fitness of the performer by assuring a reasonable probability that his survival may depend on future receipt of similar behavior.

Imagine, for instance, an early hunting band preying on big game. Situations must have arisen fairly frequently in which one member of the band would be endangered while hunting. If other members of the band could come to his aid with little risk to their own survival, there would be strong selection for genes leading to such altruistic behavior. By saving the endangered individual the others increased the probability of their own survival by reciprocal altruism in a future incident. That such altruistic behavior has both genetic and learned components is suggested in present society in which individuals observing a child about to be run down by a car may dash to the rescue with no thought for their own safety, whereas whole groups of people, given an opportunity to weigh potential costs and benefits, have allowed murders to continue without aiding the victim.

During the glaciations that periodically separated evolving subpopulations, European *H. sapiens* became differentiated into a separate ecotype, Neanderthal man. From 70,000 to 35,000 years ago this population preyed upon Europe's game herds (Klein, 1969) and evolved a large jaw and brow ridge system to which mighty chewing muscles were attached (Pfeiffer, 1969). The body build of Neanderthal man was short and stocky, which aided in heat conservation. Culture was sufficiently developed that burial and body painting were practiced (Bordes, 1961). That either warfare or hunting accidents occurred is documented by knife- or spear-abraded ribs in a Neanderthal skeleton from northern Iran (Solecki, 1957).

606

CHAPTER 20
a brief
ecological
history of
mankind

As this Iranian record indicates, Neanderthal man spread out of Europe during interglacial periods. Similarly, his ecotypic counterpart, Cro-Magnon man, spread out of the Middle East and into Europe during these same interglacials. The latter ecotype is indistinguishable from modern man by the standards of physical anthropology. In addition to physical characteristics, however, Cro-Magnon man was distinguishable from Neanderthal man by his toolmaking technique. Whereas Neanderthal man practiced the venerable tradition of shaping stones into a sort of bludgeon, Cro-Magnon man practiced the new technology of blades—stones shaped so they are at least twice as long as they are wide. The blades were soon turned into spears (Brues, 1959). The spear was a much more efficient weapon and during recessions in the Wisconsin glacial period the new technology spread rapidly through Europe. As Klein (1969) remarked, "Everywhere in Europe, this replacement seems to have happened between 40,000 and 30,000 years ago and to have taken no more than a very few thousand years." This suggests that Cro-Magnon man swept over Europe during the interglacials, resulting in the competitive exclusion of the Neanderthal ecotype (Brues, 1959). This competitive exclusion probably was accompanied by substantial gene flow so the Neanderthals rapidly converged on modern man (Brace, 1964). The world's human population at this time numbered perhaps 3 million (Deevey, 1960).

The period from the disappearance of Neanderthal man until about 10,000 years ago is the period of pronounced Pleistocene overkill of big game animals (Martin, 1967). Haynes (1966) speculated that the opening of a high protein niche by fire-driving and hafted blade tools would have diminished infant mortality considerably, thereby catalyzing the beginning of a period of rapid population growth. The first human population explosion began in Africa about 50,000 years ago, reached Australia somewhat later as man spilled down the Malayan archipelago onto that continent, and began in the Americas via migration over the Bering land bridge perhaps 25,000 years ago (Haynes, 1969). All of these migrations were accompanied by wholesale extinctions of the native fauna.

By 10,000 years ago, when man's population is estimated to have been over 5 million (Deevey, 1960), the ultimate inefficiency of overeffective hunting had forced many populations, particularly those in marginal game areas, into a return to an existence similar to that of their ground ape ancestors. Along the southern margins of the Iranian and Anatolian plateaus (Fig. 20-4) in particular, hunter-gatherer populations had turned increasingly to gathering in the absence of abundant game (Wright, 1968). These populations moved up and down the mountains with the seasons—into forested areas during the more extended dry season which was an accentuation of the drought that had expanded Africa's savannas, and then down onto the plains during the wet season. Grains of wild grasses were a food staple in many areas and grindstones were being used to prepare them for eating in the Nile Valley as early as 15,000 years ago (Wendorf et al., 1970).

Populations in Asia Minor developed an alternate ecological strategy to hunting and gathering as they learned that returning to previous camp sites guaranteed them a food source where grasses grew on the garbage heaps of the previous wet season. Among the first of these garbage grains that was cultivated, as opposed to merely

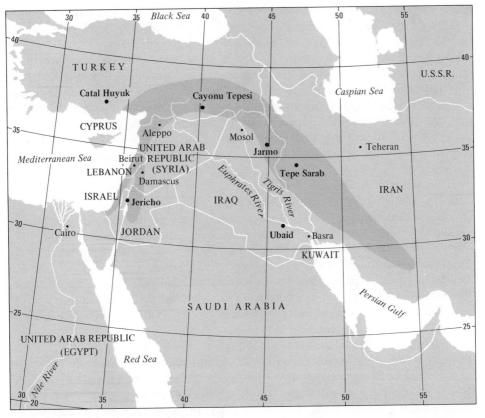

FIGURE 20-4 Map of the plateau-river basin transitional area of the mid-East superimposed on the present national boundaries. The shaded arc is where domestication of small grains and many agricultural animals occurred between 15,000 and 5,000 years ago. Principal excavated towns are indicated by dots. (After Harlan, 1971 and Boughey, 1971b.)

gathered, was wheat (Braidwood, 1960; Wright, 1968). This domestication occurred between 15,000 and 10,000 years ago. By 4200 years ago rice was being intensively cultivated by Oriental populations (Chang, 1968). These two ancient grains are still major food sources for the world's human population.

Obsidian, an important raw material in stone tools, was being traded, or at least transported, as far as 200 km from known sources by 12,000 years ago (Wright, 1968). In addition copper was being hammered into tools 9000 years ago in the Turkish village of Cayonu Tepesi (Cambel and Braidwood, 1970). The development of towns facilitated trading and a town midway between Mesopotamia and India was a busy trading center 5500 years ago (Lamberg-Karlovsky and Lamberg-Karlovsky, 1971). During the period from 12,000 to 8000 years ago there was considerable coexistence among farming and hunter-gatherer societies in such areas.

Cultivation of crops represented the expansion of the realized niches of important food sources through cultural practices. Selection from these grains for desirable

608

CHAPTER 20
a brief
ecological
history of
mankind

traits, such as grain retention, soon followed (Flannery, 1965). These practices were adaptive events of crucial importance since they expanded the food yield, though not necessarily the total yield, of the ecosystems supporting man. Domestication of plants and animals opened more abundant resources to evolving man, presenting an alternative to his previous dependence on ecological processes that had been directed largely by external forces. The contrast between operative factors in succession and in man's adaptive response to his environment is interesting. Succession proceeds from a dependence upon changes in intrinsic factors to changes in extrinsic factors. Man, in contrast, became less dependent on extrinsic factors and more dependent on intrinsic factors as he discovered agriculture and succeeding technologies.

Of equal ecological significance, particularly to man's future, was an impoverishment of diet in terms of the variety of food consumed compared with hunter-gatherers (Birdsell, 1963). This impoverishment required that man choose his diet carefully to avoid malnutrition. There is evidence to indicate that religious rituals among some early agricultural peoples may have been a device for assuring an appropriate nutritional strategy (Rappaport, 1971). And, just as man became less dependent upon extrinsic variables for his survival, dependence on maintenance of an appropriate cultural strategy became of utmost importance. It is hard to imagine a less self-sufficient society than an industrialized agricultural society if, overnight, its nonsolar energy sources were turned off.

Growth of early agricultural populations in East Asia and the Middle East resulted in the migration of excess individuals into marginal habitats. By 7000 years ago there were many villages along the Iranian Plateau with populations of 100 to 300 people (Hole, 1966). Some individuals must have been crowded into lower elevation sites along the flood plains of the Tigris-Euphrates Rivers. This competitive exclusion placed a premium on altruistic behavior as populations in the new habitat began to employ irrigation techniques to channel river water onto an otherwise parched plain. Irrigation projects required substantial present sacrifice for potential future benefit. A consequence of this requirement was the development of characteristic "hydraulic societies" in the Tigris-Euphrates, the Indus, and the Nile River valleys. Similar societies developed elsewhere, notably in China and on the Mexican plateau. As Wittfogel (1956) pointed out, "Where agriculture required substantial and centralized works of water control, the representatives of the government monopolized political power and societal leadership, and they dominated their country's economy." Such an organization represented a necessary adaptive response if the labor required for irrigation projects was to be managed effectively (Hole, 1966; Wolf and Palerm, 1955). The output of such a society was a large food excess which allowed unprecedented population growth. By 5000 B.C. Uruk, a city in southern Mesopotamia, covered over 445 ha. Assuming a population density equivalent to the early villages, over 50,000 people probably lived in Uruk. The second population explosion was beginning.

In considering the nature of such societies, Russell (1969) noted that, "To control these swarming masses, to assemble them for labor operations and organize their work, to handle the vast stores of surplus grain that began to be gathered in state granaries for feeding the labor gangs, a special kind of state evolved. At the head of it was a dictator, with absolute rights over the lives of all his subjects. However absolute his power, no one man can actually run the lives of millions of people. The ruler worked through enormous bureaucracies of state officials, supported by the surplus extracted from the peasants."

We think it is unlikely that such an organization developed from hunters and gatherers without substantial natural selection for behavioral traits. The hydraulic society represents something of the epitome of a dominance hierarchy in which some individuals command unlimited respect so that group activities may be more efficiently organized. Selection would favor both despots and subservient individuals since the yield realized from their administration and labor increased the probability of their genes contributing to the next generation. From the 5 million people in 8000 B.C. when agriculture was first developing, the human population of the earth grew to perhaps 250 million by 1 A.D. (Dorn, 1962). Much of this population was concentrated in hydraulic societies (Wittfogel, 1956).

Whereas dominance hierarchies were developing in conjunction with huge hydraulic societies on riverine plains, merchant and pastoral farming societies were developing in less productive habitats. By 3000 B.C. agriculture had spread throughout Europe (Waterbolk, 1968) and a complex trading system had developed among human societies. The discovery of smelting led to an increasingly sophisticated technology as stone implements were replaced by metal ones. The trading societies, which required a less rigid but no less participatory social system than the hydraulic societies, began to develop forms of representative government which were varyingly successful. Although the first published law code originated in the Mesopotamian hydraulic society, the concept of representative government was a product of the trading societies (McNeill, 1967).

Outside of the cultivated belt of irrigated and cleared forest lands, less rigid social structures were also developing among the herders of the great Eurasian grasslands. Horses had been used for cartage since about 4000 B.C., but herders in the mountain valleys of the Caucasus did not learn to ride them until 1300 B.C. and this practice spread rapidly north and eastward across the steppes (von Wissmann et al., 1956). This expanded the niche of animal herders into a more efficient nomadic life. By 900 B.C. the Scythians of eastern Russia had learned that their mobility had other advantages, notable in warfare, and began to raid the margins of the Near Eastern societies. Mobility was not conducive to a stable social organization and nomadic hordes tended to coalesce, separate, and coalesce again elsewhere with other bands (Russell, 1969). While a series of empires began, spread, and then contracted from the hydraulic and trading bases (the Assyrian Empire, the Median Empire, the Persian Empire, the Macedonian Empire, and the Roman Empire), the nomadic hordes made sporadic warfare on the fringes of these more organized social systems.

Hydraulic empires also were growing and then dying in Asia as a series of them attempted to expand their labor bases. Near the end of the fourth century B.C. these societies began to build a system of walls to protect themselves from each other. Finally in 221 B.C. the dictator of Ch'in unified them at terrible cost in human

610

CHAPTER 20
a brief
ecological
history of
mankind

lives (Russell, 1969). A similar unification of Indian hydraulic societies was accomplished in Asoka's empire around 250 B.C.

Eurasia, then, became ecologically differentiated into two types of ecological strategy: (a) agricultural trading societies with various formal organizational systems, the nature of which depended on the manpower needs of the resource exploitation strategy and (b) nomadic herding societies ranging across the Eurasian grasslands, with informal organizational structure dictated by the nomadic exploitation pattern, which required merely that grazing and watering rights be partitioned efficiently among tribes. As nomadic and farming populations grew, conflict among them increased.

By the second century A.D. "The nomad peoples of the Eurasian steppe may be likened to molecules of gas in a very leaky bottle" (McNeill, 1967) and both Asian and European populations were subject to periodic assault from armies of skilled horsemen. A peak of activity was reached in Europe in the fifth century A.D. as Attila's army ranged across the southern part of the continent early in the century while the Vandals spread across northern Europe and finally ranged into the eastern Mediterranean by the middle of the century. Since the forest clearings of Europe hardly constituted a favorable habitat for a nomadic grazing economy, the grazers merely raided and then fell back onto the prairie.

The nomads' search for new resources reached its culmination in the Khans, who forged them into a highly disciplined army of horsemen and infantry that conquered an empire unrivaled in extent but poor in food resources until it conquered the Asian hydraulic societies. It represented the maximum empire supportable by a society of nomadic grazers since it not only encompassed the entire Eurasian prairies, it also destroyed the cultivated lands of whole societies in Asia Minor and converted them into grazing lands. Along the system of roads built through the Khans' empire, European and Asian trading and hydraulic populations again came into contact following Marco Polo's journeys in the late thirteenth century. The earth's human population by then had grown to 400 million (Dorn, 1962).

By the early centuries of the second millineum A.D. the continent of Eurasia consisted of a system of different social systems, organized for exploiting resources differently (Fig. 20-5). Each represented a type of adaptive response to a certain habitat. Southern Asia consisted of hydraulic societies with large populations organized into totalitarian systems for efficient management of the public works required by irrigation agriculture. It seems likely that the bulk of the world's population, then as now, was concentrated in these societies (Wittfogel, 1956; Russell, 1969). The prairies of central Eurasia were exploited by nomadic herdsmen with an informal social structure which was periodically organized into armies that ranged out of these prairies onto adjacent areas. Northern Europe consisted of a number of separate feudal groups partitioning agricultural resources based on forest clearing and cultivation. Finally, there was a series of trading communities, first along the Mediterranean but later along the northern European coast, which produced manufactured goods for consumption by the ruling classes of feudal societies.

It should not be inferred that the development of distinct forms of organization was a smooth, well-ordered process. The lack of orderliness is particularly evident

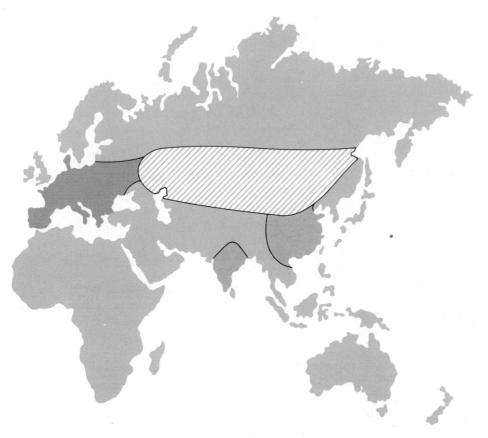

FIGURE 20-5 Map showing the distribution of feudal agricultural-manufacturing (dark brown), nomadic herding (diagonal hatching), and hydraulic (grey) societies in Europe, Eurasia, and Asia, respectively, about 1500 A.D. Other land areas were occupied primarily by hunting-gathering societies.

in the hydraulic societies in which the more radical departure from previous small group, tribal, hunting-gathering social systems occurred. Evolution of the hydraulic social systems was accompanied by periodic peasant revolts in which the laboring force of subservient individuals rose against the dominant organizing minority. In all probability such outbreaks were catalyzed by the less docile, or should we say less altruistic, members of the subservient classes. Success merely traded one group of dominant members for another while preserving the basic social structure. Regardless of their outcome these outbreaks were invariably followed by massive reprisals against the losers. The failure of a revolt, with subsequent wholesale destruction of "offenders," was an extremely effective mechanism for eliminating genes for aggressive behavior among members of the subservient class. We think it likely that such selection was an important component of the evolution of all human social systems.

The merchant and hydraulic societies of the Mediterranean area and the Far East required a bureaucracy with sufficient leisure time to develop mathematics and science, and to fuse the two in engineering projects that sustained their populations. A large bureaucracy, however, could not increase the efficiency of agriculture in such a "patchy" resource system as that in the deciduous forest lands of northern Europe. This patchiness arose out of the inability of farming techniques developed in southern Europe to exploit lowland forest soils. Farming, and consequently clearing, was confined primarily to ridgetops, generating a feudal system of partitioning resources. Sometime prior to 1000 A.D. Germanic tribes developed the mold-board plow capable of tilling the heavy soil of the European lowland (McNeill, 1967). This opened a new niche exploiting more fertile soils and decreased ecosystem patchiness. No forest soil, as we have already seen, is very rich in nutrients and sustained cultivation had not yet been achieved. The practice of rotating crops had been transported northward from southern Europe, but this practice left the ecosystem unexploited half the time. A crop rotation was gradually developed that left the soil fallow only one-third of the time. In addition the Europeans utilized a Chinese invention, the rigid wooden horse collar, to convert from an oxen power to a horse power agriculture between 800 and 1000 A.D. Because horses can work longer hours and do more work with less maintenance cost, the usable agricultural ouptut began to produce a manpower surplus.

Unneeded for the massive public works of hydraulic society by virtue of Europe's abundant rainfall, the excess population began to employ the inventions of hydraulic societies in unique ways. Watermills and windmills, for instance, invented by Greeks and Asians, respectively, were first widely employed in northern Europe for purposes other than grinding grain; England had over 8000 mills by 1086 (Russell, 1969). Europeans invented the factory by coupling mills to cranks, cams, shafts, and gears so that resources could be spun, hammered, polished, bent, and ground. Mass creation of manufactured goods catalyzed the development of a money economy and signaled the death knell of isolated feudal units. In 1650 an agricultural system in which wheat, turnips, barley, and clover were rotated over a four year cycle established a sustained-yield agriculture. As this spread outward from its Belgian origins, the productivity of ecosystems supporting European man increased dramatically. The resulting agricultural surpluses caused a precipitous decline in farm prices during the latter half of the seventeenth century and produced a shift of capital out of agriculture and into new industries (Russell, 1969). The earth's human population had grown to 500 million persons (Dorn, 1962) and the stage was set for European man, for the first time, to contribute significantly to future population growth.

Two revolutionary developments in the eighteenth century, James Watt's invention of the steam engine in 1769 and Edward Jenner's perfection of vaccination in 1796, fueled a positive feedback into human population growth that continues to the present. Coupling the steam engine to machinery created a dramatic upsurge in the per capita output of industrial societies. As McNeill (1967) remarked, "By

1789, indeed, English mills using cotton grown in India and imported to England around the Cape of Good Hope were able to undersell the Indian handweavers in India itself." After Jenner, vaccination greatly increased man's lifespan as vaccines were developed for the infectious diseases that had plagued him prior to 1800. Over the next century (Fig. 20-6), "The population of Europeans [increased] out of all proportion to that of the rest of mankind. It has been estimated that they made up about 22 percent of the world human population in 1800, about 35 percent in 1930" (Russell, 1969). This was a population explosion unprecedented since the hydraulic societies dominated population growth of early agricultural man.

"In Europe the population of the entire continent in 1800 was about 187 million. By 1900 it had increased to about 400 million, despite the fact that nearly 60 million had emigrated overseas during the . . . century and an unnumbered host had also crossed the Urals from European Russia into Siberia and central Asia" (McNeill, 1967).

What distinguished the European population was not its discoveries but its application of other population's discoveries to practical ends. Movable type, for instance, was invented in China in the eleventh century, but it remained for French copyists to combine movable type with an alphabet three centuries later. If the moldboard plow provided the tinder of the industrial revolution, the printing press lit that tinder by making knowledge widely available. This availability made a much broader segment of the population capable of participating in the innovations that support resource exploitation. A long series of intertribal wars that had surged across Europe, "Greatly increase[d] Europe's capacity to mobilize wealth, manpower, and ingenuity for political and economic purposes" (McNeill, 1967). For the technological Europeans the whole world became a resource, "They used their technological equipment mainly to control the rest of the world for purposes of trade, loot, and plantation agriculture using slave labor" (Russell, 1969). With European colonization slowing in 1900, the earth's human population stood at 1.8 billion.

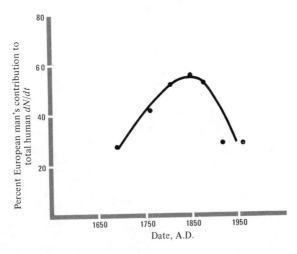

FIGURE 20-6 Relative contribution of the world's European populations to human population growth from 1650 to 1970. (Data from Hauser, 1971.)

614

CHAPTER 20
a brief
ecological
history of
mankind

The extension of the European ecological strategy, and particularly intensive agriculture and medical control of prereproductive mortality, to the rest of the world's human populations has resulted in a tremendous increase in population growth rate. European colonization (Fig. 20-7) resulted in the extermination of hunting and gathering societies throughout the Americas, Africa, and Australia. Developing agricultural and herding societies on these continents were also destroyed by the surge of European population. Another major adaptive type, central Eurasian herding, was engulfed by the European hordes sweeping over the Ural Mountains; and the world's great remaining hydraulic society (China), after colonization by industrialized populations from Europe and Japan, finally reconstituted itself in the 1950s.

In the subsequent decade, Chinese society directed the energies of its population into the most massive public works program in the history of mankind, moving an amount of dirt estimated at 450 times that moved in construction of the Panama Canal (Borgstrom, 1969). Another 40 million ha was irrigated, representing more than the total cultivated area in Canada. Today over 40 percent of the world's irrigated land is in China. More recently the Egyptian hydraulic society is attempting to recreate itself through the Aswan Dam project.

No other population even approaches European man, however, in the magnitude of resource exploitation. In 1965, for instance, North America, Europe, and the USSR accounted for over 79 percent of the human population's total energy consumption (Fisher and Potter, 1971). The rate of resource exploitation—that is, a population's "standard of living"—is closely correlated with, and presumably driven by, energy consumption (Fig. 20-8). The unprecedented ecological dominance of the European population arose out of its syntheses of scientific discoveries in the merchant and hydraulic societies into an exceedingly efficient mechanism of resource exploitation

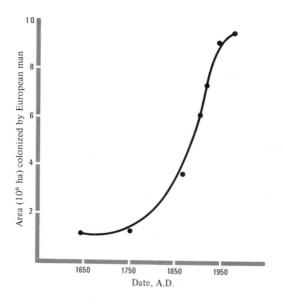

FIGURE 20-7 Total land area colonized by the earth's European population between 1650 and 1970. (Data from Hauser, 1971.)

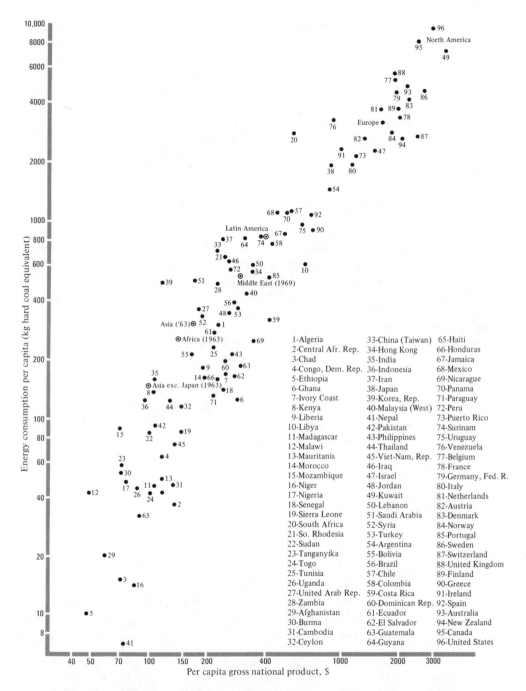

FIGURE 20-8 Data on the earth's nations showing the positive relationship between per capita energy consumption and per capita gross national product in 1965. (After Fisher and Potter, 1971.)

1-Algeria
2-Central Afr. Rep.
3-Chad
4-Congo, Dem. Rep.
5-Ethiopia
6-Ghana
7-Ivory Coast
8-Kenya
9-Liberia
10-Libya
11-Madagascar
12-Malawi
13-Mauritanis
14-Morocco
15-Mozambique
16-Niger
17-Nigeria
18-Senegal
19-Sierra Leone
20-South Africa
21-So. Rhodesia
22-Sudan
23-Tanganyika
24-Togo
25-Tunisia
26-Uganda
27-United Arab Rep.
28-Zambia
29-Afghanistan
30-Burma
31-Cambodia
32-Ceylon

33-China (Taiwan)
34-Hong Kong
35-India
36-Indonesia
37-Iran
38-Japan
39-Korea, Rep.
40-Malaysia (West)
41-Nepal
42-Pakistan
43-Philippines
44-Thailand
45-Viet-Nam, Rep.
46-Iraq
47-Israel
48-Jordan
49-Kuwait
50-Lebanon
51-Saudi Arabia
52-Syria
53-Turkey
54-Argentina
55-Bolivia
56-Brazil
57-Chile
58-Colombia
59-Costa Rica
60-Dominican Rep.
61-Ecuador
62-El Salvador
63-Guatemala
64-Guyana

65-Haiti
66-Honduras
67-Jamaica
68-Mexico
69-Nicarague
70-Panama
71-Paraguay
72-Peru
73-Puerto Rico
74-Surinam
75-Uruguay
76-Venezuela
77-Belgium
78-France
79-Germany, Fed. R.
80-Italy
81-Netherlands
82-Austria
83-Denmark
84-Norway
85-Portugal
86-Sweden
87-Switzerland
88-United Kingdom
89-Finland
90-Greece
91-Ireland
92-Spain
93-Australia
94-New Zealand
95-Canada
96-United States

616

CHAPTER 20
a brief
ecological
history of
mankind

and competitive exclusion. Constituting less than 30 percent of the world's population (Hauser, 1971), it dominates resource distribution and consumption in the biosphere.

In the last 70 years the world's human population has doubled and now exceeds 3.6 billion (Table 20-1). With an annual percentage growth rate near 2 percent, this population will *at least* double by the end of the century in the absence of a massive reversal of present reproductive or survival patterns. The conflict between pronounced concentration of resource utilization in a small portion of the population and increasing resource demand by the total population creates an ecological problem of massive proportions. It seems likely that the degree to which man has been, and can be, selected for altruistic behavior is about to be tested on an unprecedented scale.

It would be ironical, if it were not so tragic, to hear the industrialized nations cited as models of regulating human populations. During the period that European man was "regulating" through industrialization, his exploited land area increased over 900 percent to the detriment of many other of the earth's populations. Even at present the industrialized nations use their technologies to exploit resources far outside their borders. Borgstrom (1969) has calculated "ghost acreages" for many of the world's nations (Table 20-2). These ghost acreages are calculated from net imports of food, feed grains, fiber, and fish. They result from trade balances in favor of manufactured goods rather than raw materials. Compare, for instance, such developed and "regulating" countries as the Netherlands and Japan, with ghost acreages of 63.9 percent and 64.1 percent of their total acreage, with such underdeveloped and "nonregulating" countries as India and El Salvador, with 8.1 percent and 15.5 percent of their total acreage in ghost acreage. Said Borgstrom, "Holland is often put forth as an example to be followed by the entire world. Several economic handbooks and publications repeat this statement despite the fact that it is wholly unrealistic, almost absurd. If the world as a whole did follow the Dutch example . . . we would . . . need an additional satellite one and a half times the present agricultural acreage of Earth."

TABLE 20-1 Estimates of world population by regions (1650–1950)

Estimates and dates	Millions						
	World total	Africa	Northern America	Latin America	Asia (except USSR)	Europe and Asiatic USSR	Oceania
1650	545	100	1	12	327	103	2
1750	728	95	1	11	475	144	2
1800	906	90	6	19	597	192	2
1850	1,171	95	26	33	741	274	2
1900	1,608	120	81	63	915	423	6
1920	1,834	136	115	92	997	485	9
1930	2,008	155	134	110	1,069	530	10
1940	2,216	177	144	132	1,173	579	11
1950	2,515	222	166	162	1,381	571	13
1970	3,659	348	233	243	2,064	712	19

Hauser, 1971

TABLE 20-2 Ghost acreage of selected countries (1966–1967)

Area ($\frac{1}{100}$ ha) per capita

		Ghost acreage		
	Tilled land	Fish acreage	Trade acreage	Total
China	13.7	4.4	0.5	18.6
India	31.7	1.5	1.2	33.4
Pakistan	23.5	3.6	2.2	29.3
Indonesia	15.9	3.0	−0.7	18.2
Japan	5.8	22.3	15.9	44.0
Brazil	34.7	4.8	−0.5	39.0
U.A.R. (Egypt)	8.9	1.8	3.6	14.3
Nigeria	35.1	6.5	3.5	45.1
Peru	21.8	25.1	20.7	67.6
U.K.	13.5	9.1	29.1	51.7
Italy	29.0	9.2	15.8	54.0
France	40.5	4.2	1.3	46.0
Poland	48.5	6.5	5.1	60.1
Scandinavia	45.2	17.5	3.4	66.1
Netherlands	7.3	18.1	11.1	36.5
Belgium	9.4	10.3	20.9	40.6
Ceylon	16.1	5.1	1.0	22.2
Taiwan	6.8	9.7	2.6	19.1
Israel	15.4	12.1	42.9	70.4
El Salvador	19.6	0.8	2.8	23.2
Costa Rica	39.2	1.6	−12.9	27.9
Jamaica	12.8	12.0	15.4	40.2

Borgstrom, 1969

Human population growth is a classic example of a positive feedback system. Reproductive output per unit population member, in contrast to all other known animal populations, has been a positive function of number of individuals in the human population (see Chapter 9). This positive feedback has been generated by a combination of evolutionary social-technological events leading to an extended period of niche expansion for the ground ape population. The recent inputs to this positive feedback have consisted of the following (McKeown and Brown, 1955; Hauser, 1971):

1. Productivity increases produced by the agricultural, commercial, and industrial revolutions.
2. Improvements in environmental sanitation and personal hygiene, resulting from uncontaminated food and water and a decrease in the probability of infection and contagion.
3. The natural disappearance of some causes of death and disability, for example, scarlet fever.
4. The development of modern medicine.

Of these item 3 is most likely a result of evolution; that is, it is probable that the spontaneous disappearance of some of the contagious diseases that made early

618

CHAPTER 20
a brief
ecological
history of
mankind

urban life so perilous arose out of natural selection for biological resistance in the population. Items 2 and 4 undoubtedly were secondary effects of the driving force, item 1—that is, the revolution in per capita output generated by human harnessing of environmental energy sources for man's own needs. As we have already seen, the agricultural revolution in northern Europe leading to cultivation of lowland soils converted a patchy resource system into a continuous one, thereby leading to the death of feudalism. Furthermore, the employment of excess manpower in the new industries led to urbanization, with a premium consequently put on sanitation and medication as a means of preserving these dense urban populations. Simultaneously with the demand for techniques for preserving humans in a habitat that was certainly more dense than any they had previously occupied, the society was producing sufficient wealth that some members could indulge their curiosity about human sickness by, for instance, tracking down typhoid vectors. Positive feedback by man on his population growth rate arises out of evolutionary-cultural innovations, most of which consume energy stores to drive succession to early, species-poor ecosystems that are nonpersistent in the absence of human manipulation.

MODERN MAN: RESOURCES

Neither humans nor resources are evenly distributed across the earth's land surface. Population tends to be concentrated in certain oceanic coastal areas that constitute good ports, thereby facilitating trade and manufacturing, and in the basins of a few great rivers where intensive hydraulic agriculture or manufacturing is practiced (Fig. 20-9). These important rivers are the Rhine, Nile, Ganges, Yangtze, and Huang. Population growth is supported by food supply and although some populations are certainly overfed, the distribution of food is not so incongruent with population as many statements in the popular press would lead one to believe. China, for instance, has a population of about 773 million (Topping, 1971), or slightly over 21 percent of the world's total. In terms of food (Williams, 1963) China produces 33 percent of the world's rice, 30 percent of the millet and sorghum, 14 percent of the fish, 17 percent of the peanuts, 28 percent of the soybeans, and 29 percent of the swine. The world's most populous nation is quite self-sufficient at meeting current food needs. Other countries, particularly in southern Asia and South America, are not so efficient.

The continents differ conspicuously in the areal efficiency of their agriculture and in the percent of arable land that is cultivated (Table 20-3). Three continents suffer from an agricultural paradox that man's ingenuity has not yet resolved: a means of sustained cultivation of tropical forest and desert soils. The marked infertility of the former has forced most agriculture in tropical regions into a "swidden" mold, in which the forest is cut, the trees are burned to release stored nutrients into the soil, and the land is farmed for two to three years before being abandoned again. Although often referred to as a "primitive" type of agriculture, this is certainly a misnomer since "advanced" societies have been unable to perfect a more efficient adaptive response to the tropical forest biome.

Major promise for increasing agricultural yield lies with the extension of practices

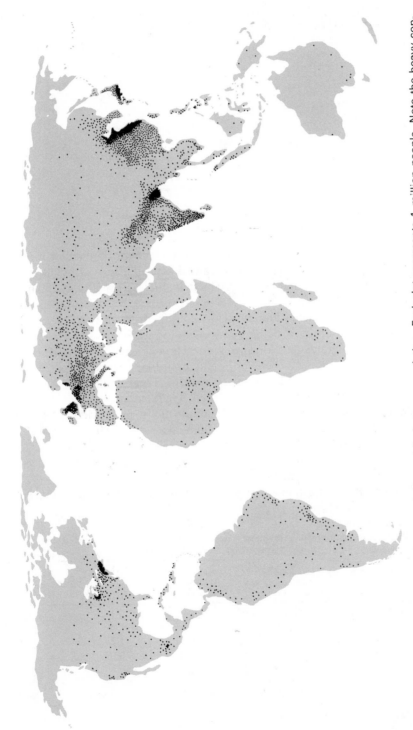

FIGURE 20-9 Geographic distribution of the world's human population. Each dot represents 1 million people. Note the heavy concentrations of people in river valleys and coastal areas. (After Williams, 1963.)

620

CHAPTER 20
a brief
ecological
history of
mankind

TABLE 20-3 Percentage of potentially arable land now cultivated, and hectares cultivated per person, on different continents

Continent	Percent cultivated	Hectares cultivated per person
Asia	83	0.28
Europe	88	0.36
South America	11[a]	0.40
Africa	22[b]	0.53
North America	51	0.93
USSR (Europe-Asia)	64	0.97
Australasia	2[c]	1.17

[a] Tropical limitation.
[b] Desert and tropical limitation.
[c] Desert limitation.
Hendricks, 1969.

perfected on maize to other crops (Table 20-4). Total maize production increased 70 percent during a 30-year period when planted acreage declined by 40 percent. This increase, which arose out of usable productivity increases of almost 200 percent during the period (Hendricks, 1969), was brought about by breeding programs producing hybrid corn varieties and by tremendous increases in fertilizer consumption. Similar practices are now being extended to the world's major grains, wheat and rice, through programs developed to a considerable extent by the Rockefeller Foundation (Borland, 1965; Rockefeller Foundation, 1967). Some people, however, question the probability of their ultimate success in balancing the food-people equation (Paddock, 1967; Ehrlich and Holdren, 1969).

In addition to an expansion of terrestrial productivity, the world's fisheries may show continued increases in yield over the immediate future (Ricker, 1969). This food source is primarily in the form of supplemental protein rather than calories. Such protein, of course, will be extremely important in supplementing the low-protein, high-caloric diet of grain.

Summarizing the best estimates of potential food yield, the National Academy

TABLE 20-4 Effects of plant breeding (hybrids), management improvement, and use of fertilizer on maize (corn) production in the United States (1933–1963)

Year	Acres harvested (millions)	Production (millions of bushels)	Yield (bushels per acre)	Percent in hybrids	Plant food used in Illinois (1000 tons)
1933	106	2,400	22.6	0.1	
1938	92	2,550	27.7	14.9	
1943	92	2,970	32.2	52.4	28
1948	84	3,600	42.5	76.0	
1953	80	3,210	39.9	86.5	294
1958	64	3,360	52.8	93.9	
1963	60	4,100	67.6	95.0+	821

Hendricks, 1969

of Sciences Committee on Resources and Man (1969) concluded that, "Forseeable increases in food supplies over the long term . . . place the earth's ultimate carrying capacity at about 30 billion people, *at a level of chronic near-starvation for the great majority.*" At a constant rate of growth equivalent to the present 2 percent, 107 years would be required to reach this size. Since, however, human population growth is itself a positive function of population size (Fig. 9-10), this population size may be reached within 65 years.

In addition to food the industrial societies need minerals. For many of these they have relied increasingly on sources outside their own borders (Fig. 20-10), as internal sources that could be exploited cheaply were exhausted. The leading producers of bauxite, for instance, are two underdeveloped countries, Jamaica and Surinam, which supply over 30 percent of this raw material for aluminum manufacture. The major world producers of aluminum, in contrast, are the United States and the USSR, which produce over 65 percent of the ingot aluminum (Foster and Harth, 1970). As the industrial nations have stripped their internal mineral supplies, the energy requirements of mining have increased substantially (Fig. 20-11); arising in part out of declines in ore quality, domestic mining energy requirements have driven capital increasingly into exploiting high-grade ore sources in underdeveloped countries. This suggests that the energy cost of development may become too high for the underdeveloped countries to pay upon depletion of their high-grade ore supplies.

The ocean has been considered as a potential source of mineral elements. How-

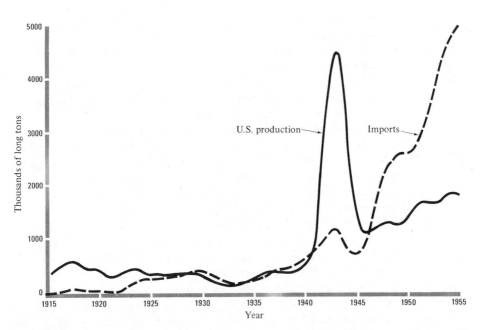

FIGURE 20-10 Relationship between domestic and imported sources of bauxite consumed in United States aluminum manufacturing. (After Cloud, 1969.)

622

CHAPTER 20
a brief
ecological
history of
mankind

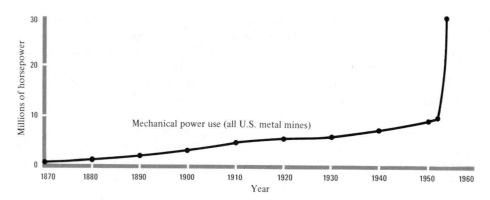

FIGURE 20-11 Increase in power consumption by United States domestic metal mines with time. (After Cloud, 1969.)

ever, if the cost of exploiting current terrestrial deposits is believed high, the cost of exploiting oceanic sources is astronomical (Cloud, 1969). Zinc is one of the minerals in limited supply in present markets and it is abundant enough in seawater to suggest recovery; current consumption in the United States economy is 122,400 tons annually. By completely stripping the zinc from 9 trillion gallons of seawater, about 9 cubic miles, we could get 400 tons of zinc. Cloud observed, "The practicality of such an operation is not impressive."

Summarizing the mineral resources of the world, the NAS Committee on Resources and Man concluded, "True shortages exist or threaten for many substances that are considered essential for current industrial society—mercury, tin, tungsten, and helium, for example. Known and now-prospective reserves of these substances will be nearly exhausted by the end of this century or early in the next." These projections suggest that it is not feasible to expect continued expansion of the present industrial economies, much less conversion of the nonindustrial economies into anything resembling those of the current developed societies.

Energy is life in the industrialized society. It drives the social system and provides the goods and services that are associated with the "standard of living." Because of this there are rather thorough theories of energy reserves and energy consumption worked out for the world (Hubbert, 1969). The theory of energy reserve discovery and exploitation proposes that rate of fossil fuel consumption is in the form of a modified normal curve against time (Fig. 20-12). This curve has been validated on small areas. The conterminous 48 states, for example, are about midway in an exploitation curve for crude oil based on reserves of 165×10^9 barrels. Over 80 percent of these energy resources will have been consumed in the single lifespan of an individual born in 1935; that is, over 80 percent of the United States domestic oil supply (exclusive of Alaska) will have been consumed in a 65-year period between 1935 and 2000. Another important source of fossil fuel energy, in fact, the fuel of choice until the mid-twentieth century, is coal. Estimated coal reserves are gigantic by comparison with oil, a total of 7.6×10^{12} metric tons (Averitt, 1969). These reserves, like most resources, are extremely heterogeneously distributed in the earth's

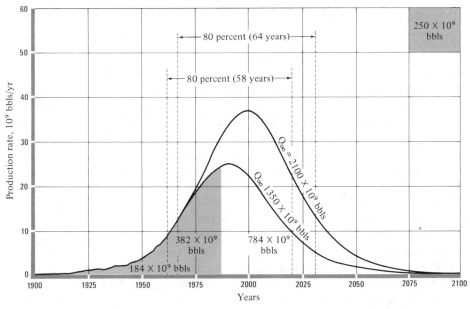

FIGURE 20-12 Estimated time distribution of world's crude oil reserves for two values of estimated total reserves ($Q\infty$). The shaded area represents known deposits, the other area is extrapolated from previous resource discovery patterns. The bulk of the world's crude oil will be consumed in the next 50 years. (After Hubbert, 1969.)

crust. Almost 57 percent of the estimated coal is in the USSR and 19 percent is in the United States. The remaining 24 percent is spread among the rest of the world's people, ranging down to 0.2 percent in South and Central America. Based on the discovery-exploitation curve, Hubbert (1969) estimated that the period of coal consumption could be sustained for a 340-year period from 2040 to 2380. It seems likely that, in terms of energy reserves, coal could sustain an extended period of industrialization.

The general problem of defining carrying capacity is emphasized by the energy problem. Clearly coal reserves are capable of sustaining a substantial industrial population on earth. These reserves are conspicuously concentrated in certain countries, however, and a more homogeneous distribution of consumption would require an equitable world trade system. In addition coal burning produces another environmental problem of first magnitude—air pollution. For example, the depletion of England's forests by a burgeoning population forced Henry II to give reluctant consent to burning coal (Ayres, 1956). The increase in coal burning thereupon forced his son, Edward I, to sign a decree imposing the death penalty for burning coal in London while Parliament was in session. As we have already mentioned the detrimental effects of fossil fuel burning in London reached something of a climax in December 1952 when 1 out of every 2000 citizens died from heavy smog. The reason many environmental problems are not a simple linear function of population size is due to the sort of positive feedback loop exemplified by air pollution:

624

CHAPTER 20
a brief
ecological
history of
mankind

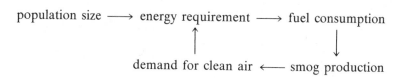

An increasing population creates an increased energy demand which increases fuel consumption. Increased fuel consumption generates smog, which generates a demand by the populace for clean air. Only if demand for clean air leads to a reduction of energy demand is the feedback loop opened.

In Edward I's time air pollution led to a ban on coal consumption. In recent times the demand resulted in fuel oil being substituted for coal and in the installation of emission control devices when further substitutions were not feasible. There has been sporadic demand by all industrial populations for curbs on air pollution, but most of these curbs are energy dependent, thereby catalyzing the positive feedback loop. The 340-year estimate of coal reserves probably would be considerably diminished in practice by positive feedback if a switch was made to a coal economy.

Another alternative source of power that has received considerable attention and research money expenditure is nuclear energy (Hubbert, 1969). Most of the research, however, has been focused solely on the question of whether it is possible to produce power economically from nuclear processes. Almost no effort has been expended on answering the more serious ecological question of what is to be done with the incredibly toxic wastes such processes produce. As Hubbert points out, "Radioactive wastes are distinguished from all other kinds of noxious wastes of chemical origin by the fact that there is no method of treating them to counteract their innate biological harmfulness." Most of the wastes of nuclear reactors are produced in the form of a viscous liquid. Present disposal practices include (a) pumping slurries into fractured rock at depth of 700 to 1000 ft at Oak Ridge, Tennessee, (b) pumping slurries into subsurface water in Idaho and Washington, (c) burying solid wastes in caverns in Tennessee and in trenches at other Atomic Energy Commission localities, and (d) discharging short-lived gaseous isotopes into the atmosphere from tall stacks.

Hubbert concluded that present disposal practices violate principles established by an advisory committee to the AEC in 1955 and that "Wastes are not being isolated from the biological environment at present, and it is questionable to what extent the same practices can be continued when the rate of waste production becomes 10 or 100 times larger than it is at present without causing serious hazard." It seems clear that the AEC reactor development program has gotten the cart before the horse. Widespread deployment of nuclear power generators in the absence of adequate disposal systems for their wastes is stupid at best and criminal at worst.

As the resource discussion demonstrates, there is considerable variation in estimates of K (carrying capacity) for the human population, depending on what one believes will be the limiting factors. Consumed resources, such as energy reserves, are amenable to straightforward, concise analysis. Nonconsumed resources, such as the ability of the biosphere to accept the toxic wastes of energy consumption, are much less tractable. Minerals, of course, are not consumed in the absolute sense, but increasing proportions of some of them, such as helium, become forever lost to the biosphere.

Water also is a nonconsumed resource, circulating continuously through the pools of the hydrologic cycle. But industrial societies use water in a very real sense, since they load it with pollutants that make it incapable of further use until purification by the hydrologic circulation system or, more rarely, by the society itself. Borgstrom (1969) estimated that if all humans used water in the same fashion in which it is used by the population of the United States, the carrying capacity of earth would be 3.65 billion persons; that is, if all the earth's people attained the same standard of living as those in the United States, by the same ecological strategy, the earth has already reached its carrying capacity for human beings as defined by water resources.

MODERN MAN: PROPERTIES ASSOCIATED WITH DENSITY

Man's uneven distribution on the surface of the earth is related to his strategy of resource exploitation. Populations of hydraulic societies tend to concentrate in the irrigated river basins they have shaped by their labors, whereas populations of industrialized societies tend to concentrate in urban centers of manufacturing and trading. The former populations are organized into a series of villages spread over the landscape to organize the work force efficiently. China's population, for example, is 80 percent rural (Foster and Harth, 1970). The industrial population is concentrated in cities so that it may be organized daily at commercial centers. An example is the United States' population which is 70 percent urban.

Many medical disorders and other causes of death and disability are correlated with urbanization (Ford, 1970). Studies of populations living in small, tribal social groups indicate a lower incidence of high blood pressure than in urban societies and no association between age and blood pressure (Cassel, 1971). Migrants from these groups to urban industrial areas, however, have higher blood pressures and the familiar positive correlation between age and blood pressure found in industrial populations.

Urbanization was associated with a series of waves of infectious diseases, some of which abated for no apparent cultural reason (Cassel, 1971). Tuberculosis was one of the first of the industrial diseases and although there were a number of cultural programs designed to control the disease, its actual decline began long before these programs were instituted. This suggests natural selection in the industrial population for tuberculosis bacillus resistance. The scourge of tuberculosis was replaced, in turn, by major malnutritional syndromes (pellagra and rickets), several diseases of early childhood, duodenal ulcer, and the current epidemic of heart disease, cancer, arthritis, and mental disorders. Some of these, such as the nutritional deficiencies and childhood diseases, were controlled by cultural modification, that is, by better diets and vaccination. Others, such as duodenal ulcer, declined for unknown reasons. Studies of lung cancer death rates after correction for smoking behavior (Haenszel et al., 1962) indicate that rural-born persons migrating to the city have significantly higher rates than city-born urban residents. The degree to which such density-related

626

CHAPTER 20
a brief
ecological
history of
mankind

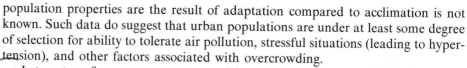

population properties are the result of adaptation compared to acclimation is not known. Such data do suggest that urban populations are under at least some degree of selection for ability to tolerate air pollution, stressful situations (leading to hypertension), and other factors associated with overcrowding.

Let us turn for a moment to a consideration of one aspect of human density that each of us can appreciate directly—family size. A number of studies have been done on the relationship between family size and properties of the family members (Wray, 1971). These provide the most direct evidence available of the damaging effect of density on human beings. In a study in Cleveland (Dingle et al., 1964) the incidence of a number of common illnesses was a positive function of family size (Fig. 20-13). Similarly, in families in Candelaria, Colombia, the percentage of malnourished children was a positive function of family size (Wray and Aguirre, 1969). Finally, intelligence as measured by standard tests was a negative function of family size in Minnesota (Reed and Reed, 1965). These studies, from a variety of cultures, provide strong correlative evidence that children pay a serious price for their parents having large families. In addition to the effects on surviving children, repeated births reduce the success of childbearing (Fig. 20-14). In a study of over 400,000 births in New York State, for example, survival of offspring declined with their birth order (Chase, 1961).

It might be argued that these effects arise out of economic conditions rather than family size—that is, that poor parents are likely to have larger families, therefore the effects above should be considered as economic rather than due to family size per se. This, however, is not an important refutation of the ecological argument; that is, decreasing resource supply to an individual is detrimental to that individual. In fact, the economic difference may only accentuate the ecological argument. When addressed on its own assumptions, however, the argument of economic correlation is not validated. Douglas et al. (1968) examined this proposition in some detail by studying British children grouped according to socioeconomic class and found that the negative relationship between intelligence test scores and family size held within families grouped into four different social classes. With social class (income level) held constant, intelligence test scores were clearly a negative function of family size (Fig. 20-15), indicating that the detrimental effects of family size are real, rather than resulting from spurious correlation with economic factors.

We have concentrated on the effects of family size on children, but there is also evidence relating family size to parental characteristics. A study of British families (Hare and Shaw, 1965) indicated that both physical and mental health deteriorated in parents as family size increased and also the effect was more marked in mothers than in fathers. Similarly a study of hypertension in Zulu families (Scotch, 1963) indicated that this illness in parents was a positive function of family size with the effect more pronounced in mothers. That overall population density is related to this effect is indicated by the fact that such a pattern was not found in rural Zulu families; it was found only in urban locations. What is not certain, of course, is the degree of cause and effect among these relationships. It might as well be argued that as the health of the parents deteriorates, their reproductive output increases.

627

MODERN
MAN:
PROPERTIES
ASSOCIATED
WITH
DENSITY

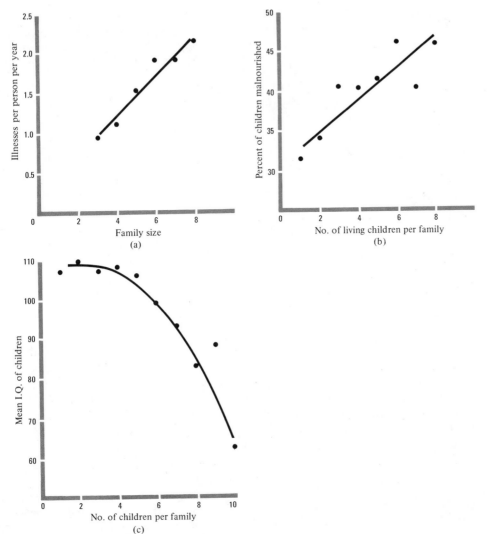

FIGURE 20-13 Relationship between family size and (a) gastrointestinal illness rate in Cleveland, Ohio, (b) malnutrition incidence in Candelaria, Colombia, and (c) I.Q. in Minnesota families. (Data from Dingle et al., 1964, Wray and Aguirre, 1969, and Reed and Reed, 1965.)

In our opinion this is a less plausible interpretation than the converse one—that overreproduction results in a deterioration in the health of the parents.

A serious problem of interpretation arises out of these studies of the effect of human density on individual humans. To what extent are such effects inherited and to what extent do they represent environmental modifications of genetic potential? For instance, do inherently less healthy parents tend to have larger families? Or

628

CHAPTER 20
a brief
ecological
history of
mankind

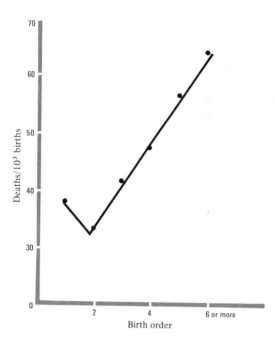

is large family size detrimental to realization of an individual's full genetic potential? Data on birth order suggest the latter. The fact that many of the same traits associated with family size are similarly associated with birth order suggests that parents insisting upon large family size are penalizing their offspring by restricting the development of their full physical and mental potential. Furthermore, the parents, and particularly the mothers, are paying a price of their own. This price is probably

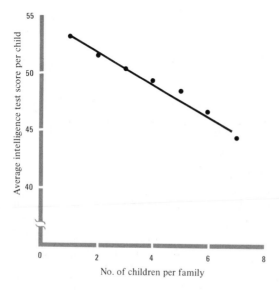

FIGURE 20-15 Decline in average intelligence test scores of family members with increase in size of British families. The social class (income level) was held constant in these studies. (Data from Douglas et al., 1968.)

accentuated with increasing urbanization. Clearly the well-being of individual humans is not served by high population density within their world, their localities, or their families.

MODERN MAN: POPULATION MOMENTUM

Many of the concepts we have been considering in this chapter have employed the technique of demography, the statistical study of human populations. The most obvious conclusion from such a study is that the world's population and resource consumption are not equally distributed among the world's nations. In fact, both are distributed according to the familiar lognormal series of ecological universes (Fig. 20-16). At the right-hand tail of the population lognormal are a series of giant nations: China, India, the USSR, the United States, Pakistan (prior to the separation of Bangla Desh), Indonesia, and Japan. These seven nations contain about 60 percent of the earth's 3.6 billion human beings. At the right-hand tail of the economic lognormal are a series of resource-dominating nations, the United States, the USSR, Japan, West Germany, and France. These five nations generate about 65 percent of the earth's total (3 trillion dollars) annual economic activity. We can offer the same sort of descriptive interpretation of these curves that was given earlier for the partitioning of resources among species populations: The most "abundant" nations combine high density or per capita economic activity with large areas or populations,

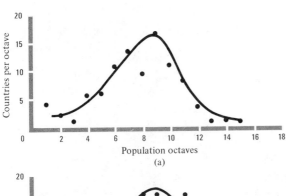

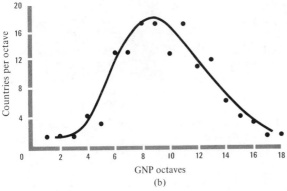

FIGURE 20-16 Approximation to lognormal distributions of (a) national population and (b) gross national product. The lowest population octave begins at 30 million persons and the largest ends at 1.092 billion persons. The smallest GNP octave begins at 1.6 million dollars and the largest ends at 1.02 trillion dollars. (Data from Foster and Harth, 1970.)

630

CHAPTER 20
a brief
ecological
history of
mankind

whereas the least "abundant" nations combine low population density or per capita income with small areas or populations. Why this should be so, of course, admits of no current functional interpretation. Perhaps the same factors govern the development of nations that govern the evolution of species; or perhaps it is merely a coincidence that the outcome of both processes fit the lognormal.

One of the most important properties of a population from the standpoint of understanding population growth and its consequences is population age structure (see Chapter 9). If rapidly growing and stable populations are compared according to the proportion of individuals in a series of age classes (Fig. 20-17), these age "pyramids" are quite different. In the stable population, individuals are relatively evenly distributed among the age classes with an abrupt tailing off at the top of the pyramid where mortality occurs. In the rapidly growing population individuals are concentrated in young age classes and the pyramid falls off at a relatively constant rate from the base.

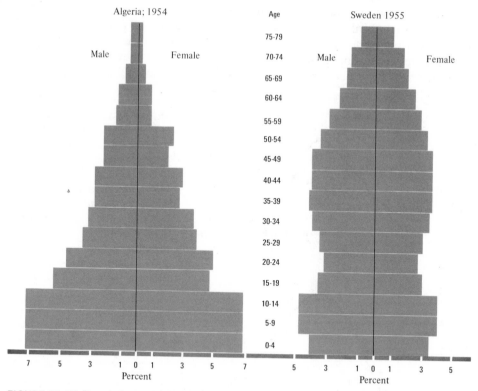

FIGURE 20-17 Population age structure for a rapidly growing nation (Algeria) and a nation approaching population stability (Sweden). The large proportion of pre-reproductives in the rapidly growing population produces considerable momentum for further growth. (After Dorn, 1962.)

Different age structures have significant consequences and, consequently, the society's future resource demands for the future course of population growth. In a typical rapidly growing (3 percent per year) modern population, in which the death rate has been curbed and the birth rate continues high, the population will have about 45 percent of its individuals younger than 15, 50 percent between 16 and 59, and 5 percent over 59 years old (Hauser, 1971). In other words almost half of the population is in a prereproductive developmental state. This generates a very high population growth momentum. Even in a typical slowly growing (1 percent per year) modern population, with only about 27 percent of the population in the prereproductive age classes, there is considerable reproductive momentum in the population. Consider, for instance, Italy, with an annual growth rate of 0.64 percent and 24.2 percent of its population under the age of 15 (Keyfitz, 1971). If the age specific birth rates dropped instantaneously to the level required to generate a stable population, the Italian population would still grow by 17.1 percent before stabilizing. This growth arises because the large prereproductive class must pass through the reproductive period. In order to stabilize at the present level, a couple would have to produce fewer children than the number required to replace themselves as they mature. If Italy has considerable momentum, however, let us consider Ecuador. This nation has an annual growth rate of 3.3 percent and 47 percent of its population is less than 15 years old. If its age specific birth rates dropped instantaneously to a replacement level, the population would grow by 66.7 percent before reaching a stable size.

This property of human population growth, its momentum, is among the most difficult to grasp and the most significant from the standpoint of its future impact. Because human population growth is a positive feedback system, there is no stable point in the population trajectory. Unlike animal populations, some of which are K responsive, the human population may not be able to depend on intrinsic regulatory mechanisms to stabilize its population size. The length of time it has taken for the human population to double is a negative exponential function of both population size (Fig. 20-18) and time.

There are, of course, intrinsic limits, such as litter size and age to sexual maturation, to the rate at which the population can double. Extrapolating the population growth trend into the future gives a date of 2026 A.D. for instantaneous doubling (von Foerster et al., 1960). This need not be interpreted literally, although it might occur as a result of widespread use of fertility drugs inducing multiple ovulations. A more reasonable interpretation would be that near this date the human population will become singularly unstable, which in turn could set off a population crash of appalling proportions, perhaps an extinctive crash.

Since neither population, nor resources, nor growth rates are equally distributed among the nations, should we assume that some populations will crash and others will not? The nations with the most stable growth characteristics and, we might assume, therefore less liable to crash are also those nations with the greatest dependence on ghost acreages to sustain their populations. The developed nations depend for their current standard of living on a flow of easily exploitable resources,

632

CHAPTER 20
a brief
ecological
history of
mankind

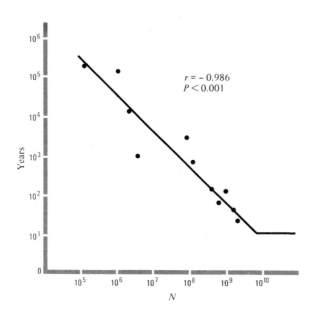

FIGURE 20-18 Decline in the time required for the world's human population to double in size with increase in size of that population. A lower limit of approximately 12 years is assumed, although this might be violated with widespread use of fertility drugs inducing multiple ovulations. The last dot is our position during the doubling time just past; we are very close to the physiological limit on doubling time. (Population data from Deevey, 1960, Dorn, 1962, and United Nations, 1969.)

many of which come from the underdeveloped world; it is difficult to imagine that widespread crashes in the underdeveloped world would spare the economies of developed nations.

One of the traditional consequences of overpopulating a resource supply has been war. This consequence was made explicit prior to World War II, "The National Socialist movement must strive to eliminate the disproportion between our population and our area. . . . We must hold unflinchingly to our aim . . . to secure for the German people the land and soil to which they are entitled" (Hitler, 1925, cited in Shirer, 1960). Whether the world's future human population can severely overpopulate resources without catalyzing increasingly severe wars can only be imagined. A more complete mechanism of population crash than extensive thermonuclear war could hardly be designed, since it would probably disorganize as well as destroy populations.

DEMOGRAPHIC TRANSITIONS AND MAN'S FUTURE

Our discussion of the population explosion suggests a sort of runaway reproduction, with couples copulating as if there were no tomorrow. Such behavior implies a general ignorance of the mechanisms of birth control even though methods of controlling population size are widely known and practiced even among nontechnological people (Boughey, 1971b): Infanticide was practiced in five hunting and gathering societies and three early agriculture societies, abortion was practiced by all but one of the groups, and half of the groups practiced intercourse avoidance to control reproductive output. It seems likely that both death rates and birth rates were relatively low in such populations and although life expectancy was comparatively short, the general level of health was good (Russell, 1969) and population size was controlled by balancing fecundity and mortality through regulating the former (Carr-Saunders, 1963).

The coming of the hydraulic revolution, which provided a burst of food resources to support population growth and a ready demand for laborers, also resulted in a substantial increase in parasitism and disease (Russell, 1969). Yellow fever, malaria, and bilharziasis are only a few of the diseases that began to afflict man concentrated into areas with substantial standing water. It seems likely that a consequence of this was an increase in the death rate and concomitant greater increase in birth rate to generate a growing population (Fig. 20-19).

In the feudal agricultural and developing trading-industrial societies of Eurasia, increasing concentration of people in towns catalyzed an increasing incidence of typhus and bubonic plague. The great pandemic, the Dance of Death, began in southern France in December 1347 and rapidly spread northward across the continent. Within five years, between one-third and one-half of Europe's population was destroyed. The number of deaths estimated for Pope Clement VI totaled almost 46 million persons. This devastatingly high death rate could only be counteracted by a high birth rate. A high birth rate, coupled with a subsequent fall in the death rate resulting from improved diet, better medical and sanitary practices, and genetic disease resistance in the population, brought about the explosive growth of post-fourteenth century European populations. Finally, three and a half centuries later the birth rates began to fall in most European countries.

Shifts in birth and death rates are products of the ecological conditions of the population and the consequent "evaluation," conscious or unconscious, by reproductive population members of their chances of reproductive success. High death rates as a part of the cultural experience probably produce pressures for high birth rates as a means of assuring "survival" into the next generation. A subsequent drop in death rates, with birth rate trends invariably lagging behind, would produce

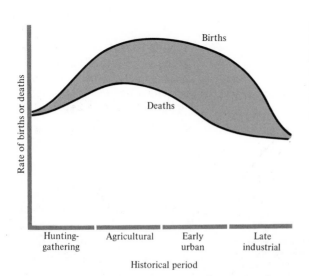

FIGURE 20-19 Hypothetical distribution of birth and death rates in human populations at different development periods. The population growth rate increases until births begin to converge on deaths in late industrial societies.

634

CHAPTER 20
a brief
ecological
history of
mankind

population explosions. The final demographic transition, once again closing the gap between birth and death rates, would then convert a population to stability.

It should be emphasized that this demographic transition is hypothetical, except for the last stages. It is well known that the growth of European nations has gradually subsided. What distinguishes these populations from current ones facing a huge gap between birth and death rates is that the European populations were able to use their technological expertise to displace other societies all over the world. Such a strategy does not appear feasible today.

More hopeful is evidence that the time taken to complete a demographic transition is constantly decreasing (Fig. 20-20). Although countries entering a demographic transition in the latter part of the nineteenth century required about 50 years to complete that transition, countries entering a transition in the 1950s completed it in 15 years (Kirk, 1971). The average annual decline in birth rates during demographic transitions beginning between 1875 and 1969 was $-0.30/1000$ population for northern European countries and the United States, -0.25 for southern European countries, -0.30 for central European countries, -0.44 for eastern European countries and Japan, and -0.91 in the underdeveloped countries which only recently entered demographic transitions. These data suggest that a convergence of birth and death rates may be accomplished much more rapidly for present populations than it was earlier.

In addition to the traditional methods of reducing fecundity, such as later marriage and sexual abstention, a variety of technological fecundity limitations have been developed over the last few decades. In varying degrees almost any method of controlling conception is more successful than chance (Table 20-5), the intrauterine device and birth control pills being the most effective. It is interesting that these two "modern" methods of birth control are both based on folk medicine, the birth control pill on the consumption of contraceptive plants by Central American Indians and intrauterine devices on the use of "magic rings" by women in the United States and perhaps elsewhere. Abortion is another technique that was commonly practiced

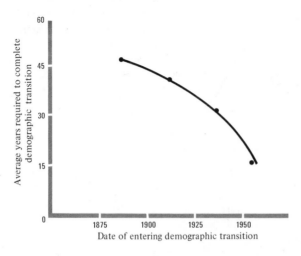

FIGURE 20-20 Decline, with time, of the period required to complete a demographic transition from 35 births per thousand persons to 20 births per thousand. (Data from Kirk, 1971.)

Average years required to complete demographic transition

Date of entering demographic transition

TABLE 20-5 Relationship between contraceptive technique employed and pregnancy rate of women of childbearing age

Method	Pregnancy rate (percent)
None	61
Douche	31
Rhythm	21
Jelly only	20
Coitus interruptus	18
Condom	14
Diaphragm	12
IUD	2.6
Sequential pills	2.0
Combination pills	0.1

Boughey, 1971b

by nonindustrial populations to limit fecundity. Most of the countries that have successfully undergone demographic transitions employed legalized abortion as a significant mechanism (Omran, 1971). Of course, the effectiveness of this method of limiting family size is 100 percent although it clearly is a much less desirable technical solution than contraception. The fact that Japan's abortion rate has declined since completion of their demographic transition suggests that the expediency of abortion will be traded for contraception if the choice and knowledge of its use is provided.

The most critical component of accomplishing a demographic transition, in the absence of governmental compulsion, is, of course, the desire of reproductive members of a society to reduce their reproductive output. An extensive survey in 1965 indicated that most of the world's human population believed that continued growth was undesirable (Table 20-6) and there was also wide approval of proposals for implementing national birth control plans. Fewer respondents, however, felt that growth of their own country was bad. Although the question apparently was not asked, we cannot help but wonder how many people would have agreed that continued growth of their own family was bad. Experience in countries in which

TABLE 20-6 Attitudes toward population growth and family planning, twenty-two countries, 1965

	Percent believe national growth a bad thing	Percent believe world growth a bad thing	Percent approve a national birth control program
Europe	67	86	76
Near East	53	78	74
Far East	53	71	82
Latin America	29	56	74
Africa	33	54	54

Stycos, 1971

636

CHAPTER 20
a brief
ecological
history of
mankind

TABLE 20-7 Opinions on population and family planning, Lima (circa 1966)

	General sample	Opinion leaders
Percent who agree that:		
Rapid population increase slows national progress.	65	51
Steps should be taken to regulate births in Peru.	79	63
To control the number of children is to contribute to the well-being of the family.	91	79
The state should teach people to limit the number of children.	81	75
The state should reward large families.	72	46
Each new child arrives with its own loaf of bread.	42	34
The Catholic woman should not use contraceptives.	54	45
Many children are needed to assure one's old age.	35	12
Number of cases	(1,000)	(100)

Stycos, 1971

birth control techniques have been made widely available and public education programs on the desirability of limiting family size have been instituted would suggest that, by and large, people are having more children than they desire. As Wray (1971) pointed out, "Family planning surveys all over the world have shown repeatedly that mothers with three or four children want no more." The problem, of course, is that three or four children will not stablize the world's populations given current death rates.

Studies of birth control attitudes in Latin American populations indicate that those who suffer most from overpopulation are also most in favor of controlling that population (Stycos, 1971). In five cities, the percentage approving of national birth control programs was 56 percent in the upper socioeconomic class, 66 percent in the middle class, and 71 percent in the lower class. An examination of birth control attitudes in Lima, Peru, compared a general sample of the population with persons characterized as "opinion leaders" by virtue of positions in government, the professions, and industry (Table 20-7). This survey indicated that opinion leaders lagged sadly behind the general population in their acceptance of the necessity of population regulation. How significant such disparity is in the failure of governments to implement effective family planning programs is not clear.

MAN'S ECOLOGY: A POSTSCRIPT

Our objective in the preceding sections has been to provide a reinterpretation of man's history in terms of the ecological principles developed in the rest of the book. Such a reinterpretation has, of necessity, been impressionistic. Our principal point is that the current status of *H. sapiens* is the product of the ecosystem resource structures which confronted him, and his adaptive responses to those resource structures. We think it is as unrealistic to regard such major cultural differences as those between industrial and hydraulic societies as solely nongenetic as it is to regard all individual differences as totally noncultural. It is apparent that present disparities between birth and death rates cannot be sustained indefinitely. Neither can the present resource exploitation strategies of the industrial societies be extended widely to other populations, or be extended far into the future. It is often argued that man is distinguished from all other organisms by his consciousness of the consequences of his actions. The extent of this consciousness is going to be severely tested over the next few generations. If man, collectively, has such consciousness, birth rates and death rates, together with resource utilization rates by different societies, will converge voluntarily. If we do not, it seems that all three will involuntarily converge on zero.

The ground ape has grown up. Or has he? It is commonly said that man "represents another order of being" (Pfeiffer, 1969). We are not so sure of that. If man is another order of being, it is because of his ability to modify his behavior nongenetically to conform with future survival needs; that is, if man is unique, it is because of his collective consciousness and his ability to modify behavior collectively under the direction of that consciousness. We are very uncertain that man possesses this ability. Whether he does or not will soon be decided by a rapid convergence of birth rate on death rate or, alternatively, a rapid transition to a huge excess of deaths over births.

GENERAL CONCLUSIONS

1. Climatic oscillations followed by directional climatic change, which modified the balance between forest and savanna biomes in Africa, placed a selective premium on traits allowing certain of the arboreal primate populations to colonize terrestrial habitats. These colonists are on the evolutionary line leading to modern man and selection for both physical and psychological traits are an important component of our evolutionary heritage.

2. The four principal strategies of habitat exploitation present in historical times are hunting-gathering, nomadic herding, hydraulic agriculture, and trading-industrial societies supported by intensive agriculture. The unprecedented growth of populations with the last technique of habitat exploitation led to extinction of many of the first two types from the seventeenth to the early twentieth centuries.

3. Although the factor, or factors, which will define the earth's carrying capacity for human beings are by no means certain, an examination of projected supplies of required resources suggests that the industrial-trading societies will deplete stocks of many resources currently regarded as essential within the next three decades, while many of the nonindustrial societies are chronically on the brink of food shortages.

4. The momentum of man's population growth, arising out of positive feedbacks on natality and negative feedbacks on mortality, has generated a constantly accelerating population growth rate as population size increases. This momentum suggests a population lacking intrinsic regulatory mechanisms sufficient to limit population size in advance of exceeding the carrying capacity.

ecology: some conclusions and projections   21

A series of generalizations in ecology can be abstracted from the general conclusions sections of preceding chapters. But while these are important hypotheses for organizing our approach to ecological phenomena, they are not yet developed into a consistent conceptual framework. Our purpose in this chapter is to abstract what we feel are a few of the most important conclusions developed in previous chapters, as a summary of the current state of the science, and to project certain questions from the conclusions that we feel may represent significant research areas of the future. Some of these generalizations are merely definitions that aid in focusing our thinking, while others are abstracted from a well-developed conceptual base, modified by empirical data. When ecology was introduced in the first chapter, we stated as our objective an understanding of how the fluxes of energy and materials into individual organisms give rise to complex communities (Fig. 21-1). A discussion of what the authors regard as some of ecology's important conclusions may help us understand how well we have succeeded in that objective, and how much is yet to be learned. In doing this, we realize we are treading on precarious ground, but the

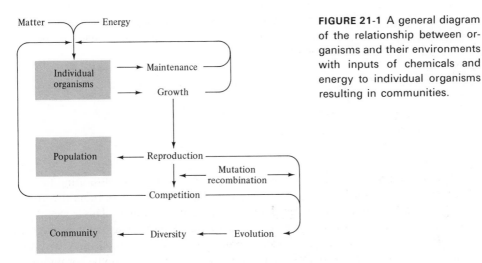

Matter ———— Energy

Individual organisms —→ Maintenance

—→ Growth

Population ←—— Reproduction

Mutation recombination

Competition

Community ←—— Diversity ←—— Evolution

FIGURE 21-1 A general diagram of the relationship between organisms and their environments with inputs of chemicals and energy to individual organisms resulting in communities.

639

TEN SUMMARY CONCLUSIONS

potential benefits seem worth the risk. The following is offered primarily to provoke the reader into reflection on the material previously covered, rather than as a definitive summary of that material.

TEN SUMMARY CONCLUSIONS

1. One of the first generalizations developed in the book was the idea of the ecosystem; that is, that organisms and their environments form a functional unit organized by fluxes of energy and chemicals. The idea that nature is organized as ecosystems is not particularly profound, and has been an implicit component of man's approach to nature for centuries. Recognizing it explicitly, however, has provided ecology with a strong organizing definition that focuses on the forces integrating organisms with their surroundings in nature. It is clear that this fundamental ecological idea is not well understood outside the science of ecology. Widespread resistance to the idea that chemicals designed to poison insects might poison other organisms demonstrates that the idea has not been well understood outside of ecological circles.

We have seen in our development of the ecosystem idea that the principal factors responsible for ecosystem organization are the following: (a) Extrinsic variables such as temperature, radiant energy supply, geology, and currents (atmospheric and aquatic circulation), which are independent of the organisms but define the general limits within which organisms must exist; (b) Intrinsic variables such as energy and nutrient pools which are modified directly by the organisms present; and (c) the genetic adaptations of the organisms present. Adaptations and extrinsic variables determine responses to intrinsic variables, and extrinsic variables may be important factors in the replenishment of depleted internal pools. Our approach to ecosystem exchanges can be refined by compartmental models (Fig. 21-2) designed to describe exchanges of energy and materials in the ecosystem through equations precisely fitted to data.

640

CHAPTER 21
ecology:
some
conclusions
and
projections

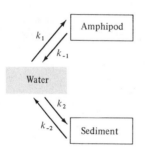

FIGURE 21-2 A compartmental diagrammatic model of exchange between ecosystem compartments regulated by rate constants (k's).

2. A second major generalization we considered was the law of limiting factors: there will be a certain factor or combination of factors which determine the ability of organisms to maintain a viable population in any ecosystem. This concept has been applied on almost every page of the book, since natural selection is directed by limiting factors, and the relationships among organisms in the same ecosystem is determined by limiting factors. From our consideration of ecotype formation in response to environmental gradients to our discussion of human population growth, the law of limiting factors has been a fundamental organizing generalization.

3. Another fundamental generalization which was introduced early in the book to direct our approach to ecology was the concept of niche: each organism has a certain set of environmental conditions within which it can exist, and its efficiency on each factor within the set will vary for different ranges of the factor. We also recognized the fact that the potential distribution of an organism may be modified by the presence of other organisms and discriminated between a genetically determined fundamental niche and a realized niche which reflected interactions with other organisms in nature.

4. Another generalization, intimately related to the law of limiting factors and the niche concept, is

$$dN/dt = r_mN(1 - N/K) \tag{21-1}$$

the logistic growth equation (Eq. 9-7). This equation examines compartment exchanges in the ecosystem in more detail through factors establishing carrying capacity; that is, limiting factors in population growth. Although we have seen that the application of this equation to real situations requires considerable qualification and refinement, it still encompasses the important idea that the rate of population growth will depend on population size and reproductive potential of population members, as modified by environmental-carrying capacity (Fig. 21-3). The precision with which it can be applied to real situations depends in part on defining N and K in functional units (such as energy flow) and in part on the degree to which the population examined has been selected for regulatory mechanisms which are density-dependent. Its fundamental basis, that no population can indefinitely outstrip the supply of limiting resources, is as valid today as when first proposed over a century ago. This, then, provides us with an understanding of the relationship between the individual and population blocks of Figure 21-1.

For plants, in which competition may be expressed in size modifications as well

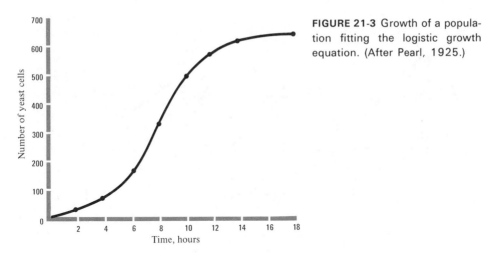

FIGURE 21-3 Growth of a population fitting the logistic growth equation. (After Pearl, 1925.)

641

TEN
SUMMARY
CONCLUSIONS

as density-dependent changes in fecundity and mortality, competition is accurately described by

$$w = Cp^{-3/2} \tag{21-2}$$

where w is mean plant weight, p is the density, and C is maximum size of an individual of the species (Eq. 11-8). This relationship holds for relating both interspecific and intraspecific competition to requirements of a single plant.

5. Fundamental to Equation 21-1 is the idea of competition among genetically similar individuals for limiting resources. Since individuals rarely occur only with individuals of the same population, the equation was expanded to include interspecific competition:

$$\frac{dN_i}{dt} = r_{mi}N_i \frac{K_i - N_i - \alpha_{ij}N_j}{K_i} \tag{21-3}$$

where α describes the degree of niche overlap between the two populations and relative exploitation efficiencies within the overlap zone (Eq. 11-1). We saw, from an examination of this equation, that competing species would coexist only if competitive relations established a balance among the competitors (Fig. 21-4). Therefore this connected populations with communities in Figure 21-1. Once again, of course, we have examined the many qualifications that must be attached to application of this simple generalization to real situations.

6. In addition to interactions among individuals at the same trophic level, inherent in Equations 21-1, 21-2, and 21-3, ecosystems are organized by relationships among trophic levels. This was considered in detail through the predator-prey interaction summarized in

$$dH/dt = r_{mH}(1 - H/K_H) - k_p HP \tag{21-4}$$

where H and P were prey and predator densities, respectively (Eq. 10-1). Constant density relationships among predators and their prey, we saw, depend on a predation

642

CHAPTER 21
ecology:
some
conclusions
and
projections

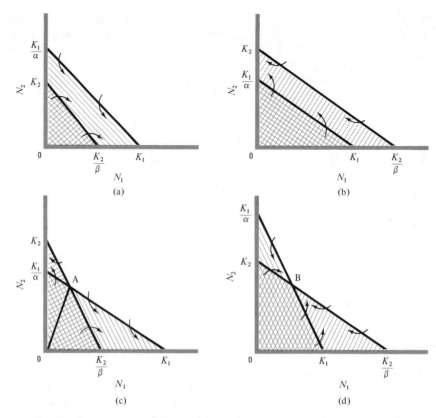

FIGURE 21-4 Generalized outcomes of competition between two species: (a) species 1 always wins, (b) species 2 always wins, (c) the winning species depends upon position of the populations around the isoclines, (d) a stable equilibrium in which the species coexist at the intersection of the equilibrium lines.

rate which is balanced with the reproductive potential of the prey population (Fig. 21-5). Since we restricted the definition of community to organisms at the same trophic level in an ecosystem, the consideration of predator-prey relations provided an approach to understanding how Figure 21-1 is expanded to connect different trophic levels. This expansion occurs through the flow of energy and materials from prey individuals to predators, with subsequent effects on predator populations and communities, as modified, of course, by all the processes of Figure 21-1.

7. An important consequence of the reciprocal relationship between organisms and their environments is a change in ecosystem organization with time. In newly established habitats, there is colonization by organisms which modify the ecosystem's intrinsic variables resulting in a sequential species replacement sequence called succession (Fig. 21-6). Eventually, an equilibrium species combination develops on the site which is relatively persistent through time, as modified by changes in extrinsic variables. A consistent pattern in succession is an increase in diversity such that the

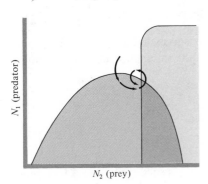

Damped cycle

FIGURE 21-5 A generalized predator-prey model in which the prey and its predator reach an equilibrium where their respective isoclines cross.

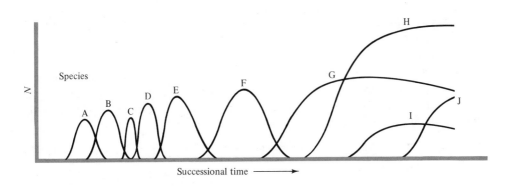

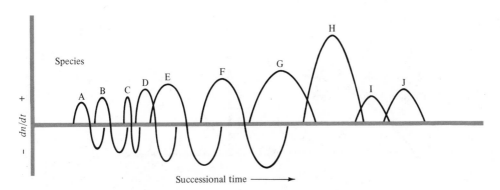

FIGURE 21-6 A generalized successional sequence with species labeled A to J. In the upper graph, population sizes are plotted against time. In the lower graph, population growth rates are plotted against time.

644

CHAPTER 21
ecology:
some
conclusions
and
projections

number of species present may be described by

$$S = a + b \log t \tag{21-5}$$

where S is the number of species present and t is successional time (see Eq. 13-1). The number of species present is a consequence of the interaction between immigration and extinction (Eq. 13-9) with

$$S = \int_0^t (I - E)dt \tag{21-6}$$

and

$$dS/dt = I - E \tag{21-7}$$

where I and E are immigration and extinction rates, respectively (Fig. 21-7). Since a change in the number of species present must be expressed through the individual population members, these equations are the sum of the population growth, competition, and predator-prey relationships considered earlier.

8. An important manifestation of the geographic distribution of extrinsic variables, particularly climate and soils on land, and depth and water circulation in freshwater and marine ecosystems, is different equilibrium species combinations, called *biomes*. These are distinguished by the growth forms of their primary producers and the feeding strategies of their consumers.

9. The balance among populations changes constantly in space, and the degree of distinctness among ecosystems is a function of the abruptness of environmental change. The steeper the environmental gradient, the more rapid the balance changes among populations. Within a small community sample, abundances seem to be divided among the species according to the geometric series

$$N_r = N[p(1 - p)^{r-1}] \tag{21-8}$$

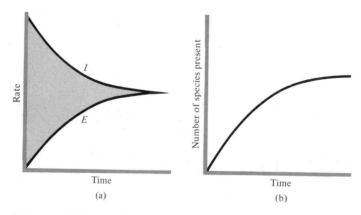

FIGURE 21-7 A generalized colonization model in which the number of species present at any time is a consequence of past immigration and extinction curves. (After MacArthur and Wilson, 1963, 1967.)

where N_r is the number of individuals in the rth species on an abundance ranking from most to least abundant species and p is a proportionality constant (Eq. 19-6). Since N_r is a measure of the local carrying capacity, as modified by predation and interspecific competition, the sums of these population processes must somehow give rise to the geometric series.

Community diversity is a function of the degree of habitat variability, as determined both by spatial gradients of limiting resource combinations, and exploitation efficiencies of component populations. High diversity suggests complex habitat gradients and fine division of those gradients among the populations present.

10. In summary, the principal factors determining ecosystem organization are:
   (a) The spatial and temporal *distribution of extrinsic variables.*
   (b) *Adaptations* of available immigrants to extrinsic and intrinsic variables.
   (c) The *supply of limiting resources,* that is, niche-carrying capacities. These will depend on site history, site location, and the evolutionary histories of available immigrants.
   (d) The *variety of limiting resources,* as determined, once again, by site history, site position, and adaptive properties of organisms.

These generalizations may be summarized as a series of five equations (page 647) from one describing energy and chemical flux in ecosystems to one describing the partitioning of energy and chemicals among species in a community. Most ecologists would throw up their hands in despair on seeing such a list of over-generalizations, and they are presented here for two reasons: (a) they have been useful as working hypotheses to guide ecological research and (b) to provoke the reader to reflect on the ecological phenomena they attempt to describe. It should be emphasized that these equations are merely an economical way of stating ideas that also can be verbalized. In fact, much of the book has consisted of stating these ideas, and their limitations, in verbal form. We are confident that a general theory of ecology is not beyond reach and that modifications of the ideas inherent in these equations, refined by careful data collection, will contribute to that theory.

A FEW PROJECTIONS

There are five basic research areas which run through ecology, and which have been reflected throughout the book. These are the study of (a) adaptations influencing the efficiency of resource utilization, (b) the relationships between efficiencies of resource utilization and competition and predation, (c) manifestations of competition and predation in energy and chemical flows, (d) factors influencing the stability of energy and materials flows, and (e) the relationships between the ecosystem's abiotic component and organization of the biotic component.

The diversity of approaches to ecology, combined with the unifying objective of understanding the functioning of organisms in nature, has provided the previous series of generalizations that reflect our current understanding of how ecological systems are organized. The ecosystem must be viewed as a dynamic entity, subject to modification by both extrinsic and intrinsic variables. Changes in energy and

646

CHAPTER 21
ecology:
some
conclusions
and
projections

chemical flow, arising out of adaptive changes in competition and predation, as well as invasion by more effective competitors and predators, result in a constantly changing dynamic balance among the populations present. The species within an ecosystem are composed of a variety of individuals that are more or less distinct genetically. The variability of the habitat occupied serves to define the fitnesses of individual genotypes. At the same time the variety of genotypes also contributes to environmental heterogeneity and operates as a further factor in natural selection. Thus the interaction between phenotype and environment is also a dynamic equilibrium that produces continuing evolutionary modifications of the genotypes present.

At the most general level we can define two approaches to ecological questions that reflect the type of problem involved. These are the questions of how ecosystems function and of why they function the way they do. The former approach may provide important insights into the dynamics of limiting factors and the relation of these to population balances. The latter focuses on the adaptive mechanisms responsible for the dynamics of limiting factors and population relationships. Clearly, a predictive theory of ecology depends on both, and the distinction between them is not clear-cut.

As an example of these two approaches with reference to the major threads of ecological research, we might examine feeding efficiency. For a bird population feeding on a certain array of prey, the question of how ecosystems function would focus on the fluxes of energy and chemicals from the prey to the predator and their relation to other ecosystem fluxes. The question of why ecosystems function the way they do would focus on such adaptive properties as bill size and structure, feeding techniques, and the relation of these to the yield of energy and limiting resources to the population.

An important development that will continue to influence ecology with more force is the growing realization that plant, animal, and microbial ecology are not separate disciplines. Many of the phenomena that have developed in studies of plant systems are equally applicable to animal systems, and vice versa. In general, the concepts of most interest in plant, animal, or microbial ecology have broad implications beyond the group where the observations were first made. This allows the formulation of more critical tests, with the organisms studied being those most susceptible to a certain analytical technique. This does not mean that ecologists will, or should, rush out and become instant generalists. There are obviously interesting and important questions that can be asked most appropriately with certain groups of organisms and not others.

One of the curses of ecology, or of any science, is the uncritical acceptance of certain fashionable ideas merely because of the force of their logic. An example from ecology is the widespread acceptance of the idea that diversity generates stability. Clearly, our survey of ecology indicates that more definitive tests are needed before this idea can be either accepted or rejected. What is often referred to as a dearth of solid theory in ecology arises more out of failure to test good ideas than a lack of such ideas. We believe that well thought out tests of the ideas in the following summary would go far toward providing a solid theory of ecology. Our most gratifying reward for writing this book would be realized if it provoked some of you who read it into designing such tests.

Resources will be taken up by ecosystem organismal compartments in a concentration-dependent fashion until uptake and loss are balanced:

$$dO/dt = k_1 E - k_{-1} O$$

If the organismal concentration of all limiting resources exceeds the amount required to meet maintenance costs, individuals will grow and reproduce, and the population will therefore grow

$$\frac{dN_i}{dt} = r_{mi} N_i \left(\frac{K_i - N_i - \alpha_{ij} N_j}{K_i} \right) - k_p N_i P$$

as regulated by carrying capacity, competition, and predation. The density of a population at any time will be a consequence of past growth trends:

$$N_t = N_o + \int_0^t \frac{dN}{dt}$$

The number of different populations (species) in a community is a function of the logarithm of the total density:

$$S_t = a + b \log \sum_{1 \to s} N_t$$

Finally, the density of a given population in a community is a product of the total density times a term describing the position of the species on a geometric series:

$$N_i = \sum_{1 \to s} N_t [p(1 - p)^{i-1}]$$

Since P_n drops about one order of magnitude at each higher trophic level, if N_i is expressed energetically, the equations above may be expanded to each trophic level to encompass the entire ecosystem.

SYMBOLS

O	concentration of a resource in an organismal compartment
E	concentration of a resource in an environmental compartment
k_1	resource uptake coefficient
k_{-1}	resource excretion coefficient
N_i	density of the ith population
N_j	density of the jth population
N_t	density of a population at a specified time
r_{mi}	maximum reproductive potential of the ith population
K_i	environmental-carrying capacity for the ith population
α_{ij}	competition coefficient describing the effect of individuals of the jth population on individuals of the ith population

648

CHAPTER 21
ecology:
some
conclusions
and
projections

k_p	predation coefficient
P	predator density
S	number of populations in a community at a given time
$\sum_{1 \to s} N_t$	density of all the populations in a community at a given time
p	a proportionality constant describing the slope of a geometric series

GENERAL CONCLUSIONS

1. Ecosystems are organized by fluxes of energy and chemicals, with the relationships among populations defined by rates of exchange and pool sizes.

2. Populations consist of genetically distinct individuals, with the variety of environments exploited both a consequence and a cause of the degree of variation among individuals. This variation is a fundamental organizer of ecosystem fluxes.

3. Populations respond to resource depletion in ways that reflect their evolutionary responses to resource predictability. Strong density-dependent regulation will occur in populations exploiting niches with resource supply predictable in time; weak density-dependent regulation will occur in populations exploiting niches with an unpredictable carrying capacity.

4. Fluxes of energy and materials arise out of the adaptive factors affecting competitive ability and predation efficiency; the stability of those fluxes will be determined by adaptive responses to both extrinsic and intrinsic variables.

5. Changes in energy and chemical flows arising out of site modification by the organisms present will result in sequential population replacements on a site, that is, succession, until a dynamic equilibrium is established.

6. Habitat variability, by defining the variety of genotypes which can effectively exploit resource combinations on a site, is an important factor in the establishment and maintenance of equilibria among populations in a given area.

7. Different equilibrium combinations, that is, biomes, occur in areas with different types of habitat.

8. Man is presently a nonequilibrium species, constantly modifying energy and materials flows in ways that he perceives to be to his benefit. No one has yet suggested a technique for maintaining this adaptive response very far into the future.

data analysis in ecology

Early ecological studies were primarily natural history observations that constituted a relatively nonquantitative base for modern ecology with its emphasis on quantification and controlled experiment. Scientists seek cause-and-effect relationships which will allow generalization from specific observations to more inclusive cases; a final product of ecological investigations is theory predicting the organization of previously unexamined ecological situations. To achieve precision in prediction requires accurate information and precise use of that information. The most precise way of describing many relationships between phenomena or objects is mathematical expressions. As a consequence many ecological relationships appear finally in mathematical form. Throughout this appendix are examples of the calculations involved in expressions commonly used in ecology.

Data are a group of measurements of phenomena or objects. A *variable* is a single phenomenon or characteristic that can take more than one value. Ecologists are interested in what determines the values of variables in ecological situations. For instance, the size of insects eaten by a bird species is a variable that might be related

649

to another variable, the length of the birds' bills. Bill size might determine the size of prey taken. Bill size then is a variable that is independent of prey size and is called an *independent variable,* but prey size, a *dependent variable,* probably depends in part on bill size. While cause-and-effect relationships generally are most easily examined by looking at the relation between independent and dependent variables, it is also possible to examine the relation between two dependent variables. A predictable relation between two dependent variables may provide insight into a common independent variable that influences both.

RELATIONSHIPS BETWEEN VARIABLES

A predictable relationship between two variables can be expressed in terms of a mathematical equation. The simplest relation is when one variable is some constant proportion of another variable. The equation for this relationship, which takes the form of a straight line (Fig. A-1), is

$$Y = a + bX \qquad \text{(A-1)}$$

where Y is the dependent variable, X is the independent variable, a is the intercept of the line on the Y-axis (the value of Y when $X = 0$), and b is the proportionality constant, the slope of the line, between Y and X. This slope, of course, may be of either positive or negative sign.

More complex relationships often can be transformed to a simpler linear relation by an appropriate transformation of variables. For example, the equation

$$Y = aX^b \qquad \text{(A-2)}$$

indicating that Y is an exponential and curvilinear function of X (Fig. A-2) can be transformed to a straight line by expressing each side of the equation as logarithms (in this case to the base 10):

$$\log_{10} Y = \log_{10} a + b \log_{10} X \qquad \text{(A-3)}$$

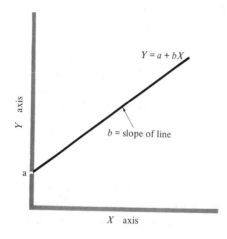

$Y = a + bX$

$b = $ slope of line

a

Y axis

X axis

FIGURE A-1 A linear relationship between two variables, X and Y, can be described mathematically by an equation of the form: $Y = a + bX$.

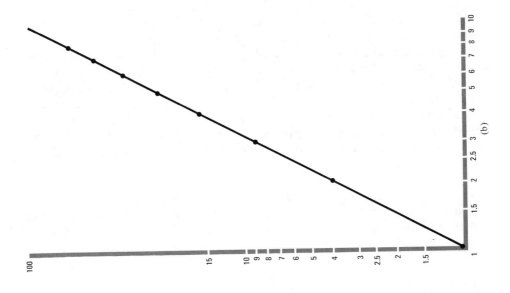

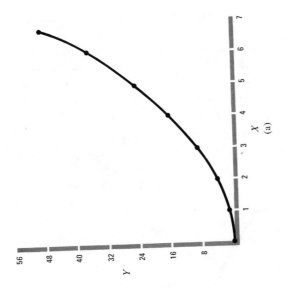

FIGURE A-2 One type of an exponential relationship between two variables can be described mathematically by an equation of the form: $Y = aX^b$. This is plotted arithmetically in (a) and on logarithmic axes in (b).

Second-order equations, such as

$$Y = a + bX + cX^2 \tag{A-4}$$

describe more complex relationships between two variables (Fig. A-3) that include both a unitary X term and an exponential function of X.

Many phenomena in ecology involve *rate functions,* changes of values of variables with time. The differential equation is a way of expressing an instantaneous (or nearly so) rate of change and is frequently employed in ecology. The notation dN/dt (the derivative of N with respect to time) is a method of noting that some variable, N, is changing a certain amount over a period of time, t. The d portions of the expression indicate that the rate of change is an instantaneous measure, but we usually can achieve only an approximation of the instantaneous rate measured over a short period of time. An example of the rate function is found in equation (A-5) of growth rate of a population for which there is no movement of individuals in or out:

$$dN/dt = B - D \tag{A-5}$$

where dN/dt denotes the rate of change of numbers in the population over a time period (t), B is the number of births in that time period, and D is the number of deaths. If births exceed deaths, dN/dt will be positive and the population will grow; if deaths exceed births, dN/dt will be negative and the population will decline.

CENTRAL TENDENCY AND VARIABILITY

The goal of precise predictions in ecology is complicated by the variability inherent in many biological data. Few organisms will respond identically to a given set of conditions; in fact response of the same organism to the same stimuli will vary at different times. To deal with variability a technique is required that defines levels of confidence in drawing conclusions. Just how confident can one be that the data are related in the way they seem to be? This requires the use of statistical methods.

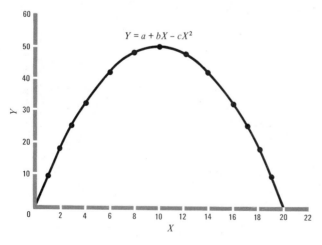

FIGURE A-3 A curvilinear relationship between two variables described by a second degree polynomial equation.

However, a problem with using statistics in making statements from biological data is that one can never make absolutely positive statements, but must always make a negative statement that a null hypothesis is probably not correct (at the same time tacitly accepting the alternative hypothesis).

A *null hypothesis* is an hypothesis of no differences (Siegel, 1956), that there is no difference in the populations from which the samples are drawn (Simpson, Roe, and Lewontin, 1960), or that the variables are not related in a specified manner. Variation in the data decreases our confidence in rejecting the null hypothesis. To cope with this variation we must employ methods of describing it and estimating its importance in our samples.

The *mean* value or arithmetic average describes the central tendency of the data. Mathematically the mean is

$$\bar{X} = \sum_{i=1}^{N} X_i/N \qquad (A\text{-}6)$$

where $\bar{X}$ signifies the mean value, $\sum_{i=1}^{N} X_i$ is the sum of all the values of the variable, and N is the total number of values. (See Example 1 for calculation of $\bar{X}$ and other indices mentioned in this section.) Once the central tendency for the values is known, the distribution of individual values around the mean should be determined. A simple measure of this variation around the mean is the *range,* the interval from the highest to lowest values (Fig. A-4). But range does not indicate how similar the

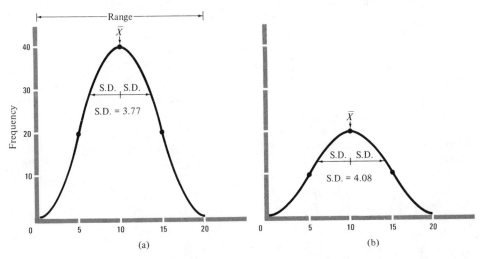

(a)

(b)

FIGURE A-4 Two curves describing frequency distributions of values of the variable X. Both curves have the same range, the same mean value, but different variances and standard deviations.

values are to each other. The two values that determine the upper and lower limits of the range may be quite different from the other values, most of which may be very close to the mean. *Variance* (σ^2) is a measure of the degree of spread of values about the mean. As the values become more widely spread about the mean, the variance will increase. The formula for the variance of a sample (s^2) is

$$s^2 = \frac{\Sigma X^2 - (\Sigma X)^2/N}{N - 1} \tag{A-7}$$

where X is each value of the variable and N is the total number of values in the sample. The square root of the variance, s, is the *standard deviation*. In a "normal" distribution of individual values around the mean, about 69 percent of the values will fall between the mean minus s and the mean plus s.

If a series of samples is taken from the same population and means calculated for each sample, the values of the means also should follow a normal distribution if the population is normally distributed. From a single sample, the standard deviation of the distribution of sample means, called the *standard error of the mean*, can be calculated from the following formula:

$$\text{S.E.} = s/\sqrt{N} \tag{A-8}$$

where s is the standard deviation of the individual values and N is the number of values in the sample.

One further problem in understanding the biological significance of variation of values around the mean relates to changing expectations of values for the variance with changing mean values. We often can assume that the variance of a trait should increase as the mean value increases. For example, it is likely that developmental control systems will produce similar-sized individuals at the same age under controlled environmental conditions. With increasing size, however, it is likely that the absolute variation will increase somewhat even though the buffering system is operating equally well at all lengths. Thus rats may vary by more than an inch in total length due to a certain degree of inefficiency in the buffering system and the input of growth stimuli. A mouse on the other hand may have the same inefficiency level but not show as much variation as the rat because the total growth on which the inefficiency acts is less than for the rat.

Biologically then a larger variation for the rat may signify essentially the same inefficiency in the developmental regulatory systems of the rat as a smaller variation signifies in the mouse. To account for these differences it is possible to calculate a value that compares the mean and standard deviation to see if the variation is of about the same degree relative to the mean for each sample. The *coefficient of variation* (V) is used for this estimate and can be calculated as follows:

$$V = 100s/\overline{X} \tag{A-9}$$

To obtain the same V for increasing mean values requires that the s increase proportionately.

PROBABILITY IN ECOLOGY

Statistical tests are available that allow one to make judgments about the validity of rejecting the null hypothesis, based on variation within the data. In the following discussion we introduce several common tests used in this text and provide examples for their calculation. This is not meant as a definitive statement regarding these tests; more detailed information on limitations of the tests, theory of the tests, and derivation of the formulas may be found in standard statistical textbooks.

The statistical test employed in a particular situation depends on the type of data, the design of the experiment or observations, and assumptions about sampling errors and randomness of the values from which the sample(s) is drawn. A statistical test allows one to judge how much the observed data depart from the null hypothesis.

As an example of how statistical tests can be used, consider the relationship between environmental temperature and respiration (a measure of energy release) in a hummingbird (Fig. A-5). Air temperature (X) is the independent variable and oxygen consumption (Y) is the dependent variable. The data points tend to fall along a straight line intersecting the Y-axis at about 13.0 ml O_2/gm/hr and the X-axis at about 40°C. The equation (see Example 2) for the straight line that best describes the relation between these two variables from the available data is

$$\text{ml } O_2/\text{gm/hr} = 13.056 - 0.326\ T_a \qquad (A\text{-}10)$$

where T_a is air temperature and ml O_2/gm/hr is a measure of the rate of oxygen use by the bird. According to the equation, as air temperature increases one degree, the rate at which oxygen is used in respiration decreases by 0.326 ml O_2/gm/hr.

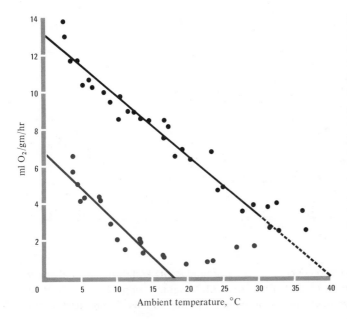

FIGURE A-5 Relationship between oxygen consumption and ambient temperature in the Purple-throated Carib hummingbird, *Eulampis jugularis*. The black dots refer to the birds in resting condition; the colored dots refer to birds in torpor. The colored line is a fit to torpor values at ambient temperatures less than 15°C. Each point represents a single determination. (Data from Hainsworth and Wolf, 1970.)

In view of the minor scatter of the data points about the line it seems reasonable to accept this equation as a good expression of the relation between X and Y. However, the "eyeball technique" often is misleading and the data should be tested statistically. Since there is an independent variable in these data, the t test is used to find out if the null hypothesis (that the slope of the line b is equal to zero) can be rejected (Example 2). If b equals zero, then the value of Y does not change with any change in X, suggesting that air temperature may not have any influence on energy release in this hummingbird. A t test, under the null hypothesis that the slope of the line is zero, gives a p value of less than 0.01, so the null hypothesis can be rejected with a high degree of confidence.

The other test that might be applied to similar data is the correlation coefficient. This is useful for data that include no independent variable. It provides confidence levels for rejecting the null hypothesis that there is no relation between the variables. A ranked correlation coefficient (Example 3) is applied if the data are not concentrated near the center of the line (Fig. A-6). A product-moment correlation coefficient (Example 4) is applied if the data are so concentrated (Fig. A-6).

A product-moment correlation coefficient, r, is the square root of an index of the variation of the points about the line, and r^2, the *coefficient of determination*, is a measure of the actual variance (Example 4). The value of r^2 may vary between 0 (no correlation) and 1.0 (perfect correlation) and represents the fraction of the variation in the Y values that is explained by the X values. If r^2 is 1.0, all variation in Y is explained by their relation to X. This is rarely the case and normally only some percentage of the variation of Y is attributable to its relationship to X, suggesting that there are either other influences on Y or errors in sampling.

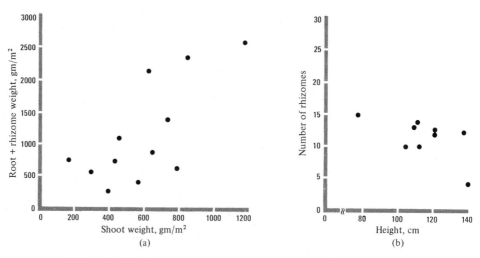

FIGURE A-6(a) Relationship between weight of underground (roots and rhizomes) and above ground (shoots) parts of cattails (*Typha*). (Data from McNaughton, unpublished.) (b) Relationship of number of rhizomes and shoot height for the cattails in (a). (Data from McNaughton, unpublished.)

Another statistical test ecologists frequently use is the *chi-square* (χ^2) test (Example 5). This is a measure of the difference between the frequencies expected under a particular null hypothesis and the observed frequencies, and provides a confidence level for rejecting the null hypothesis. The formula for χ^2 is

$$\chi^2 = \sum \frac{(O - E)^2}{E} \qquad \text{(A-11)}$$

where O is the observed frequency and E the expected frequency. If the χ^2 value is large, the observed and expected values may be different enough (depending on the degrees of freedom) that the null hypothesis can be rejected with a small chance of being wrong. Once the appropriate statistical test has been applied, the degree of departure of the observed results from the results expected under the null hypothesis will provide a measure of confidence in rejecting the null hypothesis. The null hypothesis should be rejected only if a rejection is not likely to produce a mistaken interpretation of the data. The likelihood of making a mistake in rejecting the null hypothesis is generally presented as a *probability* level (*p* value). A *p* value of 0.05 indicates that the data depart sufficiently from the null hypothesis expectation that rejecting the null hypothesis would lead to error only five times out of 100, or that sampling error could produce the observed relationship five times out of 100 even though the null hypothesis was actually correct.

The probability of being wrong in rejecting the null hypothesis depends also on the "degrees of freedom" associated with the statistical test. *Degrees of freedom* (d.f.) indicate the number of values that are not specified by known values, or the "number of items of data that are free to vary independently" (Goldstein, 1964:46). For example, if the mean value is known and all but one of the individual values are also known, then the final individual value is fixed because it must add to the others to give the mean. The degrees of freedom in this example would be $N - 1$, or one less than the total number of individual values. The fewer independent data values used in calculating the expected distributions, the greater the chances for sampling errors to cause the observed and expected values to differ. Thus as the degrees of freedom decrease the greater must be the departure of observed values from expected values before one is confident in rejecting the null hypothesis.

Once the number of degrees of freedom is known and the departure of the data from the null hypothesis calculated (using the appropriate statistical test), the *p* value can be found in tables accompanying statistics texts. In this book we do not provide statistical tables, but when tests are used we do provide *p* values along with the value of the statistical test employed. We do this so that readers may gain insight into the variability of data employed in testing ecological hypotheses.

DATA COLLECTING AND ANALYSIS

The approach employed in investigating a problem in ecology will reflect the type of problem being considered as well as the depth and precision of result that are necessary. Prior to the proliferation of computers a scientist was limited in the number of potential variables that could be considered at a single time, and the problem was compounded even more if the relationships among the variables were not constant.

The usual conceptual framework within which data are collected might be called *reductionism.* A reductionist asks a question and attempts to find the answer by examining the influence of all possible variables, one at a time, in controlled situations. This approach provides the easiest method of investigation for a human observer to comprehend the results, but it is difficult to use to investigate interactions in many ecological situations. This method of analysis is perhaps best suited to laboratory analysis of problems, but even in the laboratory it often is difficult to study the action of only a single factor on the phenomenon. In field ecology it is much more difficult to use the reductionist approach to the problem since so many factors vary despite the best attempts of the investigator to control them.

Most ecologists, by necessity, have been forced to be content with examining the influence of one or two variables on the process in which they were most interested. In most cases they had to assume that the impact of the variable factors was always of the same degree. This resulted partly from the inherent limitations of the human mind to handle a series of variables at one time. It necessitated that theoretical interpretations be as simple as possible so that the potential outcome was easy to calculate. Although simplicity of explanation is a goal always worth striving for, it is becoming more obvious that ecological relationships are not simple and that specific ecological phenomena require extensive modification of simple unifactoral explanations (Frank, 1960; Smith, 1963). To predict adequately ecological phenomena in nature may require relatively complex explanations.

By combining the reductionist theory of analysis with the mechanical abilities of the computer it has become possible to introduce the complexity in biological explanations that seemed to be required by the evidence at hand. This analytical system in ecology borrows heavily from engineering and is usually referred to as *systems analysis.* In systems analysis the variables interacting to produce a phenomenon are treated as a series of isolated factors that may have variable influences (Fig. A-7). As the number of factors considered in the solution of a problem increases, the difficulty of handling interrelated data and the necessity for computer treatment of the data increase.

A classic example of this approach to an ecological problem is found in attempts to understand predator-prey interactions (Holling, 1965, 1966; see Chapter 10). The amount and kinds of prey taken by a predator depend on a series of factors related to the predators (for example, hunger, foraging time, efficiency of foraging, time to eat prey, and other activities that require time and energy) and also on characteristics of the prey (number of prey available, energy per prey, and escape behavior of the prey). Some of the factors related to the behavior of the predator depend in part on characteristics of the prey such that, for example, the foraging time may be related to the escape behavior of the prey and prey density. Each factor can be studied singly and the influence of the other factors determined in a typically reductionist approach, but to finally understand predator-prey interactions requires putting the factors together in a cohesive whole. The experimental manipulations required to do this would be astronomical and it is at this stage that the computer

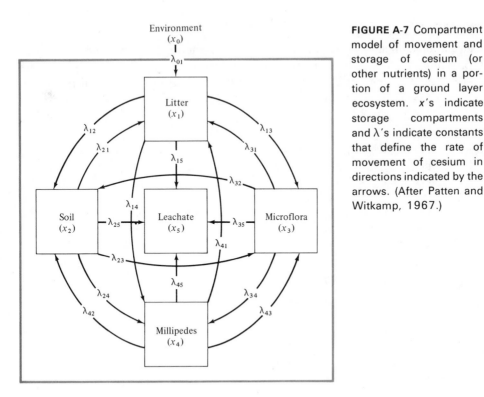

In the figure: Environment (x_0); λ_{01}; Litter (x_1); λ_{12}, λ_{21}, λ_{13}, λ_{31}, λ_{15}, λ_{32}, λ_{14}; Soil (x_2); λ_{25}; Leachate (x_5); λ_{35}; Microflora (x_3); λ_{41}; λ_{23}; λ_{45}; λ_{24}; λ_{42}; λ_{34}; λ_{43}; Millipedes (x_4).

becomes indispensable as a device for simulating effects. This simulation may allow the ecologist to discard many factors as unimportant and therefore simplify final validation through experimental testing.

It should be recognized that all ecological phenomena may not require very complex explanations although they can be explained in a complex fashion (Leigh, 1968). The quest for simplicity of explanation should remain even though the tools are now available for extremely complex explanations.

GENERAL CONCLUSIONS

1. Central tendency: $\bar{X} = \sum\limits_{i=1}^{N} X_i / N$

2. Variance: $s^2 = \dfrac{\Sigma X^2 - (\Sigma X)^2 / N}{N - 1}$

3. Equation for a straight line: $Y = a + bX$
4. Equation for an exponential: $Y = aX^b$ or $Y = ab^x$
5. Equation for second-degree polynomial: $Y = a + bX + cX^2$
6. A p value is the probability of being wrong in rejecting the null hypothesis.
7. Null hypothesis: an hypothesis of no difference or unrelated variables.
8. Correlation coefficient: a measure of the association between two variables.

EXAMPLE 1

Calculation of mean ($\bar{X}$), variance (s^2), standard deviation (s), coefficient of variation (V), standard error (S.E.), and confidence interval of the mean ($t \times$ S.E.). Data (X) = height in centimeters of cattails (*Typha latifolia*) growing in Austin, Texas.

X	X^2
80	6,400
111	12,321
115	13,225
111	12,321
106	11,236
127	16,129
135	18,225
122	14,884
121	14,641
133	17,689
$\Sigma X = 1161$	$\Sigma X^2 = 137,071$

$N = 10$

$\bar{X} = \Sigma X/N = 1161/10 = 116.1$

$$s^2 = \frac{\Sigma X^2 - (\Sigma X)^2/N}{N - 1} = \frac{137,071 - (1161)^2/10}{10 - 1} = \frac{137,071 - 134,792}{9} = 253$$

$s = \sqrt{253} = 15.9$

$V = (s/\bar{X})100 = (15.9/116.1)100 = 13.7$

S.E. $= s/\sqrt{N} = 15.9/\sqrt{10} = 5.03$

Confidence internal of mean $= t \times$ S.E.

$t_{0.95}$ (from table with degrees of freedom $= 10 - 1 = 9$) $= 2.262$

95 percent confidence limit on mean $= 2.262 \times 5.03 = 11.37$

$\bar{X} = 116.1 \pm 11.37 = 104.7$ to 127.5; so that 95 times out of 100 the mean value of a sample from this cattail population should fall within 104.7 to 127.5 cm.

EXAMPLE 2

Calculation of regression from data transformed to logarithms (base 10); t test for slope different from zero. Data on metabolic expenditure per 24 hours as a function of body weight in passerine birds. (Data from Lasiewski and Dawson, 1967.) $\bar{X} =$ mean of Xs, $\bar{Y} =$ mean of Ys.

Calculation of slope (b_{yx}) and intercept (a_y):

$$b_{yx} = \frac{\Sigma(X - \bar{X})(Y - \bar{Y})}{\Sigma(X - \bar{X})^2} \quad \text{or} \quad b_{yx} = \frac{\Sigma XY - \dfrac{\Sigma X \Sigma Y}{N}}{\Sigma X^2 - \dfrac{(\Sigma X)^2}{N}}$$

X Weight (gm)	log X	Y Metabolism (kcal/24 hr)	log Y	$(X - \bar{X})$	$(Y - \bar{Y})$	$(X - \bar{X})^2$	$(X - \bar{X})(Y - \bar{Y})$
6.1	0.7853	2.8	0.4472	−0.5571	−0.4769	0.3104	0.2657
9.0	0.9542	5.3	0.7243	−0.3882	−0.1998	0.1507	0.0776
11.2	1.0492	5.8	0.7634	−0.2932	−0.1607	0.0860	0.0471
11.7	1.0682	4.5	0.6532	−0.2742	−0.2709	0.0752	0.0743
13.0	1.1139	5.8	0.7634	−0.2285	−0.1607	0.0522	0.0367
16.6	1.2201	6.8	0.8325	−0.1223	−0.0916	0.0150	0.0112
18.5	1.2672	8.4	0.9243	−0.0752	0.0002	0.0056	−0.0000
22.0	1.3424	8.7	0.9395	0.0000	0.0154	0.0000	0.0000
22.5	1.3522	7.1	0.8513	0.0098	−0.0728	0.0001	−0.0007
23.5	1.3711	11.0	1.0414	0.0287	0.1173	0.0008	0.0034
24.5	1.3892	11.1	1.0453	0.0468	0.1212	0.0022	0.0057
26.4	1.4216	9.4	0.9731	0.00792	0.00490	0.0063	0.0039
29.4	1.4684	10.5	1.0212	0.1260	0.0971	0.0159	0.0122
31.7	1.5011	11.3	1.0531	0.1587	0.1290	0.0252	0.0205
40.0	1.6021	12.2	1.0864	0.2597	0.1623	0.0674	0.0421
43.7	1.6405	13.7	1.1367	0.2981	0.2126	0.0889	0.0634
58.0	1.7634	16.7	1.2227	0.4210	0.2986	0.1772	0.1257
71.2	1.8525	14.3	1.1553	0.5101	0.2312	0.2602	0.1179
Total	24.1626		16.6342			1.3393	0.9067

$$b_{yx} = 0.9067/1.3393$$
$$b_{yx} = 0.6770$$
$$a_y = \bar{Y} - b_{yx}\bar{X}$$
$$a_y = 0.9241 - 0.9088$$
$$a_y = 0.0153$$

Equation for relation between log metabolism per 24 hours and log body weight:

$$\log (kcal/24\ hr) = 0.0153 + 0.6770 \log (body\ wt.,\ grams)$$

Calculation of variance of Y on X (s_{yx}^2):

$$s_{yx}^2 = \frac{N - 1}{N - 2}(s_y^2 - b_{yx}^2 \cdot s_x^2)$$

$$s_{yx}^2 = \frac{17}{16}[(0.20)^2 - (0.677)^2(0.2806)^2]$$

$$s_{yx}^2 = 1.0625 \times 0.005 = 0.0053$$

$$s_{yx} = 0.0728$$

95 percent confidence interval for b_{yx}:

$$\text{Confidence interval} = \pm \frac{t_{0.05,16} \cdot s_{yx}}{s_x \cdot \sqrt{N - 1}}$$

$$= \pm \frac{2.12 \cdot 0.0728}{0.2806 \cdot 4.123}$$

$$= \pm \frac{0.1543}{1.1570} = \pm 0.1334$$

Therefore zero lies outside the 95 percent confidence interval of b_{yx} and the null hypothesis that the slope is not different from zero can be rejected.

EXAMPLE 3

Correlation analysis involving calculation of a ranked correlation coefficient (r_s). The rank is the position of the value in a sequence from low to high. X and Y are ranked independently; then the r_s value compares ranks for each Y corresponding to a particular X. If each Y is ranked exactly the same as its X, then $r_s = 1.0$; as the Y ranks differ more and more from their respective Xs, the value of r_s decreases toward zero.

Data: X = weight of shoots in gm/m^2 and Y = weight of roots and rhizomes in gm/m^2 of cattail (*Typha latifolia*) populations

X	Rank (R_x)	Y	Rank (R_y)	$R_x - R_y$	$(R_x - R_y)^2$
194	1	756	6	−5	25
310	2	548	3	−1	1
406	3	276	1	2	4
454	4	732	5	−1	1
464	5	1104	9	4	16
558	6	420	2	4	16
640	7	2204	11	−4	16
658	8	872	8	0	0
728	9	1392	10	−1	1
738	10	768	7	3	9
794	11	654	4	7	49
866	12	2396	12	0	0
1186	13	2592	13	0	0
					138

$$r_s = 1 - \frac{6(R_x - R_y)^2}{N^3 - N} = 1 - \frac{6(138)}{(13)^3 - 13} = 1 - \frac{828}{2197 - 13} = 0.621$$

P (from table with df $= N - 2 = 13 - 2 = 11$) $= 0.05$

EXAMPLE 4

Correlation analysis including the product-moment correlation coefficient (r), coefficient of determination (r^2), and prediction equation with the Y-intercept (a) and the slope (b) estimated by Bartlett's method.

Data = X = height in centimeters and Y = number of rhizomes produced by cattails (*T. latifolia*) growing in Austin, Texas. $\bar{X}$ = mean of Xs; $\bar{Y}$ = mean of Ys.

	X	$X - \bar{X}$	$(X - \bar{X})^2$		Y	$(Y - \bar{Y})$	$(Y - \bar{Y})^2$	$(X - \bar{X})(Y - \bar{Y})$
Group X_1	78	−36.4	1324.96	Group Y_1	15	7.1	50.41	−258.44
	106	−8.4	70.56		5	−2.9	8.41	24.36
	107	−7.4	54.76		12	4.1	16.81	−30.34
	111	−3.4	11.56		8	0.1	0.01	−0.34
	112	−2.4	5.76		9	1.1	1.21	−2.64
	113	−1.4	1.96		5	−2.9	8.41	4.06
	121	6.6	43.56		7	−0.9	0.81	−5.94
Group X_3	121	6.6	43.56	Group Y_3	7	−0.9	0.81	−5.94
	135	20.6	424.36		7	−0.9	0.81	−18.54
	140	25.6	655.36		4	−3.9	15.21	−99.84
	1144		2636.4		79		102.9	−393.6
	$\bar{X} = 114.4$				$\bar{Y} = 7.9$			
	$N = 10$				$N = 10$			

$$r = \frac{\Sigma(X - \bar{X})(Y - \bar{Y})}{\sqrt{\Sigma(X - \bar{X})^2 \Sigma(Y - \bar{Y})^2}} = \frac{-393.6}{\sqrt{(2636.4)(102.9)}} = \frac{-393.6}{520.85} = -0.756$$

P (with df = $10 - 2$) = 0.02

$r^2 = (-0.756)^2 = 0.571$

$$b = \frac{\bar{Y}_3 - \bar{Y}_1}{\bar{X}_3 - \bar{X}_1} = \frac{6 - 10.7}{132 - 97} = -0.134$$

$a = \bar{Y} - b\bar{X} = 7.9 - (-0.134)114.4 = 7.9 + 15.36 = 23.26$

$Y = a + bX = 23.36 - 0.134X$

EXAMPLE 5

Chi-square tests for difference between observed and expected results. Data are from Cole, 1946. Expected results are calculated from a Poisson distribution (random distribution of events when there is a very small probability of the event happening). Poisson is used on the assumption that the isopods distribute themselves at random under the boards and that the probability that an isopod will occur under a board is relatively small.

Number isopods per board	Observed frequency (O)	Expected frequency (E)	$(O - E)^2$	$(O - E)^2/E$
0	28	4.5	552.25	122.72
1	28	14.9	171.61	11.52
2	14	24.6	112.36	4.57
3	11	27.0	256.00	9.48
4	8	22.2	201.64	9.08
5	11	14.6	12.96	0.89
6	2	8.0	36.00	4.50
7	3	3.8	0.64	0.17
8	3	1.6	1.96	1.22
8	14	0.8	174.24	217.80
Total	122	122		$\chi^2 = 381.95$

$\chi^2 = 381.95$

$p = 0.001$, with 8 df

literature cited

ABELSON, P. H. 1970. Pollution by organic chemicals. *Science* **170**: 495.

ALCOCK, M. B. 1964. The physiological significance of defoliation on the subsequent regrowth of grass-clover mixtures and cereals. *Brit. Ecol. Soc. Symp.* **4**: 25–41.

ALLEE, W. C. 1926. Distribution of animals in a tropical rain forest with relation to environmental factors. *Ecology* **7**: 445–468.

ALLEE, W. C., A. E. EMERSON, O. PARK, T. PARK, and **K. P. SCHMIDT.** 1949. *Principles of Animal Ecology.* W. B. Saunders, Philadelphia.

AMES, P. L. 1966. DDT residues in the eggs of the Osprey in the northeastern United States and their relation to nesting success. *J. Appl. Ecol.* 3(Suppl.): 87–97.

ANDERSON, G. D., and **L. M. TALBOT.** 1965. Soil factors affecting the distribution of the grassland types and their utilization by wild animals on the Serengeti plains, Tanganyika. *J. Ecol.* **53**: 33–56.

ANDERSON, J., and **H. B. PRINS.** 1970. Effects of sublethal DDT on a simple reflex in brook trout. *J. Fish. Res. Bd. Canada* **27**: 331–334.

ANDERSON, K. L. 1953. Utilization of grasslands in the Flint Hills of Kansas. *J. Range Manage.* **6**: 86–93.

ANDERSON, N. H., and D. M. LEHMKUHL. 1968. Catastrophic drift of insects in a woodland stream. *Ecology* **49**: 198–206.

ANDREWARTHA, H. G. 1957. The use of conceptual models in population ecology. *Cold Spring Harbor Symp. Quant. Biol.* **22**: 219–232.

ANDREWARTHA, H. G., and L. C. BIRCH. 1954. *The Distribution and Abundance of Animals.* Univ. Chicago Press, Chicago.

ANONYMOUS. 1969. Montana warns on eating birds. *N. Y. Times* (Nov. 9).

ANONYMOUS. 1970. The mad-hatter's legacy. *Newsweek* (April 20).

ANONYMOUS. 1970. Mercury in fish. *Newsweek* (Dec. 28).

ANTONOVICS, J. 1968. Evolution in closely adjacent plant populations. V. *Heredity* **23**: 219–238.

ARMSTRONG, F. A. J., and D. W. SCHINDLER. 1971. Preliminary chemical characterization of waters in the experimental lakes area. *J. Fish. Res. Bd. Canada* **28**: 171–187.

ARMSTRONG, J. T. 1965. Breeding home range in the nighthawk and other birds; its evolutionary and ecological significance. *Ecology* **46**: 619–629.

ASTON, J. L., and A. D. BRADSHAW. 1966. Evolution in closely adjacent plant populations. II. *Heredity* **21**: 649–664.

AVERITT, P. 1969. Coal resources of the United States. *U. S. Geol. Survey Bull.* **1275**.

AYALA, F. J. 1969a. Evolution of fitness. IV. Genetic evolution of interspecific competitive ability in *Drosophila*. *Genetics* **61**: 737–747.

AYALA, F. J. 1969b. Experimental invalidation of the principle of competitive exclusion. *Nature* **224**: 1076–1079.

AYALA, F. J. 1970. Competition, coexistence, and evolution. In *Evolution and Genetics,* M. K. Hecht and W. C. Steere (eds.), pp. 121–158. Appleton-Century-Crofts, New York.

AYRES, E. 1956. The age of fossil fuels. In *Man's Role in Changing the Face of the Earth,* Vol. I, W. L. Thomas (ed.), pp. 367–381. Univ. Chicago Press, Chicago.

BAKER, H. G. 1960. Reproductive methods in speciation in flowering plants. *Cold Spring Harbor Symp. Quant. Biol.* **29**: 177–191.

BAKER, H. G. 1965. *Plants and Civilization.* Wadsworth, Belmont, California.

BAKER, H. G. 1972. Seed weight in relation to environmental conditions in California. *Ecology* **53**: 997–1010.

BAKER, H. G., and G. L. STEBBINS (eds.). 1965. *The Genetics of Colonizing Species.* Academic Press, New York.

BAKER, J. R., and J. BAKER. 1936. The seasons in a tropical rain forest (New Hebrides). *J. Linn. Soc.* (Part 2, Botany) **39**: 507–519.

BAKKER, K. 1961. An analysis of factors which determine success in competition for food among larvae of *Drosophila melanogaster*. *Arch. Neerl. Zool.* **14**: 200–281.

BAMBERG, S. A., and J. MAJOR. 1968. Ecology of the vegetation and soils associated with calcareous parent materials in three alpine regions of Montana. *Ecol. Monogr.* **38**: 127–167.

BANNER, A. H. 1952. Preliminary report on marine biology studies of Onotoca Atoll, Gilbert Isl., I. *Atoll Res. Bull.* No. 13.

BARNWELL, F. H. 1967. Daily patterns in the activity of the arboreal ant, *Azteca alfari. Ecology* **48**: 991–993.

BARRETT, G. W. 1968. The effects of an acute insecticide stress on a semi-enclosed grassland eco-system. *Ecology* **49**: 1019–1035.

BARTHOLOMEW, G. A., and **V. A. TUCKER.** 1964. Size, body temperature, thermal conductance, oxygen consumption, and heart rate in Australian Varanid lizards. *Physiol. Zool.* **37**: 341–354.

BAUMHOVER, A. H. 1955. Screw-worm control through release of sterilized flies. *J. Econ. Entomol.* **48**: 462–466.

BAZZAZ, F. A. 1968. Succession on abandoned fields in the Shawnee Hills, southern Illinois. *Ecology* **49**: 924–936.

BEALS, E. W. 1969. Vegetational change along altitudinal gradients. *Science* **165**: 981–985.

BEARD, J. S. 1942. The use of the term "deciduous" as applied to forest types in Trinidad, B. W. I. *Emp. For. J.* **21**: 12–17.

BEARD, J. S. 1944. Climax vegetation in tropical America. *Ecology* **25**: 127–158.

BEARDMORE, J. 1970. Ecological factors and the variability of gene-pools in *Drosophila.* In *Evolution and Genetics,* M. K. Hecht and W. C. Steere (eds.), pp. 299–314. Appleton-Century-Crofts, New York.

BEETON, A. M. 1965. Eutrophication of the St. Lawrence Great Lakes. *Limnol. Oceanogr.* **10**: 240–254.

BEETON, A. M. 1969. Changes in the environment and biota of the Great Lakes. *Nat. Acad. Sci. (U. S.), Eutrophication:* 150–187.

BELL, R. H. V. 1970. The use of the herb layer by grazing ungulates in the Serengeti. *Brit. Ecol. Soc. Symp.* **10**: 111–124.

BELL, R. H. V. 1971. A grazing ecosystem in the Serengeti. *Sci. Amer.* **225**: 86–94.

BENNETT, J. 1960. A comparison of selective methods and a test of the pre-adaptation hypothesis. *Heredity* **15**: 65–77.

BEYERS, R. J. 1963. The metabolism of 12 laboratory microsystems. *Ecol. Monogr.* **33**: 281–306.

BILLINGS, W. D. 1949. The shadscale zone of Nevada and eastern California in relation to climates and soils. *Amer. Midl. Nat.* **48**: 87–109.

BILLINGS, W. D. 1952. The environment complex in relation to plant growth and distribution. *Quart. Rev. Biol.* **27**: 251–265.

BIRCH, L. C. 1948. The intrinsic rate of natural increase of an insect population. *J. Anim. Ecol.* **17**: 15–26.

BIRCH, L. C. 1953. Experimental background to the study of the distribution and abundance of insects. *Ecology* **34**: 698–711.

BIRD, D. 1971. Mercury found in preserved fish. *N. Y. Times News Service* (Jan.).

BIRDSELL, J. B. 1953. Some environmental and cultural factors influencing the structuring of Australian aboriginal populations. *Amer. Nat.* **87**: 171–207.

BIRDSELL, J. B. 1963. Book review. *Quart. Rev. Biol.* **38**: 178–185.

BIRGE, E. A., and **C. JUDAY.** 1911. The inland lakes of Wisconsin; the dissolved gases of the water and their biological significance. *Bull. Wisc. Geol. Nat. Hist. Surv.* No. 22.

BISHOP, E. L. 1947. Effects of DDT mosquito larvacide on wildlife. III. *Publ. Health Rep.* (Wash.) **62**: 1263–1268.

BISWELL, H. M. 1956. Ecology of California grasslands. *J. Range Manage.* **9**: 19–24.

BJORKMAN, O., and P. HOLMGREN. 1963. Adaptability of the photosynthetic apparatus to light intensity in ecotypes from exposed and shaded habitats. *Physiol. Plant.* **16**: 889–914.

BLACK, J. N. 1958. Competition between plants of different initial seed sizes in swards of subterranean clover (*T. subterraneum* L.) with particular reference to leaf area and light micro-climate. *Aust. J. Agric. Res.* **9**: 299–318.

BLACKMAN, F. F. 1905. Optima and limiting factors. *Ann. Bot.* **19**: 281–298.

BLEDSOE, L. J., and D. A. JAMESON. 1969. Model structure of a grassland ecosystem. In *The Grassland Ecosystem: a Preliminary Synthesis,* R. L. Dix and R. G. Beidleman (eds.), pp. 410–437. Range Sci. Ser. No. 2, Colorado State Univ., Fort Collins, Colorado.

BLISS, C. I., and R. A. FISHER. 1953. Fitting the negative binomial distribution to biological data. *Biometrics* **9**: 176–200.

BLISS, L. C. 1962. Adaptations of arctic and alpine plants to environmental conditions. *Arctic* **15**: 117–144.

BLUM, H. 1962. *Time's Arrow and Evolution.* Harper Torchbooks, New York.

BLUM, U., and E. L. RICE. 1969. Inhibition of symbiotic nitrogen-fixation by gallic and formic acid, and possible roles in old-field succession. *Bull. Torr. Bot. Club* **96**: 531–544.

BOADEN, P. J. S. 1964. Grazing in the interstitial habitat. *Brit. Ecol. Soc. Symp.* **4**: 299–303.

BOFFEY, P. M. 1970. Japan: a crowded nation wants to boost its birth rate. *Science* **167**: 960–962.

BOLD, H. C. 1957. *Morphology of Plants.* Harper and Row, New York.

BONNER, J. T. 1965. *Size and Cycle: an Essay on the Structure of Biology.* Princeton Univ. Press, Princeton, New Jersey.

BOOTH, W. E. 1941. Revegetation of abandoned fields in Kansas and Oklahoma. *Amer. J. Bot.* **28**: 415–422.

BORDES, F. 1961. Mousterian cultures in France. *Science* **134**: 803–810.

BORGSTROM, G. 1969. *Too Many.* Macmillan, New York.

BORLAND, N. E. 1965. Wheat, rust and people. *Phytopath.* **55**: 1088–1098.

BORMANN, F. H. 1966. The structure, function, and ecological significance of root grafts in *Pinus strobus Ecol. Monogr.* **36**: 1–26.

BORMANN, F. H., and G. E. LIKENS. 1967. Nutrient cycling. *Science* **155**: 424–429.

BOUGHEY, A. S. 1971a. *Fundamental Ecology.* Intex Educ. Publ., Scranton, Pennsylvania.

BOUGHEY, A. S. 1971b. *Man and the Environment.* Macmillan, New York.

BOURDEAU, P. F. 1959. Seasonal variations of the photosynthetic efficiency of evergreen conifers. *Ecology* **40**: 63–67.

BOURLIERE, F., and M. HADLEY. 1970. The ecology of tropical savannas. *Ann. Rev. Ecol. Syst.* **1**: 125–152.

BOVBJERG, R. V. 1970. Ecological isolation and competitive exclusion in two crayfish. *Ecology* **51**: 225–236.

BOWEN, H. J. M. 1966. *Trace Elements in Biochemistry.* Academic Press, London.

BOWMAN, R. I. 1961. Morphological differentiation and adaptation in the Galapagos finches. *Univ. Cal. Publ. Zool.* No. 58.

BRACE, C. L. 1964. A consideration of hominid catastrophism. *Curr. Anthro.* **5**: 3–42.

BRADSHAW, A. D. 1959. Population differentiation in *Argostis tenuis.* I. *New Phytol.* **58**: 208–227.

BRADSHAW, A. D. 1960. Population differentiation in *Agrostis tenuis.* III. *New Phytol.* **59**: 92–103.

BRAIDWOOD, R. J. 1960. The agricultural revolution. *Sci. Amer.* **203**: 130–148.

BRAUN, E. L. 1950. *Deciduous Forests of Eastern North America.* Blakiston, Philadelphia.

BRAY, J. R., and J. T. CURTIS. 1957. An ordination of the upland forest communities of southern Wisconsin. *Ecol. Monogr.* **27**: 325–349.

BRAY, J. R., and E. GORHAM. 1964. Litter production in forests of the world. *Adv. Ecol. Res.* **2**: 101–157.

BROCK, T. D. 1966. *Principles of Microbial Ecology.* Prentice-Hall, Englewood Cliffs, New Jersey.

BRONSON, F. H. 1971. Rodent pheromones. *Biol. Reprod.* **4**: 344–357.

BROOKS, J. L., and S. I. DODSON. 1965. Predation, body size, and composition of plankton. *Science* **150**: 28–35.

BROWER, J. V. Z. 1960. Experimental studies of mimicry. IV. The reactions of starlings to different proportions of models and mimics. *Amer. Nat.* **94**: 271–282.

BROWER, L. P., J. V. Z. BROWER, and J. M. CORVINO. 1967. Plant poisons in a terrestrial food chain. *Proc. Nat. Acad. Sci.* **57**: 893–898.

BROWER, L. P., F. H. POUGH, and H. R. MECK. 1970. Theoretical investigations of automimicry. I. Single trial learning. *Proc. Nat. Acad. Sci.* **66**: 1059–1066.

BROWN, A. W. A. 1969. Insecticide resistance and the future control of insects. *J. Can. Med. Assoc.* **100**: 216–221.

BROWN, J. L. 1964. The evolution of diversity in avian territorial systems. *Wilson Bull.* **76**: 160–169.

BROWN, J. L. 1969a. Territorial behavior and population regulation in birds: a review and re-evaluation. *Wilson Bull.* **81**: 293–329.

BROWN, J. L. 1969b. The buffer effect and productivity in tit populations. *Amer. Nat.* **103**: 347–354.

BROWN, R. T., and J. T. CURTIS. 1952. The upland conifer-hardwood forests of northern Wisconsin. *Ecol. Monogr.* **22**: 217–234.

BROWN, R. Z. 1963. Patterns of energy flow in populations of the house mouse (*Mus musculus*). *Bull. Ecol. Soc. America* **44**: 129.

BROWN, W. L., and E. O. WILSON. 1956. Character displacement. *Syst. Zool.* **5**: 49–64.

BRUES, A. M. 1959. The spearman and the archer. *Amer. Anthro.* **61**: 458–469.

BRUNSKILL, G. J., and D. W. SCHINDLER. 1971. Geography and bathymetry of selected lake basins, experimental lakes area, northwestern Ontario. *J. Fish. Res. Bd. Canada* **28**: 134–155.

BRYANT, E. H. 1971. Life history consequences of natural selection: Cole's result. *Amer. Nat.* **105**: 75–76

BUBECK, R. C., W. H. DIMENT, B. L. DECK, A. L. BALDWIN, and **S. D. LIPTON.** 1971. Runoff of deicing salt: Effect on Irondequoit Bay, Rochester, New York. *Science* **172:** 1128–1132.

BUFFINGTON, L. C., and **C. H. HERBEL.** 1965. Vegetational changes on a semidesert grassland range. *Ecol. Monogr.* **35:** 139–164.

BURKY, A. J. 1971. Biomass turnover, respiration, and interpopulation variation in the stream limpet *Ferrissia rivularis* (Say). *Ecol. Monogr.* **41:** 235–251.

BURNETT, T. 1956. Effects of natural temperatures on oviposition of various numbers of an insect parasite (Hymenoptera, Chalcoidae, Tenthredinidae). *Ann. Entomol. Soc. Amer.* **49:** 55–59.

BURNETT, T. 1970a. Effect of temperature on a greenhouse acarine predator-prey population. *Can. J. Zool.* **48:** 555–562.

BURNETT, T. 1970b. Effect of simulated natural temperatures on an acarine predator-prey population. *Physiol. Zool.* **43:** 155–165.

BUSHLAND, R. G., and **D. E. HOPKINS.** 1953. Sterilization of screw-worm flies with x-rays and gamma rays. *J. Econ. Entomol.* **46:** 648–656.

CAIRNS, J., JR. 1964. The chemical environment of fresh-water protozoa. *Not. Naturae, Acad. Nat. Sci. Philadelphia* **365:** 1–6.

CAIRNS, J., JR. 1965. The environmental requirements of fresh-water protozoa. *Biological Problems in Water Pollution, Public Health Service Publ.* 999 WP25: 48–52.

CAIRNS, J., JR., M. L. DAHLBERG, K. L. DICKSON, N. SMITH, and **W. T. WALLER.** 1969. The relationship of fresh-water protozoan communities to the MacArthur-Wilson equilibrium model. *Amer. Nat.* **103:** 439–454.

CALHOUN, J. B. 1962. A "behavioral sink." In *Roots of Behavior,* E. L. Bliss (ed.). Harper and Row, New York.

CAMBEL, H., and **R. J. BRAIDWOOD.** 1970. An early farming village in Turkey. *Sci. Amer.* **222:** 50–56.

CAMIN, J. H., and **P. R. EHRLICH.** 1958. Natural selection in water snakes (*Natrix sipedon* L.) on islands in Lake Erie. *Evolution* **12:** 504–511.

CAMPBELL, B. 1966. The evolution of the human hand. In *Man in Adaptation: the Biosocial Background,* Y. Cohen (ed.). Aldine, Chicago.

CAMPBELL, E. 1927. Wild legumes and soil fertility. *Ecology* **8:** 480–483.

CAPERON, J. 1967. Population growth in micro-organisms limited by food supply. *Ecology* **48:** 715–722.

CAPERON, J. 1968. Population growth response of *Isochrysis galbana* to nitrate variation at limiting concentrations. *Ecology* **49:** 866–872.

CARL, E. A. 1971. Population control in arctic ground squirrels. *Ecology* **52:** 395–413.

CARLISLE, D. B., P. E. ELLIS, and **E. BETTS,** 1965. The influence of aromatic shrubs on sexual maturation in the desert locust *Schisotera gregaria. J. Insect. Physiol.* **11:** 1541–1548.

CARLSON, C. A. 1968. Summer bottom fauna of the Mississippi River, above dam 19, Keokuk, Iowa. *Ecology* **49:** 162–169.

CARRICK, R. 1963. Ecological significance of territory in the Australian Magpie, *Gymnorhina tibicen. Proc. Int. Orn. Congr.* **13:** 740–753.

CARR-SAUNDERS, A. M. 1963. *World Population: Past Growth and Present Trends.* Clarendon, Oxford.

CARSON, R. 1962. *Silent Spring.* Houghton Mifflin, Boston.

CARTER, J. L. 1967. Environmental pollution: scientists go to court. *Science* **158**: 1552–1556.

CASSEL, J. 1971. Health consequences of population density and crowding. In N. A. S.-N. R. C. publ. *Rapid Population Growth,* Vol. II, pp. 462–478. Johns Hopkins, Baltimore.

CAVERS, P. B., and **J. L. HARPER.** 1966. Germination polymorphism in *Rumex crispus* and *R. obtusifolius. J. Ecol.* **54**: 367–382.

CHADWICK, H. W., and **P. D. DALKE.** 1965. Plant succession on dune sands in Fremont County, Idaho. *Ecology* **46**: 765–780.

CHANG, K. 1968. Archeology of ancient China. *Science* **162**: 519–526.

CHARLSON, R. J., and **M. J. PILAT.** 1969. Climate: the influence of aerosols. *J. Appl. Meterol.* **8**: 1001–1002.

CHASE, H. C. 1961. The relationship of certain biologic and socio-economic factors to fetal, infant, and early childhood mortality. I. N. Y. State Dept. Health, Albany.

CHEATUM, E. L., and **C. W. SEVERINGHAUS.** 1950. Variations in fertility of white-tailed deer related to range conditions. *Trans. N. Amer. Wildl. Conf.* **15**: 170–189.

CHEMISTRY. 1968. The lead we breathe. *Chemistry* **41**: 7.

CHEW, R. M., and **A. E. CHEW.** 1970. Energy relationships of the mammals of a desert shrub. *Ecol. Monogr.* **40**: 1–21.

CHITTY, D. 1960. Population processes in the vole and their reference to general theory. *Can. J. Zool.* **38**: 99–113.

CHITTY, D. 1967. The natural selection of self-regulatory behavior in animal populations. *Proc. Ecol. Soc. Australia* **2**: 51–78.

CHRISTIAN, J. J. 1961. Phenomena associated with population density. *Proc. Nat. Acad. Sci.* **47**: 428–448.

CHRISTIAN, J. J. 1971. Population density and reproductive efficiency. *Biol. Reprod.* **4**: 248–294.

CHRISTIAN, J. J., and **D. E. DAVIS.** 1964. Endocrines, behavior and population. *Science* **146**: 1550–1560.

CLARK, J. D. 1963. The evolution of culture in Africa. *Amer. Nat.* **97**: 15–28.

CLARK, P. J., and **F. C. EVANS.** 1954. Distance to nearest neighbor as a measure of spatial relationships in populations. *Ecology* **35**: 445–453.

CLARKE, G. L. 1966. *Elements of Ecology,* 2nd printing. John Wiley, New York.

CLAUSEN, J., and **W. M. HIESEY.** 1958. Experimental studies on the nature of species. IV. *Carnegie Inst. Wash. Pub. No.* 615.

CLAUSEN, J., D. D. KECK, and **W. M. HIESEY.** 1958. Experimental studies on the nature of species. III. *Carnegie Inst. Wash. Publ.* No. 581.

CLEMENTS, F. E. 1916. Plant succession. An analysis of the development of vegetation. *Carnegie Inst. Wash. Publ.* No. 242.

CLEMENTS, F. E. 1936. Nature and structure of the climax. *J. Ecol.* **24**: 252–284.

CLEMENTS, F. E., J. E. WEAVER, and **H. C. HANSON.** 1929. Plant competition: an analysis of community functions. *Carnegie Inst. Wash. Publ.* No. 398.

CLEUGH, T. R., and **B. W. HAUSER.** 1971. Results of the initial survey of the Experimental Lakes Area, Northwestern Ontario. *J. Fish. Res. Bd. Canada.* **28:** 129–137.

CLOUD, P. 1969. Mineral resources from the sea. In N. A. S.-N. R. C. publ. *Resources and Man.* Freeman, San Francisco.

CODY, M. L. 1966. A general theory of clutch size. *Evolution* **20:** 174–184.

COHEN, J. E. 1969. Alternate derivations of a species abundance relation. *Amer. Nat.* **102:** 165–172.

COLD SPRING HARBOR SYMPOSIUM ON QUANTITATIVE BIOLOGY. 1957. No. 22, *Population Studies-Animal Ecology and Demography.* Cold Spring Harbor Biol. Lab., New York.

COLE, L. C. 1946. A study of the cryptozoa of an Illinois woodland. *Ecol. Monogr.* **16:** 49–86.

COLE, L. C. 1954a. The population consequences of life history phenomena. *Quart. Rev. Biol.* **29:** 103–137.

COLE, L. C. 1954b. Some features of random cycles. *J. Wildl. Manage.* **29:** 2–24.

COLWELL, R. K., and **D. J. FUTUYMA.** 1971. On the measurement of niche breadth and overlap. *Ecology* **52:** 567–576.

CONNELL, J. H. 1961. The influence of interspecific competition and other factors on the distribution of the barnacle *Chthamalus stellatus. Ecology* **42:** 710–723.

CONNELL, J. H. 1970. A predator-prey system in the marine intertidal region. I. *Balanus glandula* and several predatory species of *Thais. Ecol. Monogr.* **40:** 49–78.

COOK, L. M., R. R. ASKEY, and **J. A. BISHOP.** 1970. Increasing frequency of the typical form of the peppered moth in Manchester. *Nature* **227:** 1155.

COON, C. S. 1963. *The Origin of Races.* Knopf, New York.

COOPER, W. S. 1922. The broad sclerophyll vegetation of California. *Carnegie Inst. Wash. Publ.* No. 319.

COOPER, W. S. 1923. The recent ecological history of Glacier Bay, Alaska. III. *Ecology* **4:** 355–365.

COOPER, W. S. 1931. A third expedition to Glacier Bay, Alaska. *Ecology* **12:** 61–95.

COPE, O. B. 1966. Contamination of the freshwater ecosystem by pesticides. *J. Appl. Ecol.* 3(Suppl.): 33–44.

COTT, H. B. 1940. *Adaptative Coloration in Animals.* Methuen, London.

COWLES, H. C. 1899. The ecological relations of the vegetation on the sand dunes of Lake Michigan. *Bot. Gaz.* **27:** 95–117; 167–202; 281–308; 361–391.

COX, J. L. 1970. Accumulation of DDT residues in *Triphoturus mexicanus* from the Gulf of California. *Nature* **227:** 192–193.

CRISP, D. J. 1964. An assessment of plankton grazing by barnacles. *Brit. Ecol. Soc. Symp.* **4:** 251–264.

CROCKER, R. L., and **J. MAJOR.** 1955. Soil development in relation to vegetation and surface age at Glacier Bay, Alaska. *J. Ecol.* **43:** 427–448.

CURL, H., JR. 1959. The origin and distribution of phosphorus in western Lake Erie. *Limnol. Oceanogr.* **4:** 66–76.

CURTIS, J. T. 1959. *The Vegetation of Wisconsin.* Univ. Wisc. Press, Madison.

CURTIS, J. T., and **G. COTTAM.** 1950. Antibiotic and autotoxic effects in prairie sunflower. *Bull. Torr. Bot. Club* **77:** 187–191.

CUSHING, D. H. 1964. The work of grazing in the sea. *Brit. Ecol. Soc. Symp.* **4**: 207–226.

CZEKANOWSKI, J. 1913. *Zarys Metod Statystycynych, Du Grundzuge der Statischen Methoden.* Warsaw.

DART, R. A. 1953. The predatory transition from ape to man. *Int. Anthro. Ling. Rev.* 1, No. 4.

DARWIN, C. 1859. *The Origin of the Species by Means of Natural Selection.* Murray, London.

DARWIN, C., and A. R. WALLACE. 1859. On the tendency of species to form varieties; and on the perpetuation of varieties and species by natural means of selection. *J. Proc. Linn. Soc. London (Zoology)* **3**: 45–62. (Paper read before Linnean Society, July 1, 1858.)

DASMANN, R. F., and A. S. MOSSMAN. 1962. Population studies of Impala in southern Rhodesia. *J. Mammal.* **43**: 375–395.

DAUBENMIRE, R. F. 1959. *Plants and Environment.* John Wiley, New York.

DAUBENMIRE, R. F. 1966. Vegetation: identification of typal communities. *Science* **151**: 291–298.

DAUBENMIRE, R. 1968. *Plant Communities.* Harper and Row, New York.

DAVIDSON, J., and H. G. ANDREWARTHA. 1948. Annual trends in a natural population of *Thrips imaginis. J. Anim. Ecol.* **17**: 193–199.

DAVIS, C. C. 1964. Evidence for the eutrophication of Lake Erie from phytoplankton records. *Limnol. Oceanogr.* **9**: 275–283.

DAVIS, J. J., and R. F. FOSTER. 1958. Bioaccumulation of radioisotopes through aquatic food chains. *Ecology* **39**: 530–535.

DEEVEY, E. S., JR. 1947. Life tables for natural populations of animals. *Quart. Rev. Biol.* **22**: 283–314.

DEEVEY, E. S. 1960. The human population. *Sci. Amer.* **203**: 195–204.

DEEVEY, E. S. 1969. Specific diversity in fossil assemblages. *Brookhaven Symp. Biol.* **22**: 224–241.

DELIUS, J. D. 1965. A population study of skylarks *Alauda arvensis. Ibis* **107**: 466–492.

DELONG, K. T. 1967. Population ecology of feral house mice. *Ecology* **48**: 611–634.

DEVLAMING, V., and V. W. PROCTOR. 1968. Dispersal of aquatic organisms: viability of seeds recovered from the droppings of captive killdeer and mallard ducks. *Amer. J. Bot.* **55**: 21–26.

DICKMAN, M. 1968. The effect of grazing by tadpoles on the structure of a periphyton community. *Ecology* **49**: 1188–1190.

DINGLE, H. 1968. Life history and population consequences of density, photo-period, and temperature in a migrant insect, the milkweed bug *Oncopeltus. Amer. Nat.* **102**: 149–163.

DINEEN, C. F. 1953. An ecological study of a Minnesota pond. *Amer. Midl. Natur.* **50**: 349–376.

DINGLE, J. H., G. F. BADGER, and W. S. JORDAN. 1964. *Illness in the Home.* Western Reserve Univ., Cleveland.

DIX, R. L. 1960. The effects of burning on the mulch structure and species composition of grasslands in western North Dakota. *Ecology* **23**: 438–445.

DIXON, K. L. 1960. Additional data on the establishment of the chestnut-backed chickadee at Berkeley, California. *Condor* **62**: 405–408.

DOBZHANSKY, T. 1943. Genetics of natural populations. IX. Temporal changes in the composition of populations of *Drosophila pseudoobscura*. *Genetics* **28**: 162–186.

DORLAND, R. E., and **F. W. WENT.** 1947. Plant growth under controlled conditions. VIII. *Amer. J. Bot.* **34**: 393–401.

DORN, H. F. 1962. World population growth: an international dilemma. *Science* **135**: 283–290.

DOUGLAS, J. W. B., J. M. ROSS, and **H. R. SIMPSON.** 1968. *All Our Future.* Peter Davies, London.

DUBLIN, L. I., A. J. LOTKA, and **M. SPIEGELMAN.** 1949. *Length of Life.* Ronald Press, New York.

DUBOS, R. 1965. *Man Adapting.* Yale Univ. Press, New Haven, Connecticut.

DUGGAN, R. E., and **J. R. WEATHERWAY.** 1967. Dietary intake of pesticide chemicals. *Science* **157**: 1006–1010.

DUVIGNEAUD, P., and **S. DENAEYER-DeSMET.** 1970. Biological cycling of minerals in temperate deciduous forests. *Ecol. Studies* **1**: 199–225.

EASTIN, J. A. 1967. Dry matter accumulation activities of plants—their relationship to potential productivity. *Amer. Soc. Agron. Spec. Pub.* **9**: 1–19.

EBERHARDT, L. L. 1970. Correlation, regression, and density dependence. *Ecology* **51**: 306–310.

EDMONDSON, W. T. 1969. Eutrophication in North America. *Nat. Acad. Sci. (U. S.) Symp. Eutrophication:* 124–149.

EDMONDSON, W. T. 1971. Fresh water pollution. In *Environment,* W. W. Murdoch (ed.), pp. 213–229. Sinauer, Stamford, Connecticut.

EDWARDS, C. A. 1965. Effect of pesticide residues on soil invertebrates and plants. *Brit. Ecol. Soc. Symp.* **5**: 239–261.

EDWARDS, C. A., and **G. W. HEATH.** 1963. The role of soil animals in the breakdown of leaf material. In *Soil Organisms,* J. Doekson and J. van der Drift (eds.). North Holland Publ. Co., Amsterdam, Holland.

EHRLICH, P. R., and **L. C. BIRCH.** 1967. The "balance of nature" and "population control." *Amer. Nat.* **101**: 97–124.

EHRLICH, P. R., and **A. H. EHRLICH.** 1970. *Population, Resources, Environment: Issues in Human Ecology.* Freeman, San Francisco.

EHRLICH, P. R., and **P. RAVEN.** 1969. Differentiation of populations. *Science* **165**: 1228–1232.

EHRLICH, P. R., and **J. P. HOLDREN.** 1969. Population and panaceas, a technological perspective. *Bioscience* **19**: 1065–1071.

EINARSEN, A. S. 1945. Some factors affecting ring-necked pheasant population density. *Murrelet* **26**: 39–44.

EINARSEN, A. S. 1964. Crude protein determinations of deer food as an applied management technique. *Trans. N. Amer. Wildl. Conf.* **11**: 309–312.

EISENBERG, R. M. 1966. The regulation of density in a natural population of the pond snail, *Lymnea eloides. Ecology* **47**: 889–906.

EISENBERG, R. M. 1970. The role of food in the regulation of the pond snail. *Ecology* **51**: 680–684.

EISENMANN, E. 1961. Favorite foods of neotropical birds: flying termites and *Cecropia* catkins. *Auk* **78**: 636–638.

EISNER, H. E., D. W. ALSOP, and T. EISNER. 1967. Defense mechanisms of arthropods. XX. Quantitative assessment of hydrogen cyanide production in two species of millipedes. *Psyche* **74**: 107–117.

EISNER, T. 1970. Chemical defense against predation in arthropods. In *Chemical Ecology,* E. Sondheimer and J. B. Simeone (eds.), pp. 157–217. Academic Press, New York.

EISNER, T., F. C. KAFATOS, and E. G. LINSLEY. 1962. Lycid predation by mimetic adult Cerambycidae (Coleoptera). *Evolution* **16**: 316–324.

EISNER, T., and J. MEINWALD. 1966. Defensive secretions of arthropods. *Science* **153**: 1341–1350.

ELTON C., and R. S. MILLER. 1954. The ecological survey of animal communities: with a practical system of classifying habitats by structural characters. *J. Ecol.* **42**: 460–496.

EMLEN, J. M. 1966. The role of time and energy in food preference. *Amer. Nat.* **100**: 611–617.

EMLEN, J. M. 1968. Optimal choice in animals. *Amer. Nat.* **102**: 385–389.

EMLEN, J. M. 1970. Age specificity and ecological theory. *Ecology* **51**: 588–601.

ERICKSON, R. O. 1945. The *Clematis fremontii* var. *riehlii* population in the Ozarks. *Ann. Mo. Bot. Gard.* **32**: 413–460.

ERRINGTON, P. L. 1946. Predation and vertebrate populations. *Quart. Rev. Biol.* **21**: 144–177; 221–245.

ESTES, E. T. 1970. The dendrochronology of black oak, white oak, and shortleaf pine in the central Mississippi Valley. *Ecol. Monogr.* **40**: 295–316.

EVANS, F. C., and S. A. CAIN. 1952. Preliminary studies on the vegetation of an old-field community in southeastern Michigan. *Contrib. Lab. Vert. Biol. U. Michigan* **51**: 1–17.

EYRE, S. R. 1963. *Vegetation and Soils.* Aldine, Chicago.

F. A. O. 1964. *International Marine Sciences.* (Nov.): 16. UNESCO, Paris.

F. A. O. 1965. *Yearbook of Fishing Statistics.* F. A. O., Rome.

FEDERAL WATER POLLUTION CONTROL ADMINISTRATION. 1968. *The Cost of Clean Water.* (4 vols.) U. S. Gov. Print. Off., Washington.

FEENY, P. P. 1968. Effect of oak tannins on larval growth of the winter moth. *J. Insect Physiol.* **14**: 801–817.

FEENY, P. P. 1970. Seasonal changes in oak leaf tannins and nutrients as a cause of spring feeding by winter moth caterpillars. *Ecology* **51**: 565–581.

FEENY, P. P., and H. BOSTOCK. 1968. Seasonal changes in the tannin content of oak leaves. *Phytochem.* **7**: 871–880.

FELLER, W. 1943. On a general class of contagious distributions. *Ann. Math. Statist.* **14**: 389–400.

FINDENEGG, I. 1966. Relationship between standing crop and primary productivity. In *Primary Productivity in Aquatic Environments,* C. R. Goldman (ed.), pp. 271–289. Univ. Cal. Press, Berkeley.

FINNELL, H. H. 1933. The economy of soil nitrogen under semi-arid conditions. *Bull. Okla. Exp. Sta.* No. 215.

FISHER, J. L., and **N. POTTER.** 1971. The effects of population growth on resource adequacy and quality. In N. A. S.-N. R. C. publ. *Rapid Population Growth.* Johns Hopkins, Baltimore.

FISHER, R. A. 1958. *The Genetical Theory of Natural Selection.* Dover, New York.

FISHER, R. A., A. S. CORBET, and **C. B. WILLIAMS.** 1943. The relation between the number of species and the number of individuals in a random sample from an animal population. *J. Anim. Ecol.* **12:** 42–58.

FLANNERY, K. V. 1965. The ecology of early food production in Mesopotamia. *Science* **147:** 1247–1256.

FOGG, F. E. 1965. *Algal Cultures and Phytoplankton Ecology.* Athlone Press, London.

FOGG, F. E., and **W. D. WATT.** 1966. The kinetics of release of extracellular products of photosynthesis by phytoplankton. In *Primary Productivity in Aquatic Environments,* C. R. Goldman (ed.), pp. 165–174. Univ. Cal. Press, Berkeley.

FORBES, E. 1844. Report on the mollusca and radiata of the Aegean Sea. *Rep. Brit. Assoc. Adv. Sci.* (1844): 130–193.

FORBES, S. A. 1887. The lake as microcosm. *Bull. Peoria Sci. Assoc.* (1887): 77–87.

FORBES, S. A. 1907. On the local distribution of certain Illinois fishes. An essay in statistical ecology. *Bull. Ill. State Lab. Nat. Hist.* **7:** 273–303.

FORD, A. B. 1970. Casualties of our time. *Science* **167:** 256–263.

FOSTER, Z., and **M. HARTH.** 1970. *The N. Y. Times Encyclopedic Almanac. New York Times* Book Div., New York.

FRANK, P. W. 1960. Prediction of population growth form in *Daphnia pulex* cultures. *Amer. Nat.* **94:** 357–372.

FRANK, P. W., C. D. BOLL, and **B. W. KELLEY.** 1957. Vital statistics of laboratory cultures of *Daphnia pulex* De Geer as related to density. *Physiol. Zool.* **30:** 287–305.

FRANKLAND, J. C. 1966. Succession of fungi on *Pteridium* petioles. *J. Ecol.* **54:** 41–63.

FRIEDRICH, H. 1969. Marine biology. Univ. Wash. Press, Seattle.

FRITTS, H. C. 1966. Growth-rings of trees: their correlation with climate. *Science* **154:** 973–979.

FRUEDENTHAL, H. P. 1962. *Symbiodinium,* gen. Nov., and *Symbiodinium microadriacticum,* sp. N, a zooxanthella: taxonomy, life cycle, and morphology. *J. Protozool.* **9:** 45–52.

GAASTRA, P. 1963. Climatic control of photosynthesis and respiration. In *Environmental Control of Plant Growth,* L. T. Evans (ed.), pp. 113–140. Academic Press, New York.

GADGIL, M., and **W. A. BOSSERT.** 1970. Life historical consequences of natural selection. *Amer. Nat.* **104:** 1–24.

GADGIL, M., and **O. T. SOLBRIG.** 1972. The concept of r- and K-selection: evidence from wild flowers and some theoretical considerations. *Amer. Nat.* **106:** 14–31.

GASHWILER, J. S. 1970. Further study of conifer seed survival in a western Oregon clearcut. *Ecology* **51:** 849–854.

GATES, D. M. 1965. Energy, plants, and ecology. *Ecology* **46:** 1–13.

GAUSE, G. F. 1934. *The Struggle for Existence.* Williams and Wilkins, Baltimore.

GAUSE, G. F. 1935. Verifications experimentales de la theorie mathematique de la lutte pour la vie. Herman et Cie, Paris.

GAUSE, G. F. 1970. Criticism of invalidation of principle of competitive exclusion. *Nature* **227**: 89.

GEORGE, J. L. 1963. Pesticide-wildlife studies, a review of fish and wildlife studies during 1961 and 1962. *Arch. Fish. Wildl. Serv.* (Wash.) No. 167.

GEORGE, J. L., and D. E. H. FREAR. 1966. Pesticides in the Antarctic. *J. Appl. Ecol.* 3(Suppl.): 155–168.

GERKING, S. D. 1959. The restricted movement of fish populations. *Biol. Rev.* **34**: 221–242.

GERLACH, S. A. 1960. Über das tropische korallenriff als lebensraum. *Zoo Ant.* Suppl. **23**: 356–363.

GESCHWIND, N. 1964. The development of the brain and the evolution of language. *Mon. Ser. Lang. Ling.* **17**.

GIBB, J. 1954. Feeding ecology of tits, with notes on tree creeper and goldcrest. *Ibis* **96**: 513–543.

GIBB, J. 1960. Populations of tits and goldcrests and their food supply in pine plantations. *Ibis* **102**: 163–208.

GIBB, J. A. 1962. L. Tinbergen's hypothesis of the role of specific search images. *Ibis* **104**: 106–111.

GILL, F. B. 1971. Ecology and evolution of the sympatric Mascarene White-eyes, *Zosterops borbonica* and *Zosterops olivacea*. *Auk* **88**: 35–60.

GLASS, L. W., and R. V. BOVBJERG. 1969. Density and dispersion in laboratory populations of caddisfly larvae (*Cheunatopsyche,* Hydropsychiidae). *Ecology* **50**: 1082–1084.

GLASS, N. R. 1968. The effect of time of food deprivation on the routine oxygen consumption of largemouth black bass (*Micropterus salmoides*) *Ecology* **49**: 340–343.

GLEASON, H. A. 1920. Some applications of the quadrat method. *Bull. Torr. Bot. Club* **47**: 21–33.

GLEASON, H. A. 1925. Species and area. *Ecology* **6**: 66–74.

GLEASON, H. A. 1926. The individualistic concept of the plant association. *Bull. Torr. Bot. Club* **53**: 7–26.

GODWIN, H. 1923. Dispersal of pond floras. *J. Ecol.* **11**: 160–164.

GOLDSTEIN, A. 1964. *Biostatistics. An Introductory Text.* Macmillan, New York.

GOLLEY, F. B. 1960. Energy dynamics of a food chain of an old-field community. *Ecol. Monogr.* **30**: 187–206.

GOLLEY, F. B. 1965. Structure and function of an old-field broomsedge community. *Ecol. Monogr.* **35**: 113–131.

GOLLEY, F. B. 1968. Secondary productivity in terrestrial communities. *Amer. Zool.* **8**: 53–59.

GOLLEY, F. B. 1971. Energy flux in ecosystems. In *Ecosystem Structure and Function,* J. A. Wiens (ed.), pp. 69–88. Oregon State Univ. Press, Corvallis, Oregon.

GORDON, M. S., and H. M. KELLEY. 1962. Primary productivity of a Hawaiian coral reef: a critique of flow respirometry in turbulent water. *Ecology* **43**: 473–480.

GOREAU, T. F. 1963. Calcium carbonate deposition by coralline algae and corals in relation to their roles as reef-builders, *Ann N. Y. Acad. Sci.* **109**: 127–167.

GOREAU, T. F., and N. I. GOREAU. 1960. Distribution of labelled carbon in reef-building corals with and without zooxanthellae. *Science* **131**: 668–669.

GOULDEN, C. E. 1969. Temporal changes in diversity. *Brookhaven Symp. Biol.* **22**: 96–100.

GRAND, T. I. 1972. A mechanical interpretation of terminal branch feeding. *J. Mammal.* **53**: 198–201.

GRANT, K. A., and V. GRANT. 1968. *Hummingbirds and Their Flowers.* Columbia Univ. Press, New York.

GRANT, P. R. 1968. Bill size, body size, and the ecological adaptations of bird species to competitive situations on islands. *Syst. Zool.* **17**: 319–333.

GREEN, T. R., and C. A. RYAN. 1972. Wound-induced proteinase inhibitor in plant leaves; a possible defense mechanism against insects. *Science* **175**: 776–777.

GREIG-SMITH, P. 1964. *Quantitative Plant Ecology,* 2nd ed. Butterworth, Washington, D. C.

GRIME, J. P. 1966. Shade avoidance and tolerance in tree seedlings. *Brit. Ecol. Soc. Symp.* **6**: 187–207.

GRIME, J. P., and D. W. JEFFERY. 1965. Seedling establishment in vertical gradients of sunlight. *J. Ecol.* **53**: 621–642.

GRINNELL, J. 1936. Up-hill planters. *Condor* **38**: 80–82.

GWYNNE, M. D., and R. H. V. BELL, 1968. Selection of vegetation components by grazing ungulates in the Serengeti National Park. *Nature* **220**: 390–393.

HADDOW, A. J. 1961. Entomological studies from a high tower in Mpanza Forest, Uganda. VII. *Trans. Roy. Entomol. London* **113**: 315–335.

HADLEY, E. B., and R. P. BUCCOS. 1967. Plant community composition and net primary production within a native eastern North Dakota prairie. *Amer. Midl. Nat.* **77**: 116–127.

HAENSZEL, W., P. B. LOVELAND, and M. G. SIRKEN. 1962. Lung cancer mortality as related to residence and smoking histories. *J. Nat. Cancer Inst.* **28**: 947–1001.

HAINSWORTH, F. R. and L. L. WOLF. 1970. Regulation of oxygen consumption and body temperature during torpor in a hummingbird, *Eulampis jugularis. Science* **168**: 368–369.

HAIRSTON, N. G. 1959. Species abundance and community organization. *Ecology* **40**: 404–416.

HAIRSTON, N. G. 1969. On the relative abundance of species. *Ecology* **50**: 1091–1094.

HAIRSTON, N. G., and G. W. BYERS. 1954. A study in community ecology: the soil arthropods in a field in southern Michigan. *Contrib. Lab. Vert. Biol. U. Mich.* **64**: 1–37.

HAIRSTON, N. G., F. E. SMITH, and L. B. SLOBODKIN. 1960. Community structure, population control and competition. *Amer. Nat.* **94**: 421–425.

HAIRSTON, N. G., D. W. TINKLE, and H. W. WILBUR. 1970. Natural selection and the parameters of population growth. *J. Wildl. Manage.* **34**: 681–690.

HALL, A. V. 1970. A computer-based method for showing continuum and communities in ecology. *J. Ecol.* **58**: 591–602.

HAMILTON, A. L. 1971. Zoobenthos of 15 lakes in the experimental lakes area. *J. Fish. Res. Bd. Canada* **28**: 257–263.

HAMILTON, W. D. 1964. The genetical evolution of social behavior. I and II. *J. Theoret. Biol.* **7**: 1–16; 17–52.

HANSON, W. E. 1967. Cesium-137 in Alaskan lichens, caribou, and eskimos. *Health Phys.* **13**: 383–389.

HARDIN, G. 1960. The competitive exclusion principle. *Science* **131**: 1292–1297.

HARDY, G. H. 1908. Mendelian proportions in a mixed population. *Science* **28**: 49–50.

HARE, E. H., and G. K. SHAW. 1965. A study in family health. I. *Brit. J. Psychol.* **3**: 461–466.

HARLAN, J. R. 1971. Agricultural origins: centers and noncenters. *Science* **174**: 468–474.

HARPER, J. L. 1961. Approaches to the study of plant competition. In *Mechanisms in Biological Competition. Symp. Soc. Exper. Biol.* **15**: 1–39.

HARPER, J. L. 1967. A Darwinian approach to plant ecology. *J. Ecol.* **55**: 247–270.

HARPER, J. L. 1969. The role of predation in vegetational diversity. *Brookhaven Symp. Biol.* **22**: 48–61.

HARPER, J. L., P. H. LOVELL, and K. G. MOORE. 1970. The shapes and sizes of seeds. *Ann. Rev. Ecol. Syst.* **1**: 327–356.

HARPER, J. L., and I. M. McNAUGHTON. 1962. The comparative biology of closely related species living in the same area. VII. Interference between individuals in pure and mixed populations of *Papaver* species. *New Phytol.* **61**: 175–188.

HARPER, J. L., and J. OGDEN. 1970. The reproductive strategy of higher plants. I. The concept of strategy with special reference to *Senecio vulgaris* L. *J. Ecol.* **58**: 681–698.

HARPER, J. L., and G. R. SAGAR. 1953. Some aspects of the ecology of buttercups in permanent grassland. *Proc. First Brit. Weed Control Conf.* 256–265.

HARPER, J. L., J. T. WILLIAMS, and G. R. SAGAR. 1965. The behavior of seeds in soil. *J. Ecol.* **51**: 273–286.

HARRIS, L. D. 1972. An ecological description of a semi-arid East African ecosystem. Colo. State Univ. Range Sci. Ser., No. 11.

HARRIS, V. T. 1952. An experimental study of habitat selection by prairie and forest races of the deer mouse. *Contrib. Lab. Vert. Biol. U. Mich.* **56**.

HARRISON, H. L., O. L. LOUCKS, J. W. MITCHELL, D. F. PARKHURST, C. R. TRACY, D. G. WATTS, and V. J. YANNACONE. 1970. Systems studies of DDT transport. *Science* **170**: 503–508.

HARRISON, J. L. 1962. The distribution of feeding habits among animals in a tropical rainforest. *J. Anim. Ecol.* **31**: 53–63.

HARVEY, H. W. 1926. Nitrate in the sea. *J. Marine Biol. Assoc.* (U. K.) **14**: 71–88.

HARVEY, H. W. 1937. Note on selective feeding by *Calanus*. *J. Marine Biol. Assoc.* (U. K.) **22**: 97–100.

HASLER, A. D. 1947. Eutrophication of lakes by domestic drainage. *Ecology* **28**: 383–395.

HASSELL, M. P., and G. C. VARLEY. 1969. New inductive population model for insect parasites and its bearing on biological control. *Nature* **223**: 1133–1137.

HAUSER, P. M. 1971. World population: retrospect and prospect. In N. A. S.-N. R. C. publ. *Rapid Population Growth*. Johns Hopkins, Baltimore.

HAYES, W. J., JR., W. F. DURHAM, and C. CUETO. 1956. The effect of known repeated oral doses of chlorophenothane (DDT) in man. *J. Amer. Med. Assoc.* **162**: 890–897.

HAYNES, C. V. 1966. Elephant-hunting in North America. *Sci. Amer.* **214**: 104–112.

HAYNES, C. V. 1969. The earliest Americans. *Science* **166**: 709–715.

HEALEY, M. C. 1967. Aggression and self-regulation of population size in deermice. *Ecology* **48**: 377–392.

HEATH, R. G., J. W. SPANN, and J. F. KREITZER. 1969. Marked DDE impairment of mallard reproduction in controlled studies. *Nature* **224**: 47–48.

HELLER, A. N. 1968. The role of the scientist in urban ecology. N. Y. Acad. Sci. Address (May 1).

HENDRICKS, S. B. 1969. Food from the land. In N. A. S.-N. R. C. publ. *Resources and Man.* Freeman, San Francisco.

HENSLEY, M. M., and J. B. COPE, 1951. Further data on removal and repopulation of the breeding birds in a spruce-fir forest community. *Auk* **68**: 483–493.

HERBERT, D. W. M. 1965. Pollution and fisheries. *Brit. Ecol. Soc. Symp.* **5**: 173–196.

HESPENHEIDE, H. A. 1966. The selection of seed size by finches. *Wilson Bull.* **78**: 191–197.

HICKEY, J. J., J. A. KEITH, and F. B. COON. 1966. An exploration of pesticides in a Lake Michigan ecosystem. *J. Appl. Ecol.* 3(Suppl.): 141–154.

HOBBIE, J. E., and R. T. WRIGHT. 1965. Competition between planktonic bacteria and algae for organic solutes. In *Primary Productivity in Aquatic Environments,* C. R. Goldman (ed.). Univ. Cal. Press, Berkeley.

HOCKETT, C. I., and R. ASCHER. 1964. The human revolution. *Curr. Anthro.* **5**: 135–147.

HOLE, F. 1966. Investigating the origins of Mesopotamian civilization. *Science* **153**: 605–611.

HOLLAND, R. E. 1969. Seasonal fluctuations of Lake Michigan diatoms. *Limnol. Oceanogr.* **14**: 423–436.

HOLLING, C. S. 1959. The components of predation as revealed by a study of small mammal predation of the European pine sawfly. *Can. Entomol.* **91**: 293–320.

HOLLING, C. S. 1965. The functional response of predators to prey density and its role in mimicry and population regulation. *Mem. Entomol. Soc. Canada,* No. 45: 1–60.

HOLLING, C. S. 1966. The functional response of invertebrate predators to prey density. *Mem. Entomol. Soc. Canada,* No. 48: 1–86.

HOLMES, R. T. 1966. Feeding ecology of the red-backed sandpiper (*Calidris alpina*) in arctic Alaska. *Ecology* **47**: 32–45.

HOLMES, R. T. 1970. Differences in population density, territoriality and food supply of dunlin on arctic and subarctic tundra. *Brit. Ecol. Soc. Symp.* **10**: 303–319.

HOLT, C. S., and T. F. WATERS. 1967. Effect of light intensity on the drift of stream invertebrates. *Ecology* **48**: 225–234.

HOLT, S. J. 1969. The food resources of the ocean. *Sci. Amer.* **221**: 178–194.

HOPKINS, B. 1966. Vegetation of the Olokemeji Forest Reserve, Nigeria. IV. *J. Ecol.* **54**: 687–704.

HOPKINS, B. 1968. Vegetation of the Olokemeji Forest Reserve, Nigeria. V. *J. Ecol.* **56**: 97–116.

HORN, H. S. 1966. Measurement of "overlap" in comparative ecological studies. *Amer. Natur.* **100**: 419–424.

HOTCHKISS, N., and R. POUGH. 1946. Effect on forest birds of DDT used for gypsy moth control in Pennsylvania. *J. Wildl. Manage.* **10**: 202–207.

HOWARD, D. L., J. I. FREA, R. M. PFISTER, and P. R. DUGAN. 1970. Biological nitrogen fixation in Lake Erie. *Science* **169**: 61–62.

HOWELL, F. C. 1966. Observations on the earlier phases of the European Lower Paleolithic. *Amer. Anthro. Spec. Publ., Rec. Studies Paleocanthus:* 111–140.

HUBBERT, M. K. 1969. Energy resources. In N. A. S.-N. R. C. publ. *Resources and Man.* Freeman, San Francisco.

HUFFAKER, C. B. 1958. Experimental studies on predation: dispersion factors and predator-prey oscillations. *Hilgardia* **27**: 343–383.

HUFFAKER, C. B. 1970. Life against life-nature's pest control scheme. *Env. Res.* **3**: 162–175.

HUFFAKER, C. B., and C. E. KENNETT. 1956. Experimental studies on predation: predation and cyclamen-mite populations on strawberries in California. *Hilgardia* **26**: 191–222.

HUFFAKER, C. B., and C. E. KENNETT. 1969. Some aspects of assessing efficiency of natural enemies. *Can. Entomol.* **101**: 425–447.

HUNT, E. G., and A. I. BISCHOFF. 1960. Inimical effects on wildlife of periodic DDD applications to Clear Lake. *Cal. Fish. Game* **46**: 91–106.

HURD, L. E. 1972. *Stability and Diversity in Old Field Successional Ecosystems.* Ph.D. Thesis, Syracuse Univ.

HURD, L. E., M. V. MELLINGER, L. L. WOLF, and S. J. McNAUGHTON. 1971. Stability and diversity at three trophic levels in terrestrial successional ecosystems. *Science* **173**: 1134–1136.

HURLBERT, S. H. 1971. The nonconcept of diversity: a critique and alternative parameters. *Ecology* **52**: 577–586.

HURSH, C. R. 1948. Local climate in the Copper Basin in Tennessee as modified by the removal of vegetation. *U. S. P. A. Cir.* No. 774.

HUTCHINSON, G. E. 1941. Limnological studies in Connecticut. IV. The mechanisms of intermediary metabolism in stratified lakes. *Ecol. Monogr.* **11**: 21–60.

HUTCHINSON, G. E. 1957a. Concluding remarks. *Cold Spring Harbor Symp. Quant. Biol.* **22**: 415–427.

HUTCHINSON, G. E. 1957b. *A Treatise on Limnology.* John Wiley, New York.

HUTCHINSON, G. E. 1959. Homage to Santa Rosalia, or why are there so many kinds of animals? *Amer. Nat.* **93**: 145–159.

HUTCHINSON, G. E. 1961. The paradox of the plankton. *Amer. Nat.* **95**: 137–146.

HUTCHINSON, G. E. 1969. Eutrophication, past and present. *Nat. Acad. Sci. (U. S.) Symp. Eutrophication:* 17–26.

HUTCHINSON, G. E. 1970. The biosphere. *Sci. Amer.* **223**: 44–58.

HUXLEY, J. 1938. Clines: an auxiliary taxonomic principle. *Nature* **142**: 219–220.

ISAACS, J. D. 1969. The nature of oceanic life. *Sci. Amer.* **221**: 146–162.

JACKSON, P. B. N. 1961. The impact of predation, especially by the tiger-fish (*Hydrocyon vittatus* Cast.) on African freshwater fishes. *Proc. Zool. Soc. London* **136**: 603–622.

JACKSON, T. A., and W. D. KELLER. 1970. Evidence for biogenic synthesis of an unusual ferric oxide mineral during alteration of basalt by a tropical lichen. *Nature* **227**: 522–523.

JANZEN, D. H. 1967. Synchronization of sexual reproduction of trees within the dry season in Central America. *Evolution* **21**: 620–637.

JANZEN, D. H. 1969a. Allelopathy by myrmecophytes: the ant *Azteca* as an allelopathic agent of *Cecropia*. *Ecology* **50**: 147–153.

JANZEN, D. H. 1969b. Seed-eaters versus seed size, number, toxicity and dispersal. *Evolution* **23**: 1–27.

JANZEN, D. H. 1970. Herbivores and the number of tree species in tropical forests. *Amer. Nat.* **104**: 501–528.

JANZEN, D. H. 1971. Euglossine bees as long-distance pollinators of tropical plants. *Science* **171**: 203–205.

JENNY, H. 1930. A study of the influences of climate upon the nitrogen and organic matter content of the soil. *Mo. Agr. Exp. Sta. Bull. Res.* **152**.

JENNY, H. 1933. Soil fertility losses under Missouri conditions. *Mo. Agr. Exp. Sta. Bull. Res.* **324**.

JENNY, H., S. P. GESSEL, and F. T. BINGHAM. 1949. Comparative study of the decomposition rates of organic matter in temperate and tropical regions. *Soil Sci.* **68**: 419–432.

JENSEN, S., A. G. JOHNELS, M. OLSSON, and G. OITESUND. 1969. DDT and PCB in marine animals from Swedish waters. *Nature* **224**: 247–250.

JERMY, T., F. E. HANSON, and V. G. DETHIER. 1968. Induction of specific food preference in lepidopterous larvae. *Entomol. Exp. Appl.* **11**: 211–230.

JOHNSON, A., L. M. BEZEAU, and S. SMOLIAK. 1968. Chemical composition and *in vitro* digestibility of alpine tundra plants. *J. Wildl. Manage.* **32**: 773–777.

JOHNSON, E. V. 1970. Letter to the editor. *Science* **170**: 16–17.

JOHNSON, K. L., L. A. DWORETZKY, and A. N. HELLER. 1968. Carbon monoxide and air pollution from automobile emissions in New York City. *Science* **160**: 67–68.

JOHNSON, M. P., and S. A. COOK. 1968. "Clutch size" in buttercups. *Amer. Nat.* **102**: 405–411.

JOHNSTON, D. W., and E. P. ODUM. 1956. Breeding bird populations in relation to plant succession on the Piedmont of Georgia. *Ecology* **37**: 50–62.

JONASSON, P. M. 1969. Bottom fauna and eutrophication. *Nat. Acad. Sci. (U. S.) Symp. Eutrophication:* 274–305.

JONASSON, P. M., and H. MATHIESON. 1959. Measurements of primary production in two Danish eutrophic lakes. *Oikos* **10**: 137–167.

JONES, R. M. 1968. Seed production in the high veld secondary succession. *J. Ecol.* **56**: 661–666.

JORDAN, C. F. 1971. Productivity of a tropical forest and its relation to a world pattern of energy storage. *J. Ecol.* **59**: 127–142.

JORDAN, P. A., P. C. SHELTON, and D. L. ALLEN. 1967. Numbers, turnover, and social structure of the Isle Royale wolf population. *Amer. Zool.* **7**: 233–252.

JUDAY, C. 1940. The annual energy budget of an inland lake. *Ecology* **21**: 438–450.

JUKES, T. H. 1968. Letter to the editor. *Science* **159**: 695.

KEAR, J. 1962. Food selection in finches with special reference to interspecific difference. *Proc. Zool. Soc. London* **138**: 163–204.

KEEVER, C. 1950. Causes of succession on old fields of the Piedmont, South Carolina. *Ecol. Monogr.* **20**: 229–250.

KELLER, B. L., and C. J. KREBS. 1970. *Microtus* population biology III. Reproductive changes in fluctuating populations of *M. orchrogaster* and *M. pennsylvanicus* in southern Indiana, 1965–1967. *Ecol. Monogr.* **40**: 263–294.

KELLMAN, M. C. 1970. The viable seed content of some forest soil in coastal British Columbia. *Can. J. Bot.* **48**: 1383–1385.

KENDALL, R. L. 1969. An ecological history of the Lake Victoria basin. *Ecol. Monogr.* **39**: 121–176.

KERNER, A. O. 1863. *The Plant Life of the Danube Basin,* trans. by H. Conrad (1951) as the background of plant ecology. Iowa State College Press, Ames, Iowa.

KERSHAW, K. A. 1964. *Quantitative and Dynamic Ecology.* Arnold, London.

KERSTER, H. W., and D. A. LEVIN. 1968. Neighborhood size in *Lithospermum caro-liniense. Genetics* **60**: 577–587.

KETTLEWELL, H. B. D. 1956. Further selection experiments on industrial melanism in the *Lepidoptera. Heredity* **10**: 287–301.

KEYFITZ, N. 1971. Changes of birth and death rates and their demographic effect. In N. A. S.-N. R. C. publ. *Rapid Population Growth.* Johns Hopkins, Baltimore.

KHAILOV, K. M., and Z. P. BURKLOVA. 1969. Release of dissolved organic matter by marine seaweeds. *Limnol. Oceanogr.* **14**: 521–527.

KIDDER, G. W., and V. C. DEWEY. 1951. The biochemistry of ciliates in pure culture. In *Biochemistry and Physiology of Protozoa,* Vol. I. A. Lwoff (ed.), pp. 323–400. Academic Press, New York.

KING, C. E. 1964. Relative abundance of species and MacArthur's model. *Ecology* **45**: 716–727.

KING, C. E., and W. W. ANDERSON. 1971. Age-specific selection. II. The interaction between r and K during population growth. *Amer. Nat.* **105**: 137–156.

KINNE, O. 1963. The effect of temperature and salinity on marine and brackish water animals. I. Temperature. *Oceanogr. Mar. Biol. Ann. Rev.* **1**: 301–304.

KIRK, D. 1971. A new demographic transition. In N. A. S.-N. R. C. publ. *Rapid Population Growth.* Johns Hopkins, Baltimore.

KLEIBER, M. 1961. *The Fire of Life; an Introduction to Animal Energetics.* John Wiley, New York.

KLEIN, D. R. 1970. Food selection in North American deer. *Brit. Ecol. Soc. Symp.* **10**: 25–44.

KLEIN, R. G. 1969. Mousterian cultures in European Russia. *Science* **165**: 257–265.

KLOPFER, P. H., and R. H. MacARTHUR. 1961. On the causes of tropical species diversity: niche overlap. *Amer. Nat.* **95**: 223–226.

KOELLING, M. R., and C. L. KUCERA. 1965. Productivity and turnover relations in native tall grass prairie. *Iowa J. Sci.* **39**: 387–392.

KOKKE, R. 1970. DDT: its action and degradation in bacterial populations. *Nature* **226**: 977–978.

KORMONDY, E. J. (ed.). 1965. *Readings in Ecology.* Prentice-Hall, Englewood Cliffs, New Jersey.

KORMONDY, E. J. 1969. *Concepts of Ecology.* Prentice-Hall, Englewood Cliffs, New Jersey.

KOWAL, N. E. 1966. Shifting cultivation, fire, and pine forest in the Cordillera Central, Luzon, Philippines. *Ecol. Monogr.* **36**: 389–419.

KOZLOVSKY, D. G. 1968. A critical evaluation of the trophic level concept. I. Ecological efficiencies. *Ecology* **49**: 48–60.

KRAYBILL, H. F. 1969. Significance of pesticide residues in foods in relation to total environmental stress. *Can. Med. Assoc. J.* **100**: 204–215.

KREBS, C. J. 1964. The lemming cycle at Baker Lake, N. W. T. during 1959–62. *Arctic Inst. N. Amer. Tech. Papers* No. 15.

KREBS, C. J. 1970. *Microtus* population biology: behavioral changes associated with the population cycle in *M. ochrogaster* and *M. pennsylvanicus. Ecology* **51**: 34–52.

KUCERA, C. L., R. C. DAHLMAN, and M. R. KOELLING. 1967. Total new productivity and turnover on an energy basis for tall grass prairie. *Ecology* **48**: 536–541.

KUENZLER, E. J. 1961a. Phosphorus budget of a mussel population. *Limnol. Oceanogr.* **6**: 400–415.

KUENZLER, E. J. 1961b. Structure and energy flow of a mussel population in a Georgia salt marsh. *Limnol. Oceanogr.* **6**: 191–204.

LACK, D. 1933. Habitat selection in birds with special reference to the effects of afforestation on the Breakland avifauna. *J. Anim. Ecol.* **2**: 239–262.

LACK, D. L. 1954. *The Natural Regulation of Animal Numbers.* Oxford Univ. Press, London.

LACK, D. L. 1966. *Population Studies of Birds.* Clarendon Press, Oxford.

LACK, D. 1968. *Ecological Adaptations for Breeding in Birds.* Methuen, London.

LACK, D., and R. E. MOREAU. 1965. Clutch size in tropical passerine birds of forest and savanna. *Oiseau* **35** (Special): 76–89.

LAMBERG-KARLOVSKY, C. C., and M. LAMBERG-KARLOVSKY. 1971. An early city in Iran. *Sci. Amer.* **224(6)**: 102–111.

LARSEN, J. A. 1965. The vegetation of the Ennadai Lake area, N. W. T.: studies in subarctic and arctic bioclimatology. *Ecol. Monogr.* **35**: 37–59.

LASIEWSKI, R. C., and W. R. DAWSON. 1967. A re-examination of the relation between standard metabolic rate and body weight in birds. *Condor* **69**: 13–23.

LAWRENCE, D. B., R. E. SEHOUIHI, A. QUISPEL, and G. BOND. 1967. Role of *Dryas drummondii* in vegetation development. *J. Ecol.* **55**: 793–813.

LAWS, E. R., JR., A. CURLEY, and E. F. BIROS. 1967. Men with intensive occupational exposure to DDT. *Arch. Env. Health* **15**: 766–775.

LEA, E. 1930. *Rapports et Proces-Verbauz Cons. Explor. Mer.* **65**: 100–117.

LEAKEY, L. S. B. 1960. *Adam's Ancestors.* Harper and Row, New York.

LEIGH, E. J., JR. 1968. Review of K. E. F. Watt's "Ecology and Resource Management." *Science* **160**: 1326–1327.

LEMON, E., L. H. ALLEN, and L. MULLER. 1970. Carbon dioxide exchange of a tropical rain forest. Part II. *Bioscience* **20**: 1054–1059.

LEVIN, S. A. 1970. Community equilibria and stability, and an extension of the competitive exclusion principle. *Amer. Nat.* **104**: 413–423.

LEVINS, R. 1968. *Evolution in Changing Environments.* Princeton Univ. Press, New Jersey.

LEWONTIN, R. C. 1965. Selection for colonizing ability. In *The Genetics of Colonizing Species*, H. G. Baker and G. L. Stebbins (eds.), pp. 77–91. Academic Press, New York.

LIDICKER, W. Z., JR. 1962. Emigration as a possible mechanism permitting the regulation of population density below carrying capacity. *Amer. Nat.* **96**: 29–33.

LIEBIG, J. 1840. *Chemistry in its Application to Agriculture and Physiology.* Taylor and Walton, London.

LIGON, J. D. 1968. Sexual differences in foraging behavior in two species of *Dendrocopos* woodpeckers. *Auk* **85**: 203–215.

LIKENS, G. E., F. H. BORMANN, N. M. JOHNSON, and R. S. PIERCE. 1967. The calcium, magnesium, potassium and sodium budgets for a small forested ecosystem. *Ecology* **48**: 772–785.

LINDEMAN, R. L. 1942. The trophic-dynamic aspect of ecology. *Ecology* **23**: 399–418.

LINSLEY, E. G., J. W. MacSWAIN, and P. H. RAVEN. 1963. Comparative behavior of bees and Onagraceae II. *Oenothera* bees of the Great Basin. *Univ. Cal. Publ. Entomol.* **33**: 25–58.

LLOYD, J. 1965. Aggressive mimicry in *Photuris:* firefly femmes fatales. *Science* **149**: 653–654.

LLOYD, M., and H. S. DYBAS. 1966. The periodical cicada problem, I and II. *Evolution* **20**: 133–149; 466–505.

LLOYD, M., and R. J. GHELARDI. 1964. A table for calculating the equitability component of species diversity. *J. Anim. Ecol.* **33**: 421–425.

LODGE, R. W., J. B. CAMPBELL, S. SMOLIAK, and A. JOHNSTON. 1971. Management of the western range. *Can. Dept. Agric. Publ.* No. 1425.

LOOMIS, R. S., and W. A. WILLIAMS. 1963. Maximum crop productivity: one estimate. *Crop Sci.* **3**: 67–72.

LOOMIS, W. F. 1967. Skin-pigment regulation of vitamin-D synthesis in man. *Science* **157**: 501–506.

LORENZ, K. 1966. *On Aggression.* Methuen, London.

LOTKA, A. J. 1925. *Elements of Physical Biology.* Williams and Wilkins, Baltimore.

LOUCKS, O. L. 1962. Ordinating forest communities by means of environmental scalars and phytosociological indices. *Ecol. Monogr.* **32**: 137–166.

LOWRY, W. P. 1967. Biometerological inference and plant-environment interactions. In *Ground Level Climatology,* R. H. Shaw (ed.). A. A. A. S., Washington, D. C.

LUDWIG, D., and J. M. ANDERSON. 1942. Effects of different humidities, at various temperatures, on the early development of four saturniid moths (*Platysamia cecropia* Linnaeus, *Telea polyphemus* Cramer, *Samia walkeri* Felder and Felder, and *Calosamia promethea* Drury), and on the weight and water content of their larvae. *Ecology* **23**: 259–274.

LUMSDEN, W. H. R. 1966. Light, forest insects, arborviruses, and trypanosomes. *Brit. Ecol. Soc. Symp.* **6**: 209–233.

LYKKEN, L. 1970. Letter to the editor. *Science* **170**: 928.

LYONS, R. D. 1970. Mercury: no place in the world is safe. *New York Times* (Nov. 1).

MacARTHUR, R. H. 1955. Fluctuations of animal populations and a measure of community stability. *Ecology* **35**: 533–536.

MacARTHUR, R. H. 1957. On the relative abundance of bird species. *Proc. Nat. Acad. Sci.* (*U. S.*): **43**: 293–295.

MacARTHUR, R. H. 1958. Population ecology of some warblers of northeastern coniferous forests. *Ecology* **39**: 599–619.

MacARTHUR, R. H. 1960. On the relative abundance of species. *Amer. Nat.* **94**: 25–36.

MacARTHUR, R. H. 1965. Patterns of species diversity. *Biol. Rev.* **40**: 510–533.

MacARTHUR, R. H. 1966. Note on Mrs. Pielou's comments. *Ecology* **47**: 1074.

MacARTHUR, R. H. 1969. Patterns of communities in the tropics. *Biol. J. Linn. Soc.* **1**: 19–30.

MacARTHUR, R. H. 1971. Patterns of terrestrial bird communities. In *Avian Biology,* D. S. Farner and J. R. King (eds.), pp. 189–221. Academic Press, New York.

MacARTHUR, R. H., and R. LEVINS. 1967. The limiting similarity, convergence, and divergence of coexisting species. *Amer. Nat.* **101**: 377–385.

MacARTHUR, R. H., and J. W. MacARTHUR. 1961. On bird species diversity. *Ecology* **42**: 594–598.

MacARTHUR, R. H., and E. R. PIANKA. 1966. On optimal use of a patchy environment. *Amer. Nat.* **100**: 603–609.

MacARTHUR, R. H., and E. O. WILSON. 1963. An equilibrium theory of insular zoo-geography. *Evolution* **17**: 373–387.

MacARTHUR, R. H., and E. O. WILSON. 1967. *The Theory of Island Biogeography.* Princeton Univ. Press, Princeton, New Jersey.

MacFADYEN, A. 1957. *Animal Ecology. Aims and Methods.* Pitman and Sons, London.

MacFADYEN, A. 1963. The contribution of the microfauna to total soil metabolism. In *Soil Organisms,* J. Doeksen and J. van der Drift (eds.). North Holland, Amsterdam.

MACK, E. W., A. B. GARRETT, J. F. HARKINS, and F. H. VERHOEH. 1956. *Textbook of Chemistry.* Ginn, Boston.

MacKAY, R. J., and J. KALFF. 1969. Seasonal variation in standing crop and species diversity of insect communities in a small Quebec stream. *Ecology* **50**: 101–109.

MACKINTOSH, N. A. 1965. *The Stocks of Whales.* Fishing News Books, London.

MAELZER, D. A. 1970. The regression of log N_{n+1} on log N_n as a test of density dependence; an exercise with computer-constructed density-independent populations. *Ecology* **51**: 810–822.

MAGUIRE, B., JR. 1963. Passive dispersal of small aquatic organisms and their colonization of small bodies of water. *Ecol. Monogr.* **33**: 161–185.

MAGUIRE, B., JR. 1967. A partial analysis of the niche. *Amer. Nat.* **101**: 515–526.

MAJOR, J., and W. T. PYOTT. 1966. Buried, viable seeds in two California bunchgrass sites and their bearing on the definition of a flora. *Vegetatio* **13**: 253–282.

MALHOTRA, S. K., G. F. LEE, and G. A. ROHRLICH. 1964. Nutrient removal from secondary effluent by alum flocculation and lime precipitation. *Int. J. Water Poll.* **8**: 487–500.

MALLIK, M. A. B., and E. L. RICE. 1966. Relation between soil fungi and seed plants in three successional forest communities in Oklahoma. *Bot. Gaz.* **127**: 120–127.

MALONE, C. R. 1968. Determination of peak standing crop biomass of herbaceous shoots by the harvest method. *Amer. Midl. Nat.* **79**: 429–435.

MALTHUS, T. R. 1798. An essay on the principle of population as it affects the future improvement of society. Johnson, London.

MARGALEF, R. 1957. La teoria de la informacion en ecologia. *Mem. Real. Acad. Cien. Art. Barcelona* **32**: 373–449.

MARGALEF, R. 1961. Correlations entre certains caracteres synthetiques des populations de phytoplacton. *Hydrobiologia* **18**: 155–164.

MARGALEF, R. 1963. On certain unifying principles in ecology. *Amer. Nat.* **97**: 357–374.

MARGALEF, R. 1967. Laboratory analogues of estuarine plankton systems. In *Estuaries,* G. H. Lauff (ed.), pp. 515–521. Amer. Assoc. Adv. Sci., Washington.

MARSHALL, A. J. 1961. Breeding seasons and migration. In *Biology and Comparative Physiology of Birds,* A. J. Marshall (ed.), pp. 307–339. Academic Press, London.

MARSHALL, D. R., and S. K. JAIN. 1969. Interference in pure and mixed populations of *Avena fatua* and *A. barbata. J. Ecol.* **57**: 251–270.

MARSHALL, S. M., and A. P. ORR. 1964. Grazing by copepods in the sea. *Brit. Ecol. Soc. Symp.* **4**: 227–238.

MARTIN, P. S. 1966. Africa and Pleistocene overkill. *Nature* **212**: 339–342.

MARTIN, P. S. 1967. Africa and Pleistocene overkill. In *Pleistocene Extinctions*, P. S. Martin and H. E. Wright (eds.), pp. 75–120. Yale Univ. Press, New Haven, Connecticut.

MARTIN, P. S., and H. E. WRIGHT (eds.). 1967. *Pleistocene Extinctions*. Yale Univ. Press, New Haven, Connecticut.

MASON, H. L., and J. N. LANGENHEIM. 1957. Language analysis and the concept environment. *Ecology* **38**: 325–340.

MATHER, K. 1953. The genetical structure of populations. *Symp. Soc. Exp. Biol.* **7**: 66–95.

MAW, M. G. 1970. Capric acid as a larvacide and an oviposition stimulant for mosquitos. *Nature* **227**: 1154–1155.

McCORMICK, R. A., and J. H. LUDWIG. 1967. Climate modification by atmospheric aerosols. *Science* **156**: 1358–1359.

McCOY, E., and W. B. SARLES. 1969. Bacteria in lakes: populations and functional relationships. *Nat. Acad. Sci. (U. S.) Symp. Eutrophication:* 331–339.

McCREE, K. J., and J. M. TROUGHTON. 1966. Nonexistence of an optimum leaf area index for the production rate of white clover under constant conditions. *Plant Physiol.* **41**: 1615–1622.

McINTIRE, C. D. 1966. Some factors affecting respiration of periphyton communities in biotic environments. *Ecology* **47**: 918–930.

McINTIRE, C. D. 1968. Structural characteristics of benthic algal communities in laboratory streams. *Ecology* **49**: 520–538.

McINTIRE, C. D., R. L. GARRISON, H. K. PHINNEY, and C. E. WARREN. 1964. Primary production in laboratory streams. *Limnol. Oceanogr.* **9**: 92–102.

McINTIRE, C. D., and H. K. PHINNEY. 1965. Laboratory studies of periphyton and production and community metabolism in biotic environments. *Ecol. Monogr.* **35**: 237–258.

McINTOSH, R. P. 1958. Plant communities. *Science* **128**: 115–120.

McINTOSH, R. P. 1967. An index of diversity and the relation of certain concepts to diversity. *Ecology* **48**: 392–404.

McKEOWN, T., and R. G. BROWN. 1955. Medical evidence related to English population changes in the 18th century. *Pop. Stud.* **9(2)**: 119–141.

McMILLAN, C. 1959a. The role of ecotypic variation in the distribution of the central grassland of North America. *Ecol. Monogr.* **29**: 285–308.

McMILLAN, C. 1959b. Salt tolerance within a *Typha* population. *Amer. J. Bot.* **46**: 521–526.

McMILLAN, C. 1965. Grassland community fractions from central North America under simulated climates. *Amer. J. Bot.* **52**: 109–116.

McNAB, B. K. 1963. Bioenergetics and the determination of home range size. *Amer. Nat.* **97**: 133–140.

McNAUGHTON, S. J. 1966. Ecotype function in the *Typha* community-type. *Ecol. Monogr.* **36**: 297–325.

McNAUGHTON, S. J. 1968a. Autotoxic feedback in the regulation of *Typha* populations. *Ecology* **49**: 367–369.

McNAUGHTON, S. J. 1968b. Structure and function in California grasslands. *Ecology* **49**: 962–972.

McNAUGHTON, S. J., and L. L. WOLF. 1970. Dominance and the niche in ecological systems. *Science* **167**: 131–139.

McNEILL, W. H. 1967. *A World History*. Oxford Univ. Press, New York.

McNEILL, S., and J. H. LAWTON. 1970. Annual production and respiration in animal populations. *Nature* **225**: 472–474.

McNEILLY, T., and J. ANTONOVICS. 1968. Evolution in closely adjacent plant populations. IV. *Heredity* **23**: 205–218.

MECH, L. D. 1966. The wolves of Isle Royale. *Univ. Mich. Mus. Zool. Misc. Publ.* 25.

MELLINGER, M. V. 1972. *Dynamics of Plant Succession on Abandoned Hay Fields in Central New York State*. Ph.D. Thesis, Syracuse Univ.

MENDEL, G. 1866. *Versuche uber Pflanzenhybriden*. Trans. J. H. Bennet (1965). *Experiments in Plant Hybridization*. Oliver and Boyd, London.

MENZEL, D. W., J. ANDERSON, and A. RANDTKE. 1970. Marine phytoplankton vary in their response to chlorinated hydrocarbons. *Science* **167**: 1724–1726.

MILLER, R. S. 1964a. Interspecies competition in laboratory populations of *Drosophila melanogaster* and *Drosophila simulans*. *Amer. Nat.* **48**: 221–237.

MILLER, R. S. 1964b. Larval competition in *Drosophila melanogaster* and *D. simulans*. *Ecology* **45**: 132–148.

MILLER, R. S. 1967. Pattern and process in competition. *Adv. Ecol. Res.* **4**: 1–74.

MILLS, E. 1969. The community concept in marine zoology, with comments on continua and instability in some marine communities: a review. *J. Fish. Res. Bd. Canada* **26**: 1415–1428.

MILNE, A. 1961. Definition of competition among animals. *Symp. Soc. Exp. Biol.* **15**: 40–61.

MINSHALL, G. W. 1967. Role of allochthonous detritus in the trophic structure of a woodland springbrook community. *Ecology* **48**: 139–149.

MIYASHITA, K. 1963. Outbreaks and population fluctuations of insects, with special reference to agricultural insect pests in Japan. *Bull. Nat. Inst. Agr. Sci.* (Japan), No. 15.

MOBIUS, K. 1877. An oyster bank as a bioconose. In *Readings in Ecology,* E. J. Kormondy (ed.). 1965. Prentice-Hall, Englewood Cliffs, New Jersey, pp. 121–124.

MONK, C. D. 1966. Ecological importance of root/shoot ratios. *Bull. Torr. Bot. Club* **93**: 402–406.

MONK, C. D., and J. T. McGINNIS. 1966. Tree species diversity in six forest types in north central Florida. *J. Ecol.* **54**: 341–344.

MOONEY, H. A., and W. D. BILLINGS. 1961. Comparative physiological ecology of arctic and alpine populations of *Oxyria digyna*. *Ecol. Monogr.* **31**: 1–29.

MOORE, J. A. 1952. Competition between *Drosophila melanogaster* and *Drosophila simulans*. II. The improvement of competitive ability through selection. *Proc. Nat. Acad. Sci.* **38**: 813–817.

MOREAU, R. E. 1966. *The Bird Faunas of Africa and Its Islands*. Academic Press, New York.

MORSE, D. H. 1970. Ecological aspects of some mixed-species foraging flocks of birds. *Ecol. Monogr.* **40**: 119–168.

MORTIMER, C. H. 1969. Physical factors with bearing on eutrophication in lakes. *Nat. Acad. Sci. (U. S.) Symp. Eutrophication*: 340–368.

MUIR, R. C. 1965. The effect of sprays on the fauna of apple trees. I. and II. *J. Appl. Ecol.* **2**: 31–57.

MULLEN, D. A. 1968. Reproduction in brown lemmings (*Lemmus trimucronatus*) and its relevance to their cycle of abundance. *Univ. Cal. Publ. Zool.* **85**: 1–24.

MULLER, C. H. 1966. The role of chemical inhibition (allelopathy) in vegetational composition. *Bull. Torr. Bot. Club* **93**: 332–351.

MURDOCH, G. P. 1959. Evolution in social organization. In *Evolution and Anthropology: a Centennial Appraisal.* Anthro. Soc. Wash., Washington, D. C.

MURDOCH, W. W. 1966a. Population stability and life history phenomena. *Amer. Nat.* **100**: 5–11.

MURDOCH, W. W. 1966b. "Community structure, population control, and competition"—a critique. *Amer. Nat.* **100**: 219–226.

MURDOCH, W. W. 1969. Switching in general predators; experiments on predator specificity and stability of prey populations. *Ecol. Monogr.* **39**: 335–354.

MURDOCH, W. W. 1970. Population regulation and population inertia. *Ecology* **51**: 497–502.

MURDOCH, W. W. 1971. The developmental response of predators to change in prey density. *Ecology* **52**: 132–137.

MURIE, A. 1944. The wolves of Mount McKinley. *U. S. Dept. Interior Nat. Park Serv., Fauna Ser.* **5**: 1–238.

MURPHY, G. I. 1968. Pattern in life history and the environment. *Amer. Nat.* **102**: 391–403.

MUSCATINE, L., and **C. HAND.** 1958. Direct evidence of the transfer of materials from symbiotic algae to the tissues of a coelenterate. *Proc. Nat. Acad. Sci.* **44**: 1259–1263.

NASH, R. G., and **E. A. WOOLSON.** 1967. Persistence of chlorinated hydrocarbon insecticides in soils. *Science* **167**: 924–927.

N. A. S.–N. R. C. 1969. *Resources and Man.* Freeman, San Francisco.

NEALES, T. F., A. A. PATTERSON, and **V. J. HARTNEY.** 1968. Physiological adaptation to drought in the carbon assimilation and water loss of xerophytes. *Nature* **219**: 469–472.

NEMOTO, T. 1970. Feeding patterns of baleen whales in the ocean. In *Marine Food Chains,* J. M. Steele (ed.) Univ. Cal. Press, Berkeley, pp. 241–252.

NEYMAN, J., T. PARK, and **E. L. SCOTT.** 1958. Struggle for existence; the *Tribolium* model: biological and statistical aspects. *Gen. Systems* **3**: 152–179.

NICHOLSON, A. J. 1954. An outline of the dynamics of animal populations. *Aust. J. Zool.* **2**: 9–65.

NICHOLSON, A. J. 1955. Density governed reaction, the counterpart of selection in evolution. *Cold Spring Harbor Symp. Quant. Biol.* **20**: 288–293.

NICHOLSON, A. J. 1957. The self-adjustment of populations to change. *Cold Spring Harbor Symp. Quant. Biol.* **22**: 153–173.

NICHOLSON, A. J., and **V. A. BAILEY.** 1935. The balance of animal populations. *Proc. Zool. Soc. London* **1**: 551–598.

NICHOLSON, A. J., I. S. PATTERSON, and **A. CURRIE.** 1970. A study of vegetational dynamics: selection by sheep and cattle in *Nardus* pasture. *Brit. Ecol. Soc. Symp.* **10**: 129–144.

NIELSEN, A. 1950. The torrential invertebrate fauna. *Oikos* **2**: 176–196.

NIERING, W. A., R. H. WHITTAKER, and **C. H. LOWE.** 1963. The saguaro: a population in relation to environment. *Science* **142**: 15–23.

NOBUHARA, H., and M. NUMATU. 1954. Essentials of the law of geometrical progression. *Bull. Soc. Plant Ecol.* **3**: 180–185.

NORRIS, J. M., and J. P. BARKHAM. 1970. A comparison of some Cotswold beechwoods using multiple-discriminant analysis. *J. Ecol.* **58**: 603–620.

NORRIS, R., and D. W. JOHNSTON. 1958. Weights and weight variations in summer birds from Georgia and South Carolina. *Wilson Bull.* **70**: 114–129.

NUMATA, M., H. NOBUHARA, and K. SUYUKI. 1953. The quantitative composition of plant populations from the viewpoint of Motomura's law of geometrical progression. *Bull. Soc. Plant Ecol.* **3**: 89–94.

OAKLEY, K. P. 1961. On man's use of fire, with comments on tool making and hunting. In *Social Life of Early Man,* S. L. Washburn (ed.). Aldine, Chicago.

OBEID, M., D. MACHIN, and J. L. HARPER. 1967. Influence of density on plant to plant variation in fiber flax, *Linum usitatissimum* L. *Crop Sci.* **7**: 471–473.

O'BRIEN, R. D., and F. MATSUMURA. 1964. DDT: a new hypothesis of its mode of action. *Science* **146**: 657–658.

ODUM, E. P. 1959. *Fundamentals of Ecology,* 2nd ed. Saunders, Philadelphia.

ODUM, E. P. 1960. Organic production and turnover in old field succession. *Ecology* **41**: 34–49.

ODUM, E. P. 1962. Relationships between structure and function in the ecosystem. *Japanese J. Ecol.* **12**: 108–118.

ODUM, E. P. 1963. *Ecology.* Modern Biology Series. Holt, Rinehart and Winston, New York.

ODUM, E. P. 1969. The strategy of ecosystem development. *Science* **164**: 262–270.

ODUM, E. P. 1971. *Fundamentals of Ecology.* 3rd ed. Saunders, Philadelphia.

ODUM, H. T. 1957. Trophic structure and productivity of Silver Springs, Florida. *Ecol. Monogr.* **27**: 55–112.

ODUM, H. T. 1970. Summary, an emerging view of the ecological system at El Verde. In *A Tropical Rain Forest,* H. T. Odum and R. F. Pigeon (eds.), pp. I-191 to I-281. U. S. A. E. C., Div. Tech. Info., Oak Ridge, Tennessee.

ODUM, H. T., and E. P. ODUM. 1955. Trophic structure and productivity of a windward coral reef community on Eniwetok Atoll. *Ecol. Monogr.* **25**: 291–320.

ODUM, H. T., and R. F. PIGEON (eds). 1970. *A Tropical Rain Forest.* U. S. A. E. C. Div. Tech. Info., Oak Ridge, Tennessee.

ODUM, H. T., and R. C. PINKERTON. 1955. Time's speed regulator, the optimum efficiency for maximum output in physical and biological systems. *Amer. Sci.* **43**: 331–343.

OHLE, W. 1953. Phosphor als Initialfactor der Gewassereutrophierung. *Vom Wasser* **20**: 11–32.

OHLE, W. 1954. Sulfat als Katalysator des Limnischen Stuffkreislaufes. *Vom Wasser* **21**: 13–32.

OKALI, D. U. U. 1971. Rates of dry matter production in some tropical forest-tree seedlings. *Ann Bot.* **35**: 87–97.

OLD, S. M. 1969. Microclimate, fire, and plant production in an Illinois prairie. *Ecol. Monogr.* **39**: 355–383.

OLMSTEAD, C. E. 1944. Growth and development in range grasses. IV. *Bot. Gaz.* **106**: 46–74.

OLMSTEAD, C. E., and **E. L. RICE.** 1970. Relative effects of known plant inhibitors on species from first two stages of old-field succession. *Southwest. Nat.* **15:** 165–173.

OLSON, J. S. 1958. Rates of succession and soil changes on southern Lake Michigan sand dunes. *Bot. Gaz.* **119:** 125–170.

OLSON, J. S. 1963. Energy storage and the balance of producers and decomposers in ecological systems. *Ecology* **44:** 322–332.

OMRAN, A. R. 1971. Abortion in the demographic transition. In *Rapid Population Growth,* N. A. S.–N. R. C. publ., Johns Hopkins, Baltimore.

OOSTING, H. J. 1942. An ecological analysis of the plant communities of Piedmont, North Carolina. *Amer. Midl. Nat.* **28:** 1–126.

OOSTING, H. J. 1958. *The Study of Plant Communities.* Freeman, San Francisco.

ORIANS, G. H. 1961. The ecology of blackbird (*Agelaius*) social systems. *Ecol. Monogr.* **31:** 285–312.

ORIANS, G. H. 1969a. The number of bird species in some tropical forests. *Ecology* **50:** 783–801.

ORIANS, G. H. 1969b. On the evolution of mating systems in birds and mammals. *Amer. Nat.* **103:** 589–603.

ORSHAN, G. 1963. Seasonal dimorphism of desert and Mediterranean chaemyphytes. *Brit. Ecol. Soc. Symp.* **3:** 206–222.

OVINGTON, J. D., D. HEITKAMP, and **D. B. LAWRENCE.** 1963. Plant biomass and productivity of prairie, savanna, oakwood, and maize field ecosystems in central Minnesota. *Ecology* **44:** 52–63.

OVINGTON, J. D., and **D. M. LAWRENCE.** 1967. Comparative chlorophyll and energy studies of prairie, savanna, oakwood, and maize field ecosystems. *Ecology* **48:** 515–524.

PADDOCK, W. 1967. Phytopathology and a hungry world. *Ann. Rev. Phytopathol.* **5:** 375–386.

PAINE, R. T. 1966. Food web complexity and species diversity. *Amer. Nat.* **100:** 65–75.

PAINE, R. T. 1969. A note on trophic complexity and community stability. *Amer. Nat.* **103:** 91–93.

PAINE, R. T. 1971. A short-term experimental investigation of resource partitioning in a New Zealand rocky intertidal habitat. *Ecology* **52:** 1096–1106.

PAINE, R. T., and **R. VADAS.** 1969. The effects of grazing by sea urchins, *Strongylocentrotus* spp., on benthic algal populations. *Limnol. Oceanogr.* **14:** 710–719.

PALMBLAD, I. G. 1968. Competition in experimental populations of weeds with emphasis on the regulation of population size. *Ecology* **49:** 26–34.

PARK, T. 1954. Experimental studies of interspecies competition. II. Temperature, humidity, and competition in two species of *Tribolium. Physiol. Zool.* **27:** 177–238.

PARSONS, T. R., and **R. J. LEBRAUSSER.** 1970. The availability of food to different trophic levels in the food chain. In *Marine Food Chains,* J. H. Steele (ed.), pp. 325–343, Univ. Cal. Press, Berkeley.

PATALAS, K. 1971. Crustacean plankton communities in forty-five lakes in the experimental lakes area, northwestern Ontario. *J. Fish. Res. Bd. Canada* **28:** 231–244.

PATTEN, B. C., and **M. WITKAMP.** 1967. Systems analysis of cesium-134 kinetics in terrestrial microcosms. *Ecology* **48:** 813–824.

692

PAULIK, G. J. 1971. Anchovies, birds, and fishermen in the Peruvian current. In *Environment,* W. W. Murdoch (ed.), pp. 156–185. Sinauer, Stamford, Connecticut.

PAYNTER, R. A., Jr. 1966. A new attempt to construct life tables for Kent Island herring gulls. *Bull. Mus. Comp. Zool. Harvard* **133**: 489–528.

PEAKALL, D. B. 1970. Pesticide and the reproduction of birds. *Sci. Amer.* **222**(4): 73–78.

PEARL, R. 1925. *The Biology of Population Growth.* Alfred Knopf, New York.

PEARL, R. 1928. Experiments on longevity. *Quart. Rev. Biol.* **3**: 391–407.

PEARL, R., and **L. J. REED.** 1920. On the rate of growth of the population of the United States since 1790 and its mathematical representation. *Proc. Nat. Acad. Sci.* **6**: 275–288.

PEARSE, A. S., H. J. HUMM, and **G. W. WHARTON.** 1942. Ecology of sand beaches at Beaufort, North Carolina. *Ecol. Monogr.* **12**: 136–190.

PEARSON, O. P. 1954. Habits of the lizard *Liolaemus multiformis multiformis* at high altitudes in southern Peru. *Copeia* (2): 111–116.

PEARSON, O. P. 1964. Carnivore-mouse predation: an example of its intensity and bioenergetics. *J. Mammal.* **45**: 177–188.

PEARSON, O. P. 1966. The prey of carnivores during one cycle of mouse abundance. *J. Anim. Ecol.* **35**: 217–233.

PEARSON, W. D., and **D. R. FRANKLIN.** 1968. Some factors affecting drift rates of *Baetis* and Simuliidae in a large river. *Ecology* **49**: 75–81.

PENNINGTON, W. 1943. Lake sediments: The bottom deposits of the north basin of Windermere with special reference to diatom succession. *New Phytol.* **42**: 1–27.

PERRINS, C. 1965. Population fluctuations and clutch-size in the great tit, *Parus major. J. Anim. Ecol.* **34**: 601–647.

PETERLE, T. J. 1969. DDT in Antarctic snow. *Nature* **224**: 620.

PFEIFFER, J. E. 1969. *The Emergence of Man.* Harper and Row, New York.

PIANKA, E. R. 1966. Latitudinal gradients in species diversity: a review of concepts. *Amer. Nat.* **100**: 33–46.

PIANKA, E. R. 1970. On r- and K-selection. *Amer. Nat.* **104**: 592–597.

PIELOU, E. C. 1960. A single mechanism to account for regular, random, and aggregated populations. *J. Ecol.* **48**: 575–584.

PIELOU, E. C. 1966a. The measurement of diversity in different types of biological collections. *J. Theoret. Biol.* **13**: 131–144.

PIELOU, E. C. 1966b. Species-diversity and pattern-diversity in the study of ecological succession. *J. Theoret. Biol.* **10**: 370–383.

PIELOU, E. C. 1969. *An Introduction to Mathematical Ecology.* John Wiley, New York.

PIELOU, E. C., and **A. N. ARNASON.** 1966. Correction to one of MacArthur's species abundance formulas. *Science* **151**: 592.

PIMENTEL, D. 1961. Animal population regulation by the genetic feedback mechanism. *Amer. Nat.* **95**: 65–79.

PIMLOTT, D. H. 1967. Wolf predation and ungulate populations. *Amer. Zool.* **7**: 267–278.

PITELKA, F. A. 1941. Distribution of birds in relation to biotic communities. *Amer. Midl. Nat.* **25**: 113–137.

PITELKA, F. A. 1957. Some aspects of population structure in the short-term cycle of the brown lemming in northern Alaska. *Cold Spring Harbor Symp. Quant. Biol.* **22**: 237–251.

PITELKA, F. A. 1964. The nutrient-recovery hypothesis for arctic microtine cycles. I. *Brit. Ecol. Soc. Symp.* **4**: 55–56.

PITELKA, F. A., P. Q. TOMICH, and **G. W. TREICHEL.** 1955. Ecological relations of jaegers and owls as lemming predators near Barrow, Alaska. *Ecol. Monogr.* **25**: 85–117.

POLUNIN, N. 1960. *Introduction to Plant Geography.* McGraw-Hill, New York.

POMEROY, L. R., and **R. E. JOHANNES.** 1968. Respiration of ultraplankton in the upper 500 meters of the ocean. *Deep-Sea Res.* **15**: 381–391.

POMEROY, L. R., and **E. J. KUENZLER.** 1969. Phosphorus turnover by coral reef animals. In *Proc. 2nd Symp. Radioecol.,* U. S. A. E. C. TID 4500: 474–482.

PRESTON, F. W. 1948. The commonness, and rarity, of species. *Ecology* **29**: 254–283.

PRESTON, F. W. 1957. Analysis of Maryland statewide bird counts. *Bull. Md. Ornith. Soc.* **13**: 63–65.

PRESTON, F. W. 1962. The canonical distribution of commonness and rarity. *Ecology* **43**: 185–215; 410–432.

PRICE, C. A. 1970. *Molecular Approaches to Plant Physiology.* McGraw-Hill, New York.

PRICE, P. W. 1971. Niche breath and dominance of parasitic insects sharing the same host species. *Ecology* **52**: 587–596.

PROCTOR, V. W. 1959. Dispersal of freshwater algae by migratory water birds. *Science* **130**: 623–624.

PUTWAIN, P. D., and **J. L. HARPER.** 1970. Studies on the dynamics of plant populations. III. *J. Ecol.* **58**: 251–264.

PUTWAIN, P. D., D. MACHIN, and **J. L. HARPER.** 1968. Studies in the dynamics of plant populations. II. *J. Ecol.* **56**: 421–431.

RADOMSKI, J. L., W. B. DEICHMAN, E. E. CLIZER, and **A. REY.** 1968. Pesticide concentrations in the liver, brain and adipose tissues of terminal patients. *Food Cosmetic Toxicol.* **6**: 209–220.

RADWANSKI, S. A., and **G. E. WICKENS.** 1967. The ecology of *Acacia albida* on mantle soils in Zalingei, Jebel Mama, Sudan. *J. Appl. Ecol.* **4**: 569–579.

RAPPAPORT, R. 1971. The flow of energy in an agricultural society. *Sci. Amer.* **224** (3): 116–132.

RAUNKIAER, C. 1928. Dominansareal, artstaethed og formations dominanter. Kgl. Danske Videnskabernes Selskab. *Biol. Medd. Kbh.* **7**: 1.

RAUNKIAER, C. 1934. *The Life Forms of Plants and Statistical Plant Geography.* Oxford Univ. Press, Oxford.

REED, E. W., and **S. C. REED.** 1965. *Mental Retardation: A Family Study.* Saunders, Philadelphia.

REICHLE, D. E. 1968. Relation of body size to food intake, oxygen consumption, and trace element metabolism in forest floor arthropods. *Ecology* **49**: 538–542.

REICHLE, D. E. 1970. Measurement of elemental assimilation by animals from radioisotope retention patterns. *Ecology* **50**: 1102–1104.

REINERS, W. A., I. A. WORLEY, and **D. B. LAWRENCE.** 1971. Plant diversity in a chronosequence at Glacier Bay, Alaska. *Ecology* **52**: 55–70.

RIBAUT, J. 1964. Dynamique d'une population de Merles noirs *Turdus merula* L. *Revue Suisse Zool.* **71**: 815–902.

RICE, E. L., W. T. PENFOUND, and **L. M. ROHRBAUGH.** 1960. Seed dispersal and mineral nutrition in succession on abandoned fields in central Oklahoma. *Ecology* **41**: 224–228.

RICHARDS, P. W. 1952. *The Tropical Rain Forest.* Cambridge, London.

RICHERSON, P., R. ARMSTRONG, and **C. R. GOLDMAN.** 1970. Contemporaneous disequilibrium, a new hypothesis to explain the "paradox of the plankton." *Proc. Nat. Acad. Sci.* **67**: 1710–1714.

RICKER, W. E. 1969. Food from the sea. In N. A. S.–N. R. C. publ. *Resources and Man.* Freeman, San Francisco, pp. 87–108.

RICKLEFS, R. E. 1965. Brood reduction in the curve-billed thrasher. *Condor* **67**: 505–510.

RICKLEFS, R. E. 1966. The temporal component of diversity among species of birds. *Evolution* **20**: 235–242.

RIDDIFORD, L. M., and **C. M. WILLIAMS.** 1967. Volatile principle from oak leaves: Role in sex life of the polyphemus moth. *Science* **155**: 589–590.

RILEY, D., and **A. YOUNG.** 1966. *World Vegetation.* Cambridge Univ. Press, Cambridge.

RILEY, G. A. 1963. Organic aggregates in sea water and the dynamics of their formation and utilization. *Limnol. Oceanogr.* **8**: 372–381.

ROBBINS, L. H. 1972. Archeology in the Turkana District, Kenya. *Science* **176**: 359–366.

ROBERTSON, F. W. 1957. Studies in quantitative inheritance. XI. Genetic and environmental correlation between body size and egg production in *Drosophila melanogaster. J. Genet.* **55**: 428–443.

ROBINSON, J. T. 1963. Adaptive radiation in the australopithecines and the origin of man. In *African Ecology and Human Evolution,* F. C. Howell and F. Bourliere (eds.). Aldine, Chicago.

ROCKEFELLER FOUNDATION. 1967. Toward the conquest of hunger. Progress report, 1965–1966, New York.

RODHE, W. 1969. Eutrophication concepts in northern Europe. *Nat. Acad. Sci. (U. S.) Symp. Eutrophication:* 50–64.

ROGERS, J. A., and **J. KING.** 1972. The distribution and abundance of grassland species in hill pasture in relation to soil aeration and base status. *J. Ecol.* **60**: 1–18.

ROOT, R. B. 1964. Ecological interactions of the chestnut-backed chickadee following a range extension. *Condor* **66**: 229–238.

ROOT, R. B. 1967. The niche exploitation pattern of the blue-gray gnatcatcher. *Ecol. Monogr.* **37**: 317–350.

ROSENZWEIG, M. L. 1968. Net primary production of terrestrial communities: prediction from climatological data. *Amer. Nat.* **102**: 67–74.

ROSENZWEIG, M. L. 1969. Why the prey curve has a hump. *Amer. Nat.* **103**: 81–87.

ROSENZWEIG, M. L., and **R. H. MACARTHUR.** 1963. Graphical representation and stability conditions of predator-prey interactions. *Amer. Nat.* **97**: 209–223.

ROSS, B. A., J. R. BRAY, and **W. H. MARSHALL.** 1970. Effects of a long-term deer exclusion on a *Pinus resinosa* forest in north-central Minnesota. *Ecology* **51**: 1088–1093.

ROSSBY, C. G. 1941. The scientific basis of modern meteorology. In *Climate and Man,* Yearbook of Agriculture, 1941. U. S. Government Print. Off., Washington, D. C.

RUSSELL, W. M. S. 1969. *Man, Nature, and History.* Natural History Press, Garden City, New York.

RUSSELL-HUNTER, W. D. 1970. *Aquatic Productivity.* Macmillan, New York.

RYTHER, J. H. 1969. Photosynthesis and fish production in the sea. *Science* **166**: 72–82.

RYTHER, J. H., and **W. M. DUNSTAN.** 1971. Nitrogen, phosphorus, and eutrophication in the coastal marine environment. *Science* **171**: 1008–1013.

SAKAMOTO, M. 1971. Chemical factors involved in the control of phytoplankton production in the experimental lakes area, northwestern Ontario. *J. Fish. Res. Bd. Canada* **28**: 203–213.

SALISBURY, E. J. 1942. *The Reproductive Capacity of Plants; Studies in Quantitative Biology.* Bell, London.

SALISBURY, E. J. 1970. The pioneer vegetation of exposed muds and its biological features. *Philo. Trans. Roy. Soc. B. Biol. Sci. Lond.* **259**: 207–255.

SALT, G. W. 1957. An analysis of avifaunas in the Teton Mountains and Jackson Hole, Wyoming. *Condor* **59**: 373–393.

SALT, G. W. 1967. Predation in an experimental protozoan population (*Woodruffia-Paramecium*). *Ecol. Monogr.* **37**: 113–144.

SATOMI, M., and L. R. POMEROY. 1965. Respiration and phosphorus excretion in some marine populations. *Ecology* **46**: 877–881.

SATOO, T. 1970. A synthesis of studies by the harvest method. *Ecol. Studies* **1**: 55–72.

SAUER, C. O. 1952. Agricultural origins and dispersals. *Amer. Geog. Soc.,* New York.

SAWYER, C. N. 1966. Basic concepts of eutrophication. *J. Water Poll. Con. Fed.* **38**: 737–744.

SCAGEL, R. F., R. J. BANDONI, G. E. ROUSE, W. B. SCHOFIELD, J. R. STEIN, and T. M. C. TAYLOR. 1965. *An Evolutionary Survey of the Plant Kingdom.* Wadsworth, Belmont, California.

SCHINDLER, D. W. 1971. Light, temperature, and oxygen regimes of selected lakes in the experimental lakes area, Ontario. *J. Fish. Res. Bd. Canada* **28**: 157–169.

SCHINDLER, D. W., and S. K. HOLMGREN. 1971. Primary production and phytoplankton in the experimental lakes area, northwestern Ontario. *J. Fish. Res. Bd. Canada* **28**: 189–201.

SCHINDLER, D. W., and B. NOVEN. 1971. Vertical distribution and seasonal abundance of zooplankton in two shallow lakes of the experimental lakes area, northwestern Ontario. *J. Fish. Res. Bd. Canada* **28**: 245–256.

SCHMIDT-NIELSEN, K. 1964. *Desert Animals: Physiological Problems of Heat and Water.* Oxford Univ. Press, London.

SCHOENER, T. W. 1965. The evolution of bill size differences among sympatric congeneric species of birds. *Evolution* **19**: 189–213.

SCHOENER, T. W. 1968. Sizes of feeding territories among birds. *Ecology* **49**: 123–141.

SCHOENER, T. W. 1969a. Models of optimal size for solitary predators. *Amer. Nat.* **103**: 277–313.

SCHOENER, T. W. 1969b. Size patterns in West Indian *Anolis* lizards. I. Size and species diversity. *Syst. Zool.* **18**: 386–401.

SCHOENER, T. W. 1969c. Optimal size and specialization in constant and fluctuating environments: an energy-time approach. *Brookhaven Symp. Biol.* **22**: 103–114.

SCHOENER, T. W. 1970. Size patterns in West Indian *Anolis* lizards. II. Correlations with the sizes of particular sympatric species-displacement and convergence. *Amer. Nat.* **104**: 155–174.

SCHOENER, T. W. 1971. Theory of feeding strategies. *Ann. Rev. Ecol. Syst.* **2**: 369–404.

SCHOENER, T. W., and D. H. JANZEN. 1968. Notes on environmental determinants of tropical versus temperate insect size patterns. *Amer. Nat.* **102**: 207–224.

SCHOLANDER, P. F. 1955. Evolution of climatic adaptation in homeotherms. *Evolution* **9**: 15–26.

SCHULERT, A. R. 1962. Strontium-90 in Alaska. *Science* **136**: 146–148.

SCHULTZ, A. M. 1964. The nutrient-recovery hypothesis for arctic microtine cycles. II. *Brit. Ecol. Soc. Symp.* **4**: 57–68.

SCOTCH, N. A. 1963. Sociocultural factors in the epidemiology of Zulu hypertension. *Amer. J. Publ. Health* **53**: 1205–1213.

SCOTT, D. and W. D. BILLINGS. 1964. Effects of environmental factors on standing crop and productivity of an alpine tundra. *Ecol. Monogr.* **34**: 243–270.

SELANDER, R. K. 1966. Sexual dimorphism and differential niche utilization in birds. *Condor* **68**: 113–151.

SELANDER, R. K. 1970. Behavior and genetic variation in natural populations. *Amer. Zool.* **10**: 53–66.

SEVERINGHAUS, C. W., and J. E. TANCK. 1964. Productivity and growth of white-tailed deer from the Adirondack region of New York. *N. Y. Fish Game J.* **11**: 13–27.

SHANNON, C. E., and W. WEAVER. 1949. *The Mathematical Theory of Communication.* Univ. Illinois Press, Urbana.

SHAW, M. W. 1968. Factors affecting the natural regeneration of sessile oak (*Quercus patraea*) in North Wales. I and II. *J. Ecol.* **56**: 565–583; 647–660.

SHEA, K. P. 1968. Cotton and chemicals. *Sci. Cit.* (Nov.).

SHELDON, A. L. 1968. Species diversity and longitudinal succession in stream fishes. *Ecology* **49**: 193–198.

SHINOZAKI, K., and N. URATA. 1953. Apparent abundance of different species. *Res. Pop. Ecol.* **2**: 8–21.

SHIRER, W. L. 1960. *The Rise and Fall of the Third Reich.* Simon and Schuster, New York.

SHORT, Z., R. F. PALUMBO, P. R. OLSON, and J. R. DONALDSON. 1969. The uptake of iodine-131 by the biota of Fern Lake, Washington in a laboratory and a field experiment. *Ecology* **50**: 979–990.

SHURE, D. J., and P. G. PEARSON. 1969. Distribution of P-32 in *Ambrosia artemisiifolia;* its implication for trophic level studies. *Ecology* **50**: 724–726.

SIEGEL, S. M. 1956. *Nonparametric Statistics for the Behavioral Sciences.* McGraw-Hill, New York.

SILVERMAN, M. P., and MUNOZ, E. F. 1970. Fungal attack on rock: solubilization and altered infrared spectra. *Science* **169**: 985–987.

SIMBERLOFF, D. S., and E. O. WILSON. 1969. Experimental zoogeography of islands: the colonization of empty islands. *Ecology* **50**: 278–296.

SIMONS, E. L. 1964. The early relatives of man. *Sci. Amer.* **211** (1): 50–62.

SIMONS, E. L. 1972. *Primate Evolution, an Introduction to Man's Place in Nature.* Macmillan, New York.

SIMONS, E. L., D. PILBEAN, and P. C. ETTEL. 1969. Controversial taxonomy of fossil hominids. *Science* **166**: 258–259.

SIMPSON, G. G., A. ROE, and R. C. LEWONTIN. 1960. *Quantitative Zoology.* Harcourt, Brace Jovanovich, New York.

SINGH, K. P. 1968. Nutrient status of forest soils in humid tropical regions of western Ghats. *Trop. Ecol.* **9**: 119–130.

SKUTCH, A. F. 1967. Adaptive limitation of the reproductive rate of birds. *Ibis* **109**: 579–599.

SLOBODKIN, L. B. 1960. Ecological energy relationships at the population level. *Amer. Nat.* **94**: 213–236.

SLOBODKIN, L. B. 1961. *Growth and Regulation of Animal Populations.* Holt, Rinehart and Winston, New York.

SMITH, C. C. 1970. The coevolution of pine squirrels and conifers. *Ecol. Monogr.* **40**: 349–371.

SMITH, F. E. 1952. Experimental methods in population dynamics: a critique. *Ecology* **33**: 441–450.

SMITH, F. E. 1954. Quantitative aspects of population growth. In *Dynamics of Growth Processes* (E. J. Boell, ed.), Princeton Univ. Press, Princeton, N. J.

SMITH, F. E. 1963. Population dynamics in *Daphnia magna* and a new model for population growth. *Ecology* **44**: 651–663.

SMITH, F. E. 1966. *The Politics of Conservation.* Pantheon Books (Random House), New York.

SMITH, S. G. 1967. Experimental and natural hybrids in North American *Typha. Amer. Midl. Nat.* **78**: 257–287.

SNOW, D. W. 1962. A field study of the black and white manakin, *Manacus manacus,* in Trinidad. *Zoologica* **47**: 65–104.

SNOW, D. W. 1965. A possible selective factor in the evolution of fruiting seasons in tropical forest. *Oikos* **15**: 274–281.

SNOW, D. W., and B. K. SNOW. 1964. Breeding seasons and annual cycles of Trinidad land-birds. *Zoologica* **49**: 1–39.

SOLECKI, R. S. 1957. Shanidar Cave. *Sci. Amer.* **197** (5): 58–64.

SOLOMON, M. E. 1949. The natural control of animal population. *J. Anim. Ecol.* **18**: 1–32.

SOLOMON, M. E. 1964. Analysis of processes involved in the natural control of insects. In *Advances in Ecological Research,* J. Cragg (ed.), Vol. 2, pp. 1–58.

SOULE, M. 1966. Trends in the insular radiation of a lizard. *Amer. Nat.* **100**: 47–64.

SOUTHERN, H. N. 1959. Mortality and population control. *Ibis* **101**: 429–436.

SOUTHWARD, A. J. 1964. Limpet grazing and the control of vegetation on rocky shores. *Brit. Ecol. Soc. Symp.* **4**: 265–273.

SOUTHWOOD, T. R. E. 1961. The number of species of insect associated with various trees. *J. Anim. Ecol.* **30**: 1–8.

STALFELT, M. G. 1937. Der Gasaustausch der Moose. *Planter.* **27**: 30–60.

ST. AMANT, J. L. E. 1970. The detection of regulation in animal populations. *Ecology* **51**: 823–828.

STEEN, M. O. 1955. Not how much but how good. *Mo. Cons.* **16**: 1–3.

STENGER, J. 1958. Food habits and available food of ovenbirds in relation to territory size. *Auk* **75**: 335–346.

STEPHANS, G. R., and P. E. WAGGONER. 1970. Carbon dioxide exchange of a tropical rain forest. *Bioscience* **20**: 1050–1053.

STEPHANSON, T. A., and A. STEPHANSON. 1949. The universal features of zonation between tide marks on rocky coasts. *J. Ecol.* **37**: 289–305.

STEPHANSON, T. A., and **A. STEPHANSON.** 1961. Life between tide-marks in North America. IV A. Vancouver Island I.; IV B. Vancouver Island II. *J. Ecol.* **49:** 1–29; 229–243.

STERN, V. M., R. F. SMITH, R. VAN DEN BOSCH, and **K. S. HAGEN.** 1959. The integrated control concept. *Hilgardia* **29:** 81–101.

STEWART, P. A. 1952. Dispersal, breeding, behavior, and longevity of banded barn owls in North America. *Auk* **69:** 277–285.

STILES, F. G., and **L. L. WOLF.** 1970. Hummingbird territoriality at a tropical flowering tree. *Auk* **87:** 467–491.

STOCKNER, J. G., and **F. A. J. ARMSTRONG.** 1971. Periphyton of the experimental lakes area. *J. Fish Res. Bd. Canada* **28:** 215–229.

STOCKNER, J. G., and **W. W. BENSON.** 1967. The succession of diatom assemblages in the recent sediments of Lake Washington. *Limnol. Oceanogr.* **12:** 513–532.

STRICKLAND, A. H. 1966. Some estimates of insecticide and fungicide usage in agriculture and horticulture in England and Wales, 1960–64. *J. Appl. Ecol.* 3(Suppl.): 3–13.

STYCOS, J. M. 1971. Opinion, ideology, and population problems. In N. A. S.-N. R. C. publ., *Rapid Population Growth,* Johns Hopkins, Baltimore.

SWIFT, R. W. 1948. Deer select most nutritious forages. *J. Wildl. Manage.* **12:** 109–110.

TAMARIN, R. H., and **C. J. KREBS.** 1969. *Microtus* population biology. II. Genetic changes at the transferrin locus in fluctuating populations of two vole species. *Evolution* **23:** 183–211.

TANNER, J. T. 1966. Effects of population density on growth rates of animal populations. *Ecology* **47:** 733–745.

TANSLEY, A. G. 1935. The use and abuse of vegetational concepts and terms. *Ecology* **16:** 284–307.

TEAL, J. M. 1957. Community metabolism in a temperate cold spring. *Ecol. Monogr.* **27:** 283–302.

TEAL, J. M. 1962. Energy flow in the salt marsh ecosystem of Georgia. *Ecology* **43:** 614–624.

TERBORGH, J., and **J. S. WESKE.** 1969. Colonization of secondary habitats by Peruvian birds. *Ecology* **50:** 765–782.

TERNES, A. P. (ed.) 1970. The state of the species. *Nat. Hist.* **79:** 43–74.

THODAY, J. M. 1959. Effects of disruptive selection. I. Genetic flexibility. *Heredity* **13:** 187–203.

THOMAS, E. A. 1969. The process of eutrophication in central European lakes. *Nat. Acad. Sci. (U. S.) Symp. Eutrophication:* 29–49.

THOMAS, M. D. 1965. The effects of air pollution on plants and animals. *Brit. Ecol. Soc. Symp.* **5:** 11–34.

THOMAS, W. A. 1969. Accumulation and cycling of calcium by dogwood trees. *Ecol. Monogr.* **39:** 101–120.

THOMPSON, D. Q. 1952. Travel, range and food habits of timber wolves in Wisconsin. *J. Mammal.* **33:** 429–442.

THOMPSON, D. Q. 1955. The role of food and cover in population fluctuations of the brown lemming at Point Barrow, Alaska. *Trans. N. Amer. Wildl. Conf.* **20:** 166–174.

THORPE, W. H. 1945. The evolutionary significance of habitat selection. *J. Anim. Ecol.* **14:** 67–70.

TINBERGEN, L. 1960. The natural control of insects in pine-woods. I. Factors influencing the intensity of predation by songbirds. *Arch. Neerl. Zool.* **13**: 265–343.

TINBERGEN, N. 1968. On war and peace in animals and man. *Science* **160**: 1411–1418.

TINKLE, D. W. 1969. The concept of reproductive effort and its relation to the evolution of life histories of lizards. *Amer. Nat.* **103**: 501–516.

TINKLE, D. W., H. M. WILBUR, and S. G. TILLEY. 1970. Evolutionary strategies in lizard reproduction. *Evolution* **24**: 55–74.

TOPPING, A. 1971. Return to changing China. *Nat. Geogr.* **140**: 801–834.

TRANSEAU, E. N. 1926. The accumulation of energy by plants. *Ohio J. Sci.* **26**: 1–10.

TRIVERS, R. L. 1971. The evolution of reciprocal altruism. *Quart. Rev. Biol.* **46**: 35–57.

TRUOG, E. 1948. Lime in relationship to availability of plant nutrients. *Soil Sci.* **65**: 1–7.

TUKEY, H. B., Jr., H. B. TUKEY, and E. H. WITTEVER. 1958. Loss of nutrients by foliar leaching as determined by radioisotopes. *Proc. Amer. Soc. Hort. Sci.* **126**: 120–121.

TURNER, F. B., R. I. JENNRICH, and J. D. WEINTRAUB. 1969. Home ranges and body size of lizards. *Ecology* **50**: 1076–1081.

TURNER, R. M., S. M. ALCOM, G. OLIN, and J. A. BOOTH. 1966. The influence of shade, soil and water on saguaro seedling establishment. *Bot. Gaz.* **127**: 95–102.

TURRESSON, G. 1922a. The species and the variety as ecological units. *Hereditas* **3**: 100–113.

TURRESSON, G. 1922b. The genotypic response of the plant species to habitat. *Hereditas* **3**: 211–350.

UNITED NATIONS. 1969. Demographic yearbook. Stat. Off. U. N., New York.

U. N. FOOD AND AGRIC. ORG. 1967. Pesticide residues in food. *F. A. O. Agric. Studies* **73**: 1–19.

VAARTAJA, O. 1962. The relationship of fungi to survival of shaded tree seedlings. *Ecology* **43**: 547–549.

VADYA, A. P. 1961. Expansion and warfare among swidden agriculturists. *Amer. Anthro.* **63**: 346–358.

VAN DEN BOSCH, R. 1968. Comments on population dynamics of exotic insects. *Entomol. Soc. Amer.* **14**: 112–115.

VANDERMEER, J. H. 1969. The competitive structure of communities: an experimental approach with protozoa. *Ecology* **50**: 362–371.

VAN DYNE, G. M. 1969. Grassland management, research and teaching viewed in a systems context. Range Science Dept. Science series No. 3, Colorado State Univ.

VAN VALEN, L. 1965. Morphological variation and width of ecological niche. *Amer. Nat.* **99**: 377–390.

VARLEY, G. C. 1970. The concept of energy flow applied to a woodland community. *Brit. Ecol. Soc. Symp.* **10**: 389–405.

VERHULST, P. F. 1838. Notice sur la loi que la population suit dans son accroissement. *Corresp. Math. Phys.* **10**: 113–121. Trans. in Kormondy, 1965 (q.v.).

VERME, L. J. 1962. Mortality of white-tail deer fawns in relation to nutrition. *Proc. Nat. White-tailed Deer Dis. Symp.* **1**: 15–38.

VERNBERG, F. J., and W. B. VERNBERG. 1970. *The Animal and the Environment*. Holt, Rinehart, and Winston, New York.

VERNER, J. 1964. Evolution of polygamy in the long-billed marsh wren. *Evolution* **18**: 252–261.

VERNER, J. 1965. Breeding biology of the long-billed marsh wren. *Condor* **67**: 6–30.

VERNER, J. and **M. F. WILLSON.** 1966. The influence of habitats on mating systems of North American passerine birds. *Ecology* **47**: 143–147.

VESEY-FITZGERALD, D. F. 1960. Grazing succession among East African game animals. *J. Mammal.* **41**: 161–172.

VEZINA, P. E., and **D. W. K. BOULTER.** 1966. The spectral composition of near-UV and visible radiation beneath forest canopies. *Can. J. Bot.* **44**: 1267–1284.

VOIGHT, J. W., and **J. E. WEAVER.** 1951. Range condition classes of mature midwestern pasture: an ecological analysis. *Ecol. Monogr.* **21**: 39–60.

VOLTERRA, V. 1926. Variazione e fluttuazioni de numero d'individui in specie animali conviventi. *Mem. Accad. Lincei* **2**: 31–113.

VON FOERSTER, H. 1958. Basic concepts of homeostasis. *Brookhaven Symp. Biol.* **10**: 216–242.

VON FOERSTER, H., P. M. MORCA, and **L. W. AMIOT.** 1960. Doomsday: Friday, 13 November, A. D. 2026. *Science* **132**: 1291–1295.

VON WISSMANN, H., H. POECH, G. SMOLLA, and **F. KUSSMAUL.** 1956. On the role of nature and man in changing the face of the dry belt of Asia. In *Man's Role in Changing the Face of the Earth,* W. L. Thomas (ed.), Univ. Chicago Press, Chicago.

WALLACE, G. L. 1959. Insecticides and birds. *Audubon Mag.* **61**: 10–12.

WALOFF, N. 1968. Studies on the insect fauna on Scotch broom. *Adv. Ecol. Res.* **5**: 88–208.

WARMING, E. 1909. Oecology of plants, an introduction to the study of plant communities. Excerpt reprinted in *Readings in Ecology,* E. J. Kormondy (ed.), 1965. Prentice-Hall, Englewood Cliffs, New Jersey, pp. 125–129.

WARNER, R. E., K. K. PETERSON, and **L. BORGAN.** 1966. Behavioral pathology in fish: a quantitative study of sublethal pesticide toxication. *J. Appl. Ecol.* 3(Suppl.): 223–248.

WARREN, H. V., and **R. E. DELAVAULT.** 1969. Mercury content of some British soils. *Oikos* **20**: 537–539.

WATERBOLK, H. T. 1968. Food production in prehistoric Europe. *Science* **162**: 1093–1102.

WATERS, T. F. 1961. Standing crop and drift of stream bottom organisms. *Ecology* **42**: 532–537.

WATERS, T. F. 1962. Diurnal periodicity in the drift of stream invertebrates. *Ecology* **43**: 316–320.

WATERS, T. F. 1965. Interpretation of invertebrate drift in streams. *Ecology* **46**: 327–334.

WATSON, A. and **G. R. MILLER.** 1971. Territory size and aggression in a fluctuating red grouse population. *J. Anim. Ecol.* **40**: 367–383.

WATSON, A., and **R. MOSS.** 1970. Dominance, spacing behavior and aggression in relation to population limitation in vertebrates. *Brit. Ecol. Soc. Symp.* **10**: 167–218.

WATT, K. E. F. 1968. *Ecology and Resource Management: A Quantitative Approach.* McGraw-Hill, New York.

WEAVER, J. E. 1954. *North American Prairie.* Johnson, Lincoln, Nebraska.

WEAVER, J. E., and **N. W. ROWLAND.** 1952. Effects of excessive natural mulch on development, yield and structure of native grassland. *Bot. Gaz.* **114**: 1–19.

WECKER, S. C. 1963. The role of early experience in habitat selection by the prairie deer mouse, *Peromyscus maniculatus bairdi*. *Ecol. Monogr.* **33**: 307–325.

WEIBEL, S. R. 1969. Urban drainage as a factor in eutrophication. In *Nat. Acad. Sci. (U. S.) Symp. Eutrophication*: 383–403.

WELCH, H. 1968. Relationships between assimilation efficiencies and growth efficiencies for aquatic consumers. *Ecology* **49**: 755–759.

WELLS, H. G., J. S. HUXLEY, and G. P. WELLS. 1939. *The Science of Life,* book 6, part V. Garden City Publ., Garden City, New York.

WELLS, P. V. 1965. Scarp woodlands, transported soils, and concept of grassland climate in the Great Plains region. *Science* **148**: 246–249.

WENDORF, F., R. SAID, and R. SCHILD. 1970. Egyptian prehistory: some new concepts. *Science* **169**: 1161–1171.

WENT, F. W. 1955. The ecology of desert plants. *Sci. Amer.* **192** (3): 68–75.

WHITE, J., and J. L. HARPER. 1970. Correlated changes in plant size and number in plant populations. *J. Ecol.* **58**: 467–485.

WHITEHEAD, D. R., and K. W. TAN. 1969. Modern vegetation and pollen rain in Bladen County, North Carolina. *Ecology* **50**: 235–248.

WHITESIDE, M. C. 1970. Danish chydorid cladocera: modern ecology and core samples. *Ecol. Monogr.* **40**: 79–118.

WHITE-STEVENS, R. 1970. Letter to the editor. *Science* **170**: 928.

WHITTAKER, R. H. 1952. A study of summer foliage insect communities in the Great Smoky Mountains. *Ecol. Monogr.* **22**: 1–44.

WHITTAKER, R. H. 1953. A consideration of climax theory: the climax as population and pattern. *Ecol. Monogr.* **23**: 41–78.

WHITTAKER, R. H. 1954. The ecology of serpentine soils. I. Introduction. *Ecology* **35**: 258–259.

WHITTAKER, R. H. 1956. Vegetation of the Great Smoky Mountains. *Ecol. Monogr.* **26**: 1–80.

WHITTAKER, R. H. 1961. Experiments with radiophosphorus tracer in aquarium micro-systems. *Ecol. Monogr.* **31**: 157–188.

WHITTAKER, R. H. 1965. Dominance and diversity in land plant communities. *Science* **147**: 250–260.

WHITTAKER, R. H. 1967. Ecological implications of weather modification. In *Ground Level Climatology*, R. H. Shaw (ed.), A. A. A. S. publ. No. 86.

WHITTAKER, R. H. 1970a. The biochemical ecology of higher plants. In *Chemical Ecology*, E. Sondheimer and J. B. Simeone (eds.), pp. 43–70. Academic Press, New York.

WHITTAKER, R. H. 1970b. *Communities and Ecosystems*. Macmillan, London.

WHITTAKER, R. H., and P. P. FEENY. 1971. Allelochemics: chemical interactions between species. *Science* **171**: 757–770.

WHITTAKER, R. H., and G. M. WOODWELL. 1968. Dimension and production relations of trees and shrubs in the Brookhaven forest, New York. *J. Ecol.* **56**: 1–25.

WHITTAKER, R. H., and G. M. WOODWELL. 1969. Structure, production, and diversity of the oak-pine forest at Brookhaven, New York. *J. Ecol.* **57**: 155–174.

WHITTEN, R. R., and D. E. PARKER. 1945. Experimental control of shade-tree insects with DDT. *Proc. Nat. Shade Tree Conf.* **21**: 13–17.

WHITTEN, W. K., and F. H. BRONSON. 1970. Role of pheromones in mammalian reproduction. In *Advances in Chemoreception,* J. W. Johnson, D. C. Moulton, and A. Turks (eds.). Appleton-Century-Crofts, New York.

WIENS, H. J. 1962. *Atoll Environment and Ecology.* Yale Univ. Press, New Haven, Connecticut.

WILLIAMS, C. B. 1964. *Patterns in the Balance of Nature.* Academic Press, New York.

WILLIAMS, G. C. 1966. Natural selection, the costs of reproduction, and a refinement of Lack's principle. *Amer. Nat.* **100:** 687–690.

WILLIAMS, G. C., and D. C. WILLIAMS. 1957. Natural selection of individually harmful social adaptations among sibs, with special reference to social insects. *Evolution* **11:** 32–39.

WILLIAMS, J. E. (ed.). 1963. *The Prentice-Hall World Atlas,* 2nd ed. Prentice-Hall, Englewood Cliffs, New Jersey.

WILLIAMS, O. 1959. Food habits of the deer mouse. *J. Mammal.* **40:** 415–420.

WILLIAMS, W. T., G. N. LANCE, L. J. WEBB, J. C. TRACEY, and M. B. DALE. 1969. Studies in the numerical analysis of complex rain-forest communities. III. *J. Ecol.* **57:** 515–535.

WILLIS, E. O. 1963. Is the zone-tailed hawk a mimic of the turkey vulture? *Condor* **65:** 313–317.

WILSON, E. O. 1969. The species equilibrium. *Brookhaven Symp. Biol.* **22:** 38–47.

WILSON, E. O. 1970. Chemical communication within animal species. In *Chemical Ecology,* E. Sondheimer and J. B. Simeone (eds.). Academic Press, New York.

WILSON, E. O., and W. H. BOSSERT. 1971. *A Primer of Population Biology,* Sinauer, Stamford, Connecticut.

WILSON, R. E., and E. L. RICE. 1968. Allelopathy as expressed by *Helianthus annuus* and its role in old-field succession. *Bull. Torr. Bot. Club* **95:** 432–448.

WITHERSPOON, J. P. 1964. Cycling of cesium-134 in white oak trees. *Ecol. Monogr.* **34:** 403–420.

WITKAMP, M. 1966. Decomposition of leaf litter in relation to environment, microflora, and microbial respiration. *Ecology* **47:** 194–207.

WITTFOGEL, K. A. 1956. The hydraulic civilizations. In *Man's Role in Changing the Face of the Earth,* W. L. Thomas (ed.). Univ. Chicago Press, Chicago.

WOLCOTT, G. N. 1937. An animal census of two pastures and a meadow in northern New York. *Ecol. Monogr.* **7:** 1–90.

WOLF, E. R., and A. PALEM. 1955. Irrigation in the old acolhua domain, Mexico. *S. W. J. Anthro.* **11:** 265–281.

WOLF, L. L. 1970. The impact of seasonal flowering on the biology of some tropical hummingbirds. *Condor* **72:** 1–14.

WOLF, L. L., and F. R. HAINSWORTH. 1971. Time and energy budgets of territorial hummingbirds. *Ecology* **52:** 980–988.

WOLF, L. L., and F. R. HAINSWORTH. 1972. Environmental influence on regulated body temperature in torpid hummingbirds. *Comp. Biochem. Physiol.* **41A:** 167–173.

WOLF, L. L., F. R. HAINSWORTH, and F. G. STILES. 1972. Energetics of foraging: rate and efficiency of nectar extraction by hummingbirds. *Science* **176:** 1351–1352.

WOLF, L. L., and F. G. STILES. 1970. Evolution of pair cooperation in a tropical hummingbird. *Evolution* **24:** 759–773.

WOODELL, S. R. J., H. A. MOONEY, and A. J. HILL. 1969. The behavior of *Larrea divariacata* (creosote bush) in response to rainfall in California. *J. Ecol.* **57**: 37–44.

WOODWELL, G. M., and D. B. BOTKIN. 1970. Metabolism of terrestrial ecosystems by gas exchange techniques: the Brookhaven approach. In *Studies in Ecology: Analysis of Temperate Forest Ecosystems,* D. Reichle (ed.). Springer-Verlag, Berlin.

WOODWELL, G. M., and H. H. SMITH, (eds.). 1969. Diversity and stability in ecological systems. *Brookhaven Symp. Biol.,* No. 22.

WOODWELL, G. M., C. F. WURSTER, and P. A. ISSACSON. 1967. DDT residues in an east coast estuary: a case of biological concentration of a persistent insecticide. *Science* **156**: 821–824.

WRAY, J. D. 1971. Population pressure on families: family size and child spacing. In N. A. S.–N. R. C. publ., *Rapid Population Growth.* Johns Hopkins, Baltimore.

WRAY, J. D., and A. AGUIRRE. 1969. Protein-caloric malnutrition in Candelaria, Colombia. I. *J. Trop. Pediat.* **15**: 76–98.

WRIGHT, H. E. 1968. Natural environment of early food production north of Mesopotamia. *Science* **161**: 334–338.

WURSTER, C. F. 1968. DDT reduces photosynthesis by marine phytoplankton. *Science* **159**: 1474–1475.

WURSTER, D. H., C. F. WURSTER, and W. N. STRICKLAND. 1965. Bird mortality following DDT spray for Dutch elm disease. *Ecology* **46**: 488–499.

WYNNE-EDWARDS, V. C. 1962. *Animal Dispersion in Relation to Social Behavior.* Hafner, New York.

WYNNE-EDWARDS, V. C. 1965. Self-regulating systems in populations of animals. *Science* **147**: 1543–1548.

WYNNE-EDWARDS, V. C. 1970. Feedback from food resources to population regulation. *Brit. Ecol. Soc. Symp.* **10**: 413–427.

YODA, K., T. KIRA, H. OGAWA, and H. HOZUMI. 1963. Self-thinning in overcrowded pure stands under cultivated and natural conditions. *J. Biol. Osaka Cy. U.* **14**: 107–129.

YONGE, C. M., and A. G. NICHOLLS. 1931. Studies on the physiology of coral. *Sci. Rep. Brit. Mus. Nat. Hist.* **1**: 353–391.

ZINKE, P. J. 1962. The pattern of individual forest trees on soil properties. *Ecology* **43**: 130–133.

index

The index may be used as a glossary by referring to page numbers in boldface.